AF265070

Copyright 2003

by

Christopher R. Williams

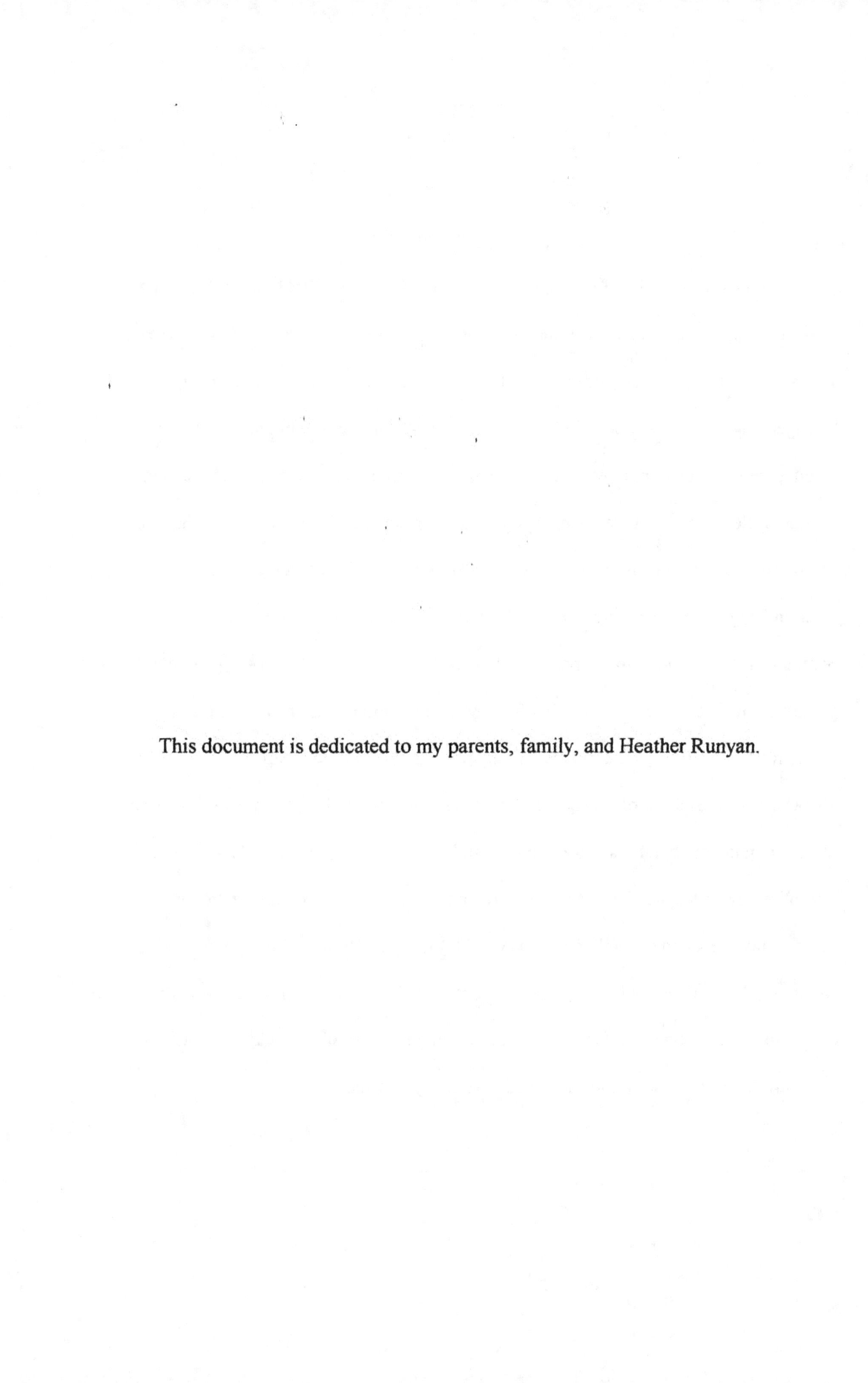

This document is dedicated to my parents, family, and Heather Runyan.

ACKNOWLEDGMENTS

Foremost I would like to thank my advisor, Linda Bloom, Ph.D., for outstanding guidance and providing an exceptional working environment in a state-of-the-art research laboratory. I thank my committee at UF, Drs. Daniel L. Purich, Arthur S. Edison, J. Bert Flanagen, and Alfred S. Lewin, for their assistance not only in guiding my project, but in guiding me to become more wise and perceptive as a scientist. I must also acknowledge my committee at Arizona State University, Drs. Neal Woodbury, Yuri L. Lyubchenko, and Kenneth J. Hoober, for their assistance in the early stages of this dissertation project. I acknowledge Manju Hingorani, Ph.D., for helpful discussions at the Keystone Symposia meetings and protein preparation, Mike O'Donnell, Ph.D. for the generous gifts of clamp loaders, Martin Webb, Ph.D., Myron F. Goodman, Ph.D., and Jeffery Bertram for the gift of the phosphate binding protein plasmid, and help in characterization and use of MDCC-PBP, and Petr Kuzmic, Ph.D., for contribution of the custom version of DynaFit and assistance with kinetic modeling. I would like to thank my following colleagues for their friendship and invaluable discussions regarding this project and science in general: Brandon Ason, Ryan Shaw, John C. Lopez, Gabriel Montaño, Ph.D., Gregory Uyeda, José Clementé, Ph.D., Joyce Feller, Ph.D. Finally, I would like to thank my parents for their unwavering support of my endevours, and for purchasing my first microscope and chemistry sets years ago.

LIST OF FIGURES

Abstract of Dissertation Presented to the Graduate School
of the University of Florida in Partial Fulfillment of the
Requirements for the Degree of Doctor of Philosophy

THE CLAMP LOADER OF *Escherichia coli* DNA POLYMERASE III:
KINETICS OF THE ATP-DEPENDENT STEPS IN THE SLIDING-CLAMP
LOADING REACTION

By

Christopher R. Williams

December 2003

Chair: Linda B. Bloom
Major Department: Biochemistry and Molecular Biology

DNA polymerase III holoenzyme, the principal enzyme responsible for *E. coli*

chromosomal replication, synthesizes stretches of DNA thousands of nucleotides long at

a rate approaching 750 nucleotides s^{-1} without dissociation. A ring-shaped DNA sliding-

clamp "β" topologically links the polymerase to DNA. A clamp loader, "γ complex",

assembles β on DNA in an ATP-dependent reaction. This dissertation project was

undertaken for investigation of the mechanism of β clamp loading by the clamp loader for

processive replication. ATP binding and hydrolysis activities of the clamp loader

promote conformational changes modulating its binding affinity for the clamp and DNA.

Using fluorescence-based steady-state and real-time stopped-flow methods, the kinetics

of these dynamic conformational changes were measured. Pre-steady-state ATP

hydrolysis assays performed in the absence or presence of β resulted in biphasic or

monophasic kinetics respectively. Biphasic kinetics suggested that the clamp loader,

equilibrated with ATP, exists in a mixture of two dominant species. Addition of β converted this mixture into a single activated population, effectively increasing its concentration. These experiments, in addition to adenosine 5'-*O*-(3-thiotriphosphate) (ATPγS)-chase assays showed that activated γ complex hydrolyzed one ATP per γ-subunit at a rate faster than ATP dissociation. The hypothesis that γ complex exists in multiple species with ATP was confirmed by measuring pre-steady-state clamp loading kinetics in DNA-binding assays. When γ complex was equilibrated with ATP, a mixture of two species formed which when mixed with DNA and β exhibited biphasic DNA binding kinetics. When equilibrated with ATP and β, rapid monophasic DNA binding kinetics resulted. Direct mixing of γ complex with DNA β and ATP displayed slow monophasic kinetics, limited by the conformational change rate. The rate of the conformational changes separating the two dominant species was determined by investigating their evolution in equilibration time with ATP. Computer modeling of these experimental data revealed a conformational change rate of ~4.5 s^{-1}. Comparison of the kinetics of a "minimal" clamp loader complex missing χ and ψ subunits, revealed that these subunits facilitate the conformational changes in γ complex required for modulation of β and DNA binding affinities during the clamp loading reaction.

CHAPTER 1
INTRODUCTION - STATEMENT OF PROBLEM

The Processive *Escherichia coli* DNA Polymerase III Holoenzyme

DNA polymerase III holoenzyme (pol III holoenzyme) is the principal enzyme

involved in replication of the *Escherichia coli* chromosome (Kornberg and Baker, 1992).

The pol III holoenzyme copies the parent chromosome with surprising speed and

processivity, moving along the DNA at a maximum velocity reaching approximately 750

nucleotides per second, without release from the template at distances of over thousands

of nucleotides. Polymerase III holoenzyme must possess these functional characteristics

to complete rapid replication of the chromosome for cell division. Rapid and processive

DNA synthesis is required for all forms of DNA metabolism and genome maintenance

including DNA repair and recombination, and is evolutionarily conserved in all branches

of life, underscoring the significance of advancing the detailed understanding of the

proteins providing these functional mechanisms.

DNA Polymerase III Holoenzyme

Pol III holoenzyme is a multi-protein complex consisting of ten distinct subunits (α,

ϵ, θ, β, τ, γ, δ, δ', χ and ψ). A core of the holoenzyme consisting of the α-5' to 3'

polymerase, ϵ-3' to 5' exonuclease, and θ subunits is paired-up by the τ_2 protein-dimer

(McHenry, 1982; McHenry and Crow, 1979). The τ_2-dimerized pol III core, termed pol

III', was originally purified from cells and isolated from pol III holoenzyme. The

replication activities of pol III core and pol III' were determined to be distributive in

kinetic and product size determination assays, whereas holoenzyme could synthesize stretches of DNA with high processivity (Fay et al., 1981; LaDuca et al., 1983). Clearly, these were incomplete forms of pol III holoenzyme, and were missing important factors necessary for complete and rapid replication of the chromosome.

Another form of pol III, termed pol III*, contained all of the subunits of pol III holoenzyme except one, the β subunit (Fay et al., 1982). Addition of this β subunit was all that was required to convert the distributive pol III* enzyme into a highly processive enzyme with DNA synthesis activity matching pol III holoenzyme. The β subunit is initially associated with the replication fork DNA at sites called preinitiation complexes where polymerase III holoenzyme assembles for DNA synthesis. Several early biochemical studies revealed that a complex consisting of the $\tau,\gamma,\delta,\delta'$, χ and ψ subunits within pol III holoenzyme comprised a complex responsible for ATP-dependent placement of the β subunit at primed sites on DNA for formation of the preinitiation complexes, and were required for conferring processivity in replication (Maki et al., 1988; Maki and Kornberg, 1988).

The β Sliding Clamp

Solution of the X-ray crystal structure of the β subunit revealed that it was a ring-shaped dimer of crescent-shaped protomers (Kong et al., 1992). β was termed a "DNA sliding-clamp" due to its ability to topologically link pol III holoenzyme to template DNA in such a way that holoenzyme remains tightly associated with DNA, yet has significant freedom of movement on the DNA. The β clamp has an inner pore lined with α-helices that act akin to "skates," traversing the clefts of the major and minor grooves in the DNA backbone as the clamp slides along. This inner pore has a diameter large enough to encircle duplex DNA as well as hybrid RNA-DNA duplex structure found at the sites of preinitiation complex formation. A combination of hydrogen bonding,

hydrophobic and ionic interactions at the dimer interfaces strengthen and hold together

the β clamp subunits. β binds directly to the core of pol III holoenzyme through the α

subunit (Stukenberg et al., 1991). The β clamp has a dissociation constant for

dimerization in the range of $\sim 6.0 \times 10^{-11}$ molar, and is stable enough to remain on

circular DNA for ~ 100 minutes (Yao et al., 1996). The extraordinary stability of the

circular β dimer in solution and on the circular chromosome stresses the need for some

mechanism to open β for its loading onto and disassembly from DNA.

The DnaX Clamp Loader – Stoichiometry and Organization of Subunits

The other processivity proteins within pol III holoenzyme ($\tau,\gamma,\delta,\delta',\chi$ and ψ) form a

complex that performs the duty of opening and loading β onto DNA for formation of

preinitiation complexes on primed DNA at the leading and lagging strands at the

replication fork (Kelman and O'Donnell, 1995). These proteins form the DnaX complex

"clamp loader," with the subunit stoichiometry $[(DnaX)_3,\delta_1\delta'_1\chi_1\psi_1]$, in which each

subunit executes a unique function (Pritchard et al., 2000). The *dna*X gene forms both τ

and γ subunits by a translational frameshift that forms a stop codon defining the γ subunit

C-terminus, approximately two-thirds the length of the complete mRNA (Flower and

McHenry, 1990; Tsuchihashi and Kornberg, 1990). Therefore, γ and τ subunits are

identical in the first two-thirds of their sequence and structure. As a consequence, the N-

terminal domain of τ acts akin to the γ subunit, and the unique C-terminal domain of τ

has a specialized function in coordinating the holoenzyme at the replication fork

(Dallmann et al., 2000). Due to the requirement of the τ subunit for dimerization of the

pol III core, the DnaX complex associated with pol III holoenzyme most likely has a stoichiometry of $\tau_2\gamma_1$ (Pritchard et al., 2000). Each of the other subunits (δ,δ',χ and ψ) is present in a single copy in this clamp loader (Onrust et al., 1995). For *in vitro* reconstitution of the DnaX clamp loader, the χ and ψ subunits bind the γ subunit through ψ-γ interaction (Glover and McHenry, 2000), and greatly increase the affinity for the $\delta\delta'$ subunits for the complex (Olson et al., 1995). The structure of the *E. coli* clamp loader recently also revealed existence of a trimer of γ (DnaX) subunits (Jeruzalmi et al., 2001a).

Clamp Loader Subunit Functions

The χ subunit interacts with single-stranded DNA binding protein (SSB) at the replication fork (Glover and McHenry, 1998). This χ-SSB interaction is thought to be involved in a primase-to-polymerase switch prior to formation of preinitiation complexes (Yuzhakov et al., 1999). The ψ subunit has no known function other than forming a structural bridging contact for χ to the clamp loader through ψ interaction with the γ subunit (Glover and McHenry, 2000; Xiao et al., 1993b). The function of the χ-SSB interaction is well characterized, however there is no known function for the χ and ψ subunits directly in loading the clamp in DNA.

It is the δ subunit that binds to, and alone has the ability to open, the β clamp (Jeruzalmi et al., 2001b; Leu et al., 2000). The surface of δ subunit which binds to β is concealed by the δ' subunit when the clamp loader is in an "inactive" state (Jeruzalmi et al., 2001a). The τ and γ subunits transduce the energy from ATP binding and hydrolysis into mechanical work within this clamp loading machine to load the clamp on DNA (Bertram et al., 2000; Onrust et al., 1991).

All of the subunits of the clamp loader, except χ and ψ are members of a diverse superfamily of AAA+ (<u>A</u>TPases <u>A</u>ssociated with a variety of cellular <u>A</u>ctivities) molecular motors (Neuwald et al., 1999). This AAA+ superfamily contains conserved sequences and structures for nucleotide binding and hydrolysis that drive conformational changes in these motor proteins. In the DnaX clamp loader ATP binding and hydrolysis by the τ (i.e., through the N-terminal γ-region) and γ subunits promote conformational dynamics that modulate the binding affinity for the β clamp as well as DNA for the loading reaction. The δ and δ' have been identified as AAA+ superfamily members based on sequence and structural homology of δ', and structural homology in the case of δ, however, neither has the ability to bind or hydrolyze ATP, although they are believed to actively participate in the conformational dynamics of the clamp loader for the loading reaction (Jeruzalmi et al., 2001a; Podobnik et al., 2003).

The Mechanism of β Clamp Loading

Figure 1-1 depicts a simplified schematic of the clamp loading mechanism (under study in this dissertation) for processive DNA synthesis by polymerase III. Initially, ATP binds to the clamp loader's τ and γ subunits. Binding of up to three ATP molecules causes conformational changes in the τ and γ subunits, and therefore changes the overall structure of the clamp loader. Although the exact nature of these ATP-dependent conformational changes is not yet known, there are two major consequences following the conformational changes. The β clamp interaction surface of the δ subunit becomes exposed, and a DNA binding surface on the clamp loader either forms or becomes exposed. Interaction of the δ subunit with β has been biochemically and structurally characterized, showing that δ induces a conformational change in the β subunit that

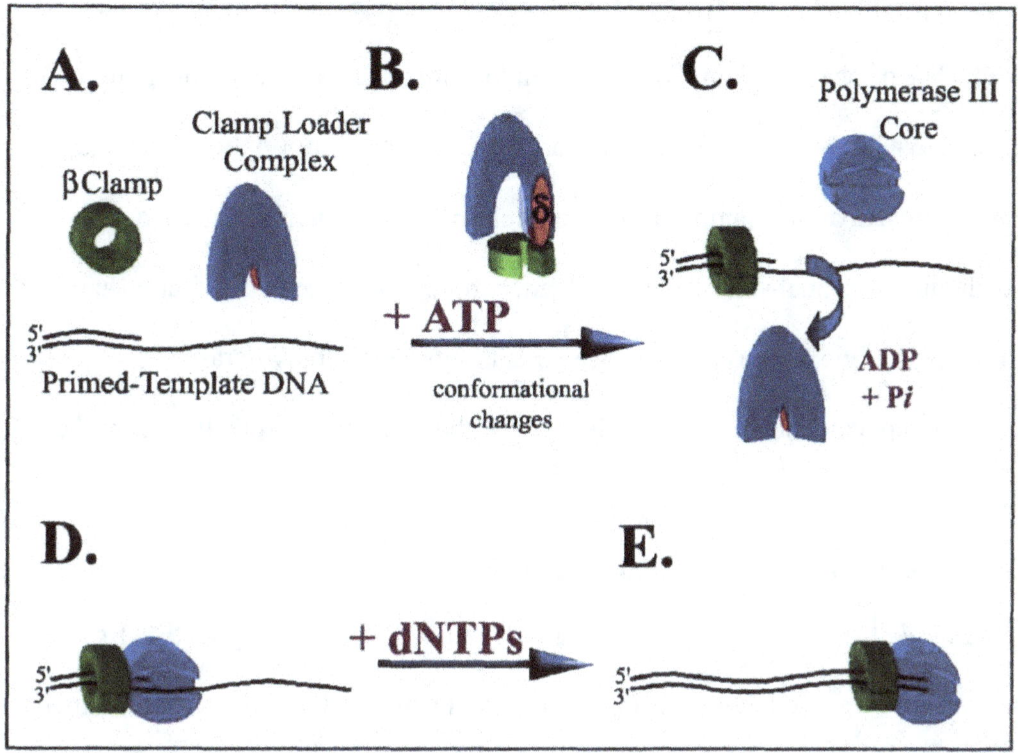

Figure 1-1. Schematic of the β clamp loading reaction for processive DNA synthesis. A) β clamp, and the DnaX clamp loader are present with a primed-template DNA substrate. B) Binding of ATP increases the affinity of the clamp loader for both β and DNA by conformational changes producing exposure of the δ subunit-β interaction surface, and formation of a putative DNA binding surface. Interaction of δ with β induces conformational changes in the clamp that open it at a single interface. C) Primed-template DNA triggers ATP hydrolysis and rapid dissociation of the clamp loader from the loaded clamp. D) DNA Polymerase then binds the same surface of β that was previously occupied by the clamp loader. E) Incorporation of deoxyribonucleotide-triphosphates (dNTPs) by polymerase then proceeds without polymerase dissociation from the template.

triggers opening at a single dimer interface (Jeruzalmi et al., 2001b; Turner et al., 1999).

The nature of the putative DNA binding surface on the ATP-bound clamp loader remains

unknown. However, it is clear that ATP binding is required for the clamp loader to bind

DNA, and further that the clamp loader preferentially places β at the 3'-end of the primer

on a replication-proficient template (Ason et al., 2003). Binding primed-template DNA

triggers the clamp loader to hydrolyze ATP. ATP hydrolysis causes release of the clamp loader from β, which then tightly closes on DNA when the δ subunit-interaction is removed. The discharged clamp loader is then in some inactive state at this point in the cycle, and most likely must undergo some as yet undefined conformational changes in order to continue loading clamps. Immediate removal of the clamp loader from the loaded clamp is essential since the clamp loader and the polymerase α subunit share the same binding surface on the clamp. Once bound to its clamp, polymerase can processively replicate thousands of nucleotides without further dissociation from the template.

Importance of the *E. coli* Model Replication System

In *E. coli*, the clamp loader and β clamp are utilized in DNA replication from initiation at the origin to DNA partitioning of parent and daughter chromosomes upon termination (Katayama, 2001; Levine and Marians, 1998). Aside from its interaction with pol III core, β clamp has been found to interact with all five *E. coli* DNA polymerases, as well as several other partner proteins in DNA metabolism such as the MutS, UvrB, and DNA ligase (O'Donnell and Lopez de Saro, 2001; Tang et al., 1999). The β clamp and DnaX clamp loading machine subunits are functionally conserved across evolution to all branches of life suggesting a fundamentally similar mechanism for processivity in all DNA metabolism (Ellison and Stillman, 2001). Increasing structural analyses also reveal that the composition of these clamps and clamp loading machines from such organisms as diverse as a bacteriophage and a human are remarkably similar, and that all clamp loaders utilize AAA+ motor proteins (Davey et al., 2002). Using *E. coli* as a model replication system has been outstandingly valuable for the determination of how complex multi-protein machines interact and work together in a well-regulated efficient manner for DNA replication. It is important to determine how the individual protein subunits of these complex biological machines carry out their own functions and communicate with each other for completion of their tasks. Additionally, understanding of these mechanisms will allow for a more complete comprehension of the important DNA metabolic tasks in which they are involved.

The General Problem Under Study and Research Questions Addressed

The general problem under study in this project is the many remaining unknown and questionable aspects of the mechanisms by the clamp loader for β clamp loading and the results and interpretations of other investigations. Conformational changes in the clamp loader are a requirement for modulation of clamp loading. However, the kinetics of clamp loader conformational changes are unknown. Different nucleotide-dependent clamp loader conformational species have been identified in proteolytic digestion experiments, but how their abundance and dynamics control the loading reaction cycle is not understood mechanistically. How do the nucleotide-dependent conformational dynamics within the clamp loader subunits drive this machine for clamp loading? Chapter 4 of this dissertation details an investigation that addresses this question. The studies describe the existence of distinct conformational species of the clamp loader and the kinetics of their activity in the clamp loading mechanism. Chapter 4 further addresses the questions concerning how β and ATP affect the kinetics of clamp loader conformational changes. An additional key question asked in chapter 4 is that of the nature of nucleotide binding to the clamp loader, and whether there is interdependency between ATP binding and the conformational dynamics. Computer modeling was applied to the experimental data for analysis of these important questions.

Sliding clamps of all organisms studied to date are known to enhance the activity of their respective clamp loader. How the β clamp affects the γ complex clamp loader for promotion of its own loading onto DNA was tested in this research project. Although it is understood that β increases the ATP hydrolysis energetics of the clamp loader, it remains unknown what the kinetic mechanism of this enhancement may be. By comparison of γ complex with the $\gamma_3\delta\delta'$ minimal clamp loader in chapter 5, along with experiments presented in chapter 4, this dissertation addresses how β clamp affects the specificity and stability of nucleotide binding by the clamp loader, and therefore the kinetics of the conformational dynamics during the clamp loading reaction for its enhancement.

In comparison to the AAA+ clamp loading machines of other organisms outlined in chapter 2, the *E. coli* clamp loader contains two additional subunits (χ and ψ). It is known that χ and ψ assist in the assembly of other subunits in the clamp loader, and the χ subunit has a role in primase-to-polymerase switching at the replication fork, but do they serve any direct function in the clamp loading mechanism by γ complex? Chapter 5 of this work outlines a detailed characterization of the $\gamma_3\delta\delta'$ minimal clamp loader. This minimal complex is missing the χ and ψ subunits, and is used to assess their roles in the clamp loading mechanism by direct experimental comparison to γ complex. Are the χ and ψ subunits "AAA+ adaptor" proteins of γ complex? Do they provide an example of a novel adaptor function through their effect on the conformational dynamics of the clamp loader? Although not originally hypothesized, the findings presented here along with findings presented elsewhere suggest that they are.

Finally, what can the details learned here about the *E. coli* clamp loading machine tell us about other molecular machines driving and regulating complex and diverse cellular tasks in other organisms? This machine is essentially a molecular switch that catalyzes a fundamental reaction in all forms of life. Can the knowledge gained through study of this clamp loading machine be used for development of novel biological or hybrid molecular machines?

Conformational Dynamics of the Clamp Loading Machine

Solution of the X-ray crystal structures of a minimal $\gamma_3\delta\delta'$ clamp loader complex and a C-terminal truncated γ subunit has recently given information revealing some of the structure-function relationships of the DnaX clamp loading machine (Jeruzalmi et al., 2001a; Podobnik et al., 2003). The subcomplex of $(\gamma_3\delta_1\delta'_1)$ forms a minimal complex with clamp loading activity similar to the complete DnaX clamp loader (Onrust et al., 1991). Each of the γ_3, δ, and δ' subunits is a AAA+ superfamily member, and is composed of three structural domains that form "C-shaped" molecules. The N-terminal domains-I and II encompass the conserved nucleotide binding site (in γ subunits only), and the third C-terminal domain-III forms an oligomerization region for this heteropentameric complex. Domain-II generally forms a mobile "hinge" between domains-I and III. ATP binding and hydrolysis by the γ-AAA+ motor subunits directly trigger conformational changes in these subunits. Although the δ and δ' subunits are AAA+ proteins, they do not have the ability to bind and hydrolyze ATP, but most likely do undergo conformational changes caused by γ subunit conformational movements. The structure of the $\gamma_3\delta\delta'$ clamp loader complex showed that the complex was arranged as a heteropentameric ring through the C-terminal oligomerization domain, and that there was extensive contact between each of the five subunits extending "down" in asymmetric orientation from the oligomerization domain. The nucleotide binding sites were found in the interfaces between the δ'-γ_1, γ_1-γ_2, and γ_2-γ_3 subunits. At least one and possibly two of these interfacial nucleotide binding sites is thought to be constitutively open whereas the third is deeply buried in an interface. This lead to a model describing a sequential series of individual γ-subunit conformational changes whereupon ATP binding to an open interface caused a conformational change opening the adjacent interface and so on (Jeruzalmi et al., 2001a), where the status of ATP binding to the nucleotide binding sites is communicated between the subunits of the clamp loader modulating its affinity for β clamp and DNA.

There are several limitations to the mechanistic inferences made in light of these structures. The clamp loader structure was determined in the absence of nucleotide. Therefore, the observed subunit interactions most likely do not represent the functional complex, as it would appear in the cell, and the overall conformation was thought to be largely due to a crystal-packing artifact. The truncated-γ subunit structure was determined in the presence of non-hydrolyzable adenosine 5'-*O*-(3-thiotriphosphate) (ATPγS), and crystallized as a tetramer (not as it would appear in a functional complex). The structure contained electron density consistent with ATPγS molecules bound to two protomers, ADP bound to a third, and no nucleotide in the fourth. ATPγS is sufficient for both β clamp and DNA binding by the clamp loader but does not confer clamp loading activity in biochemical assays (Bloom et al., 1996). Therefore, the ATPγS-bound γ subunit structure revealed important features of the conformational change in this motor subunit. However, the γ subunit was truncated and a substantial portion is missing from its C-terminus. Only the conserved AAA+ nucleotide binding domains-I and II were present. This means that the different conformations observed in each of these structures may be considerably different than the actual structure in solution, or within the cell. The experiments presented in this dissertation do not address structure specifically, however these *in vitro* studies do directly address the kinetics of the conformational dynamics of the clamp loader in solution conditions during real-time clamp loading reactions.

There are significant nucleotide-dependent conformational changes in the clamp loader that must be made during the clamp loading reaction for intersubunit communication modulating β clamp and DNA binding. Currently, only proteolytic protection assays (Hingorani and O'Donnell, 1998) have provided biochemical proof of the nucleotide-dependent conformational changes, however there is no kinetic detail known regarding the nature of these conformational changes during the clamp loading reaction. It has been previously hypothesized that the rate-limiting step of the clamp loading mechanism occurs in the clamp loader, away from DNA (Bloom et al., 1996), and could be ADP-release or additional conformational changes that "reset" the clamp loader for resuming the clamp loading cycle. The results of this dissertation address the kinetics of the nucleotide-dependent conformational dynamics of the clamp loader and add detail to what is known about the clamp loading mechanism, initially asking, how do ATP-dependent conformational dynamics drive and regulate this molecular machine for clamp loading?

The kinetics of nucleotide binding and hydrolysis by the clamp loader, DNA binding, and β clamp loading have been determined using several biochemical analyses in this project. Together, these explorations have lead to a hypothesis based on conformational dynamics for the internal workings of the clamp loading machine. Here, it is proposed that the clamp loader exists, dominantly, in two distinct conformational states that are either "activated" or "inactive" for clamp and DNA binding when in equilibrium with ATP. This project has addressed the kinetics of the ATP-dependent conformational changes separating the two clamp loader species in the clamp loading mechanism, and additionally shows that there is complexity in the conformational dynamics between these two major states. This added complexity probably arises due to the proposed asymmetry in the three nucleotide binding sites, and the possibility of having differential affinities for ATP within these sites causing a complex mixture of conformational intermediates within the clamp loader. Elucidation of the nature of the additional mixture of conformational states will require further study, but overall this project reveals, kinetically, higher populations of two major states modulating the function of the clamp loader in the β clamp loading mechanism.

Enhancement of the Clamp Loading Machine by its Clamp

It has been known for some time that the β clamp enhances, but does not trigger, the ATP hydrolysis activity of the clamp loader in the presence of DNA (Onrust et al., 1991). This feature of a sliding clamp enhancing the DNA-dependent ATP hydrolysis activity of its clamp loader is a general characteristic in bacteriophage (Pietroni et al., 2001), eukaryotic (Yoder and Burgers, 1991), and archaeal organisms (Oyama et al., 2001); thus it is a fundamental attribute in all clamp loading mechanisms. However, the mechanism of how a clamp enhances the activity of its clamp loader has remained a mystery. Part of the originally proposed research project was to determine how the *E. coli* β clamp affected the kinetics of ATP hydrolysis by the clamp loader. This dissertation addresses the mechanism of β clamp enhancement of clamp loader DNA-dependent ATP hydrolysis, and reveals several important findings interrelated with nucleotide binding and the conformational dynamics of the clamp loader. It comes as no surprise that the mechanism of β clamp enhancement of ATP hydrolysis activity is related to a conversion of the equilibrium between the dynamic conformational states of the clamp loader. It is hypothesized that the β clamp converts the conformational equilibrium of the clamp loader with ATP into a completely "active" population (i.e., a complex of ATP-bound clamp loader and β). An extension of this hypothesis is that the β clamp affects the affinity and specificity of the clamp loader for ATP, effectively "trapping" the nucleotide within the clamp loader promoting formation of the active β-bound complex poised for loading onto DNA. Selective binding and trapping the active clamp loader conformation may also stabilize a putative DNA binding surface on the clamp loader as well. An increase in the apparent concentration of this active complex poised for clamp loading would provide a direct mechanism for the increase in ATP hydrolysis turnover rate observed in kinetic assays. Due to the considerable evolutionary conservation of clamp loader function across all branches of life (Davey et al., 2002), the hypothesized mechanism of clamp enhancement of clamp loader activity presented in this dissertation could also apply to clamp loading for processive DNA synthesis in all other organisms as well.

The χ and ψ Subunits are Required for Optimum Activity of the Clamp Loader

The χ and ψ proteins do not bind DNA, do not bind or hydrolyze ATP, and are dispensable for clamp loading activity. At the replication fork *in vivo*, the χ subunit is involved in an important interaction with SSB, and a primase-to-polymerase hand-off through association with the clamp loader, but not directly involved in preinitiation complex formation by the clamp loader. A subcomplex of the clamp loader ($\gamma_3\delta\delta'$), of which the crystal structure was determined, does not contain the χ and ψ subunits, yet maintains nearly the same ATP hydrolysis and clamp loading activity as the clamp loader with χ and ψ (Onrust et al., 1991). Although the structure of the χ-ψ dipeptide is known, it is still a mystery where it binds to the clamp loader. For these reasons the $\gamma_3\delta\delta'$ clamp loader has been called the minimal clamp loader complex, and in some cases, it has been called the clamp loader itself.

This dissertation encompasses a detailed biochemical characterization of the ($\gamma_3\delta\delta'$) minimal clamp loader, and comparison of its activities to the "γ complex" clamp loader. This work shows that there are in fact several mechanistic distinctions between the minimal clamp loader and γ complex clamp loader. These differences have generally revealed that the minimal complex is slightly hindered in its nucleotide binding and hydrolysis properties and hence clamp loading activity. Therefore the χ and ψ subunits are required for optimum clamp loading activity, conceivably by strengthening the intersubunit communication within the clamp loading machine. Previous studies revealed that the χ and ψ subunits did in fact stabilize the clamp loader by increasing the affinity of τ and γ subunits for $\delta\delta'$ (Olson et al., 1995). One of the most obvious deficiencies in minimal complex function is the loss of the ability to stabilize ATP binding in an active complex with β clamp. This is hypothesized to be the result of possibly slower conformational changes in the minimal complex, and perhaps "looser" conformational dynamics in the minimal complex leading to more conformational complexity (i.e., more inactive intermediate conformational states) between the major two states hypothesized for the clamp loader. This research shows that the population of inactive conformations of the minimal complex is several-fold greater than the population of active conformation in equilibrium with ATP. β clamp does sustain the ability to enhance the ATP hydrolysis activity of the minimal complex. However, the results of kinetic analysis of ATP hydrolysis activity reveal that this enhancement is less than that of γ complex, and that β must do more "work" to convert the large population of inactive conformational states of the minimal complex into the trapped active conformation. To date, there is no other steady-state and time-resolved biochemical analysis comparing γ complex to the minimal complex in such a comprehensive manner as that presented in this dissertation, and the results have allowed a much more clear appreciation for the clamp loading mechanism catalyzed by the DnaX clamp loader.

The clamp loader is a AAA+ protein containing machine. The χ and ψ subunits are structurally unrelated and much smaller than the AAA+ subunits of the clamp loader. Since the χ and ψ subunits seem to be unique to the *E. coli* clamp loader, and most likely other gram-negative bacteria (Xiao et al., 1993a; Xiao et al., 1993b), they provide some unique functions in these prokaryotic clamp loaders compared to those of other organisms. They present a form of substrate specificity to the clamp loader for the SSB-coated lagging strand at the replication fork for assistance in preinitiation complex formation for Okazaki fragment synthesis. These characteristics of the χ and ψ subunits associate them with a new and growing list of AAA+ adaptor proteins. AAA+ adaptor proteins provide a simple and effective way for modulation of AAA+ machine function, giving the machine better control over substrate specificity and rapid redirection of its specific activity (Dougan et al., 2002). Along with their function at the replication fork, the χ and ψ subunits are proposed here to provide increased structural stability and facilitated conformational changes for intersubunit communication within the AAA+ clamp loading machine. It is thus hypothesized that they are in fact AAA+ adaptor proteins and reveal a novel adaptor function for this newly defined subset of proteins.

Application of the Analyses of the Clamp Loading Machine to Other Complex Molecular Machines

Like the clamp loader, there are many other enzyme complexes that perform complicated biological tasks such as assembly and disassembly of new protein-protein complexes, protein-cofactor complexes, and protein-DNA complexes. Orchestration of protein-protein and protein-nucleic acid interactions is likely to utilize converging mechanisms to drive the diverse functions of these multi-protein enzyme complexes. The intersubunit conformational communication observed structurally and functionally by the clamp loaders drive several essential jobs within the large replication complexes in all forms of life. The clamp loader catalyzes assembly of a new protein-nucleic acid complex without making or breaking any covalent bonds of either component. The energy for this assembly is derived from within the clamp loader by ATP-dependent conformational changes that essentially switch the clamp loader between "on" and "off" states. Understanding of subunit recognition and conformational communication driving similar switching mechanisms in molecular machines is a complicated task in science, but explorations such as that presented in this dissertation for the γ complex clamp loader could help accelerate other investigations. Not only for ATP- and GTP-utilizing energase enzymes, but also for transmembrane transport machinery, cellular cargo-transport motors, chaperone and protein modulating complexes, proteases, etc.

The investigations outlined in this dissertation are a highly focused and detailed *in vitro* examination of the mechanism of a molecular machine's mechanism of switching between active and inactive states for assembly of a protein-DNA complex at the expense of energy from ATP binding and hydrolysis. The details learned here about protein conformational changes and communication of these changes within the machine could be applied to other solution-based structure-function studies for a broad range of proteins. A key result of this and prior work shows that the information for conformational changes in these proteins is stored in the structure of the proteins themselves, akin to the storage of protein folding information encoded within secondary and tertiary peptide structure (Fersht and Shakhnovich, 1998). For example, the β clamp is not simply an unintelligent ring-shaped protein dimer, but contains structural information for release of a spring-like trigger for opening of one of its sturdy interfaces (Ellison and Stillman, 2001). In the clamp loading machines, nucleotide-dependent intersubunit communication induces the extraction of intramolecular-stored information for subunit conformational changes that, in turn, can induce changes in partner proteins.

Molecular motors and switching mechanisms are common in many energase enzymes other than the AAA+ proteins that make up the clamp loader under study here. For example, several molecular motors move along the cellular architecture of microtubule and actin cytoskeletal networks (Mehta et al., 1999; Vale, 2003). These motors power transport of organelles and other cargo throughout the cell, drive chromosomal segregation in cellular division during mitosis, give cells the capability of motility, and power the contraction of muscle fibers that give animals the extraordinary strength needed for a diverse range of movement, and let their hearts beat for entire lifetimes (Vale and Milligan, 2000). The machines powering these processes are not much unlike the clamp loading machine detailed in this dissertation. They each generally contain several domains including nucleotide binding sites, protein-protein and protein-nucleic acid binding regions, hinge-like domains, multisubunit oligomerization domains, and regulatory subunit or cofactor interfaces (Vale and Milligan, 2000). The conformational movements made by several of these nucleotide dependent energases

drive their functions, give them processive activity, and allow them to generate molecular

power that is matched on a relative scale only by man's largest macromolecular machines

(Baker and Bell, 1998; Ellison and Stillman, 2001). Several molecular switches also act

through conformational dynamics allowing protein-protein and protein-nucleic acid

interactions in signal transduction cascades (Franco et al., 2003; Phillips et al., 2003),

chromatin structural modification (Hakimi et al., 2002), transcriptional and translational

regulation (Mazumder et al., 2003), and ion channels driving action potentials for

neuronal signaling. The ras-GTPase activating protein provides an example of protein-

protein contact directly related to the subunit interfaces within the clamp loader. A

conserved "arginine-finger" residue is positioned between two proteins for catalysis of

nucleotide hydrolysis, inducing conformational changes that switch the activity status of

the protein (Ahmadian et al., 1997). Nearly all cellular processes are driven or regulated

in some way by phosphorylation / dephosphorylation mechanisms, but the activities of

energase molecular motors and switches (Purich, 2001) also appear to have great

importance in maintaining cellular function in all organisms. Therefore the significance

of investigating the inner workings of a molecular machine such as the *E. coli* clamp

loader extends far beyond the understanding of the mechanisms of DNA metabolism.

Novel Hybrid Devices Based on the Clamp Loader and Sliding Clamp

Understanding of the mechanism of the clamp loader molecular machine;

structurally and functionally how protein subunits come together and communicate with

each other based on nucleotide binding status can be exploited for development of novel

molecular hybrid devices. In even simple terms, the clamp loader is a molecular switch,

activated by ATP binding, and inactivated by DNA-dependent hydrolysis of ATP. The inactive and active states of this molecular switch are driven by conformational changes, and further affected by binding a partner protein (the clamp). Essentially, the reaction mechanism under study in this dissertation is a molecular switch that entails placement or removal of (protein) rings on (nucleic acid) rods or hoops. This activity produces a topological connection for another machine (polymerase) to the rod or hoop structure, essentially increasing its concentration at the preferred catalytic site (template DNA).

The clamp loader switching mechanism could be utilized in other systems that require regulated activation/deactivation or in systems such as memory or communications devices that require a simple (binary) on/off signal. By modification of the protein properties of the clamp loader subunits it is feasible that this energase machine/switch could be custom-tailored for guided assembly of proteins (or other materials) other than the clamp, onto specific substrates in order to control the putative-complex concentration, and activate or deactivate it specifically through NTP binding and hydrolysis.

"Off-the-Wall" Example of the Clamp and Clamp Loader in a Novel Device

There is a need for understanding and improvement of industrial catalysis systems (Coontz et al., 2003). Briefly, for homogenous catalysts, where the molecular catalyst and reactant are dispersed in the same phase, there exists a seemingly simple problem of separation of the reactants and catalysts from products. The development of heterogeneous catalysts have provided a solution by direct separation of catalyst and reactant in different phases, but in most cases do not provide the surface area for cost-efficient devices. One could envision use of a clamp loading machine to place (and remove) a ring-shaped clamp, either with the ability to bind a specific catalyst, or containing the catalyst itself, onto the reactant. This process would allow a substantial increase the heterogeneous catalytic surface area on an insoluble nano-sized rod- or hoop-shaped reactant. Perhaps the products would easily be separated in solution or gas phases. These devices would be "smart" devices where the catalytic components contain the information needed for their assembly and disassembly for disposal or recycling. For example, a spring-loaded clamp that can be placed and removed at will by a specialized machine with easily regulated activity. Improvement of industrial catalysts is only a single (imaginative) example of how the functional knowledge about sliding clamps and clamp loading machines could be used for design of a novel hybrid device. With imagination and lots of money, these and other, perhaps more significant biologically functional devices could be developed.

Design of Research Project

This research project was developed on two fluorescence-based experimental methodologies for investigation of the kinetics of DNA dependent ATP hydrolysis and β clamp loading activities of the γ complex clamp loader and the $\gamma_3\delta\delta'$ minimal clamp loader complex. The two experimental methods utilized were an anisotropy binding assay, and an *E. coli* phosphate binding protein-based MDCC-PBP ATP hydrolysis assay. The anisotropy binding assay was employed for investigation of the dynamics of the interactions between the clamp loader and fluorescent-labeled DNA or fluorescent-labeled β clamp. Studies of the DNA dependent ATP hydrolysis kinetics of the clamp loaders were performed with the MDCC-PBP ATPase assay. Both methodologies allowed for steady-state and time-resolved measurements of clamp loading kinetics.

For the anisotropy binding assay, depolarization of the emission from a fluorescent probe reports on the rotational dynamics of the fluorescent-labeled species (Lakowicz,

1999). By monitoring changes in anisotropy, one can measure binding dynamics and observe intermediates that arise in real time during a given reaction. This enables elucidation of the kinetics of discrete steps involving interactions with the labeled species in a reaction mechanism. To study binding kinetics of the clamp loader with β clamp as well as the binding kinetics of the clamp loader with DNA in the presence and absence of β clamp on steady-state, and pre-steady-state time scales, the fluorescence depolarization of X-rhodamine-labeled DNA (RhX-DNA) or pyrene-labeled β clamp (β^{pyrene}) was measured.

ATP hydrolysis kinetics were measured with the MDCC-PBP ATPase assay. This assay utilizes a site-specific mutant of *E. coli* phosphate binding protein (*phoS* gene product) that allows covalent attachment of an environmentally sensitive fluorescent probe (N-[2-(1-maleimidyl)ethyl]-7-(diethylamino)coumarin-3-carboxamide, "MDCC") near the phosphate binding cleft (Brune et al., 1994). Binding of inorganic phosphate (P_i) product (i.e. from ATP hydrolysis by the clamp loader) to the labeled-phosphate binding protein is both rapid and tight, and produces a substantial increase in fluorescence intensity readily measurable in a fluorimeter or stopped-flow real-time detection system.

Computer modeling was performed using two different programs for simulation and fitting of experimental data for investigation of the clamp loader conformational change mechanism. The KinTekSim program (Barshop et al., 1983; Dang and Frieden, 1997) was used to test model reaction mechanisms by simulation and visual comparison to the experimental data. A fitting program (DynaFit) was used to fit the experimental time course of the reaction with a least-squares regression methodology (Kuzmic, 1996).

Computer modeling of the reaction kinetics provided a way to validate, and predict

several of the mechanistic features determined in this project.

CHAPTER 2

LITERATURE REVIEW

DNA Replication in *Escherichia coli*

A complex assembly of DNA replication proteins forms at each replication fork for
synthesis of a nascent *Escherichia coli* chromosome. At the leading edge of the
replication fork is a topoisomerase, an enzyme that relieves torsional stress created by
duplex DNA unwinding. Opening up the replication fork and priming the leading and
lagging strand templates is a machine called the primosome. The primosome is made up
of two proteins, DNA helicase, which unwinds parental DNA, and primase, which
synthesizes RNA primers, the sites of initiation of replication of the parent templates.
Stabilizing the single strand template DNA that is exposed upon unwinding by the
helicase is single-stranded DNA binding protein. Associated with the primosome is a
complex of enzymes that synthesize and proofread the nascent DNA. This machine is
DNA polymerase III holoenzyme, a dimer of the synthesizing and proofreading units that
coordinate simultaneous high fidelity elongation of the leading strand, and Okazaki
fragments on the lagging strand of nascent DNA. The dimeric DNA polymerase III is
physically linked to the parental template leading and lagging strands by ring-shaped
sliding clamp processivity proteins, such that it cannot easily dissociate from the
template. A clamp loading machine, located within dimerized DNA polymerase III
utilizes the energy of ATP binding and hydrolysis to load the circular sliding clamp on
the leading strand template and on each Okazaki fragment of the lagging strand template.
DNA polymerase I and DNA ligase follow at the tail end of the replication fork, where

DNA polymerase I is poised to excise the RNA primers and replace them with DNA, and DNA ligase seals the remaining nicks produced between the Okazaki fragments on lagging strand DNA.

As a whole, this assembly of replication machines has been termed the replisome. The mass of the replisome approaches 1 million Daltons, and together, two replisomes move along the parental chromosome in opposite directions simultaneously replicating the nascent daughter chromosome from a single point of initiation and meeting at the point of termination. This process of replicating of the entire chromosome lasts about 1 hour for a pair of replisomes; however, under some conditions dichotomous replication of the chromosome can occur reducing the time of chromosomal replication to nearly 20 minutes. Up to six replisomes, at six separate replication forks, can be simultaneously synthesizing three daughter chromosomes from the single original parental chromosome during this process.

DNA Polymerase III Holoenzyme

At the heart of the replisome is the DNA polymerase III holoenzyme (pol III holoenzyme). DNA polymerase III was identified as the essential polymerase for *E. coli* chromosomal replication, distinct from the two previously discovered activities of DNA polymerase I, and II by analysis of mutants temperature sensitive for DNA synthesis and for cell viability (Gefter et al., 1971; Hirota et al., 1972).

Pol III holoenzyme replicates DNA in a semidiscontinuous manner. Pol III holoenzyme is capable of megabase processivity for continuous synthesis of the leading strand, and responsible for controlled 1 – 2 kilobase Okazaki fragment synthesis of the lagging strand. Pol III holoenzyme performs these activities simultaneously at speeds approaching 1 kilobase per second, and has an extraordinary fidelity of a single nucleotide misincorporation in 1 x 10^9 nucleotides polymerized (Kornberg and Baker, 1992).

Through purification and biochemical characterization of DNA polymerase III and accompanying proteins found to be involved in synthesis elongation, the individual subunits of this machine began to be identified. Pol III holoenzyme consists of 10 distinct subunits: α, ε, θ, τ, γ, δ, δ', χ, ψ, and β. Five of these subunits are present in two copies bringing the total number of pol III holoenzyme subunits to 20.

The core of pol III holoenzyme (pol III core) was later purified and resolved into individual subunits: α, ε, θ (McHenry and Crow, 1979). The α subunit encoded by the *dna*E gene (M_r = 140,000 Daltons), was found to be the subunit responsible for catalysis of 5'-3' DNA synthesis activity (Maki et al., 1985; Spanos et al., 1981). The ε subunit encoded by the *dna*Q gene (M_r = 25,000 Daltons) is the 3'-5' exonuclease or proofreading subunit (Scheuermann and Echols, 1984). The θ subunit encoded by the *hol*E gene (Studwell-Vaughan and O'Donnell, 1993) (M_r = 10,000 Daltons) is tightly bound to α and ε in pol III core, but a θ-specific function has not been identified. The α subunit was further characterized after overexpression and purification of the *dna*E gene product (Maki et al., 1985). Gap-filling replication assays and complementation assays were performed to show that the α subunit was responsible for the 5'-3' polymerase activity of pol III holoenzyme, and that elevated levels of α *in vivo* did not increase the amount of pol III holoenzyme in the cell.

Within pol III core, the α and ε subunits are tightly bound and complement each other's function with the overall effect of increasing the fidelity of chromosomal replication. Using purified α and ε subunits, (Maki and Kornberg, 1987) showed that a complex formed of α-ε had increased 5'-3' polymerase activity over α subunit alone, and highly increased 3'-5' ε-exonuclease activity. The affinity for DNA of the α subunit in pol III core resulted in an increase in apparent affinity of ε for the 3'-hydroxyl terminus, thus stimulating the exonuclease activity. In the cell, this proofreading activity during synthesis was found to represent a 5-fold stimulation compared to exonuclease activity uncoupled from synthesis which suggested that the fidelity of DNA replication may be controlled by the relative abundance of the α and ε subunits.

Polymerase III core is not processive in DNA replication whereas polymerase III holoenzyme is highly processive. Using a kinetic assay, along with product size determination assays it was originally shown that pol III holoenzyme was processive over thousands of nucleotides, and pol III core had distributive activity, capable of synthesizing 10-30 nucleotide stretches only (Fay et al., 1981).

It was found that pol III holoenzyme could be isolated from cells in two distinct forms other than pol III core, each form catalyzing synthesis of a characteristic length of product DNA (Fay et al., 1982; LaDuca et al., 1983). Polymerase III' was the first of these smaller forms of pol III to be purified. At the time pol III' was purified and characterized, the τ subunit was discovered and predicted to dimerize the pol III core assemblies forming pol III' (McHenry, 1982). The processivity of pol III' was increased 6-fold over pol III core, and pol III' exhibited greater ability in synthesizing long single-stranded templates coated with spermidine, a polybasic amine that stabilizes the helical structure of DNA. The Polymerase III* form was also purified, and contained all of the same subunits found in pol III holoenzyme except the β subunit-dimer. The activity of pol III* showed an increase in processivity of at least 20-fold over pol III core (LaDuca et al.; 1983). When pol III* was reconstituted with the β dimer, the characteristic processivity of the holoenzyme was restored, suggesting β as the major factor in conferring processivity on pol III holoenzyme.

Pol III holoenzyme must place the β dimer on primers formed by primase in order to form preinitiation complexes. The formation of these preinitiation complexes requires ATP binding and hydrolysis by pol III holoenzyme, and is absolutely required for initiation of processive synthesis (Burgers and Kornberg, 1982). The β subunit binds directly to the α subunit within pol III holoenzyme (Stukenberg et al., 1991), and pol III holoenzyme is dimerized by the τ subunit by direct interaction also with α (Studwell-Vaughan and O'Donnell, 1991). Through these interactions, pol III holoenzyme simultaneously and processively synthesizes DNA. Therefore, preinitiation complexes must be formed at least once on the leading strand for its continuous synthesis, and many times on the lagging strand for Okazaki fragment synthesis.

Fragmented synthesis of the lagging strand by pol III holoenzyme is in conflict with the level of processivity gained through binding the β processivity subunit. The polymerase must detach from each completed Okazaki fragment and rapidly cycle to the next preinitiation complex on the lagging strand (O'Donnell, 1987). During chromosomal replication, anywhere from 2,000 to 4,000 Okazaki fragments are synthesized at a rate of approximately 1 per second (Kornberg and Baker, 1992). Such polymerase cycling activity requires precise coordination of protein-protein and protein-DNA interactions at the replication fork. The coordinated efforts of several subunits, including processivity proteins within pol III holoenzyme are responsible for this rapid and ordered activity, and consequentially give the holoenzyme structural asymmetry, while the DNA synthesis activity of the polymerase cores remain symmetric between the leading and lagging strands.

The τ Subunit is the Coordinator of Pol III Holoenzyme Function and Processivity

The *dna*X gene codes for both the τ and γ subunits of pol III holoenzyme. The full-length *dna*X gene product is the τ subunit (M_r = 71,000 Daltons). The γ subunit (M_r = 47,500 Daltons) is created by a –1 translational frameshift adjacent to a hairpin-loop structure in the mRNA that causes a stop codon (UGA) to appear approximately two-thirds the way through the *dna*X gene (Flower and McHenry, 1990; Tsuchihashi and Kornberg, 1990). Therefore, the γ polypeptide is identical in sequence with the first two-thirds of τ. This translational frameshift occurs with about 50% yield of τ and γ polypeptides. Although they are extensively identical, the larger τ polypeptide has several important functions specific to its C-terminus, which greatly distinguishes it from γ (Gao and McHenry, 2001).

A dimer of τ subunits ($τ_2$) acts as the "glue" of pol III holoenzyme, holding it together and providing a structural scaffold for the asymmetric function of holoenzyme on the leading and lagging strands. A tight interaction between $τ_2$ and the α subunit of core occurs through the C-terminus of τ bridging two molecules of pol III core (Kim et al., 1996b; Studwell-Vaughan and O'Donnell, 1991). The C-terminus of τ also binds to the DnaB helicase, physically linking the replication machine of pol III holoenzyme with the primosome machine composed of DnaB and primase. The physical and communications link between pol III holoenzyme and the primosome enhances the DNA unwinding activity of DnaB helicase and connects pol III holoenzyme to the RNA priming activity involved in Okazaki fragment length determination and polymerase cycling on the lagging strand (Dallmann et al., 2000; Kim et al., 1996a). A single DnaB helicase couples with both the leading and lagging strands, and through interaction with the τ-bridged polymerase may help to keep both strands closely associated with the replication fork.

The τ subunit protects the β processivity protein from removal off of the leading strand indicating a direct role of τ in the high processivity of the leading strand (Kim et al., 1996c). This activity is most likely communicated through the close proximity of τ-α, and β–α binding sites on the C-terminus of α. Within this proximity, τ could potentially contact β providing direct protection from removal, or the τ-α interaction may cause a rearrangement of the β-core complex preventing removal of β. On the lagging strand, τ still protects the β processivity protein, but must be able to switch between protective and non-protective states to allow polymerase cycling. Two distinct triggers for polymerase cycling on the lagging strand have been identified (Li and Marians, 2000). The dynamic action of primase binding the replisome through DnaB helicase and synthesizing a primer is thought to trigger polymerase cycling, and collision of the lagging strand polymerase with the 5'-end of the previously synthesized Okazaki fragment also is thought to trigger cycling. With respect to the second case, a processivity switch requiring the τ subunit was identified (Leu et al., 2003; Wu et al., 1992). This "τ processivity switch" is "off" during processive synthesis, conferring protection of the β-core interaction with high processivity for completed synthesis of the Okazaki fragment. The processivity switch is turned "on" only upon incorporation of the final dNTP of the Okazaki fragment when the 5'-end of the previously synthesized primer is reached. The resulting nick in DNA activates the processivity switch, and through actions of the C-terminus of τ, core is released from β and allowed to cycle to the next preinitiation complex.

Pol III holoenzyme contains the processivity proteins necessary to form preinitiation complexes at newly primed sites on template DNA. The τ subunit is also part of this complex of processivity proteins that coordinates the protein-protein and protein-DNA interactions required for preinitiation complex formation. Unlike the functions of the τ C-terminus, τ activity in preinitiation complex formation is located in the N-terminal region of the protein identical to the γ subunit. A DnaX-complex forms within pol III holoenzyme which functions in loading β onto primed template DNA. For preinitiation complex formation, this "clamp loader" is discussed in detail below. The DnaX complex within holoenzyme contains two τ subunits (i.e., same two that dimerize core) in a complex with γ, and the δ, δ', χ, & ψ subunits (stoichiometry: $\tau_2\gamma\delta\delta'\chi\psi$) (Pritchard et al., 2000). The DnaX complex binds and hydrolyzes ATP forming preinitiation complexes on leading and lagging strand primers in what may be an ordered activity. In their report, (Glover and McHenry, 2001) showed that formation of the leading strand preinitiation complex required ATP binding but not hydrolysis by DnaX complex, and that ATP hydrolysis was required for formation of the lagging strand preinitiation complex. Further, a non-hydrolyzable analogue of ATP (ATPγS) caused removal of the polymerase presumably bound to a lagging-strand template, extending the knowledge of pol III holoenzyme as an intrinsically asymmetric dimer with distinguishable leading and lagging strand polymerases.

Structure of the β Sliding Clamp Processivity Protein

When the X-ray crystal structure of the β processivity protein was solved, it was immediately clear how this homodimer conferred processivity to polymerase III holoenzyme (Kong et al., 1992). The β sliding clamp is composed of two identical

crescent-shaped monomers arranged in head-to-tail fashion (Figure 2-1).

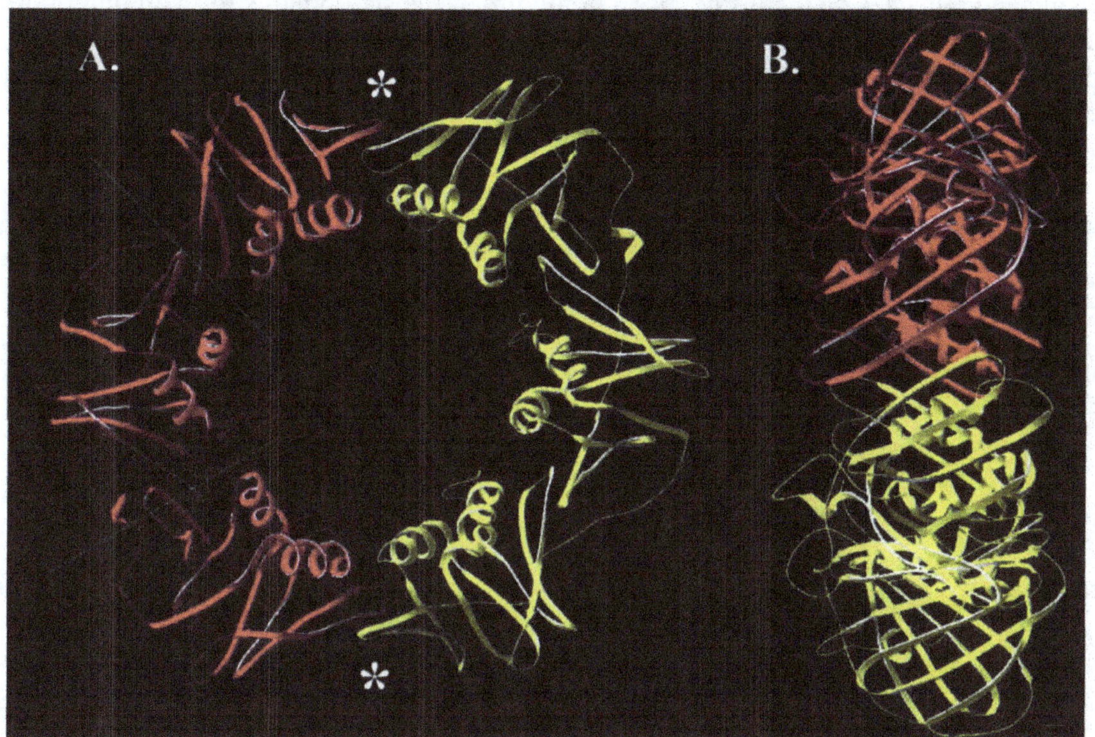

Figure 2-1. The crystal structure of β sliding clamp. Two views of the clamp composed of crescent-shaped protomers (red / yellow) are shown. A) Front view of β showing continuous β-sheet structure surrounding an inner core of α-helices. Asterisks denote the dimer interfaces. Green lines drawn through the left protomer represent imaginary boundaries between structural subdomains. From the top of the left (red) protomer, the subdomains are numbered counter clockwise: 1 (N-terminal), 2 (middle), and 3 (C-terminal). B) Side view of the β dimer showing the asymmetry of the faces. On the left face is the extended-loop region to which γ complex and pol III α subunit compete for binding. Images were composed using DeepView / Swiss-Pdb Viewer Ver. 3.7, http//www.expasy.org/spdbv/, with structural coordinates downloaded from the Protein Data Bank (Berman, Westbrook et al. 2000), http//www.rcsb.org/pbd/. (PDB code: 2POL)

The β clamp topologically links the polymerase to template DNA in a non-specific

manner providing a tight interaction with DNA while allowing essentially unlimited two-

dimensional diffusion along the template. Several structural characteristics of β clamp

confer its extraordinary ability to provide processivity to DNA polymerase. Each β monomer is formed from three structurally similar domains containing β sheets on the outside and α helices on the inside. As a homodimer, the six domains form an inner pore lined with 12 α helices that are surrounded by an essentially continuous β sheet structure. The axis of the α helices lining the inner pore are almost exactly perpendicular to the local direction of the phosphate backbone of DNA facilitating rapid movement along the duplex. For example, the α helices traverse the major and minor grooves so as not to fall into them. The height of the β clamp is ~ 80 Å. The width of β clamp (~ 34 Å) would cover about one full turn of duplex B-form DNA, or ~ 10 base pairs. It was shown biochemically that an 11-base pair primer was required by β to form a productive preinitiation complex synthesized by pol III (Yao et al., 2000). The diameter of the inner pore is ~ 35 Å, enough space to encircle either duplex B-form DNA (~ 20 Å) or a RNA – DNA hybrid duplex (i.e. a RNA-primed DNA template) (~ 21 Å). The extra space within the inner pore, between the clamp and DNA, is expected to be filled by one or two layers of water molecules.

The two faces of the β clamp are asymmetric (figure 2-1B). One face has extended loop C-terminal structures, and the other face is essentially flat giving the clamp an overall shallow cone-shaped thickness. The extended loops on the extended C-terminal face contain the binding sites for both the DnaX clamp loader and the α subunit of pol III core. The presence of overlapping binding sites for both the clamp loader and polymerase determine the proper orientation of the β clamp when loaded onto primed-template DNA and result in competition for the binding sites. DNA mediates this competition for overlapping binding sites. In the absence of DNA, β clamp binds the clamp loader, but when β is loaded onto DNA, the DNA causes the clamp loader to lose affinity for the clamp, and polymerase binds (Naktinis et al., 1996).

Calculation of the electrostatic field generated by the β clamp revealed that the protein has an overall negative electrostatic potential. However, the surface lining the inner pore has a focused positive electrostatic potential, precisely where the negatively charged DNA molecule would be. The calculated electrostatic features suggest that the inner pore of the clamp has inherent affinity for DNA. This electrostatic interaction is most likely lubricated by water molecules allowing a great degree of freedom for linear diffusion while still maintaining some affinity for DNA.

The dimer interfaces appear as a continuation of the β sheet structure from the outer surface of each monomer across a molecular boundary, and contribute at least four hydrogen bonds at each interface. In addition, two distinct sets of interactions between neighboring α-helical side chains stabilize the β dimer interface. A small hydrophobic core is formed in the center of the interface by the packing of Phe and Ile side chains with Ile and Leu side chains between monomers. Six potential intermolecular ion pairs between Glu, Arg, and Lys side chains surround the small hydrophobic core. The β dimer consequentially is very stable with a monomer-dimer dissociation constant of <60 pM, and a half-life on closed-circular DNA of ~ 100 minutes (Yao et al., 1996). Considering that the cellular concentration of β is ~250-500 nM, no monomeric β would be present *in vivo* (Kornberg and Baker, 1992). The number of specific and potentially strong interactions at the dimer interfaces underscores the requirement for a clamp loading machine to load this clamp on DNA.

The DnaX Clamp Loading Machine

Three major parts make up the pol III holoenzyme within the *E. coli* replisome. The replicative polymerase III core, the DnaX complex that contains the τ subunit –the cement of pol III holoenzyme, and the β sliding clamp processivity protein. The DnaX complex has an additional function in polymerase processivity by loading the β clamp onto primed-template DNA for preinitiation complex formation. Originally identified as the γ-δ complex or γ complex (Maki and Kornberg, 1988; O'Donnell, 1987), this DnaX complex was shown to load β onto DNA in an ATP-dependent manner (Onrust et al., 1991).

The τ and γ subunits are expressed at roughly equal levels in cells due to a translational frameshift in the DnaX gene (Flower and McHenry, 1990). Therefore it was expected that a mixed τ / γ containing DnaX complex would reside in the pol III holoenzyme. In fact, several DnaX complexes expressed from an artificial operon were isolated from cells including two forms of mixed τ / γ complex, and a γ-only complex (Pritchard et al., 2000). No τ-only complex was isolated from cells in that study. One of the two isolated forms of mixed τ / γ complex had the stoichiometry of $\tau_2\gamma_1$ of the DnaX gene products, and this form was predicted to function within the pol III holoenzyme based on the requirement of the τ subunit dimer for holoenzyme function. The separate γ_3 only complex, and / or the mixed τ_1 / γ_2 complex were predicted to function separately from holoenzyme, for example, in DNA repair activity with DNA polymerase II (Bonner et al., 1992).

Identification of all five of the subunits of γ complex (Maki and Kornberg, 1988), and the genes encoding them (Dong et al., 1993; Xiao et al., 1993a), lead to *in vitro* reconstitution and use of this γ-only complex as the model β sliding clamp loader (Onrust et al., 1995). Recently, sequence and structural analysis have identified several subunits of γ complex as AAA+ superfamily members and γ complex as a AAA+ molecular machine (Davey et al., 2002). These studies have provided significant insight into the mechanism of opening the stable ring-shaped dimer structure of β clamp and loading it onto DNA at the speed required by polymerase III catalyzed Okazaki fragment synthesis.

The γ complex contains five subunits with the stoichiometry: γ_3, δ_1, δ'_1, χ_1, and ψ_1. The χ and ψ subunits bind a γ subunit through ψ (Glover and McHenry, 2000), and greatly increase the affinity for the $\delta\delta'$ subunits to the complex (Olson et al., 1995). The γ subunit, alone in solution, was found to exist in a monomer-tetramer equilibrium, however when formed in a complex with $\delta\delta'$ became a trimer of γ_3 (Pritchard et al., 2000). The structural asymmetry derived by single copies of the δ, δ', χ, and ψ subunits of the DnaX clamp loader within holoenzyme imparts structural asymmetry in the holoenzyme, and implicates their function with the lagging strand polymerase.

The δ and δ' subunits, products of the *holA* and *holB* genes, respectively, were purified and characterized for their interactions and function in γ complex (Dong et al., 1993). These subunits have similar mass, (δ: $M_r = 38,700$ Daltons, and δ': $M_r = 36,900$ Daltons), and high affinity for each other and were usually co-purified as a consequence. The δ subunit binds β through a "β interaction element" (βIE), while δ' binds tightly to the γ subunit. During the clamp loading cycle, an "inactive" form of γ complex, the δ' subunit is thought to occlude δ from binding β and a conformational change upon binding of ATP to γ complex causes exposure of the βIE of the δ subunit (Naktinis et al., 1995). Genetic knockouts of the *holA* or *holB* genes are not viable revealing that the function of δ in clamp loading absolutely requires the δ' subunit (Song et al., 2001). Song, Pham et al. (2001) also showed that δ and δ' are required not only for ATP-dependent preinitiation complex formation, but are essential for DNA synthesis elongation as well. They participate in the DnaX clamp loader complex function in coupling to DnaB helicase at the replication fork to pol III holoenzyme for leading and lagging strand synthesis.

The δ subunit alone was found to have the ability to unload β from DNA without any requirement for ATP binding or hydrolysis (Leu et al., 2000). In this study, the approximate quantity of each subunit of γ complex was determined from whole cell lysates by protein absorbance measurements and western blot analyses. Surprisingly they discovered a ~4-fold excess of the δ subunit present separate from any within γ complex. In Okazaki fragment synthesis, about 2000 to 4000 β clamps are used for complete replication of the lagging strand. However, with only about 300 molecules of β present in the cell, there is a need to recycle the clamps left behind on the DNA as the lagging strand polymerase cycles between Okazaki fragments. To keep up with the pace of Okazaki fragment synthesis β must actively be unloaded at a rate of at least 0.01 s^{-1}. It was found that the quantity of γ complex within the cell was limited to ~140 molecules by the δ' subunit. Since pol III holoenzyme and any free γ complex are present in low quantities and perform several DNA metabolic functions such as replication and repair, it was envisioned that the excess δ subunit aided in removal of β clamps abandoned by pol III at the end of each Okazaki fragment. In their investigation (Leu, Hingorani et al., 2000) also showed that *in vitro* removal of β from circular DNA substrates by purified δ happened at a rate of 0.011 s^{-1}. This rate was comparable to measured rates of β unloading by γ complex (0.015 s^{-1}) and pol III holoenzyme (0.007 s^{-1}) in similar assays.

As all of the δ' subunit is expected to be stoichiometrically associated with γ complex, the δ' subunit would only regulate the activity of δ in clamp loading, leaving free-δ subunit to recycle leftover β clamps, keeping a sufficient level of free sliding clamp available for rapid Okazaki fragment synthesis.

Although not necessary for *in vitro* clamp loading activity, γ complex contains the χ and ψ subunits. The χ (M_r ~16,600 Daltons) and ψ (M_r ~15,000 Daltons) subunits are the products of the *holC* and *holD* genes, respectively, and have important functions adapting the lagging strand polymerase for Okazaki fragment synthesis (Onrust et al., 1995; Xiao et al., 1993a). As stated above, the ψ subunit binds to γ, bridging the interaction of the χ subunit to the complex. The χ-ψ dipeptide forms a rod-like structure (Kelman et al., 1998), however it is not known where on the γ subunit χ-ψ binds in the clamp loader complex (Jeruzalmi et al., 2001a). A sub-complex of γ−χψ, or χ-ψ alone had no independent function in promoting DNA synthesis activity by pol III (Xiao et al., 1993b).

The χ subunit through the ψ structural bridge to the γ subunit in the DnaX complex strengthens DNA polymerase III holoenzyme interactions with the single-stranded DNA binding protein (SSB)-coated lagging strand template facilitating preinitiation complex formation and processive synthesis elongation. Isolated pol III core or pol III', which are without the χ-ψ subunits, are inhibited in synthesizing DNA on SSB-coated template DNA (Glover and McHenry, 1998). Removal of χ-ψ from pol III holoenzyme results in reduced efficiency of clamp loading activity and an increase in salt sensitivity in the physiological range (40-150 mM) (Kelman et al., 1998). Investigation of a common point mutation in SSB (SSB-113) revealed that the direct interaction with the χ subunit is located at the C-terminus of SSB, and that this interaction is most likely hydrophobic in nature. Through this direct χ-SSB contact on template DNA, the χ subunit was thought to provide resistance to elevated ionic strength for proper clamp loading activity, as well as possibly keeping the lagging strand polymerase localized to the SSB-coated template for greater efficiency in synthesis elongation.

The χ subunit has been found to be required in a primase to polymerase handoff for Okazaki fragment synthesis. A "three-point switch" was proposed where the χ subunit, through direct interaction with SSB, causes primase to dissociate from a newly formed primer allowing initiation complex formation and synthesis of the Okazaki fragment to proceed (Yuzhakov et al., 1999). Primase (DnaG) interacts distributively with DnaB helicase in formation of the primosome for controlled RNA primer synthesis (Tougu and Marians, 1996). After synthesizing a primer, primase remains attached to the nascent primer, and is thought to protect it from exogenous exonuclease activity. Primase is also directly attached to SSB, and therefore blocks initiation complex formation. Through the χ-SSB interaction, but not a direct χ-primase interaction, primase is displaced, perhaps through a conformational change in SSB. The τ-DnaB helicase interaction provides a contact point to hold pol III holoenzyme to the replication fork while lagging strand polymerase cycles (Kim et al., 1996a). Contact of pol III holoenzyme to the newly primed template, and bringing the χ subunit through the DnaX clamp loader complex into the vicinity of the SSB-coated lagging strand to displace primase through SSB would allow preinitiation complex formation for polymerase cycling for rapid lagging strand Okazaki fragment synthesis.

The Dna X (τ or γ) subunits of the clamp loader bind ATP (Tsuchihashi and Kornberg, 1989). In the reconstituted γ complex clamp loader, interaction with the $\delta\delta$' subunits causes a solution-monomer-tetramer equilibrium of γ subunit to form a trimer giving the clamp loader a stoichiometry of γ_3, δ_1, δ'_1, χ_1, and ψ_1 (Pritchard et al., 2000). The γ subunits in isolation (i.e. in monomer-tetramer equilibrium) do not hydrolyze ATP at an appreciable rate with respect to activity needed for replication of the lagging strand. However, when reconstituted into a complex with the δ,δ',χ and ψ subunits, or just the δ, and δ' subunits, the rate of ATP hydrolysis is much higher, and further enhanced in the presence of β to a rate that would be rapid enough for preinitiation complex formation for Okazaki fragment synthesis (i.e. ~ 2 s^{-1}) (Onrust et al., 1991). This means that the clamp loader can bind and hydrolyze a maximum of three molecules of ATP to power the clamp loading reaction. Since the γ / τ subunits of the clamp loader are the only subunits which bind and hydrolyze ATP to power the clamp loading reaction they are known as the motor subunits of the clamp loader.

The AAA+ Superfamily of Motor Proteins

Through sequence analysis and ultimately structural analysis the γ, δ, and δ' subunits of the clamp loader have been identified as members of the AAA+ superfamily of ATPases (Davey et al., 2002; Neuwald et al., 1999). As the name of this superfamily implies: "ATPases Associated with a variety of cellular Activities", proteins of this class are involved in many cellular processes. Whole genome analyses have indicated that the AAA+ class is ancient and has undergone substantial functional divergence prior to the emergence of the major divisions of life, thus this class is large and diverse (Neuwald et al., 1999). Proteins of the class generally assist in protein transitions, such as remodeling, assembly, and disassembly. AAA+ proteins assist not only in protein-protein transitions, but also in protein-nucleic acid transitions.

AAA+ proteins including GTPase proteins, which are mechanochemical proteins that transduce the energy of nucleotide hydrolysis into useful work, are becoming recognized as a quickly growing group of "energase" enzymes (Purich, 2001). An energase can simply be defined as an enzyme that couples the change in energy of a covalent bonding state directly to changes in non-covalent substrate- and product-like states. For AAA+ machines such as the γ complex clamp loader, the chemical energy of ATP hydrolysis activity is transduced into mechanical changes in the clamp loader structure. As a AAA+ machine, γ complex clamp loader could be described as a molecular matchmaker. The γ complex modifies a protein (i.e., β clamp) such that it becomes enabled for interaction with DNA in such a way that it would not normally interact.

The AAA+ superfamily includes chaperone-like ATPases such as regulatory components of proteases, transcriptional regulators, vesicle synthesis and fusion proteins, and dynein motor proteins to name a selected few. Aside from AAA+ proteins of γ complex, other important replication proteins are members of the AAA+ class. For example, bacterial DnaA, DnaC, and RuvB proteins, and the subunits of the eukaryotic origin recognition complex, Mcm2-7 helicase, and replication factor-C subunits of the eukaryotic clamp loader (Erzberger et al., 2002; Neuwald et al., 1999).

Increasing investigation of AAA+ proteins reveal that adaptor proteins may modulate their functions. As a consequence, a growing list of these structurally unrelated adaptor proteins are being identified for many of the known AAA+ proteins (Dougan et al., 2002). Generally smaller than their AAA+ partner, AAA+ adaptor proteins provide a simple and effective way to modulate the function of the AAA+ machine. Adaptor proteins have been found to give the AAA+ protein better control over substrate specificity, allowing quick response to changing local conditions and redirection of AAA+ activity. The γ complex has two proteins, χ and ψ, that are not related in sequence or structure to the other AAA+ class subunits of the clamp loader. Questions concerning the function of χ and ψ in the clamp loading reaction by γ complex are of major importance in this dissertation, and it is attractive to propose that they are in fact adaptor proteins in complex with the AAA+ clamp loading machine. As described above, the χ subunit, through ψ, has important interactions with SSB at the replication fork for preinitiation complex formation. As will be discussed in the results of this dissertation, χ and ψ are also important for activity beyond interaction with SSB at the replication fork, and are important for optimal activity of the clamp loading machine itself. This dissertation explores the structural and conformational stability provided to γ complex by the χ and ψ proteins, and therefore may reveal a novel example of how adaptor proteins modulate the activity of their partner AAA+ machines. The X-ray crystal structures of a minimal $\gamma_3\delta\delta'$ clamp loader and a χ-ψ complex have been solved, but it still remains a mystery as to where the χ-ψ dipeptide binds to the γ subunit of the clamp loader. The limited examples of AAA+-adaptor protein complexes have shown that the adaptor proteins generally bind to the least conserved N-terminal domain of their AAA+ partner. This fact may provide direction for further structural analysis of the complete γ complex including the χ and ψ proteins.

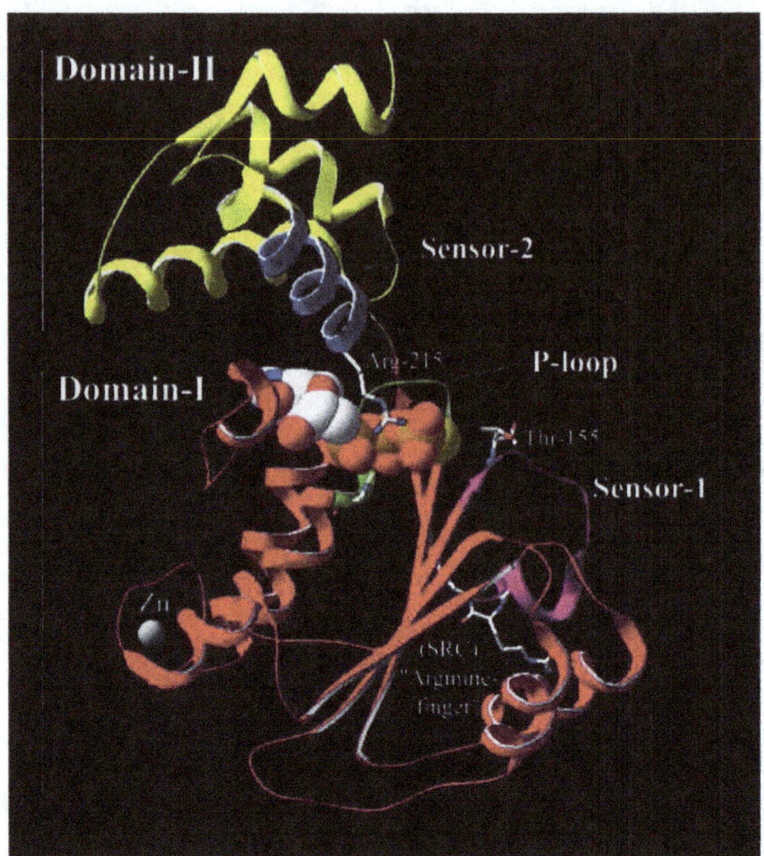

Figure 2-2 The crystal structure of a AAA+ motor protein. The γ^{1-243} crystal structure (Podobnik, Weitze et al. 2003) is shown here as a representative AAA+ motor protein. This structure was truncated at the C-terminus, and only the AAA+ domains-I (red) and –II (yellow) are shown. A molecule of ATPγS bound to the NTP-binding site is shown in space-filling representation. The conserved structural and functional features of AAA+ motors are indicated: Sensor-2 (blue), P-loop (green), and Sensor-1 (pink). Also marked are the conserved Arg (γ-215) of sensor-2 and Thr (γ-155) and (SRC) "Arginine (γ-169)-finger" of sensor-1. A Cys-coordinated Zn^{++} atom is also present in Domain-I. Images were composed using DeepView / Swiss-Pdb Viewer Ver. 3.7, http//www.expasy.org/spdbv/, with structural coordinates downloaded from the Protein Data Bank (Berman, Westbrook et al, 2000), http//www.rcsb.org/pbd/. (PDB code: 1NJF)

AAA+ proteins have several common structural features that define their function

(Figure 2-2). All AAA+ proteins have an approximately 220 amino acid conserved

structural ATPase core, although sequence is not always conserved (Jeruzalmi et al.,

2001b; Neuwald et al., 1999). The AAA+ conserved structure has two domains (I and

II), and a third oligomerization domain (III) is common in multimeric complexes of AAA+ protein subunits. AAA+ proteins generally contain the phosphate-loop (P-loop)-type NTP-binding site having the Walker-A and Walker-B motifs that are common in ATPase and GTPase proteins. The Walker-A motif (GVxxxGKT) is involved in the phosphate binding of ATP (or GTP), and the Walker-B motif (DExx) is involved in metal (Mg^{++}) binding and catalysis (Walker et al., 1982).

The P-loop α,β-fold provides a platform upon which other AAA+ structural motifs are mounted. Above the P-loop, in domain-II is the sensor-2 motif containing a highly conserved Arg residue (Arg-215 in the γ subunit). This Arg residue is predicted to interact with the β- and γ-phosphate of bound ATP (Podobnik et al., 2003). Domain II containing the sensor-2 motif is typically a hinge-like domain around which movements in domains-I and III are made.

In domain-I, flanking the P-loop from the underside is the sensor-1 motif. The sensor-1 motif contains the highly conserved Ser-Arg-Cys (SRC) sequence that is predicted to interact with the γ-phosphate of the bound ATP through hydrogen bonding. The SRC sequence includes an "arginine-finger". This arginine-finger is thought to stimulate catalysis of ATP hydrolysis by analogy to the similar, conserved switch II region of GTPase activating proteins (Ahmadian et al., 1997). In an oligomeric AAA+ complex such as γ complex, the SRC-sensor-1 motif of one AAA+ subunit interacts with the nucleotide-binding site of a neighboring AAA+ subunit, and is implicated in interfacial stimulation of ATP hydrolysis (Jeruzalmi et al., 2001a; Podobnik et al., 2003). The X-ray crystal structure of the γ complex AAA+ clamp-loading machine is described below. It has five AAA+ proteins that make up three parts of the clamp loading machine:

three γ-ATPase-motor subunits, the δ subunit, harboring the β interaction element (βIE), and the δ' "stator", upon which movement of the other subunits is supported.

X-ray Crystal Structure of the Clamp Loading Machine

The X-ray crystal structure of a reconstituted minimal clamp loader complex ($\gamma_3\delta\delta'$) was solved at a resolution of 3.0 Å using multiwavelength anomalous diffraction methodology (Jeruzalmi et al., 2001a). The subunits crystallized as a heteropentameric circle having the stoichiometry (δ'_1-γ_3-δ_1) (Figure 2-3).

While the amino acid composition of the δ and δ' subunits was complete, all three γ subunits were truncated by 57 amino acids from protease-labile C-terminii. All subunits showed a three domain "C"- shape previously observed in the structure of the δ' subunit alone (Guenther et al., 1997), giving the complex an overall form analogous to five fingers extending down from the palm of a hand. Both of the AAA+ domains I and II were observed for all five subunits. This was a surprising observation for the δ subunit, as it shared only 7-8 % sequence identity with other AAA+ proteins including those within the clamp loader. All of the ATPase motor γ subunits crystallized in nucleotide-free form, and the δ' and δ subunits were already known not to bind ATP. The X-ray crystal structure of a truncated form of the γ subunit (γ^{1-243}) including the AAA+ domains I and II only was later solved in the presence of nucleotide (Podobnik et al., 2003), and will be used for the discussion of the nucleotide binding sites and consequences of nucleotide binding the γ motor subunit.

The C-terminal domain III of all five subunits of the clamp loader forms a tight circular collar at the "top" of the complex (Figure 2-3B). The subunits are arranged where the δ' and δ subunits surround the three γ subunits.

Figure 2-3. The crystal structure of the $\gamma_3\delta\delta'$ clamp loader. The clamp loader
forms a circular heteropentameric structure, each subunit having three
domains (I II and III). A) "Front"-view: The point of view is directly
into the "open" interface between the δ' (white) and δ (blue) subunits.
The three γ subunits (different shades of red) are present behind δ' and
δ. Structural domains are indicated: Domain-III (the C-terminal
"collar"), Domain-II ("hinge"), and Domain-I (N-terminal
"functional"). B) The clamp loader has been tilted forward 90° for a
slab-view of the "top" showing the tight pentameric "collar" of the C-
terminal domain-III only. Subunit nomenclature is indicated. C) The
clamp loader has been tilted back 90° from the view in (A), and
domains-II and III have been dimmed for the "bottom" view. The N-
terminal domain-I of each subunit is colored for emphasis on their
asymmetrical disposition. The NTP binding sites of the γ subunits are
occupied with sulfate molecules from crystallization (space-filling
representation), and the Zn^{++} atoms (yellow) bound to δ', γ_1, γ_2, and γ_3
subunits are shown. The β-interaction element (βIE), and the
conserved hydrophobic-wedge residues of the βIE (δ-Leu-73 and -Phe-
74) are shown (green). Images were composed using DeepView /
Swiss-Pdb Viewer Ver. 3.7, http//www.expasy.org/spdbv/, with
structural coordinates downloaded from the Protein Data Bank
(Berman, Westbrook et al, 2000), http//www.rcsb.org/pbd/. (PDB
code: 1JR3)

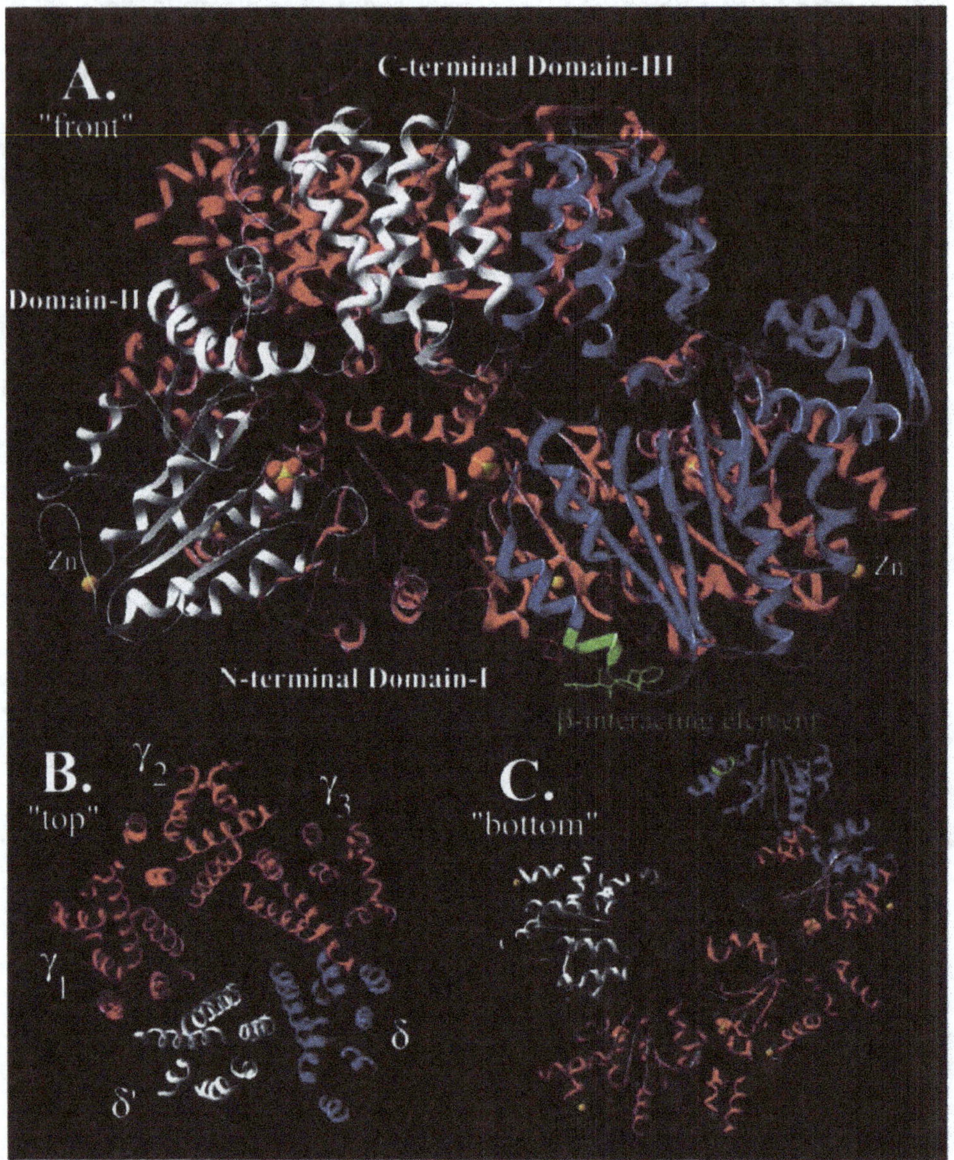

Hydrophobic interactions within a helical packing arrangement strengthen subunit interactions in this oligomerization domain. Only the C-terminal domain of the δ subunit appeared "loose" within this oligomerization domain, and was proposed to have consequences for movement of the δ subunit within the complex.

The N-terminal domains of the five subunits containing the AAA+ structural domains-I and II are highly asymmetric with respect to each other extending down from the C-terminal collar (Figure2-3C). Each N-terminal domain is increasingly splayed out with respect to the circular C-terminal collar forming an "open" σ-shape. Splaying-out of N-terminal domains in this structure resulted in considerable exposure of the δ subunit and thus the βIE. Based on biochemical studies, a nucleotide-free structure would be predicted to be more "closed" where the δ subunit-βIE would be obstructed by the δ' subunit. The authors speculate that the observed structure of the clamp loader may be an artifact of the crystal packing forces, but overall would reflect that of an "open" or activated complex.

The nucleotide binding sites of the γ subunits are buried within the subunit interfaces as is common with other oligomeric AAA+ ATPases. The AAA+ domains are arranged such that domains-I and II of one subunit cradle domain-I of the neighboring subunit providing a mechanism for conformational communication based on nucleotide binding status. Three interfacial nucleotide binding sites are between the δ'-γ_1, γ_1-γ_2, and γ_2-γ_3 subunits. The N-terminal AAA+ domains of the γ subunits were highly asymmetric in the complex, resulting in asymmetry in these interfacial nucleotide binding sites. The δ'-γ_1 interface and γ_2-γ_3 interface were open in this nucleotide-free structure, while the γ_1-γ_2 interface had the most extensive contact and therefore appeared inaccessible to nucleotide.

Structure of the Nucleotide Binding Site and the Proposed Conformational Change of the γ Subunit

The X-ray crystal structure of C-terminal truncated $\gamma^{1\text{-}243}$ was solved in the presence of ATPγS to a resolution of 2.2 Å using multiwavelength anomalous dispersion methodology (Podobnik et al., 2003). The resulting structure had four γ subunits arranged in the asymmetric crystal unit. One of these γ subunits had complete electron density corresponding to ATPγS in the nucleotide binding site, and another was missing electron density corresponding to the γ-phosphate, and therefore was taken as the structure of ADP-bound γ. Two other γ subunits in the asymmetric unit had no nucleotide bound. Slow hydrolysis of ATPγS through the crystallization procedure was thought to result in producing the ADP-bound γ. This structure allowed identification of several as yet unknown features of nucleotide binding and the conformational changes induced due to nucleotide binding. The results have important implications for the mechanism of communication between the AAA+ domains within the clamp loader and therefore the overall clamp loading mechanism.

The structure showed that the nucleotide was bound in the expected position in the cleft between domains-I and II formed by the Walker-A P-loop (see figure 2-2). The adenine ring was located in a non-polar environment, and formed hydrogen bonds from the N-6 position with the main chain carbonyl groups of Val-19 and Val-49. This feature provides a mechanism for distinguishing ATP from GTP by the γ subunit. Both the 2'- and 3'-hydroxyl oxygens of the ribose moiety were hydrogen-bonded to the carbonyl oxygen of Ala-7 of the N-terminal helix extending from domain I. Three residues of the conserved Walker-A P-loop motif coordinate the β- and γ- phosphates of the nucleotide. The Lys-51 residue forms a salt-bridge, and Thr-52 and Ser-53 form hydrogen bonds to the phosphate groups from the sensor-1 loop.

Comparison of the nucleotide-bound γ^{1-243} structure with the nucleotide-empty $\gamma_3\delta\delta'$ structure reveals that the γ subunit domain-II α-helix containing the sensor-2 motif blocked the region corresponding to the position of the ribose ring. This observation was extended to all of the empty nucleotide-binding sites in the $\gamma_3\delta\delta'$ structure. Thus it was predicted that nucleotide binding to a given γ subunit alters the position of domain-I with respect to domain-III by conformational change. The reorientation of domain-I would open up the nucleotide binding site from a collapsed state, and clearly affect the positions of the sensor-1 and sensor-2 motifs within the complex. Also, rotation of the sensor-2 region containing the conserved Arg-215 residue occurs upon nucleotide binding and provides space for the phosphate groups of the nucleotide. The sensor-1 motif does not move with respect to the P-loop, and therefore follows the general movement with the N-terminal domain-I upon nucleotide binding.

Conformational changes induced by nucleotide binding to the P-loop region are directly linked to movement of the catalytic "arginine-finger" of sensor-1, thereby coupling (i.e. communicating) changes in nucleotide binding status to the neighboring subunits of the complex. In the $\gamma_3\delta\delta'$ structure, the sensor-1 motif containing the arginine-finger, was either too far away from the neighboring nucleotide binding site that it is thought to stimulate in the case of the $\delta'-\gamma_1$ interface, or in steric-clash with the neighboring nucleotide binding site as in the case of the $\gamma_1-\gamma_2$ interface. Therefore, conformational changes in individual subunits must be made to correctly position the sensor-1 arginine-finger with the neighboring nucleotide for catalysis. These conformational changes that occur upon ATP binding not only bring the subunits into proper orientation for ATP hydrolysis, but more importantly cause exposure of the δ subunit from occlusion by the δ' stator so that it can bind β clamp.

"Crude" modeling of the β clamp bound to the clamp loader structure was performed using information from the $\delta-\beta$ structural interaction (described in detail below). This "simplistic" docking of dimeric β with the clamp loader revealed that the orientation of the $\delta-\beta$ interaction would cause the clamp to make contact with all subunits in the clamp loader (Jeruzalmi et al., 2001a). One of the key points addressed in this dissertation is the question of how β affects the clamp loader in the clamp loading mechanism. The multipoint contact suggested by the modeling of β bound to the clamp loader provides a structural basis for several conclusions concerning individual subunit conformational changes and how β may "trap" an ATP-bound "activated" clamp loader. This contact may potentially explain how the clamp loader stabilizes the open β ring while a single dimer interface remains open (see below) until a primer template junction is located.

The exact nature of how the clamp loader binds to DNA is not yet fully understood. A cysteine-coordinated zinc module is present in the δ' and all three γ subunits of the complex (figure 2-3) and may provide a interaction zone for duplex DNA, however, the function of the zinc modules in the clamp loader complex has not been investigated. Calculation of the surface electrostatic potential of the clamp loader structure invites some interesting speculation on the nature of where DNA binds and stimulates the activity of the clamp loader. Two extensive regions of positive electrostatic potential were observed on the structure. Both were seen on the N-terminal asymmetric "bottom" side of the clamp loader that is predicted to interact with the β clamp. One of these regions was on the δ subunit, adjacent to the βIE presumably where DNA would thread through the β clamp when bound to δ. The other region was observed as a partially continuous belt that tracks along the outer surface of the N-terminal domains of the γ subunits. Each of these regions of positive electrostatic potential predict that when DNA is encircled within the β clamp under the clamp loader, single-stranded DNA could be positioned to interact with the outer regions of the ATP binding domains causing additional rearrangement of sensor-2 and sensor-1 motifs stimulating hydrolysis of ATP. The requirements for formation of the DNA binding site of γ complex are further explored in the results of this dissertation.

The X-ray crystal structure of the δ subunit bound to a β monomer was also solved, and provides a possible mechanism of clamp opening by the δ subunit as well as how the β clamp may interact with γ complex through binding the δ subunit-βIE.

X-Ray Crystal Structure of the δ Subunit Bound to a β Monomer and the Mechanism for Opening the β Sliding Clamp

The X-ray crystal structures of two complexes between the δ subunit and a mutant monomeric form of β (β_1) were solved by multiwavelength anomalous diffraction (Jeruzalmi et al., 2001b). Attempts to crystallize the δ subunit with wild-type β_2 were unsuccessful, and therefore a mutant β that remained monomeric due to mutations within the dimer interface was used (Stewart et al., 2001). The δ: β_1 structures either contained full-length δ subunit, composed of all three domains observed in the γ complex structure including the AAA+ region, or truncated δ subunit (amino acids 1 – 140) δ^{1-140} containing the N-terminal domain I only. Both complexes of δ:β_1 and δ^{1-140}:β_1 crystallized in 1:1 complexes at resolutions of 2.9 Å and 2.5 Å respectively (Figure 2-4).

Examination of both structures revealed a mechanism by which β clamp is opened upon the δ subunit inducing or trapping a conformational change in β such that a stable dimeric interface cannot be formed. Some possibilities for ring opening were that δ forcibly breaks the dimer interface, but due to the fact that ATP is not required by δ to open β, this was not very probable. Another possibility was that the β clamp could oscillate between conformations and the δ subunit could have high affinity for a conformation inconsistent with ring closure.

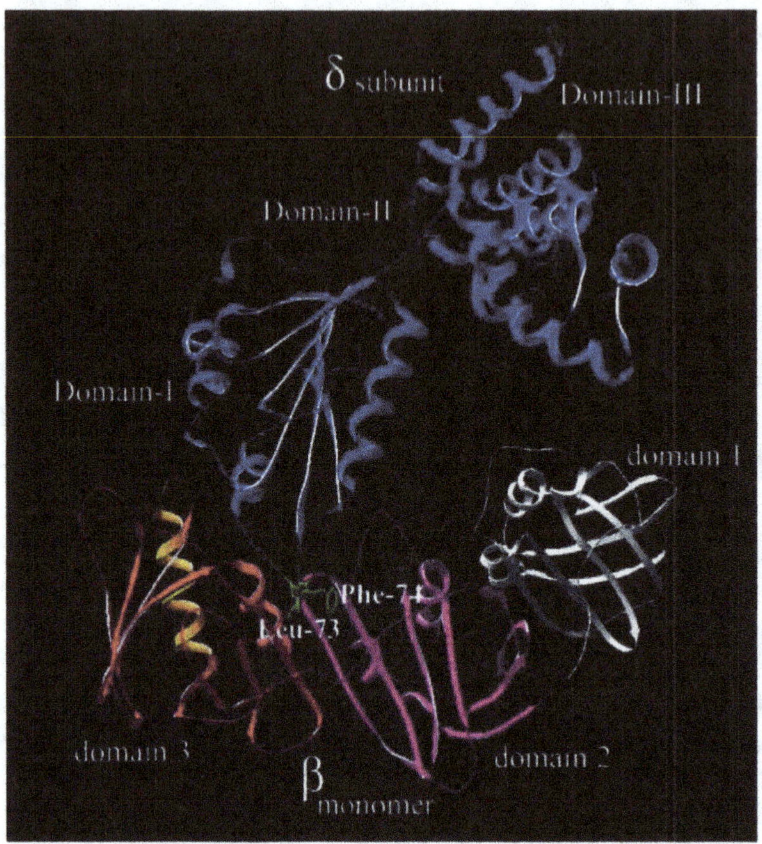

Figure 2-4. The crystal structure of the δ-β₁ complex. The full structure of the β subunit (blue) is shown bound to β₁ monomer. Domains (I II and III) of δ are indicated. The subdomains of β₁ are colored: N-terminal subdomain-1 (white), middle subdomain-2 (pink), and C-terminal subdomain-3 (red). The conserved hydrophobic-wedge, βIE (green) Leu-73 and Phe-74 of δ subunit are marked. The α-helix of the β subunit involved in the "spring-loaded" conformational change is colored yellow. Images were composed using DeepView / Swiss-Pdb Viewer Ver. 3.7, http//www.expasy.org/spdbv/, with structural coordinates downloaded from the Protein Data Bank (Berman, Westbrook et al.), http//www.rcsb.org/pbd/. (PDB code: 1JQJ)

This also was not a very probable mechanism due simply to the high stability of the β

dimer alone on circular DNA. A third possibility was that the δ subunit could induce and

stabilize a conformation in β that is inconsistent with ring closure. The results from

analysis of these structures are consistent with the third possibility, and it was proposed

that δ subunit binding induces a conformational change in the β clamp that results in straightening of a crescent-shaped β monomer such that a single dimer interface is opened to allow DNA passage into the clamp.

The βIE is on the N-terminal domain of the δ subunit, and consistent with earlier biochemical studies, makes contact with the C-terminal-extended-loop face of the β subunit (Naktinis et al., 1995). Only one δ subunit bound β_1 in the full length δ: β_1 structure, and it was observed that domains II and III of the δ subunit impeded binding of a second molecule of δ to β_1. In the δ^{1-140}:β_1 structure showed that the βIE forms a hydrophobic wedge that binds to a hydrophobic pocket on the β_1 subunit between subdomains 2 and 3 (of the β subunit) (Figure 2-4). Subdomains 2 and 3 are the middle and C-terminal structural domains of the β protomer (Kong et al., 1992). The Met-71, Leu-73, and Phe-74 residues form the hydrophobic wedge on the δ subunit. The hydrophobic pocket on β, into which this wedge is inserted, is formed by Leu-177, Pro-242, Val-247, Val-361, and Met-362 of β. The βIE hydrophobic wedge of the δ subunit is highly conserved among all bacterial δ subunit sequences examined. Likewise, the residues forming the hydrophobic pocket on β are also highly conserved across evolution, and therefore found on sliding clamps of bacteriophage, eukaryotes, and archaea (see below). Conservation of this interaction between the clamp loader and sliding clamp suggests a common mechanism for clamp opening in all known organisms.

These structures have revealed that interaction of δ with β requires conformational changes in β that weaken the dimer interface, and importantly, δ does not interact directly with the dimer interface. The structure of the β dimer interface consists of a hydrophobic core surrounded by ionic interactions (Kong et al., 1992). An α-helix participating in these interactions at the dimer interface is distorted in the wild-type dimeric β clamp such that additional hydrophobic residues contained on this distorted helix do not extend into the interface. The hydrophobic pocket binding site where the δ-subunit-βIE binds is near a 5 -residue loop adjacent to the distorted α-helix in β. The binding of δ is thought to impose strain on the distorted helix that causes its relaxation into a more ideal α-helical conformation. This results in extension of the hydrophobic residues from the previously distorted portion of the α-helix directly into the dimer interface. Removal of dimer interface stability was proposed to trigger release of spring-like tension in the crescent-shaped β monomer such that it becomes straightened. Thus the clamp is not simply an inert ring, but contains structural information for its own opening. The crescent shape of the β subunit to which δ bound was calculated to straighten by 12° at the unopened interface and by 5° at the opened interface by comparison with the structure of dimeric β.

Comparison of the δ subunit in the $\delta{:}\beta_1$ structure with the structure of the δ subunit from the $\gamma_3\delta\delta'$ structure revealed that the N-terminal domain of δ is in a considerably different position when bound to β. An α helix (α4), containing the βIE, in domain-I translates by ~5.5 Å and rotates ~45° with respect to the rest of domain-I, and the βIE or hydrophobic wedge is extended further away from the α4-helix for contact with β. The "reverse" conformational change in δ, from the β-bound form to the β-*free* form of results in exposure of specific hydrophobic residues that are important for interaction with the δ' stator of the clamp loader.

Although the two structures were solved with mutants of β that do not form dimers, the above conclusions regarding release of spring-like tension in opening an interface were consistent with molecular dynamics simulations performed on the dimeric β clamp. In these simulations, a β protomer was simply released from dimeric constraints determined from the original β_2 clamp structure. The β protomer followed a trajectory in the simulation that resulted in straightening of its crescent-shape such that it superimposed well with the structure determined from the δ-bound β monomer.

DNA structural requirements for β clamp loading by γ complex

The β clamp loading mechanism and preinitiation complex formation for processive synthesis by pol III holoenzyme has been studied on several different DNA substrates including synthetic poly-dT-primed poly-dA-templates, as well as circular phage genomes in different replicative forms (i.e. single- or multi-primed, nicked-duplex etc.). It was shown that short synthetic primed-template DNA (pt DNA) substrates (i.e. an 80-105 nucleotide template with a 30 nucleotide primer) whose sequence was based on the M13 phage where sufficient for proficient elongation by pol III core in the presence of β clamp and reconstituted γ complex, and therefore for the study of the clamp loading mechanism (Bloom et al., 1996).

Using these short synthetic pt DNA substrates in experiments along side circular phage genome substrates containing supercoiled or relaxed-structures, the DNA structural requirements for β clamp loading where investigated (Ason et al., 2000; Ason et al., 2003; Yao et al., 2000). For preinitiation complex formation at the replication fork it was known that RNA primers were required. The exact structural nature of these primers were unknown except that a 3'-hydroxyl group was annealed at the single-stranded DNA (ss DNA) / double-stranded RNA/DNA (ds RNA/DNA) junction where the preinitiation complex was formed. In experiments with supercoiled DNA and closed relaxed circular DNA substrates it was shown that reconstituted γ complex did not require a 3'-end to load β, as it was able to load the clamp onto supercoiled DNA, but not closed relaxed circular DNA (Yao et al., 2000). It was hypothesized that supercoiled DNA may produce unwound regions due to superhelical tension, and at these ds RNA/DNA / ss DNA junctions γ complex could load β forming the preinitiation complex.

Using circular single-stranded M13 phage DNA primed with synthetic DNA primers differing in specific structural features such as length and unannealed 15-nucleotide "flaps" at 5'- or 3'-ends, or both 5'- and 3'-ends, it was shown that γ complex can load β onto a wide range of these substrates (Yao et al., 2000). The minimal primer length requirement was found to be 10 base pairs, consistent with the size of the inner pore width of β, and with the known fact that *in vivo*, primers formed by primase, are ~ 10-12 nucleotides in length (Kitani et al., 1985; Kong et al., 1992). Yao, Leu et al. also showed that γ complex could load β onto primers with 5'- and/or 3'- 15-nucleotide-long unannealed flaps, supporting their conclusion that γ complex only requires a ds DNA / ss DNA junction to load the clamp.

By placing protein- or DNA secondary-structural blocks on or within the primer in replication assays, the polarity and primer spatial requirement for clamp loading were determined. Using physical (i.e. protein) blocks tightly bound at the annealed primer 5'- or 3'-ends it was shown that γ complex exhibits polarity in loading the β clamp specifically at the 3'-end of the primer. This result was expected due to the fact that polymerase synthesis elongation extends from the primer 3'-end in the 5' to 3' direction. Placement of DNA secondary structural elements at different distances upstream from the 3'-end of the primer, allowed determination of the spatial requirement of the primer for β-bound γ complex and β-bound pol III core. It was shown that with γ complex, 14-16 base pairs of primer were needed to load the clamp, and 20-22 base pairs were needed for β to bind pol III core. This result indicated that γ complex interacts directly with 4 to 6 base pairs of the primer, and that the clamp is pushed back an additional ~6 base pairs when pol III core binds β at the ds DNA / ss DNA junction (Yao et al., 2000).

Utilizing fluorescence-based steady-state and pre-steady-state kinetic anisotropy assays with a X-Rhodamine-labeled short synthetic pt DNA in solution, a DNA-triggered switch in γ complex was discovered (Ason et al., 2000). This DNA-triggered switch caused a change in affinity of γ complex for DNA, and was found to be pt DNA- and ATP hydrolysis-dependent. Initially, in steady-state assays, it was observed that γ complex had higher affinity for ss DNA than pt DNA. Even more surprising, was that the 5'-single-stranded template overhang of the pt DNA was the same sequence and length as the ss DNA substrate examined in these experiments. However, pre-steady-state real-time assays revealed DNA binding kinetics suggesting that γ complex bound transiently with high affinity to the pt DNA substrate. Use of non-hydrolyzable ATPγS in both steady-state and pre-steady-state assays with pt DNA substrates removed this dynamic switch in γ complex affinity for DNA resulting in a high affinity state only. It was therefore proposed that ATP hydrolysis by γ complex was required for cycling between high and low affinity states, and that the low affinity state may have been an ADP-bound γ complex. With ss DNA substrates γ complex with β maintained high affinity for the DNA, hydrolyzed ATP, but did not load β (Bloom et al., 1996; Turner et al., 1999).

The dynamic interaction with pt DNA and the DNA triggered switch in γ complex accomplishes two major goals in the clamp loading reaction. The primer-template, ds DNA / ss DNA junction, provides the proper site for clamp loading and preinitiation complex formation, and modulates the DNA binding affinity of the clamp loader, so as not to allow γ complex competition with pol III core bound to β at previously formed initiation complexes. The nature of the low affinity state of γ complex remains a mystery, and this dissertation addresses several possibilities [and questions] including the dynamics of conformational changes defining different states of γ complex with respect to the clamp loading reaction.

A recent investigation addressed the possibility that the pt DNA-triggered modulation of γ complex provides a dynamic mechanism for recognition of appropriate sites for β clamp specific for proficient DNA synthesis. This was accomplished by investigation of the DNA structural features required to trigger γ complex into the low affinity state and release β (Ason et al., 2003). In fluorescence-based steady-state and pre-steady-state anisotropy assays in solution, γ complex binding and clamp loading were studied with elongation-proficient DNA substrates (i.e. synthetic template DNA primed on its 3'-end or center), or elongation-deficient DNA substrates (i.e. synthetic template DNA primed such that a blunt duplex was formed at the 5'-end of the template, or unprimed-ss DNA). In steady-state titrations of γ complex to the elongation-proficient or deficient DNA substrates, results similar to the previous investigation defining the DNA-triggered switch were observed. The γ complex showed high affinity for elongation-deficient DNA substrates, but did not appear to bind well to the elongation-proficient DNA substrates. Pre-steady-state assays revealed biphasic "up-down"-kinetics (i.e., a rapid rise to a defined peak followed by a decay phase into steady-state DNA binding activity), representing transient high affinity of γ complex for the elongation-proficient DNA substrates as well as characteristic changes in anisotropy for the clamp loading reaction when β was present. For elongation-deficient DNA substrates, pre-steady-state binding kinetics were monophasic suggesting a simple high affinity bimolecular equilibrium, whether β was present or not. Additionally, a correlated DNA-binding and ATP hydrolysis assay with γ complex and elongation-proficient pt DNA clearly showed that a binary complex consisting of γ complex with pt DNA transiently formed just prior to DNA-triggered ATP hydrolysis and release of γ complex without any further binding. Overall, the results show that γ complex uses a dynamic mechanism driven by ATP binding and hydrolysis for targeting β clamp only to DNA that can serve as a template for synthesis elongation by pol III core, and prevent further interaction with polymerase-bound β.

Mechanism of the β Clamp Loading Reaction Cycle by γ Complex

Formation of a processive pol III holoenzyme requires the formation of preinitiation complexes on the leading and lagging strands at the replication fork. The ring-shaped sliding clamp must be opened and loaded onto the circular *E. coli* chromosome in order to form a topological link between pol III and DNA. The clamp loading machine performs this task through dynamic protein-protein and protein-DNA interactions. Some 26 years ago, Sue Wickner presented a model for the mechanism of DNA elongation catalyzed by DNA polymerase III that still generally holds true today (Wickner, 1976). Wickner's model described a mechanism where primed-template DNA was activated by the ATP-dependent transfer of "DNA elongation factor I" (the sliding clamp) by "DNA elongation factor III" (the clamp loader), to which DNA polymerase III then bound in an ATP-independent manner, followed by DNA elongation. Combined biochemical analysis and structural details have in recent times given us a detailed, yet still incomplete, view of the clamp loading mechanism required for processive synthesis by pol III holoenzyme.

The γ complex ($\gamma_3,\delta,\delta',\chi,\psi$), or a minimum complex of five subunits (γ_3,δ,δ') is the energase complex that transduces the energy from ATP binding and hydrolysis into mechanical work to load β onto DNA (Figure 2-5).

Within this machine, the three γ subunits serve as the motor subunits, δ has the β-interaction element (βIE), and δ' is the "stator" upon which movement of the other subunits is thought to be supported. In general terms, ATP binding and hydrolysis by the γ subunits promote conformational changes that modulate the dynamic protein-protein and protein-DNA interactions involved in clamp loading. To work at the speed and efficiency required at the replication fork within the cell, there must be precise communication between all subunits of the clamp loading machine to power and regulate its activity.

Initially, ATP binding to the γ subunits, but not hydrolysis, promotes changes in the clamp loader that modulate the binding affinity for the clamp and DNA, and therefore powers most of the steps in the clamp loading reaction (Bertram et al., 1998; Hingorani and O'Donnell, 1998). Asymmetry of the interfacial nucleotide binding sites in the clamp loader structure provided a model for activation upon ATP binding for β clamp loading. The δ'-γ_1 and γ_2-γ_3 interfacial nucleotide binding sites were open in the structure and could provide space for the first one or two molecules of ATP to bind. These conformational changes could also result in the opening of the γ_1-γ_2 ATP binding site, and further "splaying-out" of the γ subunits from the "backbone" of the δ' stator, and ultimately exposure of the δ subunit βIE (Figure 2-5B). This model would bring the maximum number of ATP molecules needed for exposure of the δ subunit βIE to three in

"open" γ complex.

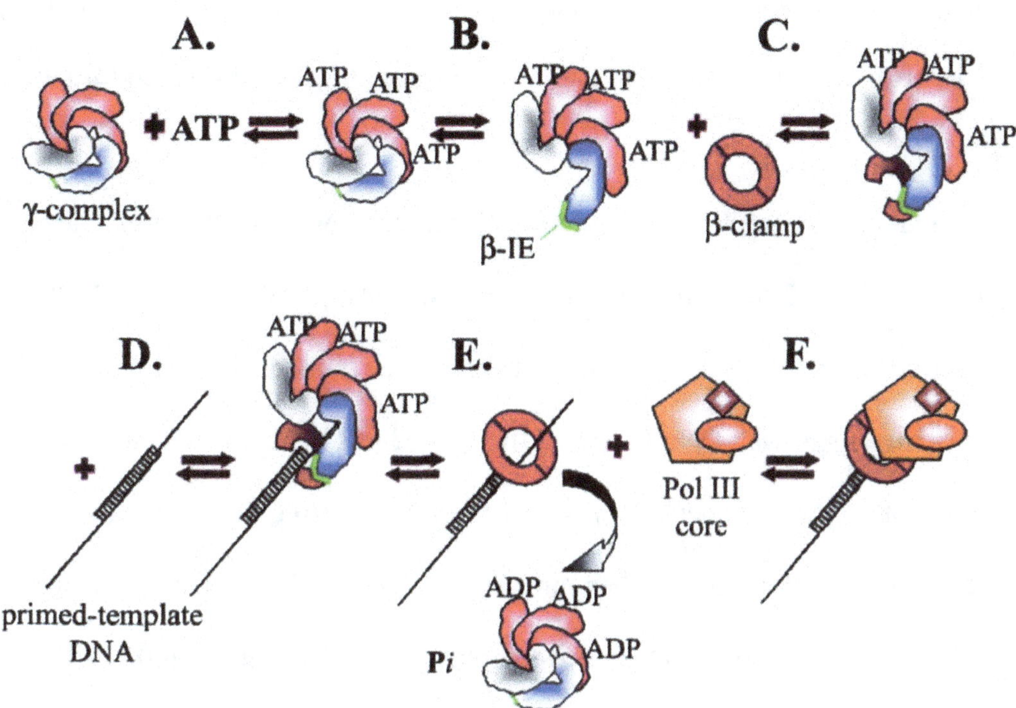

Figure 2-5. A schematic cartoon of the basic steps in the clamp loading reaction and initiation complex formation. The cartoon of the clamp loader in "closed" and "open" states is based on the γ$_3$δδ' crystal structure, and the γ complex subunits are colored as in figure 2-3. A) ATP binding to the γ subunits. B) ATP-dependent conformational changes occur within the clamp loader exposing the δ subunit βIE (green). C) β clamp binding to the clamp loader through the βIE induces a conformational change in β, opening the clamp at a single interface. D) The β-bound clamp loader complex binds pt DNA and positions the clamp at the ds DNA / ss DNA junction at the 3'-OH of the primer. E) pt DNA triggers ATP hydrolysis, the release of inorganic phosphate (P*i*), and dissociation of ADP-bound clamp loader from loaded β clamp. F) Polymerase III core (α, ε, and θ subunits) can then bind β and commence processive elongation of the template.

Originally it was shown that ATP binding caused a conformational change in γ complex

that resulted in characteristic proteolytic cleavage of the exposed δ subunit, and that this

change did not require β (Hingorani and O'Donnell, 1998; Naktinis et al., 1995). The interaction of the exposed δ subunit βIE with β then induces or stabilizes conformational changes in β that result in ring opening at a single dimer interface (Figure 2-5C). The kinetics of ATP binding and the dynamics of the conformational changes within the clamp loader are a major component of this dissertation and will be discussed further in chapters 4 and 5.

ATP binding to γ complex is also required for interaction with DNA in the clamp loading reaction (figure 2-5D) (Bloom et al., 1996; Stukenberg and O'Donnell, 1995). Studies described in this dissertation show that none of the individual subunits of the clamp loader have significant DNA binding affinity, and it is hypothesized here and that the interaction of all clamp loader subunits form the DNA binding site. With a clamp bound and opened, the clamp loader then binds DNA at a ds DNA / ss DNA junction, specifically on elongation proficient primed-template DNA (Ason et al., 2003; Yao et al., 2000). The DNA binding kinetics and positioning of β at the 3'-hydroxyl of the primer happen in less than 100 ms with an approximate bimolecular binding constant on the order of $2.0 - 4.0 \times 10^8 \, M^{-1} \, s^{-1}$ just prior to the hydrolysis of one molecule of ATP for each γ subunit in the clamp loader (Ason et al., 2000).

Primed-template DNA triggers ATP hydrolysis and dissociation of the clamp loader (Figure 2-5E). Earlier investigations of ATP hydrolysis by the clamp loader showed that mutations of the conserved lysine in the γ subunit Walker-A motif of domain-I, to alanine or arginine (K51A or K51R), resulted in complete abrogation of ATP hydrolysis activity by the clamp loader (Xiao et al., 1995). Our laboratory has also shown that these mutations of the conserved lysine in γ complex abolished both β and DNA binding in solution (unpublished), therefore it is likely that the lysine-mutants cannot bind ATP. Use of non-hydrolysable ATPγS allowed γ complex to bind β and DNA, consistent with the notion that nucleotide binding powers most of the steps in the clamp loading reaction. In these reactions, the complex of β-bound γ complex remained in a dynamic steady-state interaction with DNA for periods of time extending to 90 minutes (Bertram et al., 1998; Bloom et al., 1996). These investigations showed that although ATP binding can bring β-bound γ complex to DNA; ATP hydrolysis was absolutely necessary for release of γ complex, and completion of the clamp loading reaction. The pre-steady-state ATP hydrolysis assays showed that the ATPs were hydrolyzed at a minimal rate of 20-34 s^{-1} (Bertram et al., 2000). Closure of the β clamp on DNA was found not to require any further energy (Turner et al., 1999), therefore ATP hydrolysis by the clamp loader is not needed to close β onto DNA. The γ complex crystal structure predicted a mechanism wherein ATP hydrolysis drives final conformational changes in which β is pushed off the δ subunit βIE as δ returns to its occluded interaction with the rigid δ' subunit. This is consistent with an earlier investigation that showed ATP hydrolysis was required for release of γ complex from β on DNA, leaving the clamp behind (Bloom et al., 1996). Dissociation of γ complex from the loaded clamp is essential, and would provide space for the interaction of β with polymerase III, which is known to bind the same surface of β that γ complex had occupied (Figure 2-5F). Pre-steady-state analysis of DNA binding activity and ATP hydrolysis have shown that a single "turnover" of the clamp loading reaction takes approximately 300 ms before entering a steady-state that continues at a rate of ~2-2.5 s^{-1} (Ason et al., 2000).

Ultimately ATP hydrolysis would then allow the clamp loading machine to reset itself for continued clamp loading, most likely through nucleotide exchange and conformational changes. The γ complex released from the clamp loaded on DNA is presumably in some inactive "closed" state that would have to release ADP and become able to bind ATP again. ADP release could cause conformational changes that may transiently bring γ complex through some state in which there is no nucleotide bound. ATP binding would then cause the conformational changes that reactivate γ complex for another clamp loading reaction. In previous pre-steady-state ATP hydrolysis assays, there was a "pause" in hydrolysis activity observed during the last 200-250 ms of the first turnover of ATP during transition into the steady-state (Ason et al., 2000). Given the way these assays were initiated (i.e., with preincubation of γ complex with ATP and β before mixing with pt DNA), during this transition, the rate-limiting step of the reaction cycle was believed to occur. The exact nature of the rate-limiting step of the clamp loading reaction is not yet known. Possibilities include ADP-release, ATP binding, conformational dynamics within γ complex, and β clamp binding. The steady-state and pre-steady-state kinetics of DNA binding and ATP hydrolysis by the clamp loader are further studied in this dissertation, and reveal some novel kinetic features that predict that conformational changes within γ complex are most likely the rate-limiting step in the reaction and regulate the clamp loading mechanism.

Mutations of the β Clamp, and γ Complex δ' and γ Subunits: Effects on the Clamp Loading Mechanism

Mutations in the β clamp as well as in subunits of γ complex have been studied to provide greater detail for the interactions modulating this clamp loading mechanism. The steady-state and pre-steady-state kinetics of DNA binding and ATP hydrolysis activities were investigated using a β clamp with mutations at two positions within the dimer interface. The leucine to alanine (L273A, L108A) mutations were thought to weaken the dimer interfaces, although this mutant β was still able to form dimers in solution (Bertram et al., 1998). The β interface mutants were found to bind γ complex and DNA similar to wild-type β, but they remained bound to DNA in a ternary complex. The DNA binding assay results appeared similar to assays performed in the presence of non-hydrolyzable ATPγS where the β-bound clamp loader appeared "stuck" in a dynamic steady-state interaction with DNA, without loading β. In correlated pre-steady-state DNA binding and ATPase assays, the β interface mutants bound to DNA and hydrolyzed ATP with nearly identical kinetics as wild-type-β. However, the clamp loader steady-state ATPase activity with the β mutants was significantly decreased, coincident with extremely slow dissociation of γ complex from mutant β on DNA (Ason et al., 2000). The results confirmed the idea that the clamp loader must cycle off of β that is loaded onto DNA, and undergo a rate-limiting step that is separated from the DNA-bound state in order to continue loading clamps as previously suggested by Bloom et al. 1996. The results of these investigations also showed that although β does not require any energy from ATP hydrolysis to close on DNA, ATP hydrolysis is tightly coupled to release of the clamp on DNA. It seemed interesting that mutations within the β dimer interface so greatly affected the steady-state DNA binding and ATP hydrolysis activities of the clamp loader. Now, with better knowledge of the interaction between the clamp loader-δ subunit and β clamp it is more clear how β interface mutations would cause these effects in the clamp loader. Perhaps the conformational change of β cannot be properly promoted by the δ subunit-βIE interaction, disallowing loading onto DNA. This possibility could result in a decrease in the interaction of γ complex with DNA, and reduced DNA stimulated ATP hydrolysis activity.

To gain further insight into the AAA+ machine activity of γ complex, mutations in the sensor-1 (SRC) motif were made, and steady-state ATP binding, hydrolysis, and clamp loading activities were examined (Johnson and O'Donnell, 2003). In one of these mutations, arginine-158 of the δ' stator subunit (δ'-R158A) SRC motif was changed to alanine. This mutation removed the "arginine-finger" from the δ'-γ_1 nucleotide binding interface. The results showed that the reconstituted δ'-R158A-γ complex still bound approximately three molecules of ATP, but had reduced ATP hydrolysis activity and diminished ability to load β in DNA synthesis assays. These results implied that the "arginine-finger" contributed from the δ' subunit to the δ'-γ_1 nucleotide binding site was in fact catalytic in nature and therefore, this "stator" subunit was coupled with the motor function of the clamp loader, as well as performing as the motor support subunit. Recent results from our laboratory have shown that this reconstituted δ'-R158A-γ complex mutant has little ability to bind β, as well as significantly reduced DNA binding and clamp loading abilities (Snyder, A.K., personal communication). This is consistent with the reduced DNA synthesis activity observed in the original study.

The "arginine-finger" was removed from the γ subunits to create a second clamp loader mutant (Johnson and O'Donnell, 2003). The γ-R169A mutation removed the "arginine-finger" from the γ_1-γ_2 and γ_2-γ_3 nucleotide binding interfaces, thus only affected two nucleotide binding sites in the clamp loader. This reconstituted γ_3-R169A-complex still maintained the ability to bind three molecules of ATP and β, but had no ATP hydrolysis activity, clamp loading activity, and was unable to stimulate DNA synthesis. Our laboratory has since shown that this γ_3-R169A-complex can bind β at a level near that of wild-type γ complex in an ATP-dependent manner in solution, but showed no DNA binding activity in steady-state and pre-steady-state solution-based assays (Snyder, A.K., personal communication). Johnson and O'Donnell discussed the δ'- and γ-subunit mutants in terms of an ordered ATP binding and hydrolysis mechanism where ATP molecules are hydrolyzed in the reverse order than they were bound, requiring proper positioning of the "arginine-fingers" for catalysis. Our laboratory's results indicate that there is more than just a disruption in catalysis of ATP hydrolysis in these mutants. Taken together, all of the mutation analysis results exemplify the need for precise subunit conformational communication within the clamp loading machine for β binding and also for formation of the putative DNA binding surface on the clamp loader.

The Clamp Loading Machine Within Polymerase III Holoenzyme

Most of the biochemical analyses and all of the structural analyses of the clamp loader and clamp loading mechanism have been performed with reconstituted γ complex. How do these results apply to the clamp loader within pol III holoenzyme? The clamp loader within holoenzyme contains both the τ and γ subunits encoded by the *dna*X gene. In the τ subunit, the region of the DnaX protein structure extending from the C-terminus of the γ subunit region into the C-terminal domain of τ contains many proline residues and is thought to be unstructured and therefore a highly flexible region (O'Donnell et al., 2001). This flexible region divides the DnaX (τ) protein into the γ-AAA+ motor region, and the τ-C-terminal replisome interaction region. Within holoenzyme, there are at least two τ subunits, and a single γ subunit (Figure 2-6).

The structural arrangement of the DnaX subunits of the clamp loader within holoenzyme is likely to be [δ_1, τ_2,γ_1,δ'_1,χ_1,ψ_1] or [δ'_1,γ_1, τ_2,δ_1,χ_1,ψ_1]. The χ and ψ subunits only bind to the γ subunit through ψ, and the δ' subunit also has been shown to bind only to the γ subunit with high affinity (Glover and McHenry, 2000).

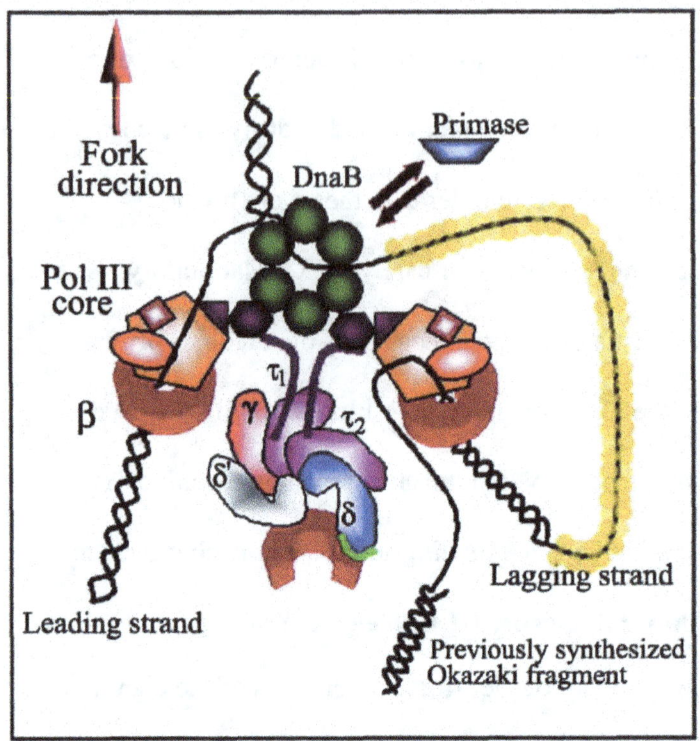

Figure 2-6. Architecture of the polymerase III holoenzyme at the replication fork organized by the DnaX clamp loading machine. The leading- and lagging-strands are unwound by hexameric DnaB helicase (green), shown bound to the SSB-coated (yellow) lagging strand. The DnaX clamp loader containing two τ subunits (violet) (δ'$_1$,γ_1, τ_2,δ_1,χ_1,ψ_1) is connected within the holoenzyme through the flexible linker regions of the τ subunit C-terminal domains that also dimerize pol III core (α ε and θ), and further cement the holoenzyme through contact with DnaB. The clamp loader χ and ψ subunits are not shown for clarity. Two β-rings (red) clamp the pol III cores to the templates, and a third is shown bound to the clamp loader. In this schematic, primase would bind and synthesize a new RNA-primer on the lagging strand. Then it is predicted that the clamp loader would "swing" over to the nascent primer, displace primase through the χ subunit, and load a fresh clamp forming a preinitiation complex. When the lagging-strand polymerase reaches the 5'-end of the previously synthesized Okazaki fragment, it would then release β and cycle to the newly formed preinitiation complex and commence elongation of DNA.

The two τ subunits cement the array of proteins at the replication fork by dimerizing the core polymerases in the holoenzyme, binding to the replicative DnaB helicase, and protecting the β subunit from removal when it is attached to the polymerase α subunit.

Through the flexible linker region, these functions of the τ subunits are connected to the γ-AAA+ motor region of the clamp loader. In this way, it is envisioned that the clamp loader associated with the holoenzyme *in vivo* has a swinging range of motion on the flexible tether loading at least one clamp on the leading strand and providing continuous clamp loading activity on the lagging strand for Okazaki fragment synthesis (O'Donnell et al., 2001).

Duplex DNA is anti-parallel, and therefore the core polymerases of the holoenzyme must work in some asymmetric fashion to simultaneously complete replication of the chromosome (Figure 2-6). The single γ subunit of the DnaX clamp loader, and consequential asymmetric distribution of the δ,δ',χ,and ψ subunits gives the holoenzyme structural asymmetry. The γ subunit is the only clamp loader subunit to which ψ and therefore χ adaptor proteins bind. This provides the holoenzyme clamp loader both structural and functional asymmetry through τ-DnaB helicase contact, and the χ-SSB interactions necessary for lagging strand synthesis.

Recently it was proposed that the asymmetry in the clamp loader provides asymmetry to the core polymerases through nucleotide binding to the clamp loader (Glover and McHenry, 2001). Use of non-hydrolyzable ATPγS was sufficient to allow the DnaX complex to load a clamp for only one polymerase, presumably the leading strand polymerase. ATP hydrolysis was absolutely required for clamp loading on the "lagging strand" polymerase in their assays, and subsequent addition of ATPγS was able to disassociate this "lagging-strand" polymerase. Whether the nucleotide binding status of the clamp loader, or simply the structural asymmetry of the clamp loader provide distinct leading and lagging strand functions to the core polymerases will require further investigation. Overall, the structurally and functionally asymmetric holoenzyme has the means to stay continuously linked to the leading strand while performing discontinuous synthesis on the lagging strand at the replication fork by the distinct structural and functional characteristics of the DnaX clamp loader. Polymerase processivity is required for DNA replication in all free living cells, and nature has provided a consistent mechanism to provide the means for polymerase processivity though conserved evolution of these replication proteins across all branches of life.

Clamps and Clamp Loaders of Bacteriophage, Eukaryotic, and Archaeal Organisms

The combined biochemical and structural investigations of the *E. coli* clamp and clamp loader have yielded many great details into the mechanism of processivity clamp assembly on DNA. These intimate details have allowed complementary studies of clamps and clamp loaders of other organisms to thrive due to the evolutionary conservation of these processivity proteins across all branches of life, and suggest a common mechanism of clamp loading in all life forms. It is well known that sliding clamps contact polymerases with DNA for processive synthesis of complete chromosomes or replicons in a fundamentally similar way (Ellison and Stillman, 2001; Hingorani and O'Donnell, 2000). Now, structural evidence is revealing the extraordinary functional similarities in clamps and clamp loading machines of viral replicases, prokaryotes, eukaryotes and archaea for processive DNA elongation (O'Donnell et al., 2001). All organisms through these evolutionary branches of life appear to utilize AAA+ proteins as the energase of their clamp loading machines (Table 2-1). Of those organisms studied, each clamp loader appears to, or is predicted to utilize a AAA+ clamp loading machine, like γ complex, that contains several ATPase motor subunits, a clamp interacting subunit, and a supporting stator subunit (Davey et al., 2002). This conservation of sequence and, to a greater extent, structural similarities of these AAA+ proteins illustrate the evolutionary importance of these clamp loading machines.

Table 2-1. Clamps and clamp loaders through evolution

Evolutionary Branch	Processivity Proteins	AAA+ protein	Function[a]	Some Characterized Organisms
Prokaryotic	β	-	-clamp	
	γ / τ	+	-motor	*Escherichia coli*[e]
	δ'	+	-stator	*Bacillus subtilus*
	δ	+	-clamp-interacting subunit[c]	*Bacillus-staerothermophilus* *Aquifex aeolicus*[f]
	χ	-	-AAA+ adaptor	*Thermus thermophilus*[f]
	ψ	-	-AAA+ adaptor	
Bacteriophage (viral)	gp45[b]	-	-clamp	
	gp44	+	-motor (stator?)	T4 phage
	gp62	+(?)	-clamp interacting subunit[c]	T7 phage RB69 phage
Eukaryotic	PCNA	-	-clamp	*Homo sapiens*
	RFC-2[d]	+	-motor	*Saccharomyces cerevisiae*
	RFC-3	+	-motor	*Schizosaccharomyces-pombe*
	RFC-4	+	-motor	*Caenorhabditis elegans*
	RFC-5	+	-stator	*Drosophila melanogaster*
	RFC-1	+	-clamp interacting subunit[c]	*Arabidopsis thaliana* *Oryza sativa* (rice)
Archaeal	PCNA	-	-clamp	*Pyrococcus furiosus*
	RFC-S	+	-motor (stator?)	*Archaeoglobus fulgidus*
	RFC-L	+	-motor / clamp interacting subunit	*Sulfolobus solfataricus* *Methanobacterium-thermoautotrophicum* ΔH

[a] by analogy to the γ complex clamp loader
[b] gp, gene product
[c] has conserved hydrophobic wedge residues of clamp interaction element
[d] nomenclature used is for yeast replication factor-C (RFC)
[e] clamp loader structure known
[f] Thermophile bacteria

Bacteriophage T4 Clamp and Clamp loader

When infecting bacteria, the bacteriophage has the necessary components to form its own replication machinery. The replisome of T4 bacteriophage contains gene product (gp) gp43 polymerase, gp41 / gp59 helicase and its accessory factor, respectively, gp61 primase, and gp32 single-stranded DNA binding protein. T4 also provides the processivity clamp gp45 and clamp loader complex gp44/62 to complete the replisome (Salinas and Benkovic, 2000). The gp45 clamp interacts with gp43 polymerase through a C-terminal region of the polymerase. This interaction between the clamp and polymerase is highly conserved across evolution (Ellison and Stillman, 2001), and the structure of a T4 clamp bound to polymerase on DNA has been modeled based on the nearly identical RB69 bacteriophage sliding clamp bound to polymerase (Alley et al., 2001; Shamoo and Steitz, 1999). To date, these are the only structural views of a polymerase bound to its clamp on DNA.

The gp45 processivity protein is a ring-shaped clamp formed of six similar structural domains like β clamp, but unlike dimeric β, is a trimeric clamp. Although there is less than 10 % sequence identity between gp45 and β, (and PCNA, see below), the clamp structure is amazingly similar (Jeruzalmi et al., 2002). The crystal structure of gp45 shows that its properties are similar to β clamp and PCNA (Moarefi et al., 2000). The gp45 clamp has an overall negative charged β-sheet outer surface surrounding an α-helical inner pore of positive electrostatic potential. The dimensions of the gp45 clamp show an inner pore diameter of ~35 Å, and a width of ~25 Å, however, the clamp has an overall triangular topology compared to the nearly circular structures β clamp and PCNA. Consistent with data that show gp45 clamp is the least stable of the known sliding clamps on DNA (Yao et al., 1996), the gp45 clamp is predicted to be slightly open in solution through a pucker which gives it a "lock-washer"-type structure (Alley et al., 1999). It is still possible that the clamp loader must open the clamp further for the loading reaction (Alley et al., 2000).

The T4 clamp loader is a pentameric complex of the gp44 and gp62 subunits with a subunit stiochiometry of four gp44 subunits to one gp62 subunit (Jarvis et al., 1989). The T4 gp44/62 complex binds and loads gp45 onto DNA in an ATP-dependent manner and requires hydrolysis of bound ATP. In earlier work, it was shown that four molecules of ATP were hydrolyzed by gp44/62 during formation of the T4 holoenzyme, and that some step following ATP hydrolysis was rate limiting in the reaction (Sexton et al., 1998). A highly detailed investigation using fluorescent resonance energy transfer techniques with fluorescent-labeled gp45 clamp, combined the pre-steady-state kinetic analysis of clamp opening and closing by the gp44/62 clamp loader with analysis of binding and hydrolysis of ATP (Alley et al., 2000). This investigation showed that hydrolysis of ATP was required to open the gp45 clamp, and then additional hydrolysis was required to load the clamp on DNA. A more recent study on the ATP hydrolysis activity of gp44/62 in clamp loading conflicted with previous work, and showed that ATP binding alone is sufficient for gp44/62 to bind gp45, and at least one ATP is required to complete the loading reaction (Pietroni et al., 2001). This study also revealed that ATP hydrolysis was not required to open the gp45 clamp, and that the rate-limiting step in the loading reaction was either ADP release from gp44/62, or a conformational change before clamp loading. This investigation is more consistent with the detailed study of the γ complex in β clamp loading.

The gp44 subunits of the complex have homology with AAA+ proteins and contain the Walker-A P-loop and sensor-1 SRC motifs needed for ATP binding and hydrolysis (Davey et al., 2002; Neuwald et al., 1999). By analogy to γ complex, the gp44 subunits are expected to be the motor subunits of the T4 clamp loader that undergo ATP-dependent conformational changes that drive gp45 binding and interaction with DNA in the loading reaction. The gp62 subunit shares no sequence homology to other AAA+ subunits, and does not bind or hydrolyze ATP. Until its structure is known it can only be predicted that gp62 is a AAA+ homologue, and the subunit that interacts with the gp45 clamp during the clamp loading reaction (i.e. like the δ subunit of γ complex). There is no support-like stator subunit in the gp44/62 complex as all of the gp44 "motor" subunits are identical. However, even in γ complex, the δ' stator subunit has been shown to be directly involved in the catalytic motor function of the clamp loader through precise positioning of its SRC motif by conformational changes (Johnson and O'Donnell, 2003). One of the gp44 subunits of the T4 clamp loader may work in a similar manner, and it may turn out that this putative, distinct, gp44 subunit may also be incapable of ATP hydrolysis like δ' in γ complex. These predictions await further biochemical and structural analyses of the T4 clamp loader.

Eukaryotic PCNA Clamp and Replication Factor-C Clamp Loader

The essential replication machinery of eukaryotic organisms was originally identified by investigations using the SV40 DNA virus as a model system. SV40 uses host cell machinery in replication of its DNA, and requires only its own "T-antigen" for replication initiation and DNA helicase activity (Waga and Stillman, 1994). Processivity proteins, proliferating cell nuclear antigen (PCNA) and replication factor-C (RFC) were identified as the sliding clamp and clamp loader, respectively, for the replicative polymerases δ /ε (Waga and Stillman, 1998). Like the γ complex clamp loader, RFC has DNA-dependent ATP hydrolysis activity that is further stimulated by the presence of the PCNA clamp (Lee et al., 1991). RFC is also directly involved in a primase to polymerase switch, similar to γ complex, and influences the length of primer synthesis by polymerase α primase (Mossi et al., 2000; Tsurimoto and Stillman, 1991).

The PCNA sliding clamp structure was solved from budding yeast *S. cerevisiae* (Krishna et al., 1994). Once again, even though sequence homology between the bacteriophage T4 clamp, prokaryotic β clamp, and PCNA was low, the structure of PCNA is nearly identical to the other clamps (Moarefi et al., 2000). Unlike the β clamp, but like T4 gp45 clamp, PCNA found to be trimeric. All three clamps have six structurally similar subdomains, dimensions (i.e. the PCNA inner pore is ~ 35 Å), and similar electrostatic characteristics on their structurally conserved outer β-sheets, and inner α-helices. The overall topology of the PCNA trimer is more hexameric or circular than gp45, remains closed in solution, and is more stable on DNA than gp45, but not as stable as dimeric β clamp (Yao et al., 1996).

RFC has been extensively studied from yeast and human origin, and has been identified as a heteropentameric complex consisting of one large subunit and four small subunits of different mass. The subunit nomenclature is: RFC- (yeast/human): 1/p140, 2/p37, 3/p36, 4/p40, and 5/p38. For simplicity the yeast RFC nomenclature will be used here. Deletion analysis of RFC subunits showed that the C-terminus of each subunit was required for formation of the pentameric complex, and that this complex must form to allow DNA stimulated ATP hydrolysis activity (Uhlmann et al., 1997). It was also shown that none of the individual subunits alone had DNA binding activity (Uhlmann et al., 1996). Using surface plasmon resonance (SPR) and filter magnetic-bead binding assays, it was shown that ATP mediated all interactions of RFC with PCNA and DNA (Gomes and Burgers, 2001). This study also showed that ATP binding but not hydrolysis was required for the interaction with DNA, and that use of non-hydrolyzable ATPγS resulted in a "stuck" complex of PCNA-bound RFC on DNA. These results are consistent with complementary investigations performed with γ complex. The DNA structural requirements for PCNA loading by RFC were also tested and showed that RFC could load PCNA on ds DNA / ss DNA junctions not requiring an annealed 3'-OH similar to the DNA structural requirement for γ complex catalyzed clamp loading activity.

Recent studies with yeast RFC (Schmidt et al., 2001a), as well as previous investigations with human RFC (Cai et al., 1998) identified the subunits involved in ATP hydrolysis activity and those essential for DNA recognition by mutational analysis of the conserved Walker-A lysine residue in each subunit. The results of these mutational analyses revealed that RFC subunits 2,3, and 4 were required for ATP binding and hydrolysis in efficient PCNA loading. Mutations in RFC-1, the large subunit, had no effect on PCNA binding and loading abilities. RFC-5 has modified Walker-A and B sequences suggesting that it does not bind or hydrolyze ATP, similar to the δ' subunit of γ complex (Cai et al., 1998; Schmidt et al., 2001a). Consequentially, these Walker-A mutations of RFC-5 did not effect PCNA binding and loading activities.

As in the T4 clamp loading reaction cycle, the use of ATP by RFC in PCNA loading is also not fully understood. This may be due to limitations of the SPR and filter binding analyses that were being performed along side steady-state ATP hydrolysis analysis of these reactions. Despite these limitations, a pathway of multiple stepwise ATP binding events is thought to be required for RFC loading PCNA onto DNA, and that this proposed reaction resembles the well studied *E. coli* clamp loading reaction pathway (Gomes et al., 2001). In the yeast PCNA loading reaction, 2 ATP molecules initially bind RFC, PCNA then binds RFC, and RFC gains affinity for a third ATP. Conformational changes within RFC are most likely taking place as each ATP binds an RFC subunit. Finally, RFC-PCNA binds pt DNA, and apparently one more molecule of ATP binds RFC, perhaps to the RFC-1 subunit. Upon interaction with pt DNA, PCNA is released, coincident with hydrolysis of up to all four bound ATP molecules. It remains unknown how many ATP molecules are used for this reaction.

The four small subunits of RFC (2-5) form a functional core, and are used in other cellular processes. The large RFC-1 subunit swaps with another protein, specific for functions such as replication termination (Kouprina et al., 1994), and cell cycle checkpoint control (Green et al., 2000). Differing the nature of ATP binding affinities and putative conformational changes in the RFC-2-5 core may provide differential reaction mechanisms for these diverse functions. Perhaps, the first two ATP molecules bind and cause conformational changes that allow a presentation of general binding site for PCNA and DNA only when RFC-1 is present, or form and expose different binding sites when another protein takes the place of RFC-1. Then, other ATPs could bind and produce conformational changes specific for each function. Another possibility could be that when a specific protein such as PCNA binds, it alters the affinity for ATP at additional sites in RFC specific for, in this case, clamp loading.

All five RFC subunits are AAA+ homologues, and are related in sequence to the γ and δ' subunits of γ complex (Neuwald et al., 1999; O'Donnell et al., 1993). The heteropentameric stiochiometry of the RFC subunits predicts that this eukaryotic clamp loader will have structural similarity to γ complex. The RFC-1 subunit is the least conserved of the RFC subunits, and its specificity in clamp loading with the RFC-2-5 core, and dispensability from the RFC-2-5 core in functions other than clamp loading suggest that it may be the PCNA clamp interacting subunit of the RFC complex by analogy to the δ subunit of γ complex, (see Table 2-1). Natural modifications of the Walker-A and B nucleotide-binding domain of the RFC-5 subunit suggest that this subunit may not bind or hydrolyze ATP, but it still has a sensor-1 SRC motif that could interact with the other RFC subunits (Cullmann et al., 1995). Like the δ' subunit of γ complex, RFC-5 is predicted to function as the stator in this eukaryotic clamp loading machine (Davey et al., 2002). The remaining RFC-2, 3, and 4 subunits, when mutated in the conserved Walker-A sequence for ATP binding and hydrolysis show the most severe *in vitro* defects and *in vivo* phenotypes (Schmidt et al., 2001a; Schmidt et al., 2001b). Therefore RFC-2, 3, and 4 were proposed to be the motor subunits of this machine. In fact, as heterotrimeric complexes, both yeast and human RFC-2-3-4 complexes had DNA-dependent ATPase activity (Ellison and Stillman, 1998).

This functional analogy of the different RFC AAA+ subunits to the γ complex subunits for prediction of RFC clamp loader structure and mechanism of clamp loading awaits solution of an atomic structure of the RFC clamp loader. Although, as described below, the atomic structure of part of the archaeal RFC clamp loader which has significant homology to eukaryotic RFC, has been solved, and closely resembles the

structure of γ complex motor subunits.

Human RFC alone and in a complex with PCNA has been viewed by transmission

electron microscopy (TEM) and atomic force microscopy (AFM) (Shiomi et al., 2000).

The TEM results showed that RFC forms a "closed-U-form" structure in the absence of

ATP. When viewed in the presence of ATP, the RFC molecules appeared as a more

"open-C-form" structure demonstrating that ATP caused RFC to open. Partial

proteolysis results with ATP, non-hydrolyzable ATPγS or ADP confirmed that there was

induction of a distinct structural change in RFC only in the presence of ATP. Additional

TEM imagining also showed that ATPγS or ADP had virtually no effect on the structure

of RFC (i.e. RFC remained in "U-form" similar to that without nucleotide). Although the

images were low resolution and did not allow specific distinction of RFC or PCNA

molecules in the imaged RFC-PCNA complexes, the authors concluded that the "open"

ATP-bound RFC "C-form" was most likely bound to PCNA in the TEM images.

Archaeal PCNA Clamp and Replication Factor-C Clamp Loader

Biochemical analyses and recent structural analysis of the clamps and clamp

loaders of several archaeal organisms are providing important information that is helping

bridge the gap between the prokaryotic structural and functional clamp loading

characteristics and eukaryotic structural and functional clamp loading characteristics.

Archaeal PCNA clamps and RFC clamp loaders share ~30-40 % sequence

homology to their eukaryotic counterparts (Seybert et al., 2002). Archaeal organisms

have a PCNA clamp and a RFC clamp loader complex (Table 2-1). Archaeal RFC and

PCNA have been biochemically characterized for *Sulfolobus solfataricus* (De Felice et

al., 1999; Pisani et al., 2000), *Methanobacterium thermoautotrophicum* ΔH (Kelman and

Hurwitz, 2000), *Archaeoglobus fulgidus* (Seybert et al., 2002), and *Pyrococcus furiosus* (Oyama et al., 2001).

The RFC clamp loading machine from each of these archaeons is composed of two subunits, RFC-L (large), and RFC-S (small), and each has AAA+ homology (Davey et al., 2002). Biochemical analyses of RFC and PCNA of these organisms have identified several general characteristics, and these characteristics are consistent with the clamp loading mechanisms of both eukaryotic and prokaryotic organisms. Each archaeal RFC must form a complex to acquire DNA-dependent ATP hydrolysis activity. This ATPase activity is associated with the RFC-S subunits, and has shown preference for ds DNA / ss DNA primer-template junctions (Kelman and Hurwitz, 2000; Pisani et al., 2000). For the *Archaeoglobus fulgidus af*RFC clamp loader, it was further shown that the ATP hydrolysis activity was enhanced 2 to 3-fold by *af*PCNA, and that ATP binding, not hydrolysis was required to stimulate the interaction of *af*PCNA with DNA (Seybert et al., 2002). Each archaeal RFC clamp loader was also shown to support processive DNA synthesis with its respective PCNA and pol B-family archaeal DNA polymerase. The major difference in the characterization of these clamp loaders was their subunit stiochiometry. All have been shown to form pentameric complexes similar to the eukaryotic and prokaryotic clamp loaders, except the *Methanobacterium thermoautotrophicum* ΔH RFC, which has been shown to form a hexameric complex of two RFC-L subunits with 4 RFC-S subunits. The general pentameric subunit stoichiometry resembles the T4 bacteriophage clamp loader and the eukaryotic RFC, having four RFC-S subunits with 1 RFC-L subunit.

About the same time the γ complex clamp loader crystal structure was solved, the crystal structure of a RFC-S subunit complex of *Pyrococcus furiosus* was solved at 2.8 Å resolution (Oyama et al., 2001). The RFC-S subunit formed a "dimer-of-trimers", not necessarily the functional assembly of the clamp loader. Each RFC-S subunit showed significant AAA+ structural homology. Superimposition of a RFC-S subunit with the γ complex δ' subunit structure had a root-mean-square displacement of only 3 Å for the two conserved AAA+ domains (I and II), and showed marked overall similarity. The RFC-S subunits had the Walker-A P-loop and Walker-B motifs flanked by sensor-2 and sensor-1 motifs forming the AAA+-structurally conserved nucleotide binding site, and also had the SRC "arginine-finger" motif. Each RFC-S subunit had an α-helical C-terminal Domain-III corresponding to the circular "collar" oligomerization domain of the γ complex clamp loader. The subunit interfaces within each trimer of RFC-S subunits was arranged such that each nucleotide binding site was in contact with the neighboring subunit, and the interfaces differed between each subunit, suggesting asymmetry among the subunits in the crystal, a feature distinguishing the γ complex structural-functional characterization (Jeruzalmi et al., 2001a).

The RFC-S subunits are AAA+ proteins, and based on the extraordinary structural similarity with the γ AAA+ subunits of the prokaryotic clamp loader, suggest that the mechanism of ATP-dependent conformational changes that drive the γ complex is most likely the same mechanism driving the archaeal clamp loader. The RFC-S subunit has the highest sequence similarity to the eukaryotic RFC-3 AAA+ subunit (Cann and Ishino, 1999), and therefore two predictions can be made. First, the RFC-S subunit of archaeal organisms is the "motor" subunit of the clamp loader, and second, that the mechanism of the eukaryotic clamp loader may be closer to the mechanism of the prokaryotic clamp loader than originally expected. To add to this, the RFC-L subunit generally lacks an SRC motif, and shows highest sequence similarity to the eukaryotic RFC-1 subunit (Cann and Ishino, 1999). Thus it has been proposed that like the eukaryotic RFC-1 and prokaryotic δ subunit, RFC-L is the clamp interacting subunit of the archaeal clamp loading machine (Davey et al., 2002).

The functional conservation of the sliding-clamps and clamp loaders in all branches of life is readily understandable considering the importance of their functions in DNA replication and therefore in cell division for propagation of any given organism. These are ancient proteins that have not changed considerably for billions of years. The use of the *E. coli* model system for studying the mechanisms of clamp loading for processive replication has thus given us the fundamental basis for study of the clamp loading mechanisms in all forms of life. In this dissertation project, the conformational dynamics of the *E. coli* clamp loader were under investigation, and it is hoped that the results can be applied for understanding of other AAA+-based clamp loaders and provide additional functional insights for other AAA+ motors.

MATERIALS AND METHODS

NOTE: The anisotropy binding assays and MDCC-PBP ATPase assays were performed with both γ complex and the minimal complex clamp loader, usually in "back-to-back" (i.e. hours apart) assays. Therefore, the term "clamp loader" will be used throughout this chapter to describe both, except where different concentrations of each were used or where only the activity of γ complex was studied.

Proteins, Reagents, and Oligonucleotide Substrates

DNA Polymerase III Proteins

DNA polymerase III proteins were a generous gift from Mike O'Donnell's laboratory at The Rockefeller University, New York, NY. Proteins were stored in 20 mM Tris-HCl pH7.5, 2 mM DTT, 0.5 mM EDTA, and 10% glycerol. Assay buffer for all experiments contained 20 mM Tris-HCl pH 7.5, 50 mM NaCl, 40 µg/mL Bovine serum albumin (Invitrogen Corp., Carlsbad, CA), 5 mM DTT, and 8 mM $MgCl_2$ (Table 3-1).

Concentrations of γ complex were determined from absorbances at 280 nM in 6 M guanidine hydrochloride and the calculated extinction coefficient. This concentration was verified by amino acid analysis following acid hydrolysis of the protein (performed by the Protein Chemistry Core Facility, Biotechnology Program, University of Florida). A sample for amino acid analysis was prepared by dialyzing γ complex against 20 mM sodium phosphate buffer pH 7.5 at 4 °C to remove glycerol and other buffer components that interfere with the analysis. The concentration of this dialyzed protein was also determined by three other methods for comparison; 1) its absorbance at 280 nM in 6 M guanidine hydrochloride pH 6.5, 2) its concentration in a Bradford assay using prepared reagents from BioRad and a BSA standard, and 3) its concentration in a Bradford assay using an IgG standard. The amino acid analysis and absorbance measured under denaturing conditions yielded concentrations of, 1.6 µM and 1.7 µM, respectively, within experimental error. Both Bradford-type assays overestimated the concentration of protein by a factor of 1.2 using the BSA standard and a factor of 3.0 using the IgG standard. The concentration of β was determined from its absorbance at 280 nm and the extinction coefficient for native protein of $(17\,900\ M^{-1}cm^{-1})$ (Johanson et al., 1986).

Reagents

ATP was from Sigma-Aldrich Co. (St. Louis, MO) and 5'-O-(3-thiotriphosphate) (ATPγS) was from Roche Molecular Biochemicals. Bacterial purine nucleoside phosphorylase (PNPase) and 7-methylguanosine (MEG) were from Sigma-Aldrich and stored as solutions at –80 °C.

Table 3-1. Assay and protein buffers

Standard assay buffer	Protein storage	Protein dilution
20 mM Tris-HCl pH 7.5 50 mM NaCl 40 µg/mL BSA 5 mM DTT 8 mM MgCl$_2$	20 mM Tris-HCl pH 7.5 0.5 mM EDTA 2 mM DTT 10 % glycerol	20 mM Tris-HCl pH 7.5 40 µg/mL BSA 5 mM DTT 0.5 mM EDTA 10 % glycerol

Oligonucleotide Substrates

Synthetic oligonucleotides were made on an ABI 392 DNA/RNA synthesizer using standard β-cyanoethylphosphoramidite chemistry and reagents from Glen Research Corp. (Sterling, VA). Denaturing polyacrylamide gel electrophoresis was used to purify new DNA. The 30-mer primer has the sequence: 5'-GAG CGT CAA AAT GTA GGT ATT

TCC ATG AGC GTT TTT CCT GTT GCA ATG GC-3' and the 105-mer template has

the sequence: 5'-GAG CGT CAA AAT GTA GGT ATT TCC ATG AGC GTT TTT

CCT GTT GCA ATG TCG GGC GGT AAT ATT GTT CTG GAT ATT ACC AGC

AAG GCC GAT AGT TTG AGT TCT TCT-3'. A 5'-C2-dt amino linker (Glen Research

Corp., Sterling, VA) was incorporated onto the 105-mer during synthesis. Following

synthesis and purification, the 5'-amino group was covalently labeled by addition X-

rhodamine isothiocyanate (RhX) (Molecular Probes Inc., Eugene, OR) as described

(Bloom et al., 1996). Recently, it was demonstrated that the position of the RhX

fluorescent probe on the 105-mer template had no effect on anisotropy measurements in

pre-steady-state experiments (Ason et al., 2003). For the MDCC-PBP ATPase assay, a

105-mer of the same sequence was synthesized, but was not functionalized with the 5'-

C2-dt amino linker, as it did not require fluorescent labeling. The 30-mer primer was

annealed to the X-rhodamine-labeled 105-mer (or unlabeled 105-mer) by incubating 1.2

mol equivalent of the 30-mer with 1 mol equivalent 105-mer in a water bath at 80 °C then

allowed to slowly cool to room temperature. Excess 30-mer was not further purified

from the p/t DNA since it has been previously shown that it had no effect on clamp

loading reactions in anisotropy assays (Ason et al., 2000).

Purification of *Escherichia coli* Phosphate Binding Protein

E. coli phosphate binding protein (PBP) was purified and labeled with N-[2-(1-

maleimidyl)ethyl]-7-(diethylamino)coumarin-3-carboxamide (MDCC) (Molecular Probes

Inc., Eugene, OR) using the method described by (Brune et al., 1998) without the use of

the "enhanced P_i-mop". Using this procedure, approximately 90 mg of the mutant

A197C phosphate binding protein was recovered from a 2-L culture. LB medium (10

mL) containing 12.5 mg/L tetracycline was inoculated with *E.coli* strain ANCC75 harboring the plasmid pSN5182/7 with the mutated *pho*S gene (a gift from Martin

Webb). This culture was diluted to 100 mL in the same medium after 6 h at 37 °C. This

culture was then grown for 16 h at 37 °C, and used for inoculation (i.e. 10 mL added) of

four 500-mL cultures containing high-phosphate TG-plus medium with supplements

(Table 3-2) containing 12.5 mg/L tetracycline (Brune et al., 1994).

Table 3-2. TG-plus media contents

Buffer	Essential amino acids	Salts
120 mM Tris-HCl pH 7.2	20 mg/mL L-arginine-HCl	120 mM NaCl
4 g/L glucose	10 mg/mL L-leucine	30 mM KCl
0.640 mM KH_2PO_4 pH 7.5[a]	8 mg/mL L-histidine-HCl	4.5 mM Na_2SO_4
	4 mg/mL L-methionine	0.3 mM $MgSO_4$
0.064 mM KH_2PO_4 pH 7.5[b]	4 mg/mL L-tryptophan	0.3 mM $CaCl_2$
	4 mg/mL L-adenosine	1.5 μM $FeSO_4$
	2 mg/mL L-thiamine	

[a] for high-potassium phosphate
[b] for low-potassium phosphate

For optimum growth of cells it is advantageous to use 2.8-L Fernbach culture flasks

for the 500-mL cultures with vigorous shaking throughout to obtain maximum aeration of

the cultures. The cultures were incubated at 37 °C for ~ 6 h at which point the optical

density at 600 nm was measured. The optical density at 600 nm should approach 2.0 for

optimum growth in these cultures. Cells were then pelleted by centrifugation at 1,670 x g

(JLA-10.500 rotor) at room temperature in autoclaved centrifuge bottles for 30 min.

Cells were resuspended in 500 mL TG plus containing the same supplements except low

phosphate was used (Table 3-2). It was important to wash the pellets in the low

phosphate medium to remove any carryover of the high phosphate from the original TG

plus medium. Cells were grown for another 16 h at 37 °C and then harvested by

centrifugation at 3,000 x g (JLA-10.500 rotor) for 25 min at room temperature.

An osmotic shock procedure was used to release PBP from the cells' periplasm (Willsky and Malamy, 1976). Supernatant was discarded and pellets were resuspended in 650 mL total volume containing 10 mM Tris-HCl pH 7.6 and 30 mM NaCl at room temperature. This was pooled into two 500-mL centrifuge bottles and spun at 3,000 x g (JLA-10.500 rotor) for 25 min at room temperature. The supernatant was discarded and the cells were resuspended in 350 mL of the same buffer. The cells were centrifuged as above, the supernatant removed, and the pellet weighed. Typically 6 – 10 g of cells were obtained from the 2-L culture. Cells were then resuspended in 100 mL 33 mM Tris-HCl pH 7.6 at 25 °C in a centrifuge bottle containing a magnetic stir bar. While stirring rapidly on a magnetic stir plate, 100 mL of 40% sucrose, 0.1 mM EDTA, and 33 mM Tris-HCl pH 7.6 was rapidly added at 25 °C. After the mixture stirred for 10 min at 25 °C cells were spun at 18 600 x g (JLA-10.500 rotor) for 20 min at 4 °C. After discarding the supernatant, the cells' outer membranes were burst by osmotic shock by rapid resuspension in 200 mL ice cold 0.5 mM $MgCl_2$. After rapid stirring for 15 min on an ice bath, the cells were spun at 18 600 x g (JLA-10.500 rotor) for 10 min at 4 °C. The supernatant ("shockate") was then dialyzed against the column equilibration buffer: 10 mM Tris-HCl pH 7.6, and determined by absorbance at 280 nm to be ~ 200 mg unpure periplasmic protein. The Osmotic shock procedure was repeated once again on the pelleted cells by another resuspension in ice cold 0.5 mM $MgCl_2$. Generally this second osmotic shock produced an additional significant quantity of PBP from the cells for purification (~ 70 mg).

PBP was purified from the shockate by FPLC anion exchange purification on a 75-mL High-Trap Q-Sepharose column (Amersham Biosciences Corp., Piscataway, NJ) equilibrated with 10 mM Tris-HCl pH 7.6 and 1 mM $MgCl_2$. The column was loaded with ~ 200 mg protein, and washed with 5-column volumes of equilibration buffer. A 1-L linear gradient of 0 to 200 mM NaCl was initiated at a flow-rate of 8 mL / min. The protein eluted in a single broad peak at ~ 30 mM NaCl. The peak fractions were pooled and concentrated using centricon-centriplus YM-10 centrifuge tubes (10 mL each) (Millipore Corp., Bedford, MA). Protein concentration was determined by absorbance readings at 280 nm using an extinction coefficient for PBP ($\varepsilon^{0.1\%}$ at 280 nm, 60 880 M^{-1} cm^{-1}). The identity of purified PBP was confirmed using denaturing polyacrylamide gel electrophoresis analysis with molecular weight standards and previously purified PBP as markers. Purified A197C PBP (~ 69 mg) was stored at –80 °C.

Labeling of Phosphate Binding Protein with MDCC

The purified A197C PBP protein was labeled in a solution (4 mL total volume) of 20 mM Tris-HCl pH 8.1 containing ~100 μM PBP, 200 μM MDCC in dimethylformamide, 168 μM MEG, 0.2 U/mL PNPase, and 5 μM $MnCl_2$. The labeling reaction was incubated for 30 min at room temperature while gently turning on a rotisserie with protection from light. After the labeling reaction PBP was concentrated in a Centricon-Centriplus YM-10 centrifuge tube (Millipore Corp., Bedford, MA) to a volume of approximately 1 mL. Free MDCC and other low molecular weight impurities were removed a with P6 gel desalting column (39-mL) (Bio-Rad Laboratories Inc., Hercules, CA). Labeled-protein separated from free MDCC, and eluted in the void volume of the column. MDCC-PBP was pooled and dialyzed against equilibration buffer (10 mM Tris-HCl pH 8.0) overnight. MDCC-PBP (typically 11-12 mg in ~ 9-13 mL) was then loaded onto a 5 X 5 mL High-Trap Q-Sepharose column (Amersham

Biosciences Corp., Piscataway, NJ) equilibrated with 10 mM Tris-HCl pH 8.0. After washing with 5-column volumes of equilibration buffer, a 200-mL linear gradient from 0 to 50 mM NaCl in equilibration buffer was initiated. This purification typically gave ~ 70 % recovery of >80 %-labeled / responsive MDCC-PBP. Pooled fractions were concentrated in an Amicon stirred-cell pressure concentrator (Millipore Corp., Bedford, MA) using a Biomax-(PB) polyethersulfone membrane (NMWL 5000 Daltons) (Millipore) at ~ 44 PSI, 4 °C. Concentrated MDCC-PBP was then dialyzed overnight against Mono-Q column equilibration buffer (10mM Tris-HCl pH 8.0, 100 μM KH_2PO_4). In general ~ 9 mg MDCC-PBP was loaded onto the equilibrated 8-mL Mono-Q column (Amersham Biosciences Corp., Piscataway, NJ). The column was then washed with 5-column volumes of equilibration buffer, and an 80-mL linear gradient from 0 to 30 mM NaCl was initiated. Two peaks were eluted; the earliest peak fractions contained P_i-responsive protein, and the latter, non-responsive protein. Both peaks were pooled separately and concentrated using an Amicon stirred-cell pressure concentrator (Millipore Corp., Bedford, MA) using a Biomax^{tm}-(PB) polyethersulfone membrane (NMWL 5000 Daltons) (Millipore) at ~ 44 PSI, 4 °C. The concentrated P_i-responsive protein was dialyzed overnight against 1 L 10 mM Tris-HCl pH 8.0 (changing the dilution buffer two times), and then stored in –80 °C. There is some remaining P_i contamination in the purified MDCC-PBP, but this was removed with the "P_i-mop" before each experiment (see below).

Characterization of MDCC-Labeled Phosphate Binding Protein

MDCC-PBP was analyzed by three methods, 1) UV/VIS absorbance for concentration calculation and determination of the efficiency of labeling, 2) a responsivity test to determine the extent of molar response to P_i, and 3) active site titration with P_i standards.

Removal of Free-Inorganic Phosphate (P_i) Contamination with the "Pi-mop"

The "Pi-mop" utilizes purine nucleoside phosphorylase (PNPase) to catalyze the phosphorolysis of 7-methylguanosine (MEG) for removal of P_i from buffers, quartz cuvettes and the stopped-flow instruments. Phosphorolysis of 7-methylguanosine is essentially irreversible toward formation of ribose-1-phosphate product (Nixon et al., 1998). To minimize free-P_i contamination, disposable plasticware was used, and rinsed only in deionized water to clean. At room temperature this "P_i-mop" removes the free-P_i contamination to levels below the threshold of sensitivity of MDCC-PBP (<0.1 μM) (Brune et al., 1994). Generally, quartz cuvettes and the stopped-flow apparatus were preincubated in buffer containing the "P_i-mop": PNPase (typically 0.1-0.2 U/mL) and MEG (typically ~ 160 μM).

Concentration and Efficiency of Labeling of MDCC-PBP

The concentration of MDCC-PBP was determined by its absorbance at 280 nm after correcting for the absorbance of MDCC at 280 nm (0.164X of the MDCC absorbance peak at 430 nm was subtracted from the absorbance at 280 nm) using an absorbance scan from 250 to 460 nm. Using the molar extinction coefficient for MDCC at 430 nm (46 800 $M^{-1}cm^{-1}$) (Brune et al., 1994) the concentration of MDCC was determined. Using the value of MDCC concentration, the percent labeled-PBP (i.e. efficiency of labeling) was determined (typically 86-90%). This percent-labeled quantity was used as the concentration of MDCC-PBP all MDCC-PBP ATPase assays since it is only this labeled-PBP that undergoes an increase in fluorescence upon binding P_i.

Therefore, the only fluorescence response observed in the MDCC-PBP ATPase assays always corresponds to the fraction (percent) labeled-PBP. It is noteworthy that the dissociation constant for the labeled-PBP is an order of magnitude lower (i.e. has greater affinity for P_i) than for any unlabeled-PBP which may have come through the purification (Brune et al., 1998).

Characterization of the fluorescence-molar response of MDCC-PBP to P_i

A "P_i -responsivity test" was performed by scanning the fluorescence emission of MDCC-PBP in standard assay buffer, (Table 3-1), using the QuantaMaster QM-1 fluorimeter described below with excitation and emission monochromator band passes of 2 nm. Approximately 4 µM MDCC-PBP (either P_i -responsive or non-responsive) in assay buffer containing MEG was initially scanned, then "P_i-mop" PNPase was added to begin the removal of P_i contamination. Emission scans were performed at 5 min intervals until the peak fluorescence intensity reached a minimum value (typically took 15-20 min), and was completely red-shifted from ~ 464 nm to ~ 472 nm. The minimum peak intensity at 472 nm was recorded as the intensity value for Pi-free MDCC-PBP. To this was added a solution containing potassium phosphate (KH_2PO_4) giving a final concentration of 200 µM in order to observe the saturated intensity of completely P_i - bound MDCC-PBP. The fold-difference between the P_i -free and P_i -bound intensities is noted as the Pi-response, and was generally 7- to 10-fold for P_i -responsive MDCC-PBP, and 1- to 2-fold for non-responsive MDCC-PBP. The observed intensities of the P_i-free and P_i-bound peaks for non-responsive MDCC-PBP were characteristically at least an order of magnitude lower than the observed intensities for responsive MDCC-PBP.

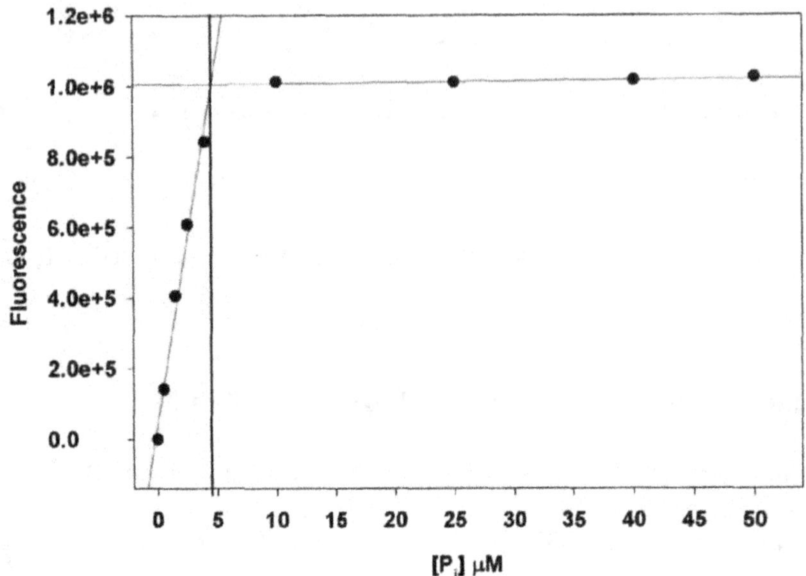

Figure 3-1. P$_i$ titration analysis of MDCC-PBP. 5 µM MDCC-PBP was subjected to a range of potassium phosphate (P$_i$) from 0 to 50 µM in P$_i$-mopped assay buffer. The average fluorescence emission intensity values at 464 nm from 30 s measurements are plotted as a function of concentration of potassium phosphate standards. Linear regressions for the rising and saturated data points intersect at a point reflecting the apparent concentration of the fraction of labeled-MDCC-PBP (~4.5 µM) (vertical line).

Active Site Titration of MDCC-PBP

MDCC-PBP was titrated with P$_i$ of known concentrations. Potassium phosphate was used as the source of P$_i$ in the titration. The experimental setup of this analysis was designed to mimic the steady-state MDCC-PBP ATPase assay described below. Fluorescence intensities at 464 nm were recorded for solutions containing 5 µM MDCC-PBP subjected to a range of potassium phosphate from 0 to 50 µM. The potassium phosphate (P$_i$) standards were added to separate assays in random order to prevent any systematic error in the analysis.

A 76-µL sample containing 5 µM MDCC-PBP in assay buffer (Table 3-1) was placed in a quartz microcuvette (internal dimensions 3 X 3 mm), (Hellma Cells Inc.,

Plainview, NY). Fluorescence emission intensity of Pi-free MDCC-PBP was recorded for 30-s. Then a 4-μL solution containing varied concentrations of P_i was mixed into the cuvette, and another 30 s measurement of P_i-bound emission intensity was recorded. The averaged values of the 30-s measurements of P_i-free and P_i-bound fluorescence intensities were used to analyze the nature of P_i binding to MDCC-PBP. Figure 3-1 shows a typical P_i titration.

This characterization has great significance for the quantitation of P_i in the MDCC-PBP ATPase assays. The rising and saturated data points of this titration were separately subject to linear regression and the point of their intersection relates to the concentration of MDCC-PBP in the experiment since the fluorescence of MDCC-PBP in response to P_i saturates at stiochiometric quantities of P_i (Brune et al., 1994). Control titrations containing an equimolar amount of unlabeled-PBP showed no change in the fluorescence response (not shown). Together, the active site-P_i titration shows that only the labeled-fraction of PBP (~ 4.5 μM) actually binds P_i. Thus, it is proper to use the concentration of the fraction of MDCC-labeled PBP in the calculation of molar P_i in the MDCC-PBP ATPase assays.

Fluorescence Anisotropy Binding Assays

Calculation of Anisotropy

In each anisotropy binding assay, vertically and horizontally polarized emission intensities were measured and corresponding background intensities were measured for solutions containing assay buffer only. The background intensities were subtracted from experimental intensities for background correction. To correct for any polarization bias a "G-factor" was determined by measuring vertical and horizontal emission intensities

95

when the fluorescent-labeled species was excited with horizontally polarized light

(Lakowicz, 1999). Anisotropy values were calculated using background corrected

intensities and the equation, $r = (I_{vv} - gI_{vh})/(I_{vv} + 2gI_{vh})$, where r is anisotropy, and I_{vv} &

I_{vh} are the background corrected intensities of vertically and horizontally polarized

emission, respectively, measured when samples are excited with vertically polarized light

and g is the G-factor. The anisotropy values determined in titration assays were generally

converted to the fraction-bound of fluorescent-labeled substrate, and fit to the quadratic

equation for estimation of the binding constant (K_d): Fraction bound X_b = [AB].

$$[AB] = \frac{(K_d + [A] + [B] - \sqrt{(K_d + [A] + [B])^2 - 4[A][B]}}{2[B]} * (r_{bound} - r_{free}) + r_{free}$$

Where [A] and [B] are the concentrations of the clamp loader and fluorescent-labeled

DNA or β, respectively, and r_{bound} and r_{free} are the characteristic steady-state anisotropy

values for bound- and free-labeled DNA or β, respectively.

Steady-state Measurement of Clamp Loader – RhX-DNA Binding Kinetics

Steady-state fluorescence-polarized emission intensity measurements for clamp

loading assays were recorded using a Quantamaster QM-1 fluorimeter (Photon

Technology International Inc., Lawrenceville, NJ) equipped with a 75-W xenon arc lamp,

an excitation monochromator, dual emission monochromators, photon-counting detectors

with Hamamatsu R928 PMTs (Hamamatsu Photonic Systems, Bridgewater, NJ), and

Glan-Thompson polarizers. Samples were excited with vertically polarized light at 580

nm (5 nm band pass) and both vertically and horizontally polarized emission at 610 nm

(5 nm band pass) were simultaneously measured in a "T-format". Reaction final volumes

were 80 μL in a quartz microcuvette (internal dimensions 3 X 3 mm, (Hellma Cells Inc.,

Plainview, NY).

Determination of clamp loader – RhX-labeled DNA binding constants

In fluorescence-based titration assays for the calculation of RhX-pt DNA

anisotropy, the Quantamaster fluorimeter setup described above was used. A solution of

RhX-labeled DNA, either single-stranded (ss DNA) or primer-template (pt DNA)

substrates, at a concentration of 50 nM in assay buffer (Table 3-1) was placed in a

cuvette. Initially, background polarized emission intensities were recorded with the free

RhX-DNA substrate. When β was included in titration assays, it was added to the RhX-

DNA solution after background fluorescence measurements. The final concentration of β

was 500 nM in these assays. Next the RhX-DNA was titrated by addition of a constant

volume (8 μL) of a solution containing clamp loader, at differing concentrations, in assay

buffer. ATP was present either originally in the assay buffer, or added later giving a final

concentration of 0.3 mM or 0.5 mM. The amount of ATP was high enough such that its

concentration was essentially constant over the short time-course of the titration

experiment. Typically, 30 s measurements of polarized emission intensities were

averaged for each step in the titration assays; including background measurements, after

addition of β (when present), and after addition of each solution of clamp loader for

calculation of anisotropy values. In some assays, the steady-state polarized emission

intensities were observed over longer time-courses, and anisotropy was calculated per

each second of the experiment. Steady-state anisotropy values are a function of both the

lifetime and rotational correlation time of a fluorophore (Lakowicz, 1999; Otto et al.,

1994). It was previously determined that the fluorescence lifetime of X-rhodamine-labeled DNA did not change when clamp loader and β clamp were present in assays with RhX-DNA, demonstrating that the measured changes in steady-state anisotropy were the result of changes in rotational dynamics of the fluorescent probe on the DNA substrate (Bloom et al., 1996).

Determination of the effect of ADP on clamp loader - RhX-labeled DNA binding

The steady-state fluorescence emission intensities of RhX-DNA (either ss DNA or pt DNA) in assay buffer were measured for calculation of anisotropy as described above. Initially, background measurements were recorded for a solution containing either ss or pt RhX-DNA in the presence of differing concentrations of ADP. ADP was varied over a range of 0.2 mM to 0.8 mM in this solution. When present, β clamp (final concentration, 500 nM) was then added. This was followed by addition of a solution containing clamp loader (final concentration, 500 nM). ATP was finally added to the cuvette giving a final concentration of 0.3 mM. At all steps, reactions were monitored for 30 s and resulting polarized emission intensities were averaged for the calculation of anisotropy.

Determination of clamp loader - β^{pyrene} clamp binding constants

Fluorescence-based titration experiments were performed using the Quantamaster fluorimeter in a similar manner to those above for RhX-DNA except that the excitation light was at 345 nm and the polarized emission intensities were measured at 375 nm (5 nm band pass). Initially, background intensity measurements were recorded for a solution containing pyrene-labeled β clamp (β^{pyrene}) in assay buffer (final concentration, 50 nM). Then, a constant volume (8 μL) of a solution containing clamp loader was added to the cuvette. Concentrations of the clamp loader were varied in separate assays. After addition of the clamp loader solution, ATP-independent β^{pyrene}-clamp loader binding activity was measured. Finally, ATP (final concentration, 0.3 mM) was added to the reaction to induce ATP-dependent binding of β^{pyrene} and clamp loader. At each step in the assay, a 60-s measurement of polarized emission intensities was recorded, and the average intensity values for these 60-s measurements was used in the calculation of anisotropy.

Determination of the apparent binding constant of ATP to the clamp loader

In fluorescence-based titration assays with β^{pyrene}, similar to those described above, the apparent binding constant of ATP to the clamp loader was estimated. Initially, background intensity measurements were recorded for a solution containing pyrene-labeled β clamp (β^{pyrene}) (final concentration, 65 nM) in assay buffer. To this was added a solution of clamp loader (final concentration, 400 nM) in assay buffer. To induce binding of β^{pyrene} and clamp loader, a constant volume (4 μL) of a solution containing ATP was added to the reaction. The concentration of ATP was varied in separate assays. At each step in the assay, a 60-s measurement of polarized emission intensities was recorded, and the average intensity values for these 60-s measurements was used in the calculation of anisotropy. The concentrations of ATP stocks were determined from the average of at least three absorbance measurements at 259 nm using the molar extinction coefficient at 259 nm (15 400 $M^{-1}cm^{-1}$).

Determination of the effect of ADP on clamp loader - β^{pyrene} binding

A constant volume (8 μL) of a solution containing clamp loader was added to a cuvette containing a solution of β^{pyrene} in assay buffer following initial background intensity measurements. Concentrations of clamp loader were varied while each separate assay contained 55 nM β^{pyrene} in assay buffer. Next, a solution of ATP (final concentration, 0.3 mM) was added to induce clamp loader interaction with β^{pyrene}.

Finally, a solution containing an excess amount of ADP (final concentration, 0.8 mM) in

a small volume (2 µL) was added to the cuvette. At each step in the assay, a 60-s

measurement of polarized emission intensities was recorded, and the average intensity

values for these 60-s measurements was used in the calculation of anisotropy.

Pre-Steady-State Measurement of Clamp Loader – RhX-DNA Binding Kinetics

A Biologic SFM-4 stopped-flow (Molecular Kinetics, Pullman, WA) equipped with

four independently driven reagent syringes and a 31-µL cuvette (model # FC-15) with a

1.5 mm path length was used to record polarized emission intensities in real-time.

Dichroic sheet polarizers (380 – 770 nm, model # 27346), (ThermoOriel Corp., Stratford,

CT) were mounted directly onto the cuvette. A QuantaMaster QM-1 fluorimeter,

described above, was used as an excitation source by focusing the output from the

excitation monochromator (580 nm, 1.0 – 1.5 nm band pass) onto a fused silica fiber

optic consisting of a bundle with dimensions of 0.25 X 2.56 mm containing nine 250-µM

fibers (Fiberguide Industries, Stirling, NJ). The excitation fiber optic was mounted

against the stopped-flow cuvette "sandwiching" the dichroic sheet polarizer. Vertically

and horizontally polarized emission were collected simultaneously in a "T-format"

through UV/VIS liquid light guides with a 5-mm core diameter (ThermoOriel Corp.,

Stratford, CT) mounted directly against the stopped-flow cuvette "sandwiching" the

dichroic sheet polarizers. Fluorescence emission was detected with a photon-counting

detection system consisting of two channels each with a filter holder containing a 610 nm

cut-on filter (CVI Laser Corp., Albuquerque, NM), mounted onto an ambient

photomultiplier housing (Products for Research, Inc., Danvers, MA) with a R4457P PMT

(Hamamatsu Corp., Bridgewater, NJ). Signals from both PMTs were detected via a SR455 preamplifier, a SR400 gated-photon counter (Stanford Research, Sunnyvale, CA), and MCS-II multichannel scalar cards (Oxford Research, Oakridge, TN). The external synch from the SFM-4 stopped-flow unit triggered data acquisition by the MCS-II cards.

All assays were performed at 20 °C. For each assay run, vertically and horizontally polarized emission intensities were recorded at 1-ms intervals, and 18 – 20 stopped-flow runs were signal averaged. The cuvette was flushed out prior to each run with assay buffer (Table 3-1). Following each assay, the steady-state anisotropy of RhX-DNA free in assay buffer (r_{free}) was measured using the setup described above for steady-state anisotropy calculation. Since the excitation polarizer is "sandwiched" between the cuvette and fiber optic in the Biologic SFM-4 stopped-flow, it is not easily rotated for horizontal excitation to determine the G-factor polarization bias of the system. However, since there are no emission monochromators in this system, there is only modest polarization bias. The G-factor was solved for with the steady-state anisotropy value of free RhX-DNA left over from the assays, using the equation: $G = [(I_{vv})/(I_{vh})] / [(1 - r_{free})/(2r_{free} + 1)]$, where r_{free} is the steady-state anisotropy of free RhX-DNA, and I_{vv} & I_{vh} are the background corrected intensities of vertically and horizontally polarized emission, respectively, measured when samples are excited with vertically polarized light.

Pre-steady-state kinetics of clamp loading initiated at different reaction steps

Three different mixing schemes for pre-steady-state anisotropy binding assays were achieved by changing the reagents in the stopped-flow drive syringes. All reactions were initiated by mixing 80 μL each from two syringes at a 10-mL/s flow rate. The stopped-flow reaction dead time under these conditions was 3.7 +/- 0.7 ms determined using the method of (Peterman, 1979). See below for description of the dead-time determination.

For scheme-1 mixing, one syringe contained 500 nM clamp loader, 0.5 mM ATP, and 1.2 μM β in assay buffer. The second syringe was loaded with 100 nM RhX-pt DNA, and 0.5 mM ATP in assay buffer. For scheme-2 mixing, one syringe was loaded with 500 nM clamp loader, and 0.5 mM ATP in assay buffer. The second syringe was loaded with 100 nM pt DNA, 1.2, 2.4, or 4.8 μM β, and 0.5 mM ATP in assay buffer. For scheme-3 mixing, one syringe contained 500 nM clamp loader only. The second syringe was loaded with 100 nM pt DNA, 1.2 or 4.8 μM β, and 0.2, 0.5, or 1.0 mM ATP in assay buffer. In control reactions without enzyme, one syringe was loaded with assay buffer only, and the second syringe was loaded with 100 nM pt DNA, 1.2 μM β, and 0.5 mM ATP in assay buffer.

Pre-steady-state kinetics of clamp loading on different RhX-DNA substrates

The activity of γ complex with RhX-pt DNA versus RhX-ss DNA was compared by analysis of pre-steady-state kinetics using anisotropy binding assays. For these experiments, scheme-1 mixing was utilized. One syringe was loaded with 480 nM γ complex, 0.5 mM ATP, and 1 μM β (when present) in assay buffer. The second syringe was loaded 100 nM RhX-pt DNA or RhX-ss DNA and 0.5 mM ATP in assay buffer. Reactions were initiated as above.

Fluorescence-based MDCC-PBP ATP Hydrolysis (ATPase) Assay

Steady-State Kinetics of ATP hydrolysis

The steady-state change in fluorescence of MDCC-PBP was measured using a Quantamaster QM-1 fluorimeter (Photon Technology International Inc., Lawrenceville,

NJ). Samples were excited at a wavelength of 425 nm (2 nm band pass), and emission was selected at 464 nm (2 nm band pass).

A 76-μL sample of clamp loader, β clamp (when present), pt DNA, and MDCC-PBP in standard assay buffer (Table 3-1) was placed in a quartz microcuvette (internal dimensions 3 X 3 mm, (Hellma Cells Inc., Plainview, NY). A 10-minute timebased scan was initiated, and approximately 14-s later a 4-μL solution of ATP was added to initiate the reaction. The P_i-mop (described above) was used to remove P_i contamination separately from the assay buffer (0.06 U/mL PNPase, 160 μM MEG), and the solution containing MDCC-PBP, clamp loader, pt DNA, and β (when present) (0.1 U/mL PNPase, 160 μM MEG). Both mopped solutions were diluted 4-fold when added to the cuvette before starting the reaction. This dilution effectively reduced the PNPase concentration such that it had only a minimal effect on the steady-state measurement of P_i-binding MDCC-PBP in these assays. Initial velocities for reactions containing 41 nM γ complex, 50 nM pt DNA, 50 nM β (when present), and 3.4 μM MDCC-labeled phosphate binding protein in assay buffer were measured for reactions containing a range in molarity of the substrate ATP (0.6, 1.1, 2.6, 7, 10, 20, 40, and 77 μM). For analysis of the minimal complex, reactions contained 100 nM γδδ', 100 nM pt DNA, 312.5 nM β. In all experiments, the emission intensity measured prior to the addition of ATP was recorded as the intensity of P_i-free MDCC-PBP. Following each reaction, 1 μL of potassium phosphate (16.2 μM) was added to give a final saturating excess concentration of P_i (200 μM). The saturated emission intensity was recorded as the intensity of completely P_i-bound MDCC-PBP.

The concentrations of ATP stocks were determined from the average of at least 3 absorbance measurements at 259 nm using the molar extinction coefficient at 259 nm (15 400 $M^{-1}cm^{-1}$).

The mole fraction of MDCC-PBP bound to P_i $X_b(t)$ was calculated by subtracting the intensity of free MDCC-PBP (I_f) from observed intensities at time t $I_{obs}(t)$ and dividing this difference by the difference of fully saturated P_i -bound MDCC-PBP (I_b) and free MDCC-PBP $X_b(t) = [I_{obs}(t) - I_f] / [I_b - I_f]$. Reactions were performed 3 times and the initial velocities were calculated as the linear slope of P_i release (μM) from clamp loader ATPase activity ($\leq$ 10% product). Initial velocities were plotted versus ATP concentration then analyzed using the Michaelis-Menten equation to determine steady-state ATP hydrolysis parameters, where [S] is the concentration of ATP.

$$v = \frac{V_{max}[S]}{K_m + [S]}$$

Michaelis-Menten equation.

Effect of β concentration on the steady-state ATP hydrolysis activity of the clamp loader

The steady-state MDCC-PBP ATPase assay was performed as described above with different concentrations of β. For analysis of γ complex, a range of 50, 100, 200, and 740 nM β was used. For analysis of $\gamma_3\delta\delta'$ minimal complex, a range of 200, 310, 600, 1500 nM β was used. The concentrations were 41 nM γ complex, 50 nM p/t DNA, and 3.4 μM MDCC-labeled phosphate binding protein in assay buffer, or for the minimal complex, reactions contained 100 nM $\gamma_3\delta\delta'$, 100 nM pt DNA, and 3.4 μM MDCC-PBP.

Steady-state ATPγS-chase assay

Steady-state ATP hydrolysis activity of the clamp loader was chased by addition of non-hydrolyzable ATPγS after initiation of the MDCC-PBP ATPase assay. The ATPγS-chase assay was performed by initiating a timebased fluorescence emission measurement of a solution containing 50 nM clamp loader, 150 nM β (when present), 150 nM pt DNA, and 3.4 μM MDCC-PBP in assay buffer that was P_i-mopped as described above for the steady-state MDCC-PBP ATPase assay. After 14 s, 4-μL of a solution containing ATP was added to this solution giving a final concentration of 77 μM ATP. A constant volume of ATPγS (4 μL) was then added after ~10-s into the reaction time course. The ATPγS concentration was varied to produce differing final concentrations over the range of: 20, 40, 80, 160, and 800 μM in separate chase reactions.

Pre-Steady-State Kinetics of ATP Hydrolysis

Measurement of the change in fluorescence of MDCC-PBP in real-time was carried out on a SX.18MV Stopped-Flow Reaction Analyzer (Applied Photophysics Ltd., UK). Excitation light from a 150-W xenon arc source was at 425 nm through a monochromator with a 0.5 mm band pass. Fluorescence emission intensity was recorded through a 455 nm cut-on filter (CVI Laser Corp., Albuquerque, NM) by an analogue PMT-detection device. Data points were collected every 1 ms for the first 2 s of the reaction time course, then every 5 ms for the remaining 10 s using a split-timebase. All experiments were performed at 20 °C. The stopped-flow apparatus and all reagents were P_i-mopped (0.14 U/mL PNPase, 168 μM MEG) for ~ 45 min prior to each experiment.

Generally, a *sequential-mix "three-syringe"* experiment was performed where 80 μL each of the contents of two syringes were mixed and preincubated for 1 s. The preincubated solution (80 µL) was then rapidly mixed with 80 µL of the contents of a third syringe. This action both initiated the reaction and triggered the detection system. The stopped-flow reaction dead time under these conditions was 1.47 ms determined using the method of (Peterman, 1979) as described below. Each assay typically had three steps. 1) An initial measurement of the fluorescence intensity of P_i-free MDCC-PBP in the presence of pt DNA, clamp loader, and β (when present), and no ATP. 2) One or more experimental "shots" in the presence of ATP. 3) A final shot under conditions of saturating excess P_i for measurement of the fluorescence intensity of completely P_i-bound MDCC-PBP.

Standard pre-steady-state MDCC-PBP ATPase assay

In a standard sequential-mix "three-syringe" assay, one syringe was loaded with 1.08 µM clamp loader, and 4 µM β (when present) in assay buffer (Table 3-1). A second syringe was loaded with 400 µM ATP in assay buffer. For initial measurement of the fluorescence intensity of P_i-free MDCC-PBP as described in step one above, this second syringe initially contained assay buffer only, then was replaced with the solution containing ATP for the experimental shots. The contents of these two syringes were mixed and preincubated for 1 s as above, then mixed with the contents of a third syringe loaded with 5.4 µM MDCC-PBP, and 2 µM pt DNA in assay buffer. Final reaction concentrations were 270 nM clamp loader, 1 µM β (when present), 1 µM pt DNA, 2.7 µM MDCC-PBP, and 100 µM ATP. For the final saturated emission intensity measurement of completely P_i-bound MDCC-PBP, the syringe containing ATP was removed, and 14 µL potassium phosphate (16.2 mM) was added to the syringe, mixed with the remaining ATP solution, and then flushed into the stopped-flow. A final experimental shot added ~ 200 μM excess P_i to the reaction mixture.

Pre-steady-state MDCC-PBP ATPγS-chase assay

"Back-to-back" pre-steady-state MDCC-PBP ATPase assays were performed in the presence and absence of β where ATP hydrolysis activity was chased by mixing with ATPγS at initiation of the reaction. For these sequential-mix "three-syringe" ATPγS-chase assays, one syringe was loaded with 1.08 μM clamp loader, and 4 μM β (when present) in assay buffer. After the initial measurement of the fluorescence intensity of P_i-free MDCC-PBP as described in step one above, the second syringe, originally containing assay buffer only, was exchanged with a syringe loaded with 400 μM ATP in assay buffer. The contents of these two syringes were mixed and preincubated for 1 s, then mixed with the contents of a third syringe loaded with 2 mM ATPγS, 5.4 μM MDCC-PBP, and 2 μM pt DNA in assay buffer. Final reaction concentrations were 270 nM clamp loader, 1 μM β (when present), 1 μM pt DNA, 2.7 μM MDCC-PBP, 100 μM ATP, and 1 mM ATPγS (a 10-fold excess over ATP). For the final saturated emission intensity measurement of completely P_i-bound MDCC-PBP, the syringe containing ATP was removed, and 14 μL potassium phosphate (16.2 mM) was added to the syringe, mixed with the remaining ATP solution, and then flushed into the stopped-flow. A final experimental shot added ~ 200 μM excess P_i to the reaction mixture.

Measurement of the effect of γ complex concentration in the pre-steady-state MDCC-PBP ATPase assay

The above standard pre-steady-state sequential-mix "three-syringe" assay was performed with a range of concentrations of γ complex in the presence of β. In three separate assays, one syringe was loaded with 0.4, 1.08, or 1.8 µM γ complex and 6.4 µM β in assay buffer. After the initial measurement of the fluorescence intensity of P_i-free MDCC-PBP as described in step one above, the second syringe, originally containing assay buffer only, was exchanged with a syringe loaded with 1.6 mM ATP in assay buffer. The contents of these two syringes were mixed and preincubated for 1 s, then mixed with the contents of a third syringe loaded with 5.4 µM MDCC-PBP and 3.2 µM pt DNA in assay buffer. Final reaction concentrations were 100, 270, or 450 nM γ complex, 1.6 µM β, 1.6 µM pt DNA, 2.7 µM MDCC-PBP, and 400 µM ATP. For the final saturated emission intensity measurement of completely P_i-bound MDCC-PBP, the syringe containing ATP was removed, and 14 µL potassium phosphate (16.2 mM) was added to the syringe, mixed with the remaining ATP solution, and then flushed into the stopped-flow. A final experimental shot added ~ 200 µM excess P_i to the reaction mixture.

Pre-steady-state MDCC-PBP ATPase assay with varying preincubation periods

A standard pre-steady-state sequential-mix "three-syringe" assay was performed for γ complex in the absence of β where the preincubation period after mixing the contents of the first two syringes was varied. One syringe was loaded with 1.08 µM γ complex in assay buffer. After the initial measurement of the fluorescence intensity of P_i-free MDCC-PBP as described above, the second syringe, originally containing assay buffer only, was exchanged with a syringe loaded with 1.6 mM ATP in assay buffer. In separate experimental shots, the contents of these two syringes were mixed and preincubated for 15, 50, 100, 250, 500, or 1000 ms, then mixed with the contents of a third syringe loaded with 5.4 µM MDCC-PBP, and 2 µM pt DNA in assay buffer. Final reaction concentrations were 270 nM γ complex, 1 μM pt DNA, 2.7 μM MDCC-PBP, and 400 μM ATP. For the final saturated emission intensity measurement of completely P_i-bound MDCC-PBP, the syringe containing ATP was removed, and 14 μL potassium phosphate (16.2 mM) was added to the syringe, mixed with the remaining ATP solution, and then flushed into the stopped-flow. A final experimental shot added ~ 200 μM excess P_i to the reaction mixture.

Pre-steady-state MDCC-PBP ATPase assay with no preincubation period

This assay is a single-mix experiment utilizing only two syringes of the stopped-flow apparatus. By using this *simple "two-syringe"* experimental setup, there was essentially no preincubation period of clamp loader with ATP prior to reacting with pt DNA in the presence of MDCC-PBP. The stopped-flow reaction dead time under these conditions was 1.6 ms determined using the method of (Peterman, 1979) described below.

One syringe was loaded with 540 nM γ complex in assay buffer, and the other syringe was loaded with 5.4 μM MDCC-PBP, 2 μM pt DNA, and 800 μM ATP in assay buffer. The contents of the two syringes were rapidly mixed to initiate the reaction and trigger the detection system. Final reaction concentrations were 270 nM γ complex, 1 μM pt DNA, 2.7 μM MDCC-PBP, and 400 μM ATP. For the final saturated emission intensity measurement of completely P_i-bound MDCC-PBP, the syringe containing ATP was removed, and 14 μL potassium phosphate (16.2 mM) was added to the syringe, mixed with the remaining ATP solution, and then flushed into the stopped-flow. A final experimental shot added ~ 200 μM excess P_i to the reaction mixture.

Determination of the ATP concentration dependence on γ complex ATP hydrolysis

activity

The single-mix "two-syringe" MDCC-PBP ATPase assay with no preincubation period was also used to explore the concentration dependence of ATP on the kinetics of ATP hydrolysis by γ complex. In separate assays, one syringe was loaded with 540 nM γ complex in assay buffer, and the other syringe was loaded with 5.4 μM MDCC-PBP, 2 μM pt DNA, and 200, 500 μM, or 1 mM ATP in assay buffer. The contents of the two syringes were rapidly mixed to initiate the reaction. Final reaction concentrations were 270 nM γ complex, 1 μM pt DNA, 2.7 μM MDCC-PBP, and 20, 80, 100, 200, 250, 400, or 500 μM ATP. For the final saturated emission intensity measurement of completely P_i-bound MDCC-PBP, the syringe containing ATP was removed, and 14 μL potassium phosphate (16.2 mM) was added to the syringe, mixed with the remaining ATP solution, and then flushed into the stopped-flow. A final experimental shot added ~ 200 μM excess P_i to the reaction mixture.

Computer Modeling of ATP Hydrolysis Kinetic Data

Kinetics data in Figure 4-9 were fit to the model to shown in Figure 4-10A using the program DynaFit (Kuzmic, 1996). The DNA binding rate has been measured previously (Ason et al., 2000; Ason et al., 2003). The forward rate constant for DNA binding was set to a constant value of 235 $μM^{-1}s^{-1}$ determined by fitting the previously published data to our model. This value is consistent with the $300 - 400$ $μM^{-1}s^{-1}$ rates that were reported based on simulation of these data. The concentration of MDCC-PBP•P_i as a function of time was calculated by including experimentally determined rate constants for MDCC-PBP binding P_i (Brune et al., 1994) as constants. The total concentration of γ complex was treated as an adjustable parameter to account for the difference between total and active protein concentrations as well as the experimental accuracy of concentration and volume measurements. The fit yielded a concentration of 0.25 μM (93% active) γ complex for assays containing an experimentally determined of 0.27 μM γ complex. Some steps were modeled using a single forward rate constant because these data do not provide sufficient information to calculate a unique pair of forward and reverse rate constants. This does not mean to imply that these steps are irreversible.

Data in Figure 4-10C were fit to a version of the model in Figure 4-10A that included an additional initial step converting an enzyme species PTTT to the species ETTT. For this fit, all of the rate constants shown in Figure 4-10A were held constant; the rate of conversion of PTTT to ETTT, k_{pe}, and the total concentration of γ complex were the only parameters fit. This fit yielded a concentration of 0.26 μM γ complex. For simulation of the different species of γ complex in a 1-s equilibration (Figure 4-10B) using the model in Figure 4-10A, the program KinTekSim was used (Barshop et al., 1983; Dang and Frieden, 1997). Appendix A contains all DynaFit program scripts, and analyses indices used in the computer modeling, as well as the KinTekSim simulations.

Correlated Pre-Steady-State MDCC-PBP ATPase Assays and Fluorescence Anisotropy Binding Assays

The standard sequential-mix "three-syringe" assay described above for the SX.18MV apparatus was used for pre-steady-state MDCC-PBP ATPase assays as well as pre-steady-state fluorescence anisotropy measurements under nearly identical experimental conditions. The measurement of anisotropy on the SX.18MV stopped-flow apparatus required the use of the FP-1 attachment that housed optics for the polarization of the excitation light at 580 nm, and measurement of polarized emission intensities from RhX-pt DNA. Also fitted on the apparatus were 610 nm cut-on filters (CVI Laser Corp., Albuquerque, NM) specific for the X-rhodamine fluorescent probe. Other than these instrumental modifications required for performing anisotropy measurements, the standard "three-syringe" reaction was unchanged between the MDCC-PBP ATPase assay and anisotropy binding assay.

For each assay, one syringe was loaded with 900 nM clamp loader, and 2 μM β (when present) in assay buffer. A second syringe contained 800 μM ATP in assay buffer. The contents of these two syringes were mixed and preincubated for a period of 1 s, then mixed with the contents of a third syringe, loaded with either 10 μM MDCC-PBP, and 900 nM pt DNA in assay buffer for the MDCC-PBP ATPase assays, or 900 nM RhX-pt DNA in assay buffer for anisotropy binding assays. Final reaction concentrations were 225 nM γ complex, 500 nM β (when present), 450 nM pt DNA, 200 μM ATP. The MDCC-PBP ATPase assays contained a final concentration of 5 μM MDCC-PBP.

In the MDCC-PBP ATPase assays, the second syringe initially contained assay buffer only, for the initial measurement of the fluorescence intensity of P_i-free MDCC-PBP. This second syringe was then exchanged with a syringe loaded with 400 μM ATP in assay buffer. For the final saturated emission intensity measurement of completely P_i-bound MDCC-PBP, the syringe containing ATP was removed, and 14 μL potassium phosphate (16.2 mM) was added to the syringe, mixed with the remaining ATP solution, and then flushed into the stopped-flow. A final experimental shot added ~ 200 μM excess P_i to the reaction mixture.

For the anisotropy binding assays, RhX-pt DNA with assay buffer only was initially placed into the cuvette for setup of the PMTs, and software calculation of the G-factor. The stopped-flow software ultimately calculates anisotropy in the real-time output over the time courses of each experimental reaction shot. Up to 14 separate experimental recordings of calculated anisotropy were signal averaged to reduce the signal-to-noise ratio.

Stopped-Flow Dead Time Determinations

Determination of the Dead Time for the Biologic SFM-4 Stopped-Flow

The dead time for the Biologic SFM-4 stopped-flow described above was determined using the method of (Peterman, 1979). The quenching of N-acetyl-tryptophanamide (N-AcTrpNH$_2$) fluorescence by N-Bromosuccinamide (NBS) in sodium phosphate buffer (pH 7.0) was used as a reaction for dead time determination. Rapid mixing of differing concentrations of NBS quencher results in single-exponential decays in N-AcTrpNH$_2$ fluorescence at different amplitudes and rates. N-AcTrpNH$_2$ fluorescence was excited at 280 nm, and emission intensity was measured through 320 nm cut-on filters (CVI Laser Corp., Albuquerque, NM). Thin-film polarizers were removed from the stopped-flow apparatus in this experimental setup.

In each reaction, one syringe contained 20 μM N-AcTrpNH$_2$ in sodium phosphate buffer (monobasic) (pH 7.0). The second syringe was loaded with varying concentrations of NBS in sodium phosphate buffer (monobasic) (pH 7.0). The contents of each syringe were rapidly mixed in 80-μL shots at a 10-mL / s flow-rate. Generally, 13 reaction shots were signal averaged for each concentration of NBS. In between reactions differing in concentration of NBS quencher, the stopped-flow syringe containing NBS was thoroughly rinsed with deionized water. The final reactant concentrations were 10 μM

N-AcTrpNH$_2$, 50 mM sodium phosphate (monobasic) pH 7.0, and a range of NBS: 312.5,

500, 625, 750, 1000, 1250, 1500, and 2000 μM.

Each fluorescence decay curve was fitted as a single exponential decay in the

program SigmaPlot (SPSS Inc., UK) for determination of the experimental decay

amplitudes and rates. Figure 3-2 shows a fit defining the experimentally determined rate

constants plotted as a function of NBS concentration. This graph shows that the

increasing rate of the quenching reaction is linear as a function of NBS concentrations

used in this analysis.

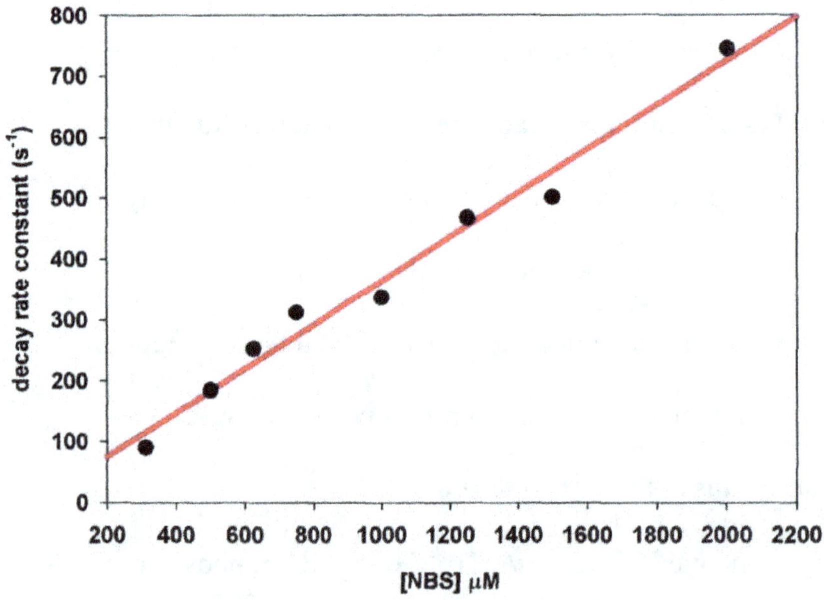

Figure 3-2. A plot of observed reaction decay rates as a function of NBS concentration. The experimentally determined fluorescence decay amplitudes as a function of observed decay rates were fitted to the equation: $A = A_o * exp^{(-kt)}$, for the calculation of dead time in Figure 3-3, where t is the dead time, A represents the amplitude of the reaction, A_o is the calculated initial amplitude, and k is the experimentally observed rate constant.

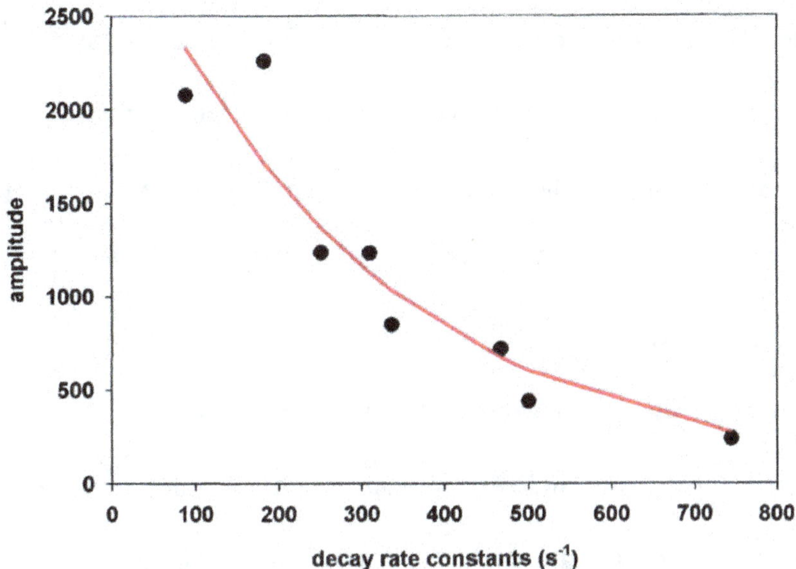

Figure 3-3. Fluorescence decay amplitudes plotted as a function of experimentally observed decay rate constants to determine dead time of the SFM-4 stopped-flow using a 10 mL / s flow rate. The red curve is the fitted data to the equation: $A = A_o*exp^{(-kt)}$. The calculated dead time (t) is 3.3 ms in this example.

Determination of the Applied Photophysics SX.18MV Stopped-Flow Reaction Analyzer Sequential- and Single-mix Dead Times

The dead time for the Applied Photophysics SX.18MV stopped-flow described above was determined using the method of (Peterman, 1979). The quenching of N-acetyl-tryptophanamide (N-AcTrpNH$_2$) fluorescence by N-Bromosuccinamide (NBS) in sodium phosphate buffer (monobasic) (pH 7.0) was used as a reaction for dead time determination. N-AcTrpNH$_2$ fluorescence was excited at 280 nm, and emission intensity was measured through a 320 nm cut-on filter.

The sequential-mix "three-syringe" setup dead time was determined in an experiment where one syringe contained a solution of 40 µM N-AcTrpNH$_2$ in sodium phosphate buffer (pH 7.0). A second syringe was loaded with a solution of sodium phosphate buffer only (pH 7.0). The contents of these two syringes were mixed and preincubated for 1 s, then mixed with the contents of a third syringe loaded with varying concentrations of NBS quencher in sodium phosphate buffer (pH7.0). The third syringe and stopped-flow apparatus were rinsed thoroughly with deionized water in between reactions differing in NBS concentration. The final reactant concentrations were 10 µM N-AcTrpNH$_2$, 50 mM sodium phosphate pH 7.0, and a range of NBS: 312.5, 500, 625, 750, 1000, 1250, 1500, and 2000 µM.

Using the data analysis function of the SX.18MV software, the observed fluorescence decay reactions were fit to a single exponential equation for determination of the experimental amplitudes and decay rates. The "X-datum" initial data point was set to 2.4 ms in this analysis, as suggested by the manufacturer. Changing this initial data point to 2.2 or 2.3 ms had no significant effect on the dead time calculation. Figure 3-4 shows the experimentally determined fluorescence decay amplitudes as a function of observed decay rates were fitted to the equation: $A = A_o * \exp^{(-kt)}$, for the calculation of dead time (t). As expected, analysis of the observed rate of the quenching reaction was linear as a function of NBS concentrations used (not shown). The dead time calculated using this method for the sequential-mix setup was 1.47 ms. Analysis of the reaction data recorded by the stopped-flow software for all MDCC-PBP ATPase experiments performed over the course of two years revealed that the average software-calculated dead time was 1.48 (+/- 0.16 ms) for 241 "shots". This is in excellent agreement with the reaction dead time experimentally calculated by the Peterman method.

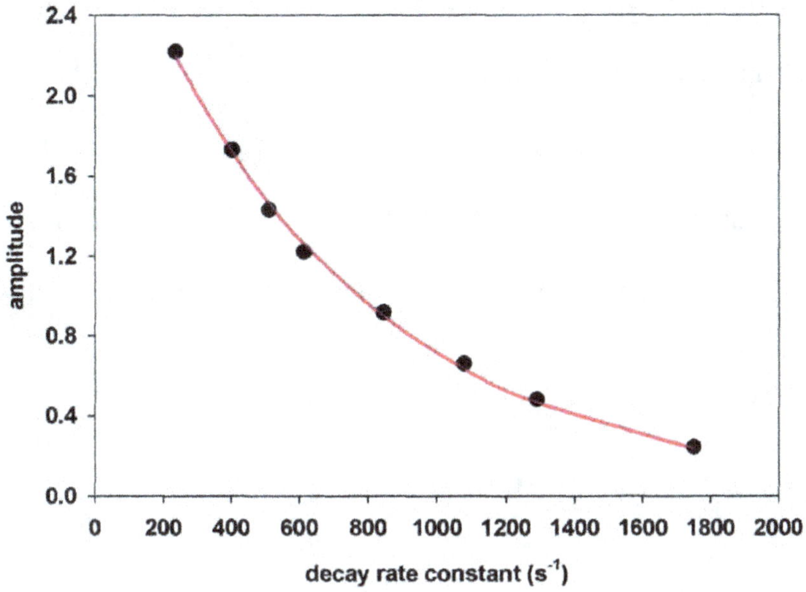

Figure 3-4. Sequential-mix reaction fluorescence decay amplitudes plotted as a function of experimentally observed decay rate constants to determine dead time of the SX.18MV stopped-flow. The red curve is the fitted data to the equation: $A = A_o * \exp^{(-kt)}$. The calculated dead time (t) is 1.47 ms.

The single-mix "two-syringe" setup dead time was determined in a set of experiments where one syringe contained a solution of 20 μM N-AcTrpNH$_2$ in sodium phosphate buffer (pH 7.0). The second syringe was loaded with varying concentrations of NBS quencher in sodium phosphate buffer (pH7.0). The second syringe and stopped-flow apparatus were rinsed thoroughly with deionized water in between reactions with differing NBS concentrations. The final reactant concentrations were 10 μM N-AcTrpNH$_2$, 50 mM sodium phosphate pH 7.0, and a range of NBS: 312.5, 500, 625, 750, 1000, 1250, 1500, and 2000 μM.

The experimental fluorescence quenching decay amplitudes and rates were determined as described above using the SX.18MV stopped-flow software. Figure 3-5 shows the experimentally determined fluorescence decay amplitudes as a function of observed decay rates fitted to the equation: $A = A_o * \exp^{(-kt)}$, for the calculation of dead time (t). The dead time calculated using this method for the single-mix setup was 1.6 ms.

Analysis of the reaction data recorded by the stopped-flow software for all single-mix

MDCC-PBP ATPase experiments performed over the course of two years revealed that

the average software-calculated dead time was 1.30 (+/- 0.13 ms) for 23 "shots". This is

consistent with the reaction dead time experimentally calculated by the Peterman method.

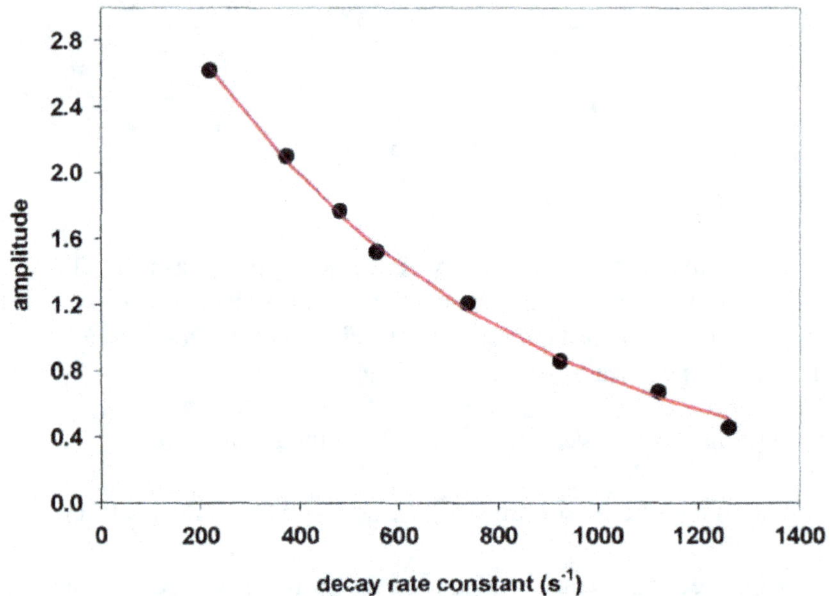

Figure 3-5. Single-mix reaction fluorescence decay amplitudes plotted as a function of experimentally observed decay rate constants to determine dead time of the SX.18MV stopped-flow. The red curve is the fitted data to the equation: $A = A_o * \exp^{(-kt)}$. The calculated dead time (t) is 1.6 ms.

CHAPTER 4

ATP-DEPENDENT CONFORMATIONAL CHANGE IN THE CLAMP LOADER

Introduction

The *Escherichia coli* DNA polymerase III γ complex clamp loader assembles the ring-shaped β sliding-clamp onto DNA. Once bound to DNA, the β clamp topologically links the core polymerase to the template that it is copying to increase the processivity of DNA synthesis. ATP binding and hydrolysis promote conformational changes within the γ complex that modulate its affinity for the clamp and DNA allow it to accomplish the task of assembling clamps on DNA. The mechanical task of assembling clamps on DNA requires that protein•protein and protein•DNA interactions change during a single cycle of a clamp loading reaction. The clamp loader must initially have a high affinity for clamps and DNA to bring the clamp to DNA, but then must have a decreased affinity to release the clamp onto DNA and avoid competing with the polymerase for loaded clamps. (Jeruzalmi et al., 2001a; Naktinis et al., 1995). ATP binding to the γ subunits induces conformational changes (Podobnik et al., 2003) that expose this region of δ allowing the clamp loader to bind with high affinity and open the β clamp. ATP binding also increases the affinity of the clamp loader for DNA (Ason et al., 2000; Bloom et al., 1996), although a discrete DNA binding site has not yet been identified. Subsequent binding to a primed DNA template triggers ATP hydrolysis reducing the affinity of the clamp loader for β and DNA (Ason et al., 2000; Ason et al., 2003). This reduced affinity is likely to be produced by conformational changes that mask β and DNA binding

118

domains. Dissociation of the ADP-bound clamp loader from β and DNA allows the clamp to close and gives the polymerase access to the newly loaded clamp.

The precise nature of the ATP-induced conformational changes within the clamp loader is not yet known. Successive binding of ATP to each of the three-γ subunits could potentially promote a series of conformational changes that ultimately produces an activated complex. Structural data for a minimal *E. coli* clamp-loading complex ($\gamma_3\delta\delta'$) suggest that conformational changes that occur in each γ subunit are translated to the δ subunit pulling it away from δ' and exposing the β binding domain (Jeruzalmi et al., 2001a). The conformational flexibility that must be inherent in this complex to produce these structural changes and the presence of three distinct ATP binding sites suggest that several conformational states of the complex could exist in solution.

In this chapter, ATP hydrolysis and clamp loading reactions were measured under steady-state conditions to study the affect β has on the clamp loader affinity for DNA and ATP, and also to analyze the clamp loader "Michaelis-Menten" kinetics in the presence or absence of β. ATP hydrolysis and clamp loading reactions were also measured under pre-steady-state conditions to identify kinetic phases associated with conformational transitions in the γ complex ($\gamma_3\delta\delta'\chi\psi$) on binding ATP. A novel ATP hydrolysis assay was used to measure the kinetics of ATP-dependent conformational changes where γ complex was incubated with ATP for a defined period of time, 0 to 1000 ms, to differentially populate conformational states that are formed on binding ATP. DNA was then added to trigger ATP hydrolysis and probe the relative populations of species present. Computerized simulation and fitting of the experimental kinetic data were applied for estimation of the rates of conformational changes between three species of γ complex presented in a novel model.

Steady-State Characterization of γ Complex ATP Hydrolysis, DNA Binding and Clamp Loading Activities

Enhancement of Steady-State ATP Hydrolysis Kinetics of γ Complex by β clamp

The DNA-dependent ATPase activity of γ complex is stimulated by the β-sliding clamp (Onrust et al., 1991; Stukenberg et al., 1991). The steady-state kinetics of ATP hydrolysis triggered by primed-template (pt) DNA were re-examined here with a newly designed methodology (i.e. for study of γ complex) to compare the hydrolysis activity by γ complex with pt DNA alone or in the presence of β. The fluorescence-based MDCC-PBP ATPase assay was utilized for measurement of the initial velocities of ATP hydrolysis activity. In MDCC-PBP ATPase assays, the kinetics of ATP hydrolysis were measured as a function of inorganic phosphate release from the clamp loader and therefore establish a minimum rate for ATP hydrolysis (Bertram et al., 2000; Brune et al., 1998). Separate steady-state assays were initiated by addition of a range of molarity of the substrate ATP (from 0.6 to 77 μM) to a solution in a cuvette containing γ complex, pt DNA, and MDCC-PBP in the presence or absence of β. The initial reaction velocities under these conditions were measured as a function of an increase in fluorescence of MDCC-PBP upon binding inorganic phosphate. Three separate reactions were performed for each concentration of ATP, and resulting averaged initial velocities were plotted against ATP concentration, and analyzed by fitting to the Michaelis-Menten equation, giving the steady-state parameters shown in Table 4-1.

Table 4-1. Steady-state ATP hydrolysis kinetics of γ complex in the absence and presence of β[a]

	γ complex with β	γ complex without β
V_{max} (μMs^{-1})	0.07 (0.02)[b]	0.02 (0.003)
$k_{cat}{}^{complex}$ (s^{-1})[c]	2.0 (0.4)	0.6 (0.07)
K_m (μM)	11 (6.7)	6.0 (3.0)
$k_{cat}{}^{complex}$ / K_m (μM^{-1}s^{-1})	0.2 (0.05)	0.1 (0.04)

[a] Steady-state MDCC-PBP ATPase assays were performed by addition of a range in molarity of ATP from 0.6 to 77 μM to a solution of 41 nM γ complex with 50 nM ptDNA and 50 nM β (when present). All assays contained 2.4 μM MDCC-PBP and P_i-mopped assay buffer containing 20 mM Tris-HCl, 50 mM NaCl, 5 mM DTT, 40 μg/mL BSA, and 8 mM MgCl$_2$.
[b] Standard deviations were calculated from three separate steady-state MDCC-PBP ATPase assays.
[c] Since the clamp loader has three subunits capable of hydrolyzing ATP, the turnover number ($k_{cat}{}^{complex}$) reflects the activity of the combined activity of the complex.

These data are consistent with previous analyses using autoradiography ^{32}P-based methods for measuring the DNA-dependent ATP hydrolysis activity of γ complex in the presence and absence of β (Bertram et al., 2000; Hingorani et al., 1999). The $k_{cat}{}^{complex}$ was increased approximately 3-fold, and the K_m value roughly doubled in the presence of β. The $k_{cat}{}^{complex}$ / K_m "specificity constant" value from this analysis can be understood as an apparent second-order binding constant for ATP. Interestingly, this value is not what would be expected for diffusion-controlled binding of a small nucleotide substrate such as ATP. Therefore, this result suggests that ATP binding was followed by a slow step in the reaction before hydrolysis activity. The ATP specificity constant $k_{cat}{}^{complex}$ / K_m value was increased in the presence of β clamp about 1.8-fold. This measure of enhancement of ATP binding specificity in the presence of β demonstrates that β partitions a fraction of γ complex towards a more active clamp loading form, possibly increasing or bypassing the rate of the slow step preceding ATP hydrolysis. The results of steady-state ATP hydrolysis activity are outlined and discussed in additional detail in the following chapter where γ complex activity is compared to the activity of the minimal clamp loader complex (Table 5-1).

Steady-State DNA Binding and Clamp Loading Activities of γ Complex

Examination of the steady-state DNA binding and clamp loading kinetics of γ complex was accomplished using the fluorescence-based anisotropy DNA binding assay. The γ complex was studied in the presence and absence of β, to observe the DNA binding activity in solution based assays. In original studies, measurement of DNA replication activity was used as a gauge of clamp loading activity by γ complex (Maki and Kornberg, 1988; Stukenberg et al., 1991). The solution-based anisotropy DNA binding assay offers study of γ complex activity under equilibrium conditions, and was originally used for investigation of γ complex by (Bloom et al., 1996), and later extended for clamp loading assays with mutant β clamps (Bertram et al., 1998). In some labs, gel filtration assays under non-equilibrium conditions, are still used to assay the clamp loading activity of γ complex (Johnson and O'Donnell, 2003).

Here, the anisotropy DNA binding assay was used to monitor the reaction of γ complex with X-Rhodamine (RhX)-labeled pt DNA in the absence or presence of β over extended periods of time. In these assays, the observed steady-state anisotropy of the RhX-probe on DNA is a population weighted average of the combined anisotropies for free DNA and protein-bound DNA. Therefore, the observed anisotropy increases as the population of protein-bound DNA increases (Bloom et al., 1996). Polarized emission intensities of the RhX-probe covalently attached to pt DNA were measured and used in calculation of anisotropy over time. The steady-state measurement of DNA binding activity or clamp loading activity was observed in assays where ATP was added to a solution of RhX-ptDNA, β clamp (when present) and γ complex to initiate the reaction.

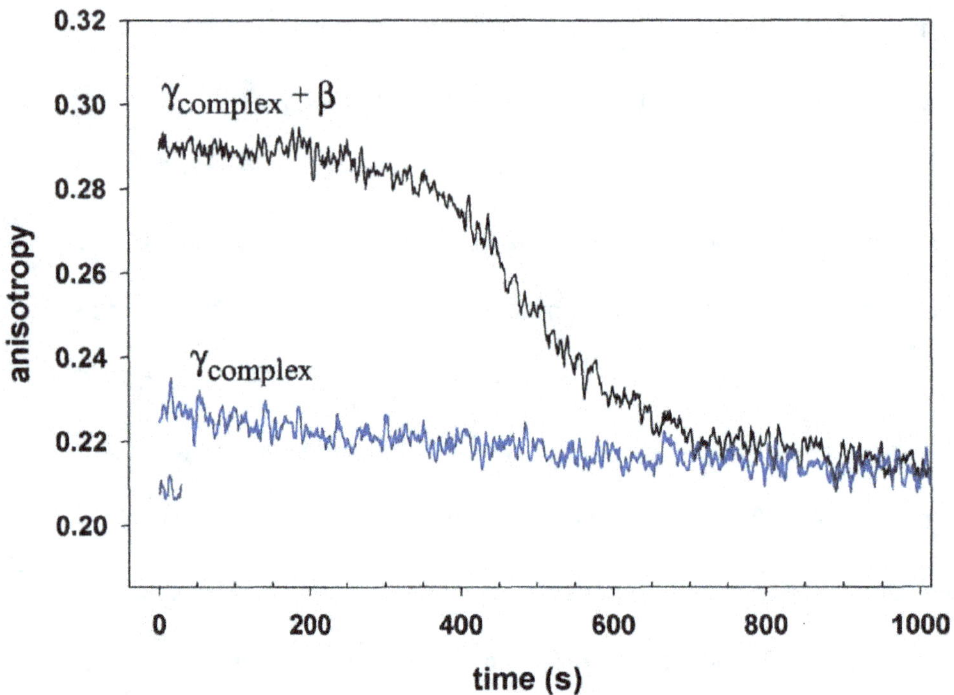

Figure 4-1. Steady-state DNA binding and clamp loading activities of γ complex. The change in steady-state anisotropy of a 30-nucleotide-primed 105-nucleotide template DNA labeled with RhX at the 5' end is plotted against time for γ complex (blue), and for the clamp loading reaction; γ complex in the presence of β (black). Reactions contained 50 nM RhX-ptDNA, 500 nM γ complex, 500 nM β (when present), and 0.3 mM ATP in a solution of assay buffer: 8 mM MgCl$_2$, 20 mM Tris-HCl pH 7.5, 50 mM NaCl, 5 mM DTT, 40 μg/mL BSA. Steady-state polarized intensities from the RhX probe were recorded once per second for 30 minutes. The plot shows a slightly expanded timescale, since both reactions no longer showed activity for the longer time period. A 30-second plot of anisotropy of RhX-pt DNA, free in solution is shown (gray).

Figure 4-1 shows the resulting time courses for steady-state anisotropy of DNA binding by γ complex in the presence or absence of β.

The steady-state anisotropy of RhX-pt DNA was approximately 0.21, and was increased slightly (~0.23) in the DNA binding reaction with γ complex in the absence of β. This low activity slowly decayed back to the level of free RhX-pt DNA over the course of the assay. In the presence of β, γ complex catalyzed the clamp loading reaction on this RhX-pt DNA substrate (Ason et al., 2000). A large increase in anisotropy was initially seen (~0.29), consistent with a dynamic interaction of γ complex and β with the RhX-pt DNA. The loaded β clamps easily slide off of this short DNA substrate keeping their concentration essentially in constant equilibrium. This high anisotropy value was sustained for more than 200 s, before decaying to the level of free RhX-pt DNA over and additional ~450 s, indicating that the clamp loading reaction continued at a constant rate until all of the ATP was used up by γ complex. It was also possible that a large amount of the ADP product of hydrolysis built up and began to inhibit the reaction.

The difference observed in these assays between γ complex DNA binding and clamp loading activities in the presence or absence of β, indicates that β enhanced that activity of the clamp loader, consistent with the steady-state hydrolysis assays described above. The sustained clamp loading reaction suggests that β clamp increased the affinity for ATP binding γ complex, or possibly increased the substrate (ATP) specificity of the clamp loader.

The steady-state anisotropy binding (DNA binding/clamp loading) assay was used for further analysis of the clamp loading reaction. Assays were performed to identify the concentration of ATP at which the clamp loading reaction was no longer sustained. Figure 4-2 shows the results of ten separate clamp loading reactions where the final concentrations of ATP were decreased from 450 to 25 μM. The steady-state clamp loading reactions were again recorded over an extended time course after addition of varied concentrations of ATP to a solution of RhX-pt DNA, γ complex and β. Polarized emission of the RhX-pt DNA was measured for calculation of anisotropy, and the assays were allowed to run until the reaction anisotropy reached a level equivalent to free RhX-

pt DNA indicating that all of the ATP was used up.

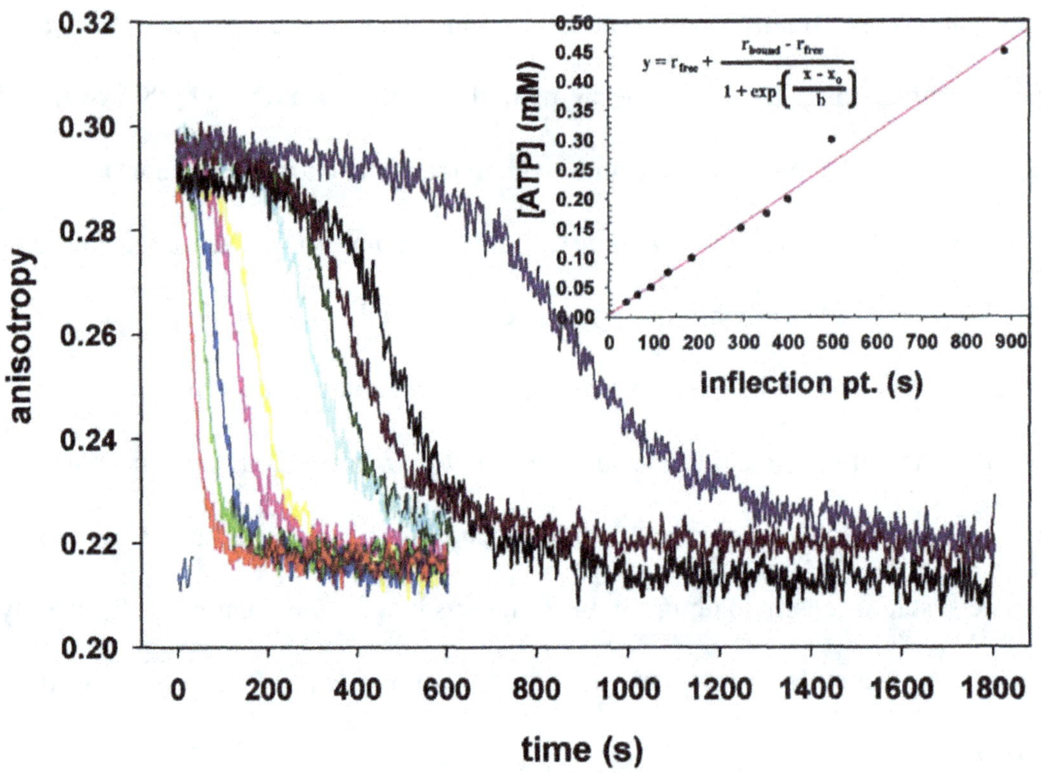

Figure 4-2. Steady-state kinetics of clamp loading as a function of ATP concentration. Varied concentrations of ATP were added to a solution of RhX-ptDNA, γ complex, and β to initiate clamp loading reactions. ATP concentrations were (left-to-right in the plot) (μM): 25 (red), 38 (green), 50 (blue), 75 (pink), 100 (yellow), 150 (cyan), 175 (dark green), 200 (dark red), 300 (black), and 450 (magenta). Final reactant concentrations were, 50 nM RhX-ptDNA, 500 nM γ complex, and 500 nM β in solutions of assay buffer: 8 mM $MgCl_2$, 20 mM Tris-HCl pH 7.5, 50 mM NaCl, 5 mM DTT, 40μg/mL BSA. Steady-state polarized intensities from the RhX probe were recorded once per second for 10 min or 30 min, depending on the ATP concentration. A 30-second plot of anisotropy of RhX-pt DNA, free in solution (r_{free}) is shown (gray). The curves were fitted to a 4-parameter sigmoid logistic equation in SigmaPlot 8.0 software, shown in the inset. The inset shows the calculated points of inflection of each curve (i.e. where tangents of the curves meet at the center of the sigmoidal decay) plotted against the concentration of ATP added to each assay. The red line is a linear fit to the data, with the y-intercept equal to 4.88 μM ATP.

It was observed that an ATP concentration between 25 and 38 μM (Figure 4-2, red and green, left-most curves) was unable to sustain the clamp loading reaction, and the anisotropy quickly decayed to the level of unbound RhX-pt DNA. In contrast, with increasing concentration of ATP, the steady-state clamp loading activity was sustained for increasing amounts of time. This indicated that the point at which the anisotropy transitions into a decay phase, there was no longer enough ATP available for γ complex to continue loading β. At the highest concentration of ATP (450 μM), the clamp loading reaction was maintained up to roughly 12 min (~750 s).

Each of the clamp loading reaction curves was fitted to a four-parameter sigmoid-logistic function (see equation, Figure 4-2, inset). Fitting the curves to this equation revealed that the slope of the decay increased with decreasing concentrations of ATP (b value), and that the point of inflection during the decay phase increased in time (x_o value) with increasing ATP concentrations. The inflection point represents the midpoint anisotropy value of the decay phase for each reaction, presumably where there is no longer enough ATP available for clamp loading activity. By plotting the calculated inflection points as a function of starting ATP concentration (Figure 4-2, inset), and fitting these data to a linear regression, the concentration of ATP present at zero time (y-intercept) was estimated to be 4.88 μM. Since ATP binding is required for γ complex to bind β with high affinity, this value indicates the concentration of ATP bound to γ complex at initiation of the clamp loading reaction, and represents an apparent dissociation constant for ATP binding. This apparent K_d is consistent with previously the previously measured value (~2 μM) (Hingorani and O'Donnell, 1998), and also with the apparent K_d measured in this dissertation project with anisotropy binding assays using fluorescent-labeled β clamp (~2 μM) (see Figure 5-4). This apparent Kd for ATP was used in this research project for determination of a appropriate molar-range of ATP in the design of the steady-state MDCC-PBP ATPase assay presented above, and again in the following chapter. These assays were performed early in this dissertation project, and the resulting DNA binding and clamp loading kinetics were later addressed in additional detail by pre-steady-state analysis.

Pre-Steady-State Kinetics of DNA-Dependent ATP Hydrolysis by γ Complex

Pre-Steady-State MDCC-PBP ATPase Assays for γ Complex in the Absence and Presence of β Clamp

Pre-steady-state kinetics of ATP hydrolysis by γ complex were measured in assays with and without the β clamp to determine how β alters the kinetics of ATP hydrolysis. A real time fluorescence-based assay containing *E. coli* phosphate binding protein (PBP) covalently labeled with MDCC was used to measure the amount of inorganic phosphate produced on hydrolysis of ATP (Bertram et al., 2000; Brune et al., 1994). When MDCC-PBP binds inorganic phosphate in a 1:1 stoichiometry, an increase in the fluorescence of MDCC occurs. Sequential mixing assays were performed by incubating γ complex (0.27 μM) with ATP (100 μM) in the presence or absence of β (1 μM) for 1 s prior to adding a solution of DNA (1 μM) to trigger hydrolysis and MDCC-PBP (2.7 μM) to give the final concentrations indicated in parentheses. The DNA substrate used in all experiments consisted of a 30-nt primer annealed to a 105-nt template and supports loading of the β clamp and processive synthesis by pol III core (Ason et al., 2000; Bloom et al., 1996; Hingorani et al., 1999).

In assays with γ complex alone (Figure 4-3, black trace), the first turnover of ATP is biphasic and is followed by a slower steady-state phase beginning in about 700 ms.

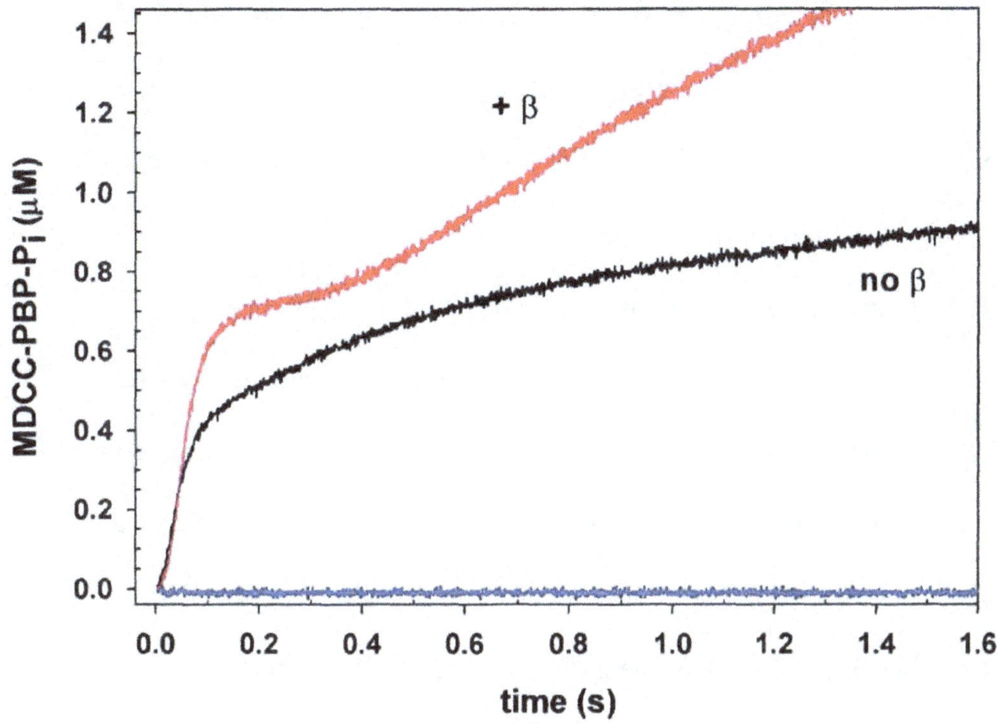

Figure 4-3. Kinetics of ATP hydrolysis by γ complex in the presence and absence of β. The ATP hydrolysis kinetics for γ complex in the absence (black) and presence of β (red) were measured in real time. Using a sequential-mix "three-syringe" stopped-flow assay, one syringe loaded with a solution containing γ complex and β (when present) was mixed with the contents of a second syringe loaded with ATP. The resulting mixture was preincubated for a period of 1 s, and then mixed with the contents of a third syringe loaded with a solution-containing pt DNA and MDCC-PBP. All solutions were prepared in P$_i$-mopped assay buffer: 20 mM Tris-HCl, 50 mM NaCl, 5 mM DTT, 40 μg/mL BSA, and 8 mM MgCl$_2$. Final reactant concentrations were 270 nM γ complex, 1.0 μM β (when present), 100 μM ATP, 1.0 μM pt DNA, and 2.7 μM MDCC-PBP. The flat trace (gray) at the bottom of the figure showing no change was a negative control assay performed in the absence of ATP. Raw fluorescence data were transformed into the concentration of P$_i$-bound MDCC-PBP as a function of time using the equation, $X_b(t) = [I_{obs}(t)-I_f] / [I_b - I_f]$ to solve for the fraction of P$_i$-bound MDCC-PBP $X_b(t)$, then multiplying this value by the concentration of MDCC-PBP present in the assay to get the value MDCC-PBP-P$_i$ (μM) plotted. I_f was the fluorescence intensity of P$_i$-free MDCC-PBP in assay buffer, and I_b was the saturated fluorescence intensity of completely P$_i$-bound MDCC-PBP in the presence of 200 μM potassium phosphate (P$_i$ source).

The rapid pre-steady-state phase of ATP hydrolysis is complete within about 100 ms whereas slower phase lasts about 600 ms. The amplitude of the rapid phase represents about 60 % of the total burst amplitude. In contrast to assays with γ complex alone, a single pre-steady-state phase of ATP hydrolysis is present in reactions with β and the second slower phase is not. Addition of β (Figure 4-3B, red trace) increases both the overall rate of the first turnover of ATP and the steady-state rate of ATP hydrolysis. The burst of ATP hydrolysis in reactions with β occurs at the same rate as the rapid phase in assays without β, and its amplitude is equal to the sum of the amplitudes of the rapid and slow phases in assays without β (Figure 4-3, black trace). Burst amplitudes indicate that 2.7 molecules of ATP are hydrolyzed per molecule of γ complex in the presence and absence of β. The presence of β increases the overall pre-steady-state rate of ATP hydrolysis by increasing the relative proportion of the rapid reaction.

Pre-Steady-State MDCC-PBP ATPase Assays at Different Concentrations of γ Complex in the Presence of β

The pre-steady-state MDCC-PBP ATPase assay is essentially an active-site titration of the clamp loader enzyme, allowing quantification of the number of ATP molecules hydrolyzed during the first turnover of the clamp loading reaction. During development of the assay, it was questionable if the real time fluorescence response by MDCC-PBP in this assay was reliably measuring the molar release of inorganic phosphate product from hydrolysis of ATP by γ complex. At the time, only one other group was using this pre-steady-state assay to quantify the number of molecules of ATP hydrolyzed by an enzyme (PcrA helicase) during its reaction (Dillingham et al., 2000).

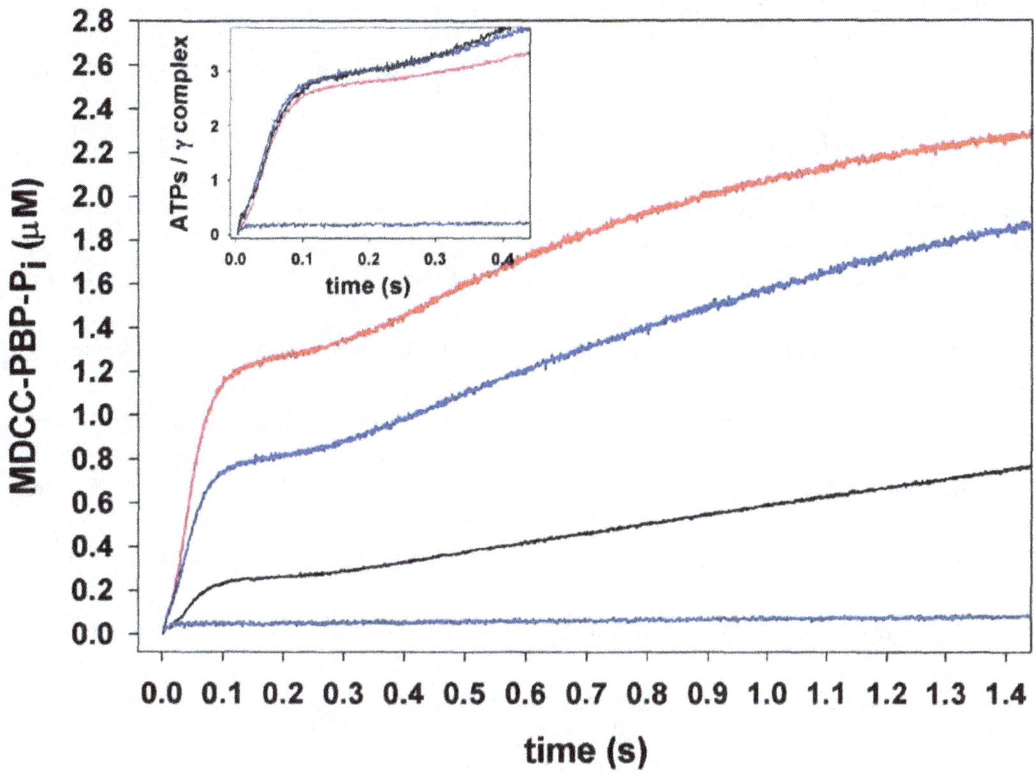

Figure 4-4. Quantification of the number of ATP molecules hydrolyzed by γ complex in the first turnover of β clamp loading. The pre-steady-state ATP hydrolysis activity during clamp loading by γ complex was measured at concentrations of 87 nM (black trace), 270 nM (blue trace, middle), and 450 nM (red trace) using the sequential-mix "three-syringe" stopped-flow MDCC-PBP ATPase assay. One syringe loaded with a solution containing varying concentrations of γ complex with β was mixed with the contents of a second syringe loaded with ATP. The resulting mixture was preincubated for a period of 1 s, and then mixed with the contents of a third syringe loaded with a solution-containing pt DNA and MDCC-PBP. All solutions were prepared in P_i-mopped assay buffer: 20 mM Tris-HCl, 50 mM NaCl, 5 mM DTT, 40 µg/mL BSA, and 8 mM $MgCl_2$. Final reactant concentrations were 87, 270, or 450 nM γ complex, 1.6 µM β, 400 µM ATP, 1.6 µM pt DNA, and 2.6 µM MDCC-PBP. The flat trace (gray) at the bottom of the figure showing no change was a negative control assay performed in the absence of ATP. The inset shows the moles of ATP hydrolyzed per mole of γ complex during the "burst" phase for all three clamp loading reactions was the same (trace colors as indicated above). Raw fluorescence data were transformed into the concentration of P_i-bound MDCC-PBP using the equation described in figure 4-3.

To determine if our MDCC-PBP ATPase assay was responding specifically to the

number of ATP molecules hydrolyzed by γ complex during the clamp loading reaction, three separate assays were performed at different concentrations of γ complex while the MDCC-PBP concentration was held constant. Sequential mixing assays were performed where varying concentrations of γ complex in the presence of β were mixed with ATP, then preincubated for 1 s before mixing with pt DNA and MDCC-PBP. Figure 4-4 shows that the pre-steady-state "burst" amplitude of the molar MDCC-PBP-P$_i$ response changed in proportion to the concentration of γ complex. This result demonstrates that the same number of ATP molecules were hydrolyzed in the first turnover of the clamp loading reaction when the relative concentration of γ complex to MDCC-PBP was varied. Figure 4-4, inset, shows that approximately three molecules of ATP were hydrolyzed per molecule of γ complex at a similar rate during the first turnover in each assay. This result is consistent with the clamp loader structure containing three AAA+ γ (ATPase) subunits (Jeruzalmi et al., 2001a), suggesting that each γ subunit hydrolyzed one molecule of ATP during these assays.

ATPγS-Chase of Pre-Steady-State ATP Hydrolysis Activity by γ Complex

ATPγS "chase" experiments were performed to determine the fate of ATP bound by γ complex on addition of DNA. These chase assays were done under identical sequential mixing conditions to pre-steady-state ATPase assays except that ATPγS (1 mM final concentration) was added to the solution of DNA and MDCC-PBP. ATPγS is hydrolyzed 4 orders of magnitude more slowly than ATP by γ complex (Hingorani and O'Donnell, 1998) and therefore is effectively non-hydrolyzable on the time scale of these experiments. In these assays, γ complex is initially bound to ATP and addition of an excess (10-fold) of ATPγS to assays will limit γ complex to a single round of ATP hydrolysis. If securely bound, ATP is hydrolyzed more rapidly than it dissociates, then the pre-steady-state kinetics of ATP hydrolysis should be unaffected.

Figure 4-5. Pre-steady-state kinetics of ATP hydrolysis by γ complex in the presence and absence of the β clamp in assays with and without an ATPγS chase. Sequential mixing "three-syringe" MDCC-PBP ATPase assays were performed where γ complex was incubated with ATP in the A) absence (black trace) or B) presence (black trace) of β for 1 s prior to the addition of solution containing DNA and MDCC-PBP. ATPγS chase experiments were done under identical conditions except that ATPγS was included in the solution of DNA and MDCC-PBP (red labeled + ATPγS). The increase in fluorescence of MDCC-PBP when it bound the P_i product of ATP hydrolysis was measured in real time. Fluorescence intensities were converted to the concentration of MDCC-PBP bound to P_i using the equation described in figure 4-3, and are plotted as a function of time. Control experiments done in the absence of ATP are also shown (gray traces, labeled no ATP). Final reaction concentrations were 0.27 μM γ complex, 100 μM ATP, 1.0 μM β clamp (when present), 1.0 μM DNA, and 2.7 μM MDCC-PBP in assay buffer containing 20 mM Tris$\cdot$HCl pH 7.5, 50 mM NaCl, 8 mM $MgCl_2$, 5 mM DTT, and 40 μg/ml BSA. In ATPγS chase assays, ATPγS at a final concentration of 1 mM was added to the solution of DNA and MDCC-PBP.

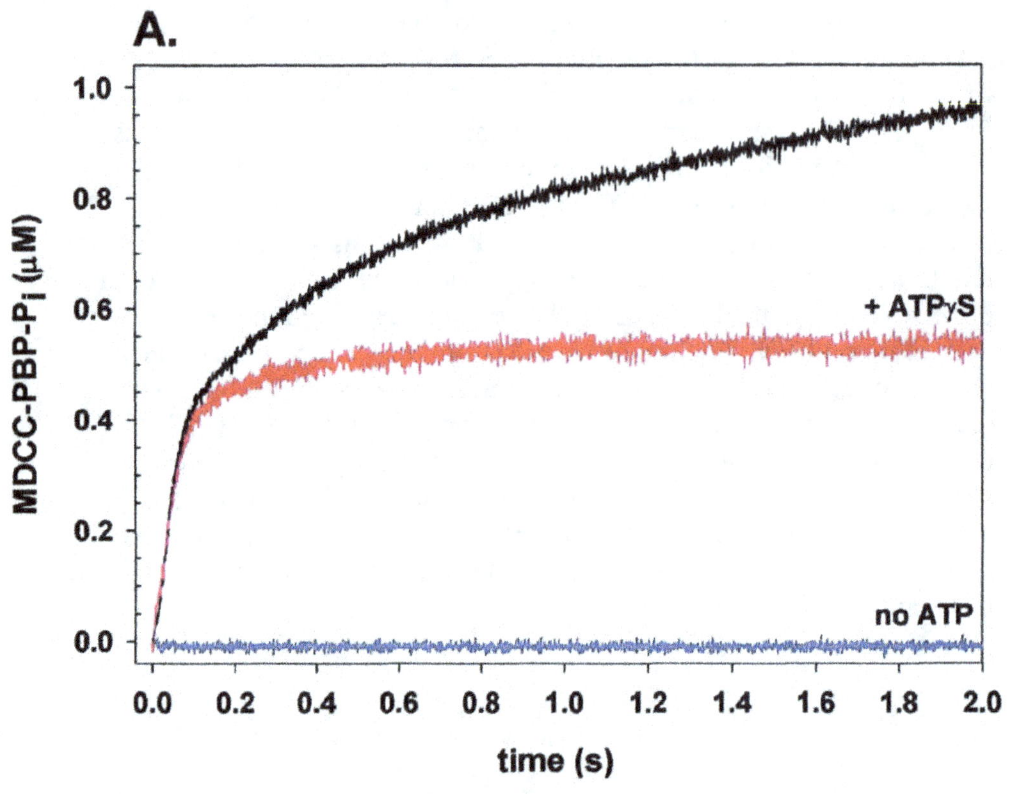
A.
1.0
0.8
0.6
0.4
0.2
0.0
MDCC-PBP-P$_i$ (µM)
+ ATPγS
no ATP
0.0 0.2 0.4 0.6 0.8 1.0 1.2 1.4 1.6 1.8 2.0
time (s)

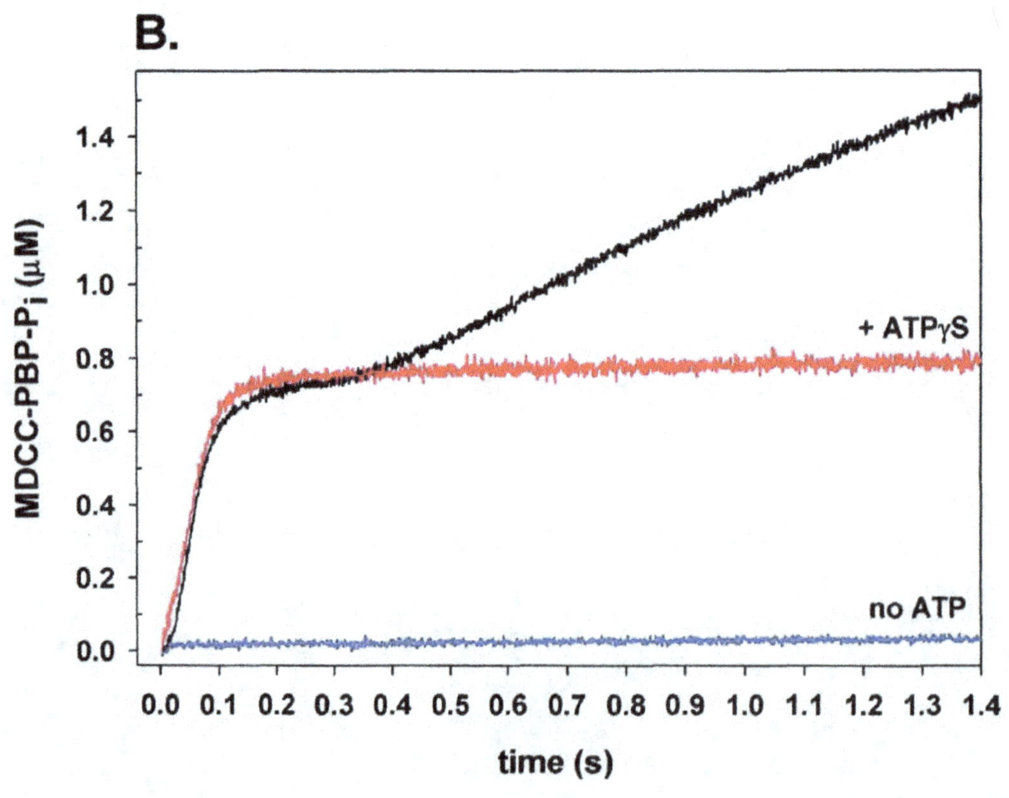
B.
1.4
1.2
1.0
0.8
0.6
0.4
0.2
0.0
MDCC-PBP-P$_i$ (µM)
+ ATPγS
no ATP
0.0 0.1 0.2 0.3 0.4 0.5 0.6 0.7 0.8 0.9 1.0 1.1 1.2 1.3 1.4
time (s)

If ATP dissociation competes with hydrolysis, then bound ATP can be exchanged for ATPγS and the amount of ATP hydrolyzed will reflect relative rates of hydrolysis and dissociation.

In preliminary steady-state ATPγS-chase assays, ATP hydrolysis activity was completely abrogated following addition of a 10-fold excess of ATPγS to reactions both in the absence and presence of β. The steady-state ATP hydrolysis activity of γ complex in the absence of β showed some ability to continue in the presence of lower ATPγS:ATP ratios, however, the ATP hydrolysis activity in the presence of β was poisoned even at substoichiometric ATPγS:ATP ratios. Further details of these steady-state chase reactions are outlined and discussed in chapter 5 of this dissertation.

In assays without β (Figure 4-5A, red trace, + ATPγS), addition of ATPγS reduces the amount of ATP hydrolyzed by γ complex in the first turnover. The rapid phase of ATP hydrolysis remains, but most of the ATP that is hydrolyzed in the slow phase is no longer hydrolyzed. This result shows that ATP hydrolyzed in the rapid phase is hydrolyzed more rapidly than it dissociates whereas ATP dissociation is more rapid than hydrolysis in the slow phase. In assays with β (Figure 4-5B, red trace, + ATPγS), addition of ATPγS does not affect the amplitude of the burst of ATP hydrolysis indicating that bound ATP is hydrolyzed more rapidly than it dissociates.

Together these ATPase assays demonstrate that the β clamp increases both the pre-steady-state and state-state rates of ATP hydrolysis by γ complex. The sliding clamp appears to help γ complex bypass a slower reaction step that forms product more slowly than it exchanges substrate. A key question is what gives rise to the biphasic kinetics of ATP hydrolysis in assays without β. The two most likely possibilities are that individual γ subunits within the γ complex hydrolyze ATP at different rates, or that two populations of γ complex exist, with one binding DNA and hydrolyzing ATP more rapidly than the second.

Pre-Steady-State Kinetics of β Clamp Loading by γ Complex Initiated at Different Steps of the Reaction Cycle

To distinguish between the two possible mechanisms resulting in biphasic ATP hydrolysis kinetics in assays without β, the kinetics of γ complex loading β on DNA were measured in assays that were initiated at different steps in the loading cycle. The rationale for this set of experiments is that if two species of γ complex give rise to biphasic ATPase kinetics then different populations of these species may be present at different stages of the loading reaction. A real time fluorescence anisotropy-based assay was used to measure clamp loading where the anisotropy of an X-rhodamine (RhX) probe covalently attached to DNA (RhX-pt DNA) increases when the DNA is bound by protein (Bloom et al., 1996; Otto et al., 1994). Polarized intensities of the RhX probe were measured during the time course of loading reactions and used to calculate the anisotropy of the probe as a function of time (Lakowicz, 1999). Clamp loading reactions were performed using different mixing schemes to add γ complex, ATP and β to DNA, but all reactions contained final concentrations of 0.25 μM γ complex, 500 μM ATP, 0.6 μM γ, and 0.05 μM RhX-DNA.

Kinetics of Clamp Loading when γ Complex is Equilibrated with ATP

In the first mixing scheme, γ complex was pre-incubated with ATP prior to adding β and DNA. Because DNA binding is rapid and a pre-requisite for the burst of ATP hydrolysis (Ason et al., 2003), the DNA binding/clamp loading kinetics are likely to mirror the kinetics of ATP hydrolysis. If biphasic kinetics of ATP hydrolysis were due to individual γ subunits within the γ complex hydrolyzing ATP at different rates, then a single species of γ complex should form on equilibration with ATP and give rise to relatively rapid monophasic kinetics of clamp loading. On the other hand, if two species of γ complex were formed on equilibration with ATP one of which rapidly bound β and DNA relative to the other, then biphasic kinetics of clamp loading should occur. Clamp loading reactions done using this mixing scheme, produce a biphasic increase in anisotropy consistent with a mechanism where two species load clamps on DNA at different rates (Figure 4-6A). The increase in anisotropy reaches a maximum value of about 0.31 in 200 ms and is followed by a small decrease to a steady-state value of about 0.3. The small decrease in anisotropy at the transition from the pre-steady-state to steady-state regimes is most likely due to dissociation of the γ complex from the clamp and DNA (Ason et al., 2000; Bertram et al., 1998). The experimental conditions used in this anisotropy DNA binding assay were mimicked in an analogous pre-steady-state MDCC-PBP ATPase assay for examination of the kinetics of ATP hydrolysis during this clamp loading reaction (see below, Figure 4-7).

It is possible that when reactions are initiated in this manner, γ complex can either bind β first or bind DNA first. If this were the case then, the two species that give rise to the biphasic DNA binding/clamp loading kinetics would be free γ complex and a $\beta \cdot \gamma$ complex. To test this possibility, the experiment was repeated at two additional concentrations of β, 1.2 and 2.4 μM. If free γ complex were present, increasing the concentration of β should increase the rate at which γ complex binds β relative to DNA. This increase in binding rates would be reflected in increases in the rate and amplitude of a kinetic phase due to $\beta \cdot \gamma$ complex binding DNA and a decrease in the amplitude of a phase due to free γ complex binding DNA. However, increasing the concentration of β in these reactions does not affect the kinetics of clamp loading (Figure 4-6A). This result demonstrates that the biphasic clamp loading kinetics may not be the result of free γ complex and a β•γ complex binding DNA. Instead, two different populations of γ complex may form on equilibration with ATP, one of which loads clamps more rapidly than the other. Another important result of these experiments is that the rates of the clamp loading reaction were largely unaffected by increasing the concentration of β indicating that loading kinetics are independent of the rate of a bimolecular association reaction between β and γ complex.

Kinetics of Clamp Loading when γ Complex is Equilibrated with ATP and β

In ATPase assays, the slow phase of ATP hydrolysis is not present when β, γ complex, and ATP are pre-incubated prior to addition of DNA. To determine whether the slow phase can also be eliminated in clamp loading reactions, a second experiment was done where γ complex, ATP, and β were present in one stopped-flow syringe where they could establish equilibrium and RhX-DNA was in the second syringe. This reaction was done using the same concentrations of proteins and DNA as in the first mixing scheme. Initiating reactions after γ complex binds ATP and β produces a single phase of increasing anisotropy to a value of about 0.33 in the first 50-100 ms (Figure 4-6B, black trace) of the time course. This rapid increase in anisotropy was followed by a slow decrease over about 500 ms to a steady-state value of about 0.3. Again, the increase is most likely due to a complex of β and γ complex binding DNA and the decrease due to dissociation of γ complex from the clamp that has been loaded on DNA (Ason et al., 2000; Bertram et al., 1998). The anisotropy (0.3) remains higher than free DNA (0.21) because a steady-state population of β remains bound to DNA (Bloom et al., 1996).

Figure 4-6. Kinetics of clamp loading measured in reactions that were initiated at different stages of the loading cycle. The increase in anisotropy of an RhX probe covalently bound to DNA is plotted as a function of time for clamp loading reactions. The anisotropy for free DNA (black) is also plotted and shows no change. Assays contain final concentrations of 0.25 μM γ complex, 500 μM ATP, 0.05 μM DNA and several concentrations of β, as indicated, in assay buffer containing 20 mM Tris•HCl pH 7.5, 50 mM NaCl, 8 mM $MgCl_2$, 5 mM DTT, and 40 μg/ml BSA. A) Reactions were initiated by adding a solution of γ complex and ATP to a solution of β and DNA and contained 0.6 (blue), 1.2 (red), and 2.4 (green) μM β. B) Reactions were initiated by addition of a solution of γ complex, ATP, and β to a solution of DNA (black trace) and contain 0.6 μM β. For comparison of kinetics, data using the mixing scheme shown in A is also plotted (blue trace). C) Reactions were initiated by adding γ complex to a solution of ATP, β, and γ complex and contained 0.6 μM (black) or 2.4 μM β (red). D) The clamp loading reactions in C with 0.6 μM β, were also repeated with different final concentrations of ATP: 100 μM (blue), 250 μM (green), and 500 μM (red; same plot as the black trace in C). The plot in (D) is on expanded scale of time and anisotropy.

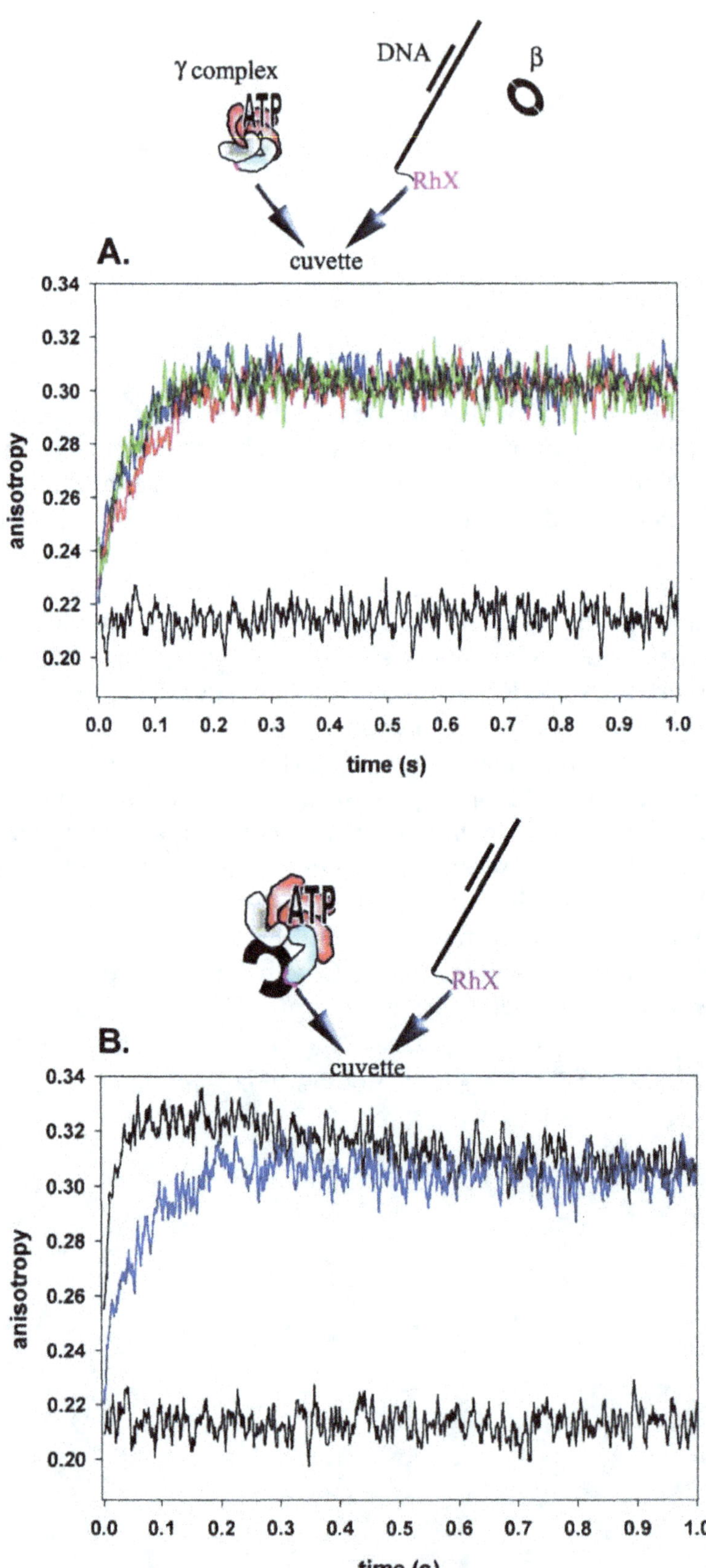
γ complex
DNA
β
ATP
RhX
A.
cuvette
0.34
0.32
0.30
0.28
0.26
0.24
0.22
0.20
anisotropy
0.0 0.1 0.2 0.3 0.4 0.5 0.6 0.7 0.8 0.9 1.0
time (s)
ATP
RhX
B.
cuvette
0.34
0.32
0.30
0.28
0.26
0.24
0.22
0.20
anisotropy
0.0 0.1 0.2 0.3 0.4 0.5 0.6 0.7 0.8 0.9 1.0
time (s)

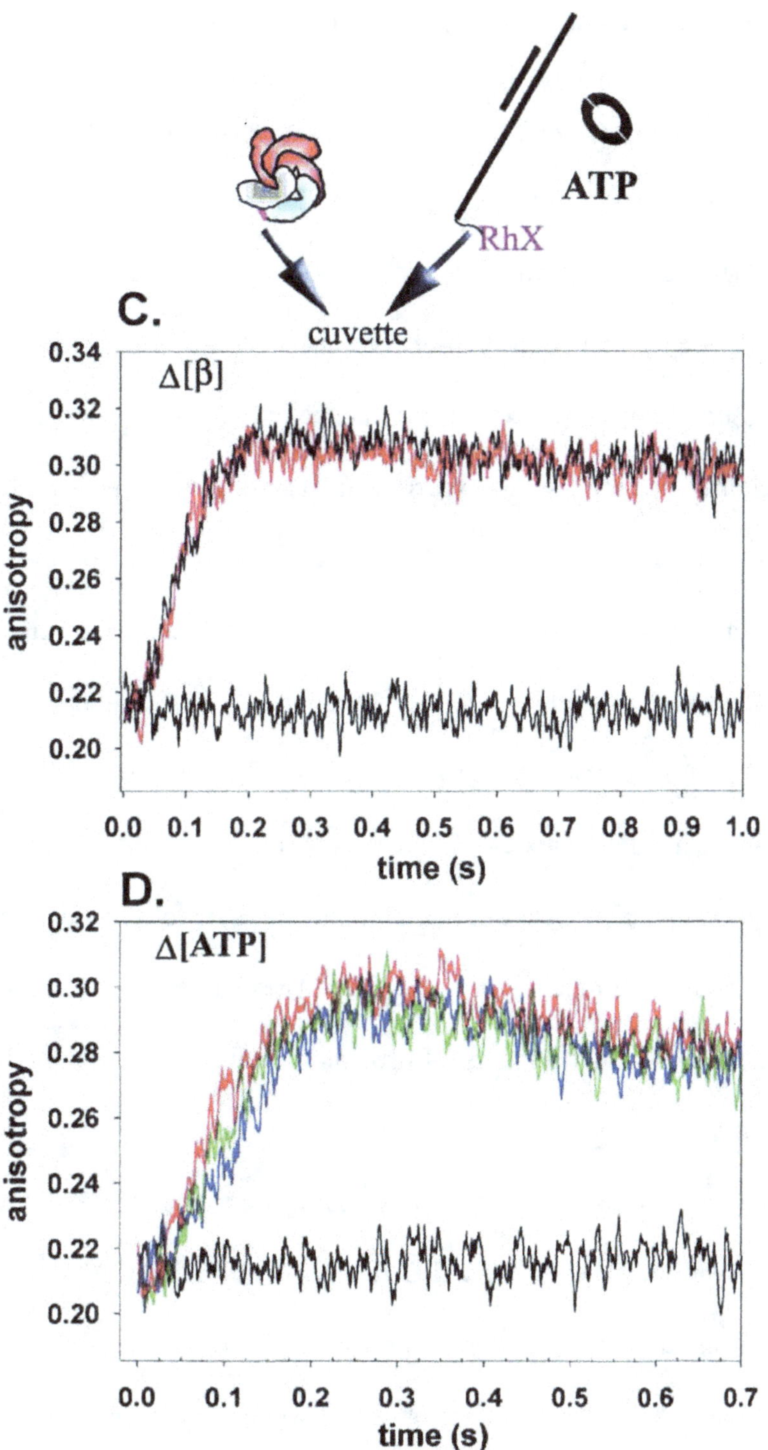

Figure 4-6. Continued.

Kinetics of Clamp Loading when γ Complex is Added Directly to a Solution of ATP, β, and DNA

Equilibration of γ complex with ATP produces a biphasic increase in anisotropy whereas equilibration with ATP and β produces only a rapid phase (Figure 4-6B, compare blue and black traces, respectively). If the slow step in clamp loading reactions were due to some ATP-dependent conformational changes, then it is possible that slow kinetics would be observed if γ complex were not incubated with ATP prior to initiating clamp-loading reactions. In a third set of experiments, reactions were initiated by adding γ complex to a solution of β, RhX-DNA, and ATP. Reactions contained 0.25 μM γ complex, 0.05 μM RhX-DNA, 0.6 μM β and saturating ATP (500 μM, see below) as before. When reactions are initiated prior to ATP binding, a substantial lag of about 30 ms in the kinetics of clamp loading is produced (Figure 4-6C). This lag is followed by an increase in anisotropy to a value of about 0.31 that is complete in about 250 ms. This increase in anisotropy is complete in about the same time it takes for the slow phase of reactions using the first mixing scheme (Figure 4-6A). These kinetics do not change when the concentration of β is increased to 2.4 μM and thus are not limited by the rate of γ complex binding β.

The reactions described above, initiated by adding γ complex to a solution of β, RhX-DNA, and ATP were repeated at three different concentrations of ATP to examine if ATP binding contributed to the slow step in these clamp loading reactions. There was no apparent change in clamp loading kinetics observed when the ATP concentration was lowered from the amount used in the original assay (Figure 4-6D). The lag preceding slow clamp loading kinetics remained at approximately 30 ms at each concentration of ATP, and was followed by an increase in anisotropy to a value of about 0.31 that was complete in about 250 ms. Taken together, these results demonstrate that the clamp loading reaction initiated by adding γ complex to ATP, β, and DNA is limited by the rate of an intramolecular reaction, possibly ATP-induced conformational changes in the γ complex

Kinetics of ATP Hydrolysis During Clamp Loading when γ Complex is Equilibrated with ATP

A pre-steady-state MDCC-PBP ATPase assay was performed under mixing conditions that mimicked those used in the pre-steady-state clamp loading anisotropy assay shown in figure 4-6A. In that assay, biphasic DNA binding/clamp loading kinetics were observed. The biphasic clamp loading kinetics related to the biphasic ATP hydrolysis kinetics originally seen in the MDCC-PBP ATPase assay where γ complex was equilibrated with ATP, and then mixed with pt DNA (Figure 4-3, black trace). Together, these results indicated that there were two species of γ complex present upon equilibration with ATP prior to reacting with DNA and β in the clamp loading assay. To investigate the kinetics of ATP hydrolysis when γ complex is equilibrated with ATP, and mixed with a solution of pt DNA and β, a sequential mix MDCC-PBP ATPase assay was performed. In this assay, one syringe containing a solution of γ complex (0.27 μM) was mixed with a solution containing ATP (400 μM) from a second syringe, and preincubated for a duration of 2 s before mixing with a solution from a third syringe containing pt DNA (1.0 μM), β (1.0 μM), and MDCC-PBP (2.7 μM). Final reaction concentrations are as indicated in parentheses. Figure 4-7 shows a direct comparison of the ATP hydrolysis kinetics for this clamp loading reaction with the biphasic hydrolysis kinetics originally observed with pt DNA only.

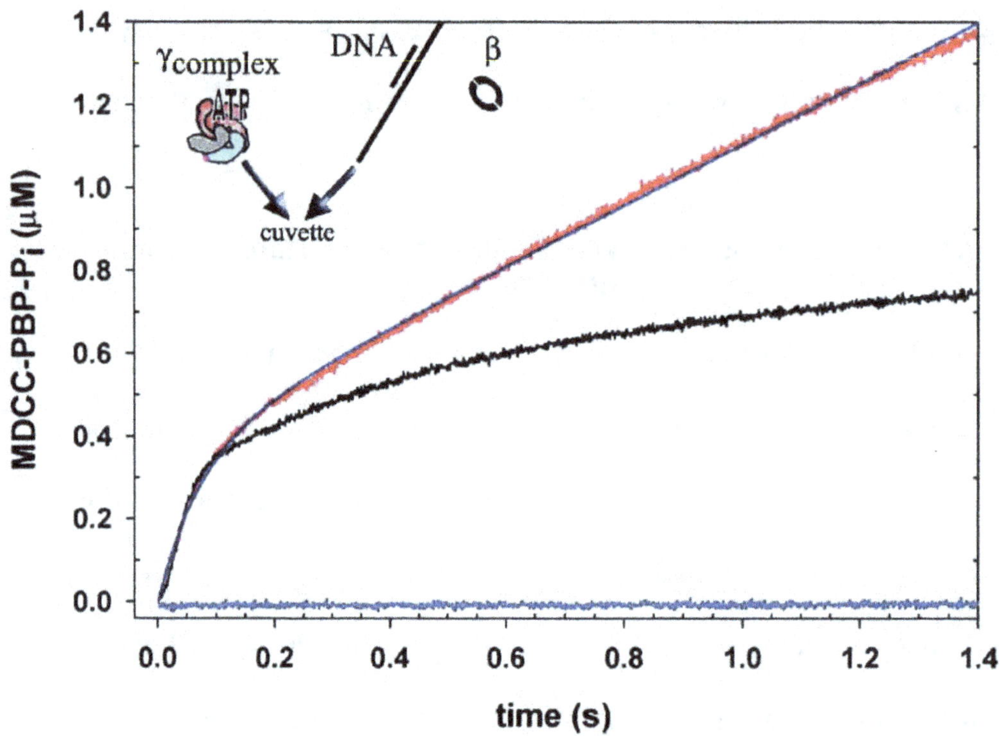

Figure 4-7. Kinetics of ATP hydrolysis during the clamp loading reaction when γ complex is equilibrated with ATP. Using a sequential-mix "three-syringe" MDCC-PBP ATPase assay, with a mixing scheme (inset) analogous to that used in the anisotropy DNA binding assay (figure 4-6A), the ATP hydrolysis kinetics of the clamp loading reaction where measured (red trace). One syringe loaded with a solution containing γ complex was mixed with the contents of a second syringe loaded with ATP. The resulting mixture was preincubated for a period of 2 s, and then mixed with the contents of a third syringe loaded with a solution-containing pt DNA, β, and MDCC-PBP. The mixing scheme shown in the inset represents the contents of the second and third syringes. All solutions were prepared in P_i-mopped assay buffer: 20 mM Tris-HCl, 50 mM NaCl, 5 mM DTT, 40 μg/mL BSA, and 8 mM $MgCl_2$. Final reactant concentrations were 270 nM minimal complex or γ complex, 1.0 μM β, 400 μM ATP, 1.0 μM pt DNA, and 2.7 μM MDCC-PBP. The reaction of γ complex with pt DNA only was performed under identical conditions, except that there was no β present in the third syringe (black trace). The flat trace (gray) at the bottom of the plot showing no change was a negative control assay performed in the absence of ATP. The solid curve (blue) is a fit of the clamp loading ATPase data (red) to an empirical expression representing a single exponential rise followed by a linear rise: $y = A_o[1-exp^{(-kt)}] + C(t)$. Raw fluorescence data were transformed into the concentration of P_i-bound MDCC-PBP using the equation shown in figure 4-3.

The kinetics for the clamp loading reaction (red trace) showed a rapid single exponential rise that led directly into a steady-state linear rise (0.74 μMs^{-1}). The steady-state phase of the clamp loading reaction was identical to that previously observed for reactions when γ complex was equilibrated with ATP and β before mixing with DNA, (compare to Figure 4-5B, black trace). The single rapid phase matched similarly in amplitude and rate with the pre-steady-state rapid phase of the biphasic reaction with pt DNA only (Figure 4-7, black trace). These results suggest that the predicted population of γ complex active species, (~40 %) in equilibrium with ATP, rapidly bound to and loaded β, then entered a steady-state clamp loading reaction as the smaller population of inactive species was converted to an active species, analogous to the biphasic clamp loading anisotropy data above (Figure 4-6A). This result also suggests that the second phase observed in the biphasic anisotropy DNA binding assays is possibly equivalent to the rate of conformational change(s) within γ complex, and that this rate limits the steady-state clamp loading reaction.

Kinetics of ATP Hydrolysis when γ Complex is not Equilibrated with ATP

Slow kinetics of clamp loading were observed when γ complex was not pre-incubated with ATP prior to initiating reactions. To determine whether this was also the case in ATPase assays and whether the rate of ATP binding was contributing to the kinetics of clamp loading, single mixing ATPase assays were done where γ complex was added directly to a solution of ATP, DNA, and MDCC-PBP. Reactions contained 0.27 μM γ complex, 1 μM DNA, 2.7 μM MDCC-PBP, and 20, 80, 200, 400, μM ATP. As in clamp loading assays, a substantial lag in product formation preceded a burst of ATP hydrolysis (Figure 4-8).

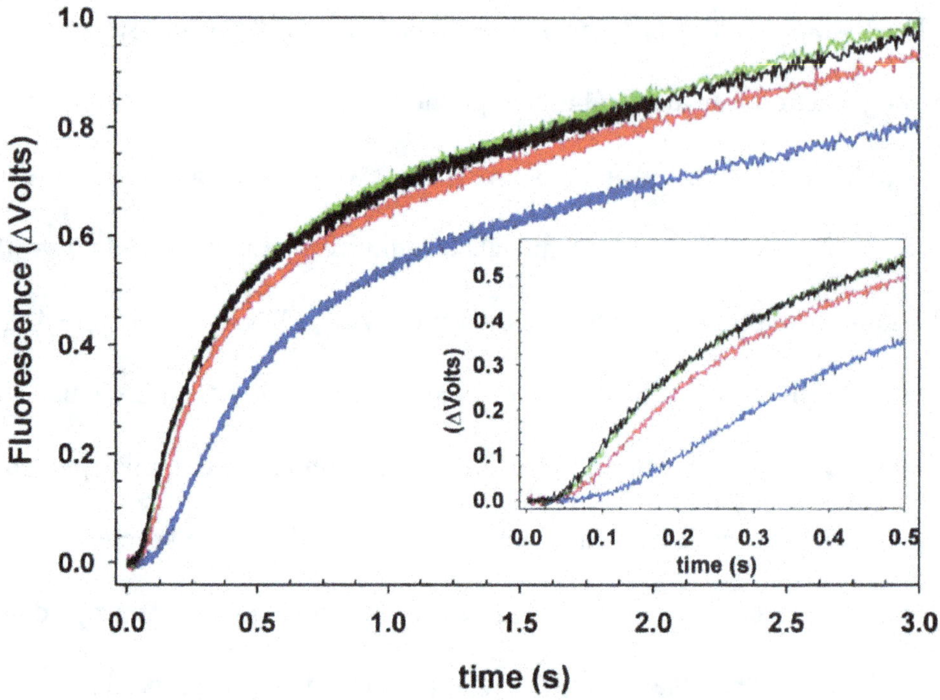

Figure 4-8. Kinetics of ATP hydrolysis in reactions initiated by the addition of γ
complex to ATP and DNA. Single mixing experiments were performed
where γ complex was added directly to a solution of ATP, pt DNA, and
MDCC-PBP. Kinetics of ATP hydrolysis were determined by measuring the
increase in fluorescence of MDCC-PBP on binding P_i and are plotted as raw
MDCC fluorescence (Δ Volts) as a function of time. The time base changes
from 1 ms to 5 ms intervals at 2 s. The inset shows the first 0.5 s on an
expanded scale. Reactions contained 20 (blue), 80 (red), 200 (green), 400
(black), μM ATP. A lag in product formation is present at each ATP
concentration. Final concentrations of other reagents remained constant and
were 0.27 μM γ complex, 1.0 μM DNA, and 2.7 μM MDCC-PBP in assay
buffer containing 20 mM Tris•HCl pH 7.5, 50 mM NaCl, 8 mM $MgCl_2$, 5 mM
DTT, and 40 μg/ml BSA. Experimental data for these assays were
contributed by Anita K. Snyder.

The length of the lag phase decreased with increasing concentrations of ATP and ranged

from about 25 to 30 ms for concentrations of ATP greater than 200 μM to 70 ms for 20

μM ATP. At concentrations above 200 μM ATP, the length of the lag phase and pre-

steady-state rates of ATP hydrolysis are independent of ATP concentration. At these

higher concentrations, ATP binding was not limiting the rate of hydrolysis. A significant result of these experiments is that the biphasic nature of the burst of ATP hydrolysis persists even when γ complex is not equilibrated with ATP. The rate of the rapid phase of ATP hydrolysis is reduced relative to reactions where γ complex was equilibrated with ATP making the biphasic nature of the burst more difficult to identify by visual inspection alone.

Kinetics of Formation of the Two Populations of γ Complex

The results of the experiments presented above demonstrate that two populations of γ complex are formed when γ complex is equilibrated with ATP and that an intramolecular reaction, possibly ATP-induced conformational changes, limits the observed rate of clamp loading reactions when γ complex is not pre-incubated with ATP. To more clearly define the kinetics of the conformational transitions that occur when γ complex binds ATP, a series of sequential mixing reactions were performed where the length of time that γ complex was incubated with ATP was varied. If an ATP-induced conformational change limits the observed rates of clamp loading and ATP hydrolysis in assays where γ complex is not pre-incubated with ATP, then the populations of the conformational states should vary as a function of the incubation time with ATP. After a defined preincubation time, DNA was added to trigger ATP hydrolysis and probe the relative populations of γ complex species that have formed. In these experiments, γ complex was incubated with ATP for 0, 15, 50, 100, 250, 500, or 1000 ms prior to addition of DNA and MDCC-PBP. Final concentrations were 0.27 μM γ complex, 400 μM ATP, 2.7 μM MDCC-PBP, and 1 μM DNA. These data show the evolution of the two major populations of γ complex species that are formed on equilibration with ATP. As the length of the incubation period increases, the length of the lag phase decreases and

Figure 4-9. Kinetics of ATP hydrolysis when γ complex is incubated with ATP for a defined period of time prior to addition of DNA. Sequential mixing experiments were performed where γ complex was incubated with ATP for 15 (red), 50 (blue), 100 (green), 250 (cyan), 500 (purple), or 1000 (black) ms prior to addition of DNA and MDCC-PBP. Kinetics of ATP hydrolysis were determined by measuring the increase in fluorescence of MDCC-PBP on binding inorganic phosphate and are plotted as the concentration of MDCC-PBP•P_i complex as a function of time on time scales of A) 2.0 s and B) 0.4 s. A control reaction with no ATP (gray) shows the signal for free MDCC-PBP as a function of time. The solid black curves through the data were calculated by fitting the data to the model shown in figure 4-10A. Final concentrations were 0.27 μM γ complex, 400 μM ATP, 1.0 μM DNA, and 2.7 μM MDCC-PBP in assay buffer containing 20 mM Tris•HCl pH 7.5, 50 mM NaCl, 8 mM $MgCl_2$, 5 mM DTT, and 40 μg/ml BSA. Raw fluorescence data were transformed into the concentration of P_i-bound MDCC-PBP using the equation shown in figure 4-3.

the rate of the initial phase of product formation increases (Figure 4-9). Incubation of γ

complex with ATP for 250 ms gives nearly the same kinetics as for 500 and 1000 ms

indicating that the conformational transitions have reached equilibrium by this point in

time.

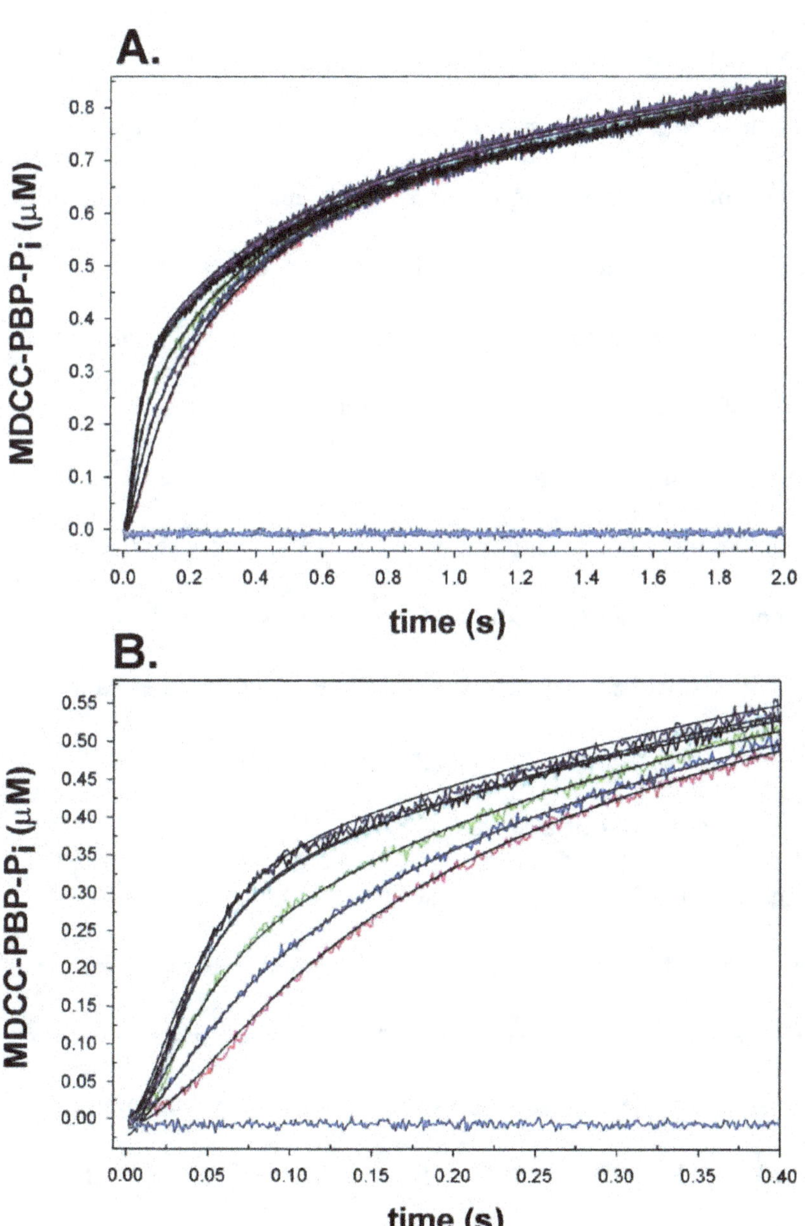

Computer Modeling of ATP Hydrolysis Reaction Kinetics

To determine the kinetic relationship between γ complex species that give rise to the biphasic kinetics of ATP hydrolysis and clamp loading, data from sequential mixing experiments shown in Figure 4-9 were modeled using DynaFit (Kuzmic, 1996). Modeling demonstrated that at least three different species of γ complex were required to produce biphasic kinetics of ATP hydrolysis and that these three species could not all exist on the linear pathway to products. The minimal model that best fit the data is shown in Figure 4-10A (see computer modeling methods in chapter 3 for details). In this model, γ complex is initially present in a state, ETTT, that can convert to two different species, one that is active for DNA binding, ATTT, one that is inactive, ITTT, and does not directly form products. The rates of interconversion of these three states are relatively slow with rates in the range of 2.4 to 4.5 s^{-1}. In this minimal model, it is assumed that each of these species is bound to three molecules of ATP because ATP hydrolysis rates were independent of the ATP concentration at this saturating concentration (400 μM ATP). However, it is possible that the E and/or I states have fewer molecules of ATP bound and that very rapid ATP binding steps and conformational changes exist between the three states.

The γ complex in the activated state, ATTT, rapidly binds DNA. This DNA•γ complex then hydrolyzes each of its three molecules of ATP sequentially at the same rate. The ATP hydrolysis steps are modeled using single forward rate constants because the presence of MDCC-PBP limits the reverse reaction such that it cannot accurately be determined. Previous work has shown that the DNA dissociation step is likely to be very rapid relative to the steady-state recycling of the clamp loader, and because these data do not directly measure this step, DNA dissociation was set to be concurrent with hydrolysis

Figure 4-10. Kinetic modeling of ATP hydrolysis reactions. A) The minimal
kinetic model used to fit the ATP hydrolysis data in Figure 4-9. Three
states of the γ complex, ETTT, ATTT and ITTT are related by the
kinetic constants shown (cartoons are based on the structure of the
clamp loader, see chapter2, figure 2-3). The ATTT form binds DNA
(N) rapidly and hydrolyzes its three molecules of ATP (T) in
succession producing ADP (D) and inorganic phosphate (P_i). The γ
complex dissociates from DNA on hydrolyzing its third molecule of
ATP and is recycled back to the ETTT state. Experimentally
determined rates of γ complex binding DNA (Ason et al., 2000;
Bertram et al., 2000), k_{onN}, and MDCC-PBP (PBP) binding phosphate
(Brune et al., 1994), k_{on} and k_{off}, were set constant in the fit. All other
rate constants were determined by fitting the data to the model using
DynaFit (Kuzmic, 1996). B) The concentrations of each of the three γ
complex species were simulated with KinTekSim (Barshop et al.,
1983; Dang and Frieden, 1997) and calculated based on the rate
constants shown in A and are plotted as a function of time. C) An
additional step (PTTT $\rightarrow$ ETTT) is required in the kinetic model
shown in A to fit single mixing ATPase reactions as in Figure 4-8
where the pre-incubation time is effectively 0 ms. Concentrations in
this assay were 0.27 μM γ complex, 500 μM ATP, 1.0 μM DNA, and
2.7 μM MDCC-PBP in assay buffer containing 20 mM Tris•HCl pH
7.5, 50 mM NaCl, 8 mM $MgCl_2$, 5 mM DTT, and 40 μg/ml BSA. All
DynaFit program scripts and analysis for these models, and
KinTekSim simulations are presented in the appendix.

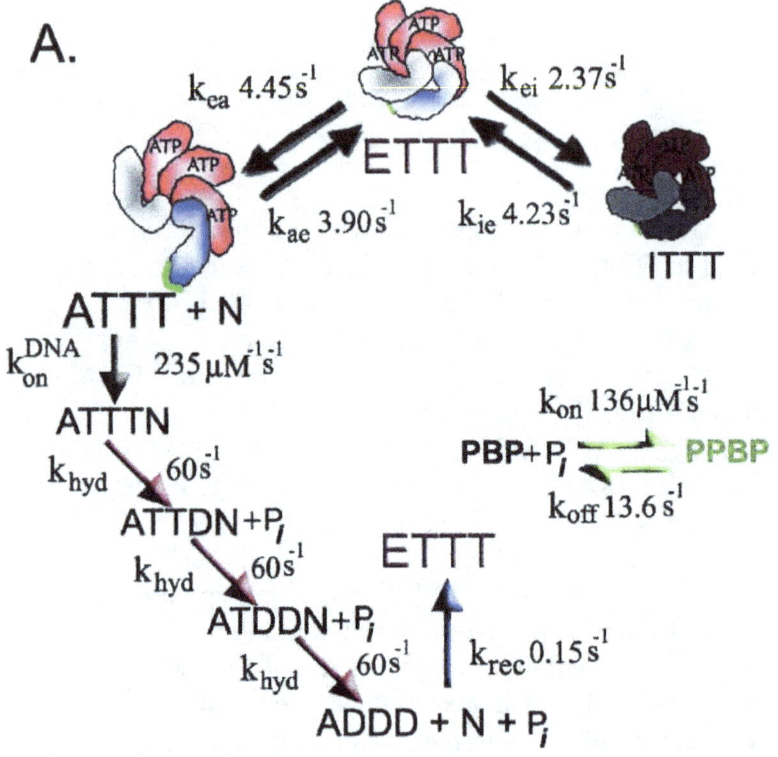

A.
k_ea 4.45 s⁻¹
k_ei 2.37 s⁻¹
ETTT
k_ae 3.90 s⁻¹
k_ie 4.23 s⁻¹
ITTT
ATTT + N
k_on^DNA   235 μM⁻¹s⁻¹
ATTTN
k_hyd   60 s⁻¹
ATTDN + P_i
k_hyd   60 s⁻¹
ATDDN + P_i
k_hyd   60 s⁻¹
k_on 136 μM⁻¹s⁻¹
PBP + P_i   PPBP
k_off 13.6 s⁻¹
ETTT
k_rec 0.15 s⁻¹
ADDD + N + P_i

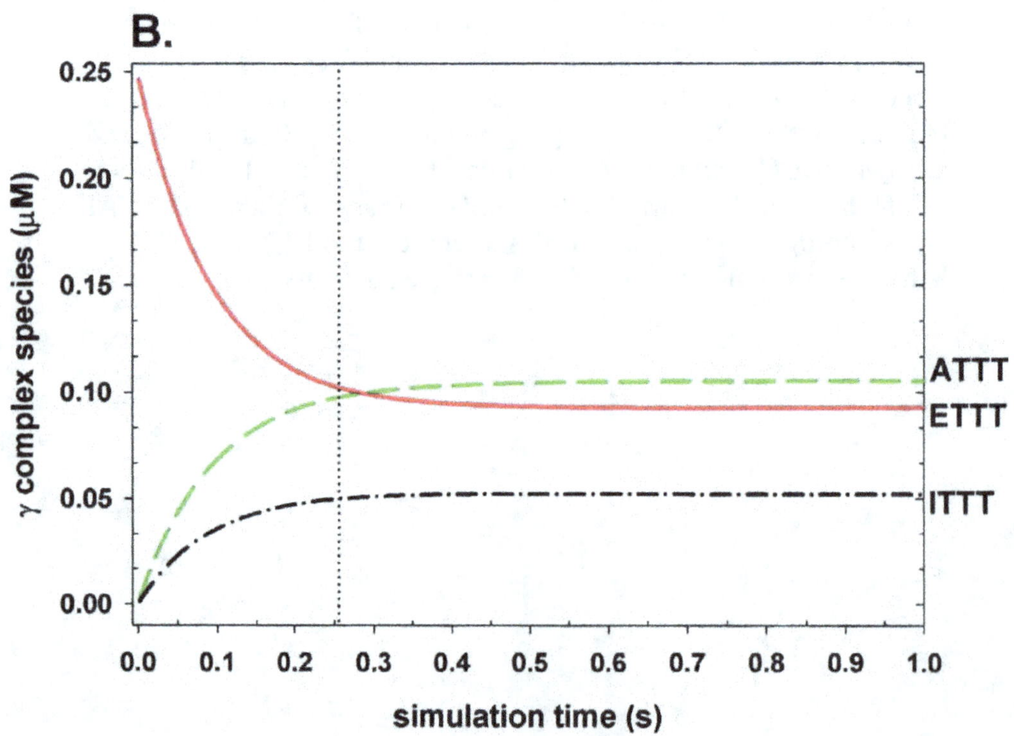

B.
0.25
0.20
0.15
0.10
0.05
0.00
γ complex species (μM)
ATTT
ETTT
ITTT
0.0 0.1 0.2 0.3 0.4 0.5 0.6 0.7 0.8 0.9 1.0
simulation time (s)

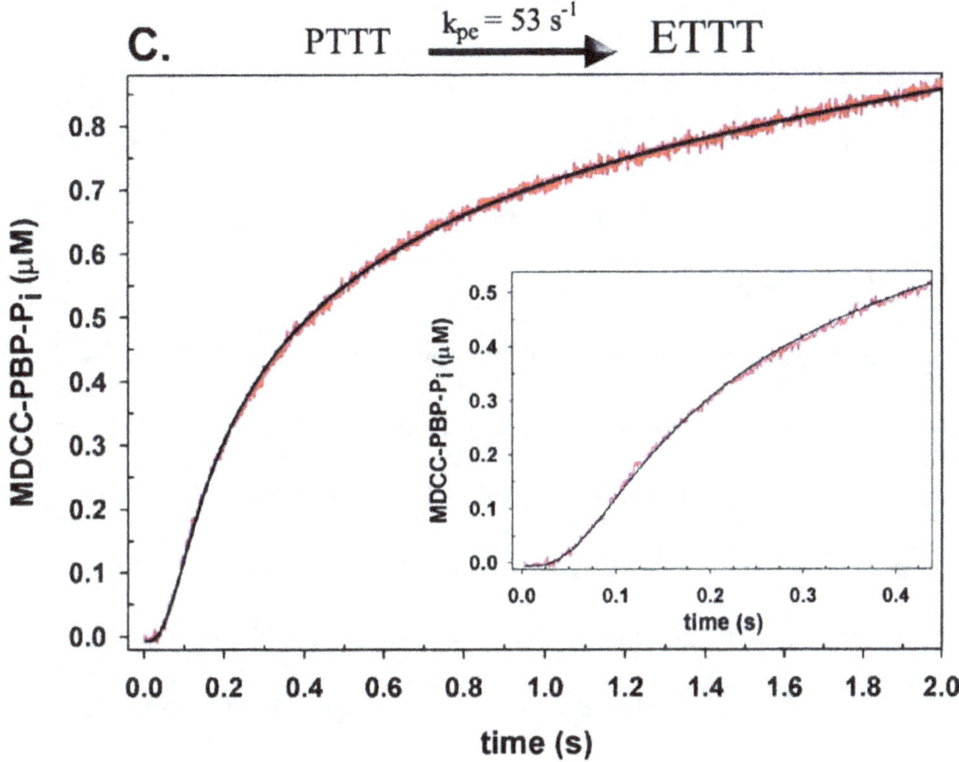

Figure 4-10. Continued.

of the third molecule of ATP. Finally, the steady-state portion of the reaction was

modeled as a pseudo-first order conversion of the ADP-bound γ complex (ADDD) back

to the initial ATP-bound state (ETTT). This pseudo-first order approximation is

reasonable because only a small fraction of the ATP substrate is consumed on the time

scale of the reaction.

The three species of γ complex, ETTT, ATTT, and ITTT, which are in equilibrium

in the presence of ATP generate the two populations that give rise to the biphasic kinetics

of ATP hydrolysis and clamp loading. The rapid phase of ATP hydrolysis is produced

largely from the population of γ complex in the ATTT state and the ETTT and ITTT

states produce the slow phase. The populations of each of these species was simulated

and calculated as a function of time using the kinetic constants in Figure 4-10A, and are plotted in Figure 4-10B. The concentration of each changes over the first 250 ms and then remains constant just as the kinetic data for ATP hydrolysis do not change at pre-incubation times greater than 250 ms. At equilibrium, concentrations are 0.093 µM ETTT, 0.105 µM ATTT, and 0.052 µM ITTT based on a total concentration of 0.25 µM γ complex calculated from the fit.

In ATPase assays where γ complex was not pre-incubated with ATP (Figure 4-8), a pronounced lag (25 to 30 ms) in product formation is observed. The minimal model in Figure 4-10A does not generate a lag that is long enough to fit these data well. An additional species of γ complex (PTTT) that converts relatively rapidly ($k_{pe} = 53$ s^{-1}) to the ETTT state is required; these data could be fit to the model using the rate constants shown in Figure 4-10A when this single step was included. Figure 4-10C shows the data for an assay where γ complex was mixed with DNA, MDCC-PBP and ATP. The resulting fit from the minimal model (Figure 4-10A) with the additional step PTTT ➔ ETTT is shown over the experimental data.

Discussion

The *E. coli* γ complex functions as a molecular machine to assemble ring-shaped β clamps on primers where DNA synthesis is slated to begin (reviewed in (Davey et al., 2002). This mechanical task requires that the affinity of clamp loader for β and DNA be modulated during the course of the clamp loading cycle. Initially, the γ complex must have a high affinity for both β and DNA to efficiently bring them together. ATP binding produces a high affinity binding state (Ason et al., 2000; Naktinis et al., 1995; Turner et al., 1999) where the δ subunit can bind the clamp and "crack" it open (Leu et al., 2000; Stewart et al., 2001; Turner et al., 1999). Thus, ATP binding allows the γ complex to form a ternary clamp loader•clamp•DNA complex (Bertram et al., 1998) where an open clamp is most likely encircling DNA. At this point, the affinity of the clamp loader for the clamp and DNA must be reduced so that the clamp can be closed around DNA and the γ complex can dissociate so as not to interfere with the polymerase binding the newly loaded clamp on DNA. A primed DNA template provides the trigger to convert the γ complex to a low affinity binding state (Ason et al., 2000; Ason et al., 2003). Upon binding a primer end, the γ complex hydrolyzes ATP and releases the clamp and DNA (Ason et al., 2003).

Binding and hydrolysis of ATP most likely induce conformational changes that produce the high and low affinity binding states, respectively (Ason et al., 2000; Jeruzalmi et al., 2001a; Naktinis et al., 1995). Each clamp loader can hydrolyze up to three molecules of ATP, as each γ subunit has an ATP binding site. These binding sites are located at the interface of two domains within a γ subunit and also at the interface between two adjacent subunits within the complex (Jeruzalmi et al., 2001a; Podobnik et al., 2003). ATP binding by each γ subunit is likely to promote a conformational change within that subunit that is then communicated to adjacent subunits. Thus, multiple conformational changes within the complex are likely to be associated with the transition to a high affinity binding state. Ultimately, this combination of conformational changes exposes a region on the N-terminal domain of the δ subunit that binds the clamp. A DNA binding domain is also created or exposed by these conformational changes (Ason et al., 2000). Because there are three ATP binding sites within the clamp loader, different conformational states are likely to exist depending on the occupancy of these sites. In addition, the complex nature of the interactions between protomers and the conformational flexibility that must exist in the complex suggest that different conformational states may exist even among complexes where the occupancy of the ATP binding sites is the same. In this dissertation project, the ATP-dependent conformational dynamics of the γ complex have been investigated by measuring the kinetics of ATP hydrolysis and clamp loading under conditions where these changes contributed to the observed kinetics.

When γ complex is completely equilibrated with ATP (pre-incubation times of ≥ 250 ms), the first turnover of ATP is biphasic (Figure 4-3, black trace). Clamp loading assays done under parallel conditions, initiated by adding an equilibrated solution of γ complex and ATP to β and DNA, a biphasic increase in anisotropy is also observed indicating that two species were loading clamps at different rates (Figure 4-6A). If the biphasic kinetics of ATP hydrolysis were due to individual γ subunits within the γ complex hydrolyzing ATP at different rates, then a single phase of clamp loading resulting from a single population of γ complex loading clamps would have been observed. The ATPγS-chase assay of ATPase activity also showed different populations of γ complex when equilibrated with ATP before adding DNA, and showed that ATP is securely bound to one of these populations when directly compared to the original biphasic hydrolysis kinetics (Figure 4-5A). An ATP-bound population hydrolyzed ATP faster than it dissociates in a pre-steady-state rapid step that directly related to the rapid phase of the originally observed biphasic hydrolysis kinetics. A pre-steady-state ATP hydrolysis assay mimicking the mixing conditions of the pre-steady-state clamp loading assay (Figure 4-7) showed that when γ complex was equilibrated with ATP before mixing with DNA and β, a single population rapidly hydrolyzed ATP, then appeared to enter a steady-state clamp loading reaction. The population of clamp loader that rapidly hydrolyzed ATP in this assay directly relates to the initial rapid phase observed for clamp

loading kinetics, as well as the population of γ complex which rapidly hydrolyzed ATP in

the biphasic reaction with DNA only. Therefore, this rapid phase may be due to some γ

complex alone binding to pt DNA and hydrolyzing ATP. The rate of the slower second

phase of the biphasic hydrolysis kinetics (i.e. some inactive species population) was

increased to a rate identical to the steady-state rate of ATP hydrolysis for clamp loading

in this assay designed to mimic the clamp loading assay (Figure 4-7). These ATP

hydrolysis data for the clamp loading reaction in comparison to the reaction with DNA

only are consistent with the ability of β to enhance the steady-state ATP hydrolysis

activity of the clamp loader. In the mimicked pre-steady-state clamp loading reaction

(Figure 4-6A), the slower second phase may relate to conformational changes required

for γ complex to gain affinity for β and load it on DNA. A steady-state rate of ATP

hydrolysis for clamp loading was observed during this time in the clamp loading ATP

hydrolysis assay, suggesting that the rate of the ATP-dependent conformational changes

giving the clamp loader affinity for β and DNA is a rate limiting step in the clamp

loading reaction.

In ATPase assays where γ complex is incubated with β and ATP, a single pre-

steady-state phase of ATP hydrolysis is produced at the same rate as the rapid phase in

assays without β (Figure 4-3, red trace). The amplitude of the single phase of ATP

hydrolysis in assays with β is equal to the sum of the amplitudes of the two phases in

assays without β and indicates that about 2.7 molecules of ATP are hydrolyzed per γ

complex. This number is interpreted as one molecule of ATP hydrolyzed by each γ

subunit. This experimental value is less than the theoretical value 3 most likely because of differences between the concentrations of active enzyme and total protein in the assay. ATPγS-chase of the pre-steady-state hydrolysis activity in the presence of β showed that the molecules bound to the γ subunits were hydrolyzed faster than they dissociate, indicating that the β clamp has an effect on the stability of the clamp loader. Specifically, in the presence of β, the bound-ATPs appear to be secure or "trapped" in the complex until DNA triggers hydrolysis upon clamp loading. The observation that the sum of the amplitudes of the rapid and slow phases of ATP hydrolysis in assays without β is also equivalent to 2.7 molecules of ATP indicates that both populations of γ complex are hydrolyzing all their ATP but at different rates.

The pre-steady-state ATP hydrolysis assays performed in the presence of β at three different γ complex concentrations (Figure 4-4) demonstrate further that the clamp loader hydrolyzed approximately three ATP molecules at nearly the same observed rate in the single rapid phase of each assay, and importantly, that the molar response to hydrolysis by MDCC-PBP was adequately measured. This result has significance for any study of an energase enzyme (i.e., ATPase, GTPase), in which the investigator is using the MDCC-PBP NTP hydrolysis assay to quantify the number of nucleotide molecules hydrolyzed under pre-steady-state "burst" conditions.

To determine the kinetic relationship between γ complex species that give rise to the two phases of product formation in ATPase and clamp loading assays, the kinetics of formation of these species was measured in pre-steady-state ATPase assays. In these assays, γ complex was pre-incubated with ATP for a defined period of time, 15 to 1000 ms, during which different populations of the various species formed. Kinetic data for ATP hydrolysis were dependent on the relative concentrations of these species.

Modeling of these data showed that the biphasic kinetics of ATP hydrolysis can be generated by three different γ complex species that we have denoted as ATTT, ETTT, and ITTT (Figure 4-10A and B). In other words, three different states of the γ complex produce two different kinetic populations that give rise to two phases of product formation. In these reactions, γ complex is initially present in the ETTT state and this state produces two new states, ATTT and ITTT. Modeling further demonstrated that one of these species, ITTT, does not form products directly. It is believed that these three species represent three different conformational states of the γ complex.

ATP hydrolysis was also measured in assays where γ complex was not preincubated with ATP, effectively a 0 ms time point (Figure 4-8 and Figure 4-10C). These reaction time courses showed an extended lag of about 25 to 30 ms that was not present in assays where the pre-incubation period was 15 ms or greater. This extended lag was also reflected in time courses for clamp loading where reactions were initiated by the addition of γ complex to ATP, β, and DNA (Figure 4-6C). The model used to fit ATPase data for pre-incubation times of 15 – 1000 ms does not generate a lag that is long enough to adequately fit the "0 ms" preincubation data. However, these data can be fit by the inclusion of an early step, PTTT → ETTT, which generates the ETTT state where the model in Figure 4-10A begins. A simulation of the model including the PTTT → ETTT step shows that formation of ETTT peaks at approximately 30 to 40 ms, consistent with the extended lag observed in the assays where there was an effective 0 ms preincubation time (see Appendix A, figure A-1). The rate of this early step is relatively rapid, 53 s-1. Data for preincubation times of 15-1000 ms do not adequately define this kinetic step because it is rapid, a half-life of about 11 ms, relative to the shortest delay time. To define this step in these assays, shorter pre-incubation times would be required which are not experimentally possible with our instrumentation. Although an extra step was required to generate the lag in the 0 ms pre-incubation data, the rest of the time course was fit very well by the model and rate constants shown in Figure 4-10A. The quality of this fit indicates that the model adequately represents both data sets for the 0 ms and 15-1000 ms pre-incubation reactions.

What does this early step that generates the ETTT state represent? One likely possibility is that this step reflects the rate of ATP binding to the complex. These experiments were done under conditions where ATP binding was not rate-limiting and that are effectively "V_{max}" conditions. Therefore, an ATP binding step could be represented as a pseudo-first order reaction. Although it is not rate limiting, it would still take some finite amount of time for three ATP molecules to bind the complex and this time may be reflected in the lag. Note that the length of the lag phase did increase substantially to about 70 ms in an ATPase reaction at a lower sub-saturating concentration, 20 µM, of ATP (Figure 4-8, blue trace). This observation is consistent with an ATP binding step contributing to the lag in product formation. Another possibility is that this lag represents local conformational changes within a single γ subunit that occur when a single molecule of ATP binds the complex.

What is the physical nature of the three kinetic states, ATTT, ETTT, and ITTT that give rise to the biphasic kinetics of ATP hydrolysis? The model (Figure 4-10A) describes the ATTT state as one that is activated for DNA binding suggesting that the ETTT to ATTT transition reflects conformational changes that either expose or create a DNA binding site. This transition may also reflect conformational changes that expose the δ subunit and activate the complex for β binding. The pre-steady-state rate of ATP hydrolysis in assays with β was identical to the rapid phase of hydrolysis in assays without β. If β selectively bound this activated species of γ complex, then during the 1000 ms equilibration period (Figure 4-3, red trace) β could ultimately convert all γ complex to an activated β•ATTT complex. This would then produce a single phase of ATP hydrolysis that occurred at the same rate as the rapid phase in assays without β. The high affinity of ATP-bound γ complex for β, Kd of 3-4 nM (Naktinis et al., 1995) (see also chapter 5 Figure 5-3), suggests that β would be able to drive the equilibrium in this direction under our assay conditions. Previous work has shown that when γ complex is pre-incubated with β and ATP for only 80 ms, the pre-steady-state formation of ADP is biphasic (Hingorani et al., 1999) as we have shown in assays without β. This observation is consistent with selective binding of β to an activated ATTT state, although more pre-incubation time points need to be taken to provide stronger support for this model. An alternative explanation for the effect of β on ATPase kinetics is that β could bind the ETTT state and this interaction could promote the formation of the ATTT state at the expense of the ITTT state. This type of mechanism would require that β change the kinetics of this step rather than simply shifting the equilibrium of γ complex states by selectively binding one of the three species. These two possibilities could be differentiated in ATPase experiments where γ complex was pre-incubated with ATP in the presence of β for different lengths of time.

A key feature of our kinetic model (or any model that would adequately describe these data) is that a nonproductive branch in the pathway to products must be present to generate the kinetic data shown in Figures 4-8 and 4-9). In other words, one species of γ complex must form that does not produce products directly. These kinetic data could not be adequately described by a series of states that exist on the linear pathway to products (e.g. E1TTT → E2TTT → ATTT → products). In our minimal model, we have represented this branch as an interconversion of ETTT and ITTT. We do not yet know what the physical nature of this ITTT state is. One possibility is that the ITTT state simply represents an alternative ATP-bound form of the γ complex that is thermodynamically stable but not kinetically active. Another interesting possibility is that the ITTT state is not initially formed form the ETTT state, but instead, ATP loading into binding sites in a different order forms the ETTT and ITTT states independently. It is possible that ordered binding of ATP is required to generate the most rapid and efficient series of conformational changes to produce the high affinity binding state of γ complex (Johnson and O'Donnell, 2003). Structural data for the $\gamma_3\delta\delta'$ minimal complex suggests that two of the three ATP binding sites may be open in the nucleotide-free complex so that two different orders of loading ATP may be possible (Jeruzalmi et al., 2001a). Perhaps the ETTT state is formed when ATP binds in the preferred order and the ITTT state is formed when ATP binds in an alternative order. The ITTT state then must convert to the ETTT state before it can go on to products. It is possible that ITTT converts directly to ETTT, or that ITTT converts to ETTT by dissociation and rebinding of ATP and the rates defined by our minimal model reflect the overall rate for the process.

Regardless of what the physical nature of these kinetic states is, the data show that the rates of conformational changes within the γ complex are relatively slow and limit the rate of the pre-steady-state ATP hydrolysis and clamp loading reactions when γ complex is not pre-equilibrated with ATP. The rates of transitions between ETTT and ATTT and ETTT and ITTT states are relatively slow on the order of 2-4 s^{-1} with the rate of conversion of ETTT to ATTT being about 4.5 s^{-1}. The magnitude of these values is similar to the steady-state rate constant or turnover number, $k_{cat}^{complex}$ of 2 to 3 s^{-1} (Table 4-1) (Bertram et al., 2000; Hingorani et al., 1999), for reactions with β and may also reflect the rate-limiting step under these conditions. Additionally, $k_{cat}^{complex}/K_m$ values for steady-state reactions are relatively slow, about 0.8 to 1.8 x 10^5 $M^{-1}s^{-1}$ (Table 4-1) (Bertram et al., 2000; Hingorani et al., 1999), to represent a bimolecular binding reaction between ATP and γ complex. The magnitude of $k_{cat}^{complex}/K_m$ is likely to be reduced relative to the theoretical upper limit (a diffusion controlled binding rate) by a two-step binding reaction consisting of relatively rapid ATP binding followed by slow conformational changes induced by ATP binding.

CHAPTER 5

CHARACTERIZATION OF THE MINIMAL CLAMP LOADER COMPLEX AND
COMPARISON TO GAMMA COMPLEX

Introduction

The *E. coli* β sliding-clamp loader is a complex molecular machine made up of a heteropentameric core of AAA+ subunits $(\gamma/\tau)_3\delta\delta'$ along with two additional subunits (χ and ψ) (Davey et al., 2002; Neuwald et al., 1999; Onrust et al., 1995). The core sub-complex of the clamp loader has the necessary components for loading the sliding clamp onto DNA for promotion of processive synthesis (Onrust et al., 1991). The X-ray crystal structure of this sub-complex $(\gamma_3\delta\delta')$ of the clamp loader was the first solved of any of the evolutionarily conserved clamp loaders (Jeruzalmi et al., 2001a), and has had a broad impact across the entire replication field (Ellison and Stillman, 2001; Jeruzalmi et al., 2002; O'Donnell et al., 2001).

The mechanism and energetics of clamp loading has been extensively explored for γ complex $(\gamma_3\delta\delta'\chi\psi)$ in both steady-state and pre-steady-state analyses (Bertram et al., 2000; Hingorani et al., 1999; Kelman and O'Donnell, 1995). However, the analysis of the sub-complex $(\gamma_3\delta\delta')$ has only been sparingly examined in steady-state assays only (Dong et al., 1993; Onrust et al., 1991), with the generalized conclusion that it has activity similar to γ complex (Jeruzalmi et al., 2001a). The previous chapter (4) of this dissertation continues the detailed study of the clamp loading mechanism by γ complex, and shows the kinetics of the ATP-dependent conformational transitions within the clamp loader using an array of ATP hydrolysis and DNA binding assays along with computer modeling and fitting the kinetic data. Here, many of the same analyses are presented for the "minimal complex" clamp loader $\gamma_3\delta\delta'$ in direct comparison to those of γ complex with the goal of further understanding the activity of this conserved "core" clamp loader, and additionally understanding the need for the χ and ψ subunits in the *E. coli* clamp loader. The work presented in this chapter encompasses the most detailed study of the $\gamma_3\delta\delta'$ clamp loader to date, and includes steady-state and pre-steady-state analyses of DNA binding/clamp loading, β clamp binding, and ATP hydrolysis activities of this minimal complex. The results show that the activity of the minimal complex is nearly the same as γ complex, as had been previously known; however, the data show that the missing χ and ψ subunits are a necessity for γ complex, and confer conformational stability to the clamp loader leading to increased optimal activity. Another key result of the comparison of the $\gamma_3\delta\delta'$ minimal complex to γ complex is that the β clamp is responsible for the similarities observed for these clamp loaders. It is well known that the sliding clamps enhance the ATP hydrolysis activity of their clamp loader, and therefore the clamp loading activity of γ complex (Onrust et al., 1991; Stukenberg et al., 1991), and also clamp loaders of other organisms (Ellison and Stillman, 2001). The comparison of $\gamma_3\delta\delta'$ and γ complex steady-state and pre-steady-state kinetics in the absence and presence of β lead to a possible clamp loader enhancement mechanism wherein the clamp, by binding the clamp loader, converts it to a completely activated population and secures the bound molecules of ATP such that this ATP-bound-clamp loader-β clamp complex is committed for the loading reaction as soon as DNA is located.

$\gamma_3\delta\delta'$ is the Minimal Clamp Loader Complex Which Can Bind DNA

Analysis of DNA Binding Activity of the Individual Subunits or Sub-complexes of the Clamp Loader

The steady-state anisotropy of RhX-labeled ss DNA was analyzed to measure

binding interactions between purified individual subunits of γ complex (γ, δ, δ', $\chi-\psi$), as

well as several sub-complexes of subunits ($\gamma\delta$, $\delta\delta'$, $\gamma\chi\psi$, $\gamma\psi\delta'$, $\gamma\delta\delta'$). The τ subunit

dimerizes the core polymerase *in vivo*, and is also a constituent of the clamp loader in the

holoenzyme. Therefore the steady-state binding activity of the τ subunit alone as well as

in a sub-complex with $\delta\delta'$ ($\tau\delta\delta'$) was also measured with the RhX-ss DNA. The

observed steady-state anisotropy of the RhX-probe on DNA is a population weighted

average of the combined anisotropies for free DNA and protein-bound DNA. Therefore,

the observed anisotropy increases as the population of protein-bound DNA increases

(Bloom et al., 1996). The change in anisotropy of RhX covalently attached to a single-

stranded 50-mer DNA was measured from calculated anisotropy from titration assays

with individual subunits or sub-complexes. Increasing concentrations of the purified

individual subunits, sub-complexes, or γ complex were titrated into a solution containing

50 nM RhX-ss DNA and 0.5 mM ATP. The ATP was added last in these assays such

that the change in anisotropy in the absence, then presence of ATP could be compared.

None of the individual subunits, sub-complexes, or γ complex bound ss DNA in the

absence of ATP. Figure 5-1 shows that none of the individual subunits and several of the

combinations of subunits in sub-complexes have ss DNA binding activity under these

conditions. Only a specific combination of subunits, the sub-complex $\gamma_3\delta\delta'$, showed

ATP-dependent ss DNA binding activity. The corresponding $\tau\delta\delta'$ sub-complex also

showed ss DNA binding activity.

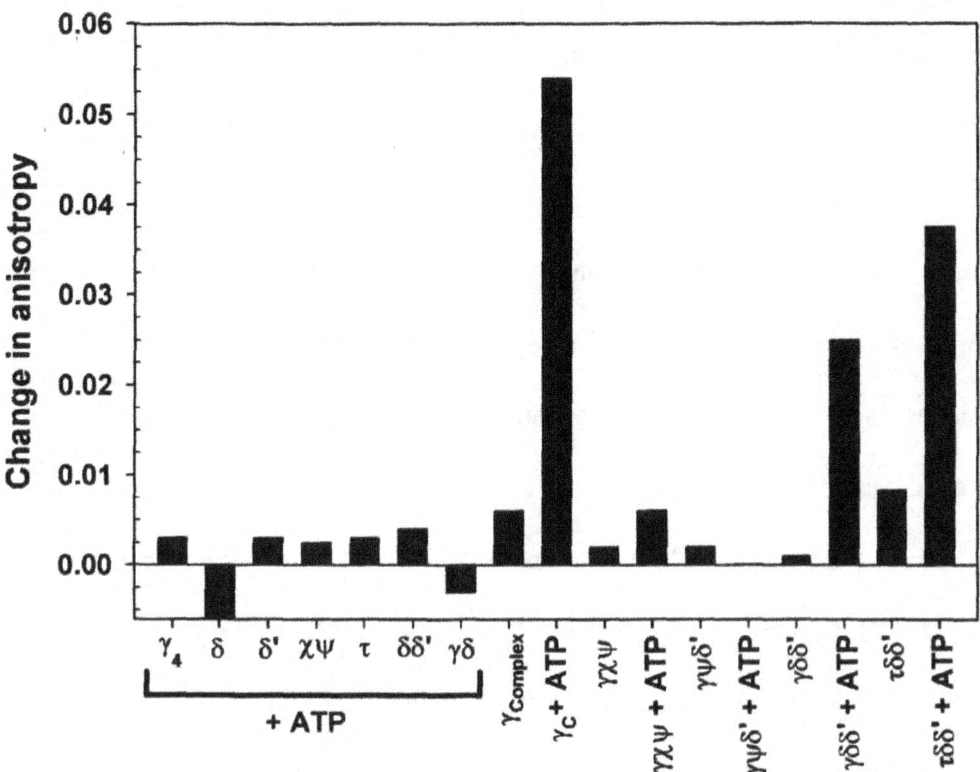

Figure 5-1. Change in steady-state anisotropy for RhX-ss DNA in the presence of individual subunits, sub-complexes, and γ complex with or without ATP. Changes in the steady-state anisotropy of a 50-nucleotide ss DNA labeled with RhX at the 5'-end is plotted against the value for a single data point from titration experiments representing saturated binding of the protein to Rhx-ss DNA (i.e., when there was binding). The γ_4, δ, δ',$\chi\psi$, τ_2, $\delta\delta'$, $\gamma\delta$ subunits were all in the presence of ATP, and γ complex or various sub-complexes were in the absence or presence of ATP as indicated. The steady-state polarized intensities from the RhX probe were measured upon the addition of a constant volume of a solution containing protein to a solution in the cuvette containing 50 nM RhX-ss DNA. A constant volume of ATP was subsequently added to the cuvette giving a final concentration of 0.5 mM, and the polarized intensities were again measured for calculation of anisotropy. All proteins were titrated over a range of 50 to 500 nM, and beyond 2 μM in some cases. The anisotropy value for free RhX-ss DNA (r_{free}) was subtracted from the anisotropy for RhX-ss DNA in the presence of protein (r_{bound}) to give the value of change in anisotropy plotted. Data from titration experiments for the γ_4, δ, δ',$\chi\psi$, τ_2, $\delta\delta'$, $\gamma\delta$ subunits were contributed by Xu Liu, Ph.D.

Analysis of the DNA Binding Activity of $\gamma_3\delta\delta'$ Minimal Complex in the Absence and Presence of β

To further characterize the $\gamma_3\delta\delta'$ "minimal complex", titration assays were performed with RhX-pt DNA in the presence and absence of β clamp. The change in steady-state anisotropy of RhX-pt DNA was examined upon the addition of the minimal complex to determine the minimal complex-pt DNA binding activity as well as the β clamp loading activity when β was present in the assay. The pt DNA substrate consisted of a 105-nucleotide template with a 30-nucleotide primer, annealed approximately in the center of the template, creating a 50-nucleotide-long ss DNA 5'-overhang of the template. The sequence of the 50-nucleotide overhang was identical to the sequence of the 50-mer ss DNA used in the above titration assays (see chapter 3, materials and methods for DNA sequences). It was previously shown that the 30/105-mer pt DNA substrate supports DNA synthesis by the polymerase III core in the presence of γ complex, β, and single-stranded DNA binding protein (SSB) (Hingorani et al., 1999). Increasing concentrations of the minimal complex (100 to 1200 nM) were added to 50 nM RhX-pt DNA in the presence and absence of 500 nM β in a solution containing 8 mM $MgCl_2$ and 0.5 mM ATP.

In the absence of β, the minimal complex appears to have low binding affinity for pt DNA, similar to γ complex (Figure 5-2). Even though the sequence of the 50-nucleotide ss DNA overhang of this pt DNA substrate is identical to the ss DNA 50-mer to which both the minimal complex and γ complex clearly bind in figure 5-1, they do not appear to have much affinity for it within the context of the pt DNA substrate. It was previously shown for γ complex that there was a transient interaction with pt DNA, and this interaction triggered rapid dissociation of γ complex leaving it in a form that was

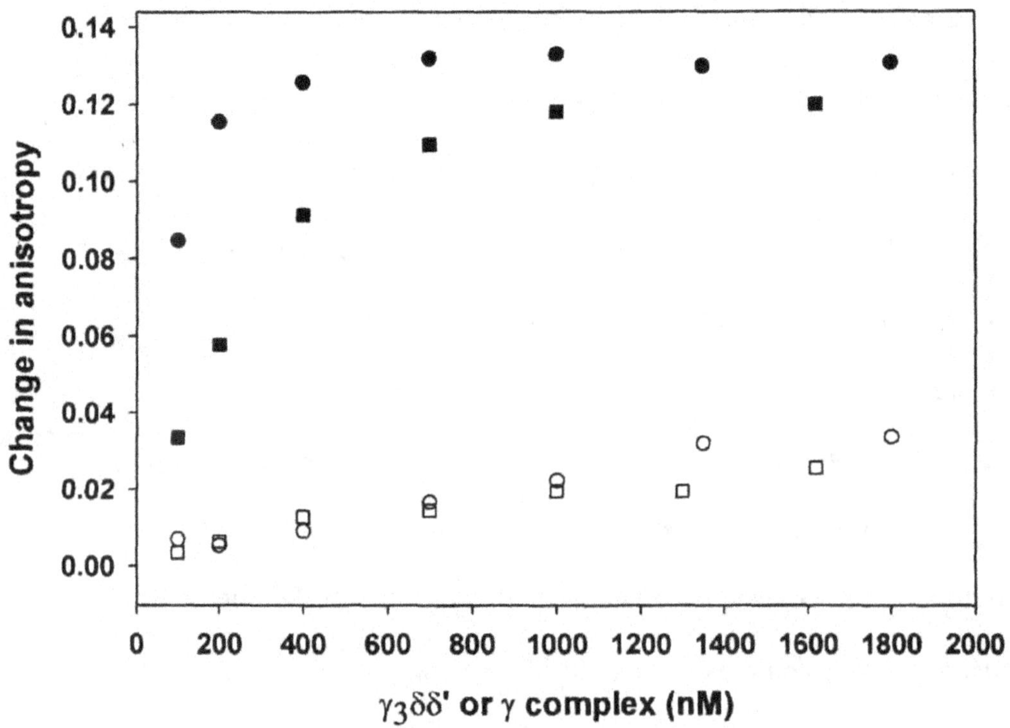

Figure 5-2. Steady-state binding of the minimal complex or γ complex with RhX-pt
DNA with and without β. The steady-state polarized emission intensities of
RhX-DNA were measured for calculation of anisotropy after solutions
containing increasing concentrations of either the minimal complex alone
(open squares) or γ complex alone (open circles) were added to a cuvette
containing 50 nM RhX-pt DNA, 8 mM $MgCl_2$, 20 mM Tris-HCl pH 7.5, 50
mM NaCl, 5 mM DTT, 40μg/mL BSA and 0.5 mM ATP. Titration assays
were also performed for the minimal complex or γ complex in the presence of
2.5 μM β (filled squares or filled circles, respectively). The concentration of
minimal complex or γ complex in the absence or presence of β is plotted
against the change in anisotropy from separate titration assays. The
anisotropy value for free RhX-pt DNA (r_{free}) was subtracted from the
anisotropy for RhX-pt DNA in the presence of protein (r_{bound}) to give the
value of change in anisotropy plotted.

modeled to undergo a slow step to regain affinity for DNA, resulting in the apparent low

affinity for DNA in the steady-state (Ason et al., 2000). The comparable result here, with

the minimal complex, indicates that the minimal complex undergoes a similar slow step

to regain affinity for DNA.

For assays performed in the presence of β, the minimal complex gained high affinity for pt DNA, and loaded β (Figure 5-2). When $\gamma_3\delta\delta'$ was originally reconstituted and purified by others (Onrust et al., 1991), it was shown that like γ complex, $\gamma_3\delta\delta'$ was capable of loading β and conferring processive DNA synthesis in assays with polymerase III core, β, and SSB. The minimal complex with β showed a slightly lower affinity for pt DNA than γ complex, however, both appeared to saturate the pt DNA above a concentration of 1 μM.

Comparison of β Clamp Binding Affinity of the Minimal Complex and γ Complex

Equilibrium β^{pyrene} Binding Activity of the Minimal Complex and γ Complex

After determining that the steady-state DNA binding activity of the minimal complex is similar to that of γ complex, the equilibrium binding of γ complex and the minimal complex to β clamp was examined in solution. β^{pyrene} binding assays were performed by measuring the steady-state polarized emission intensities of the pyrene probe covalently attached to β for calculation of anisotropy as a function of minimal complex or γ complex concentration (Figure 5-3). Like that for the RhX-labeled DNA substrates above, the observed steady-state anisotropy of the pyrene-probe on β is a population weighted average of the combined anisotropies for free-β and protein-bound β in solution. Therefore, the observed anisotropy increases as the population of protein-bound β increases.

Results show that minimal complex and γ complex had essentially identical affinity for β^{pyrene} both in the presence or absence of ATP. Dissociation constants for this binding interaction (K_d) were calculated based on a reversible second-order kinetic model: $A+B \leftrightarrows AB$, where A is the ATP-bound clamp loader, and B is β^{pyrene}.

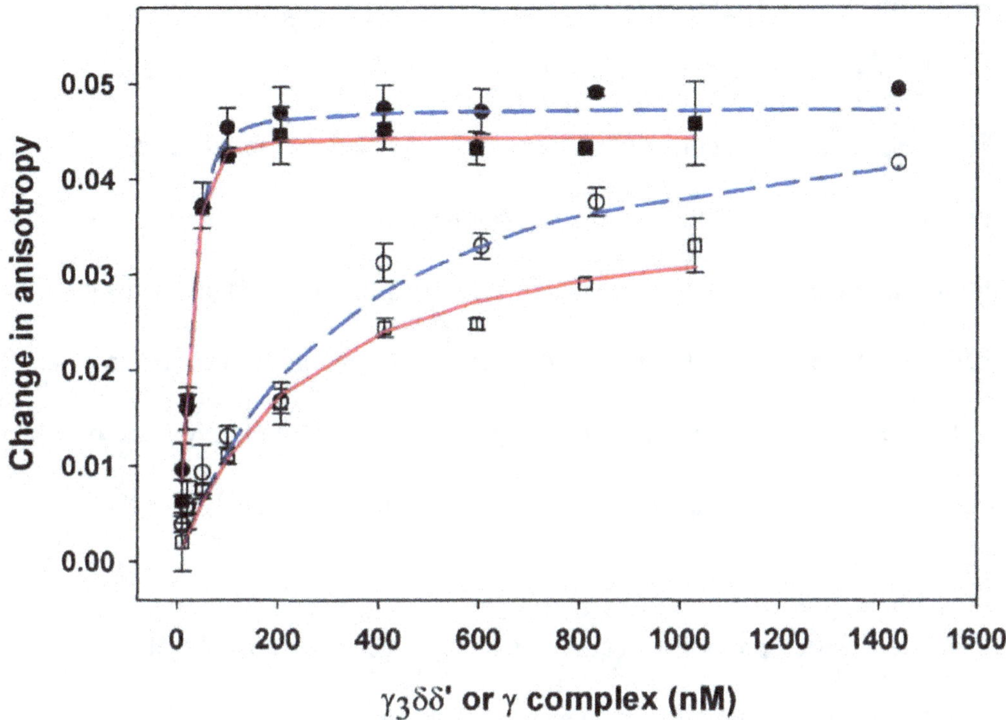

Figure 5-3. Steady-state anisotropy binding activity of γ complex or the minimal complex with β^{pyrene} in the presence or absence of ATP. The steady-state polarized emission intensities of β^{pyrene} were measured and used to calculate anisotropy for solutions containing increasing concentrations of either the minimal complex in the presence of ATP (closed squares) or absence of ATP (open squares) or γ complex in the presence of ATP (closed circles), or absence of ATP (open circles). Solutions containing the minimal complex or γ complex were added to a cuvette containing 50 nM β^{pyrene}, 8 mM MgCl$_2$, 20 mM Tris-HCl pH 7.5, 50 mM NaCl, 5 mM DTT, and 40 µg/mL BSA. The anisotropy value for free β^{pyrene} (r$_{free}$) was subtracted from the anisotropy for β^{pyrene} in the presence of different concentrations of γ complex or minimal complex (r$_{bound}$) to give the value of change in anisotropy plotted. Mean and standard deviation values are plotted for two independent measurements. Nonlinear fits to the quadratic equation (see text) are plotted for minimal complex (red, solid) and γ complex (blue, dashed).

The empirical solution for the quantity AB was fitted to the experimental data using the following quadratic equation in terms of the change in anisotropy of β^{pyrene} for estimation of the K$_d$.

$$[AB] = \frac{(K_d + [A] + [B]) - \sqrt{(K_d + [A] + [B])^2 - 4[A][B]}}{2[B]} * (r_{bound} - r_{free}) + r_{free}$$

In the presence of ATP the minimal complex had a K_d of 3 nM, and γ complex had a K_d of 4 nM for β^{pyrene}. Both the minimal complex and γ complex remained stably bound to β^{pyrene} for a minimum of 10 minutes (not shown). Surprisingly, both the minimal complex and γ complex bound β^{pyrene} in the absence of ATP with dissociation constants of 200 and 240 nM, respectively. Even though the dissociation constants for binding in the absence of ATP were ~ 50-fold higher than the K_d values measured in the presence of ATP, they still reflect significant binding activity for the clamp loader with β^{pyrene} in the absence of ATP.

A K_d of 3 nM was reported previously, (Naktinis et al., 1995) for γ complex and β in the presence of ATP using surface plasmon resonance (SPR) methods. The investigators also reported a K_d for interaction of γ complex and β in the absence of ATP that was 1000-fold higher in the absence of ATP. The dissociation constants measured using the steady-state anisotropy binding titration assays here, with β^{pyrene}, were estimated in solution equilibrium conditions (i.e., γ complex was immobilized using the SPR method) and the first measured for the $\gamma_3\delta\delta'$ minimal complex.

Apparent Dissociation Constant for ATP Binding the Minimal Complex or γ Complex

The dissociation constants for ATP binding the minimal complex or γ complex were determined indirectly as a function of ATP-dependent clamp loader binding to β^{pyrene}. ATP is required by the clamp loader to bind the β clamp with high affinity, and therefore the steady-state anisotropy assay was used to measure the minimal complex or γ complex binding activity with β^{pyrene} as a function of ATP concentration.

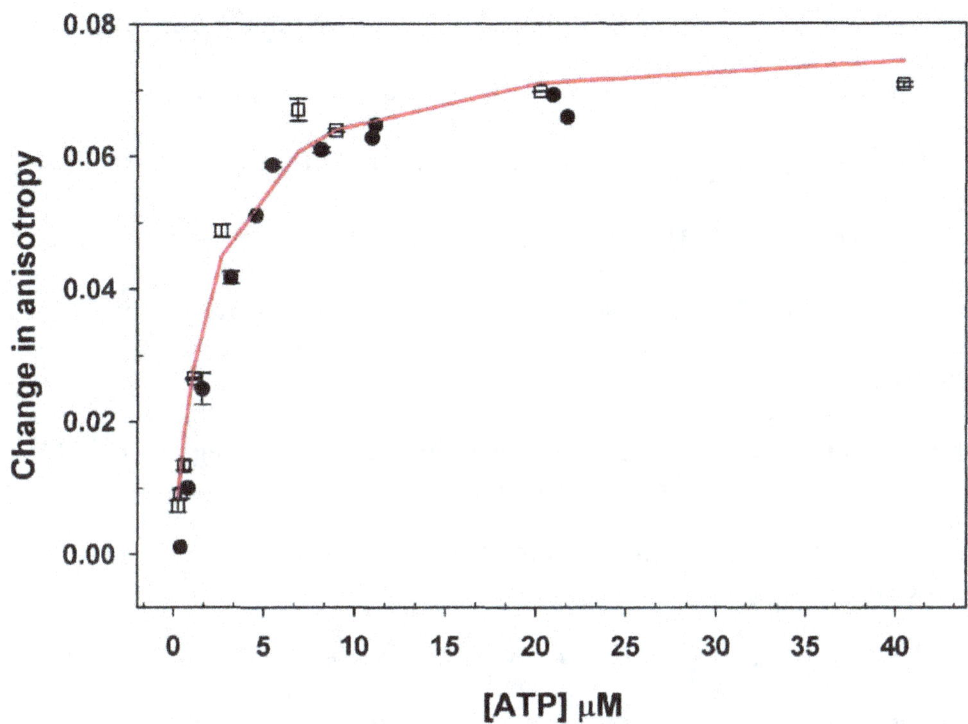

Figure 5-4. Minimal complex or γ complex binding β^{pyrene} as a function of ATP concentration. The steady-state polarized emission intensities of β^{pyrene} with the minimal complex or γ complex were measured and used to calculate anisotropy following addition of a solution of a constant volume containing increasing concentrations of ATP in separate assays. Concentrations were: β^{pyrene} (65 nM), minimal complex or γ complex, (400 nM) in 8 mM MgCl$_2$, 20 mM Tris-HCl pH 7.5, 50 mM NaCl, 5 mM DTT, and 40 µg/mL BSA. The change in anisotropy of β^{pyrene} in the presence of minimal complex (squares) or γ complex (circles) is plotted against the concentration of ATP. The anisotropy value for free β^{pyrene} (r$_{free}$) was subtracted from the anisotropy value for β^{pyrene} with minimal complex or γ complex (r$_{bound}$) in the presence of varied concentrations of ATP to give the value of change in anisotropy plotted. Mean and standard deviation values are plotted for two independent measurements. A nonlinear fit to the quadratic equation (see text) (red curve) is plotted for minimal complex only for clarity.

Since this assay only measures the binding of clamp loader to β^{pyrene}, and not ATP binding directly to the clamp loader, it is an indirect measure of the affinity of ATP for the clamp loader. The K$_d$^{apparent} for ATP binding was estimated using the reversible second-order kinetic model: A + B ⇆ AB (described above), where A is the concentration of ATP-bound clamp loader at each concentration of ATP and B is β^{pyrene}.

The results showed that the minimal complex bound ATP with essentially identical

affinity as γ complex ($K_d^{apparent}$ = 2-3 μM). This value is consistent with previous

measures of ATP binding to the polymerase III holoenzyme (~2 μM) in UV crosslinking

assays (Biswas and Kornberg, 1984), assays with purified γ subunit (~2.1 μM)

(Tsuchihashi and Kornberg, 1989), and with γ complex (~2 μM) (Hingorani and

O'Donnell, 1998).

Kinetics of ATP Hydrolysis by the Minimal Complex Measured Using the MDCC-PBP ATPase Assay

Steady-State ATP Hydrolysis Kinetics of $\gamma_3\delta\delta'$ Minimal Complex in the Absence and Presence of β Clamp

It was previously determined for the minimal complex, along with γ complex, that

β increases the steady-state rate of DNA dependent ATP hydrolysis (Onrust et al., 1991).

To investigate the DNA-dependent ATP hydrolysis activity of the minimal complex

further, measurement of the initial velocities of ATP hydrolysis activity was performed in

steady-state conditions with the MDCC-PBP ATPase assay.

Steady-state reactions were initiated by addition of a range of ATP concentrations

approximately 10-fold less than or greater than an apparent dissociation constant for ATP

binding (~5 μM), estimated from steady-state clamp loading assays as a function of ATP

concentration described in chapter 4 (Figure 4-2). Initial reaction velocities were

measured in three separate assays each for the minimal complex with pt DNA only and

the minimal complex with pt DNA and β. Initial velocities were averaged and plotted

against ATP concentration and the resulting rectangular hyperbolic relationship was fitted

with the Michaelis-Menten equation: (where v is the initial velocity, and [S] is ATP

concentration),

$$v = \frac{V_{max}[S]}{K_m + [S]}$$

describing a system where the observed rate depends on the concentration of the enzyme-substrate complex. The resulting steady-state kinetic parameters for ATP hydrolysis are presented in Table 5-1.

Table 5-1. Steady-state ATP hydrolysis kinetic parameters for the minimal complex and γ complex in the absence and presence of β[a]

	γ complex		$\gamma_3\delta\delta'$ minimal complex	
	No β	+ β	No β	+ β
V_{max} (μMs^{-1})	0.02 (0.003)[b]	0.07 (0.02)	0.04 (0.01)	0.08 (0.01)
$k_{cat}^{complex}$ (s^{-1})[c]	0.6 (0.07)	2.0 (0.4)	0.4 (0.13)	0.8 (0.14)
K_m (μM)	6.0 (3.0)	11 (6.7)	9.0 (3.0)	6.0 (4.0)
$k_{cat}^{complex} / K_m$ (μM^{-1}s^{-1})	0.1 (0.04)	0.2 (0.05)	0.05 (0.02)	0.2 (0.09)

[a] Steady-state MDCC-PBP ATPase assays were performed by addition of ATP ranging in concentrations from 0.6 to 77 μM to a solution of 41 nM γ complex with 50 nM ptDNA and 50 nM β (when present), or 100 nM $\gamma_3\delta\delta'$ minimal complex with 100 nM ptDNA and 100 nM β (when present). All assays contained 2.4 μM MDCC-PBP and P$_i$-mopped assay buffer containing 20 mM Tris-HCl, 50 mM NaCl, 5 mM DTT, 40 μg/mL BSA, and 8 mM MgCl$_2$.
[b] Standard deviations were calculated from three separate steady-state MDCC-PBP ATPase assays.
[c] Since the clamp loader has three subunits capable of hydrolyzing ATP, the turnover number ($k_{cat}^{complex}$) reflects the combined activity of the complex.

The steady-state rate of ATP hydrolysis ($k_{cat}^{complex}$) by the minimal complex increased 2-fold. For γ complex, β increased the rate of ATP hydrolysis activity approximately 3-fold. Changes in the K_m values in the presence or absence of β were not clear in this analysis. Previously a turnover rate of 1.8 s^{-1} and a K_m of ~22 μM for γ complex in the presence of β as reported (Hingorani et al., 1999). The K_m values derived from this analysis with the minimal complex and γ complex are all higher than the apparent K_d of 2 μM determined for ATP with the β^{pyrene} anisotropy binding assay, suggesting that the K_m value is not a simple binding affinity constant, but some other combination of rates. The enhancement of the turnover rate ($k_{cat}^{complex}$) for both the minimal complex and γ complex by β indicates the possibility that β may partition a more active form of each, essentially increasing the clamp loaders' concentrations, towards the clamp loading pathway which requires ATP hydrolysis.

The specificity constant for ATP ($k_{cat}^{complex}/K_m$) can be an apparent bimolecular binding rate constant. This value increased 4-fold for the minimal complex and 2-fold for γ complex when β was present. Interestingly, these $k_{cat}^{complex}/K_m$ values are all roughly 1000-fold lower than would expected for diffusion-controlled bimolecular binding of a small nucleotide cofactor such as ATP. This result indicates that rapid ATP binding was followed by a slow step leading to hydrolysis for both the minimal complex and γ complex. Overall, the results of the steady-state analysis show that in the absence or presence of β the minimal complex hydrolyzes ATP with slightly lower activity than γ complex.

The absence of sigmoidal curves in the analysis of initial velocity of ATP hydrolysis activity versus ATP concentration for the minimal complex or γ complex with pt DNA only, argues against β being an allosteric effector. The mechanisms of positive cooperativity predict that without its allosteric effector, an enzyme generally displays sigmoidal initial velocity kinetics (Fersht, 1999). Whether or not β was present, all plots of initial velocity against ATP concentration from the steady-state MDCC-PBP ATPase assays were rectangular hyperbolas that fit to the Michaelis-Menten equation, not sigmoidal.

Pre-Steady-State Kinetics of ATP Hydrolysis by $\gamma_3\delta\delta'$ Minimal Complex in the Absence and Presence of β Clamp

Pre-steady-state analysis was performed to characterize and determine if the DNA-dependent ATP hydrolysis activity of the minimal complex required to load β on pt DNA differs from the activity of γ complex. Does the minimal complex hydrolyze ATP in a different clamp loading reaction mechanism resulting in the small differences observed in the steady-state ATP hydrolysis parameters? The χ and ψ subunits of γ complex are not present in the minimal clamp loader complex. Do the χ and ψ subunits change the way the clamp loader hydrolyzes ATP in the clamp loading reaction?

For the pre-steady-state MDCC-PBP ATPase assay, a sequential-mixing "three-syringe" stopped-flow experiment was performed. One stopped-flow syringe contained a solution of $\gamma_3\delta\delta'$ minimal complex and β (when present), and a second syringe contained ATP. The contents of these syringes were mixed and preincubated for 1 s. After preincubation, $\gamma_3\delta\delta'$, ATP, and β (when present) were rapidly mixed with a solution from a third syringe containing pt DNA and MDCC-PBP. Mixing with the third syringe triggered the detection system allowing real-time measurement of the fluorescence change due to inorganic phosphate binding MDCC-PBP upon pt DNA stimulated ATP hydrolysis activity of the minimal complex. The DNA substrate used in all experiments was the 30-nucleotide-primed/105-nucleotide template DNA described above. The 1-s period of preincubation time was sufficient for equilibration of a complex of $\gamma_3\delta\delta'$ and ATP, or a complex of $\gamma_3\delta\delta'$, ATP and β (see simulations of equilibration in the appendix). The preincubation time was also increased up to 2 s, and there was no further increase in ATP hydrolysis activity observed in the first turnover demonstrating that a 1-s period of preincubation is sufficient for formation of the $\gamma_3\delta\delta'$-, ATP, +/- β complex.

The results of pre-steady-state MDCC-PBP ATPase assays for the minimal

complex in the absence or presence of β compared to identical assays for γ complex are

shown in Figure 5-5 A and B. The reaction time course for ATP hydrolysis for the

minimal complex in the presence of β showed a slight lag, a single rapid phase of ~ 100

ms, another lag phase of ~ 150-200 ms constituting the first turnover. The single rapid

phase in this assay is essentially an enzyme active-site titration depicting the amount of

active minimal complex present in the assay. Although the concentration of γ complex or

the minimal complex used in these assays calculated to be the same, the activity in the

first turnover was lower for the minimal complex than for γ complex. From the single

pre-steady-state rapid phase amplitude, an estimated ~1.5 molecules of ATP were

hydrolyzed. This result suggested that the concentration of "live" minimal complex was

roughly half that calculated to be present in the assay. This was not always the case

during the course of this study. In earlier assays the amplitudes of the rapid phase were

in good agreement with the γ complex activity. Therefore, this difference may be due to

error in calculation of clamp loader concentrations, or the minimal complex may be

unstable in storage, and consequentially lose some activity. The instability of the

minimal complex has been noted previously (Olson et al., 1995). However, the

mechanistic features of the pre-steady-state ATP hydrolysis reaction were preserved in

relation to γ complex, but the pause at the end of the first turnover was less defined for

the minimal complex.

In the assay for the minimal complex without β (figure 5-5A, blue trace), there was

a small lag (~ 20-30 ms), followed by two kinetic phases. Fitting the data to a double

exponential expression indicated that there were either two pre-steady-state phases

Figure 5-5. Kinetics of ATP hydrolysis by the minimal complex or γ complex in the presence and absence of β. A) ATP hydrolysis kinetics for the minimal complex in the presence (black) and absence of β (blue). B) ATP hydrolysis kinetics for γ complex in the presence (black) and absence of β (blue). Using a sequential-mix "three-syringe" stopped-flow assay, one syringe loaded with a solution containing the minimal complex or γ complex and β (when present) was mixed with the contents of a second syringe loaded with ATP. The resulting mixture was preincubated for a period of 1 s, and then mixed with the contents of a third syringe loaded with a solution-containing pt DNA and MDCC-PBP. All solutions were prepared in P_i-mopped assay buffer: 20 mM Tris-HCl, 50 mM NaCl, 5 mM DTT, 40 μg/mL BSA, and 8 mM $MgCl_2$. Final reactant concentrations were 270 nM minimal complex or γ complex, 1.0 μM β (when present), 100 μM ATP, 1.0 μM pt DNA, and 2.7 μM MDCC-PBP. The flat trace (gray) at the bottom of each figure showing no change was a negative control assay performed in the absence of ATP. Raw fluorescence data were transformed into the concentration of P_i-bound MDCC-PBP using the equation, $X_b(t) = [I_{obs}(t) - I_f] / [I_b - I_f]$ to solve for the fraction of P_i-bound MDCC-PBP $X_b(t)$, then multiplying this value by the concentration of MDCC-PBP present in the assay to get the value MDCC-PBP-P_i (μM) plotted. I_f was the fluorescence intensity of P_i-free MDCC-PBP in assay buffer, and I_b was the saturated fluorescence intensity of completely P_i-bound MDCC-PBP in the presence of 200 μM potassium phosphate (P_i source).

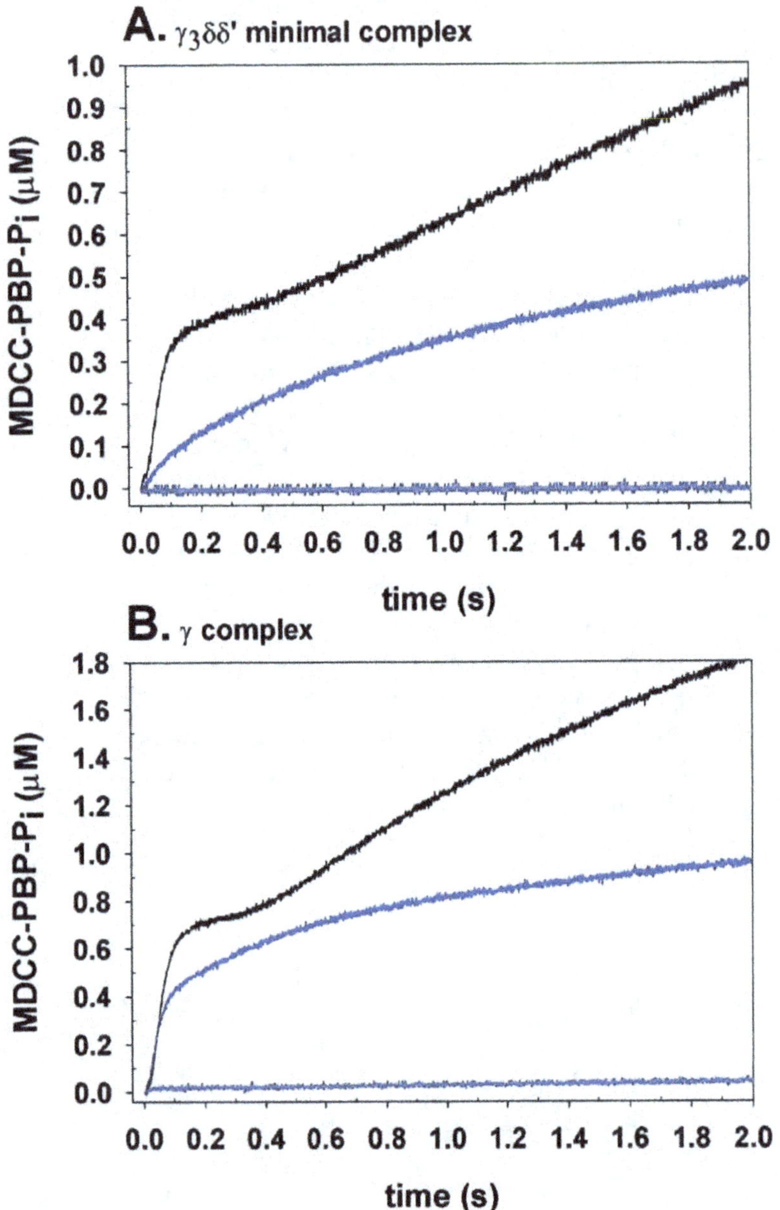

preceding transition to the steady-state, or that there was only a single rapid pre-steady-state phase that transitioned into steady-state activity since the end of a putative second pre-steady-state phase was not well defined both visually or mathematically. In either case, the amplitude of the small pre-steady-state rapid phase was ~ 25% of the rapid phase activity observed for the reaction in the presence of β (i.e., 25% of total active clamp loader). For γ complex in the absence of β (Figure 5-5B, blue trace), pre-steady-state biphasic kinetics are clearly visible with the amplitude of the initial rapid phase approximately 40-50 % of the first turnover amplitude in the assay with β (Figure 5-5B, black trace).

Kinetics of ATP Hydrolysis when the Minimal Complex is not equilibrated with ATP

While investigating the dependence of the pre-steady-state ATP hydrolysis kinetics of γ complex on the preincubation period with ATP, assays were performed in which there was no preincubation period (Figure 4-8). In that study, the possibility that ATP binding to γ complex was limiting the hydrolysis kinetics was addressed. The results showed that the rate of ATP binding could be contributing to the slow conformational change kinetics in the clamp loader. By performing the assay with different concentrations of ATP, it was determined that ATP concentrations above 200 μM ($\sim$100-fold $K_d^{apparent}$) were high enough such that ATP binding was not limiting the rate of hydrolysis, but slow conformational changes may have been limiting.

For comparison of the ATP hydrolysis kinetics of the minimal complex to the kinetics by γ complex when neither is equilibrated with ATP, pre-steady-state MDCC-PBP ATPase assays excluding equilibration with ATP were performed. Figure 5-6 shows the resulting kinetics of ATP hydrolysis by the minimal complex compared to γ complex with 500 μM ATP. Using a single-mix setup, a solution of the minimal complex or γ complex was added directly to a solution containing ATP, DNA, and MDCC-PBP in the stopped-flow. An extended lag phase preceded biphasic ATP hydrolysis activity. The lag phase displayed by the minimal complex activity was approximately 50–60 ms, at least twice the duration observed when the minimal complex was equilibrated with ATP for 1-s in the previous assays (Figure 5-5A, blue trace).

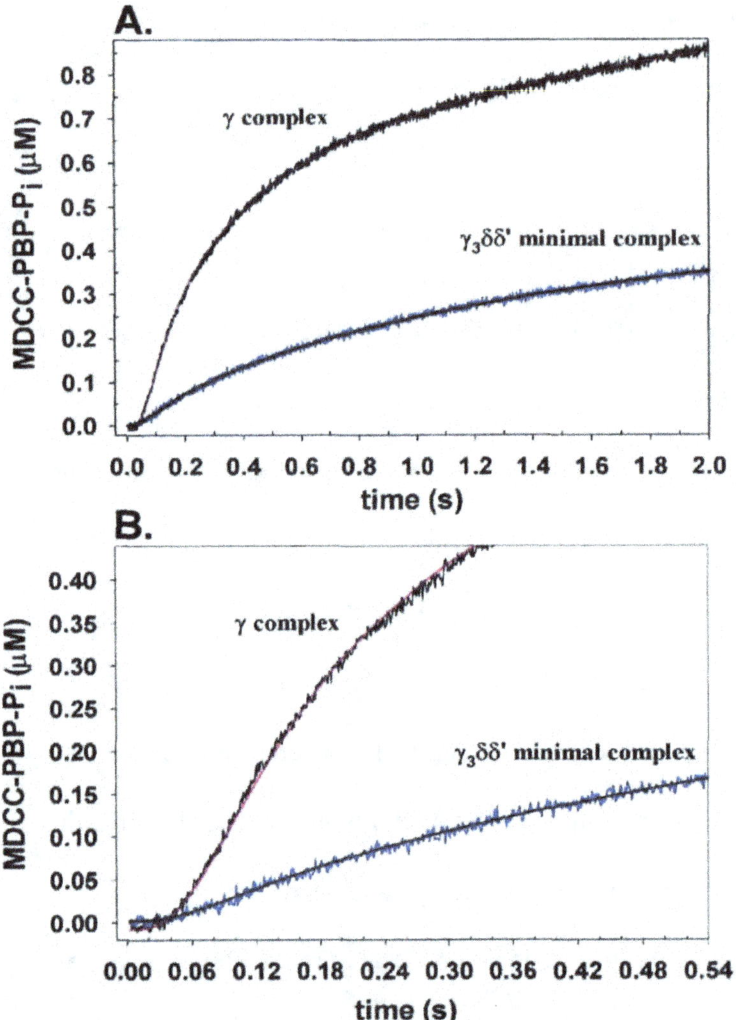

Figure 5-6. Kinetics of ATP hydrolysis of the minimal complex or γ complex directly mixed with pt DNA and ATP. A) In the stopped-flow, a simple "two-syringe" assay was performed where a solution from one syringe loaded with the minimal complex (blue trace) or γ complex (black trace) was mixed with the contents of a second syringe loaded with pt DNA, ATP, and MDCC-PBP. B) MDCC-PBP-P_i is plotted against time on an expanded scale for better observation of the lag phase for γ complex (black) and the minimal complex (blue). Solid lines through the data (red, γ complex and black, $\gamma_3\delta\delta'$ minimal complex) are fits using the model described in chapter 4, figure 4-10 with DynaFit (Kuzmic, 1996). All solutions were prepared in P_i-mopped assay buffer: 20 mM Tris-HCl, 50 mM NaCl, 5 mM DTT, 40 μg/mL BSA, and 8 mM $MgCl_2$. Final reactant concentrations were 270 nM minimal complex or γ complex, 500 μM ATP, 1.0 μM pt DNA, and 2.7 μM MDCC-PBP. Raw fluorescence data were transformed into the concentration of P_i-bound MDCC-PBP using the equation shown in figure 5-5.

Notably, biphasic ATP hydrolysis kinetics following the initial lag persisted even without equilibration of the minimal complex with ATP. This was a key result in the investigation of γ complex in similar assays. Raw fluorescence data for the minimal complex reaction were fitted with a double exponential expression, excluding the long lag phase, revealing rates that were roughly 2 s^{-1} for the first phase and 0.2 s^{-1} for the second phase. The minimal complex data were also fitted using the computer model presented in chapter 4 (Figure 4-10) (here, figure-5-11) that includes the PTTT $\rightarrow$ ETTT step prior to the ATP equilibration steps. The resulting fit to that model is shown in Figure 5-6 (solid black trace). In modeling the minimal complex data, the lowered concentration due to "dead" enzyme was accounted for, and the forward and reverse rates between the "activated" ATTT and "inactivated" ETTT states were determined from fitting experimental data from a 1-s equilibration with ATP (figure 5-5A, blue) as well as the rates describing the ETTT $\leftrightarrows$ ITTT transition (appendix). The results of the experimental analysis and fitting suggest that the minimal complex has a slower forward conformational change towards the ATTT species in the model. The model estimate of this forward rate was 2.6 s^{-1}, consistent with the empirical fit of the raw fluorescence data (2 s^{-1}). This estimated rate was about 2-fold slower than the estimated rate for the formation of "activated" γ complex. Qualitatively, the comparison indicates that the conformational changes giving the clamp loaders affinity for DNA were much slower for the minimal complex than γ complex. It is possible that without the χ and ψ subunits, the minimal complex has more inherent conformational instability than γ complex (Olson et al., 1995), and causes a slower transition towards an "activated" complex. In the future this assay should be repeated with different concentrations of ATP to study the rate of ATP binding to the minimal complex.

ATPγS-Chase of Pre-Steady-State ATP Hydrolysis Activity by the $\gamma_3\delta\delta'$ Minimal Complex

An analogue of ATP, adenosine 5'-O-(3-thiotriphosphate) (ATPγS) was used in MDCC-PBP ATPase assays to qualitatively study the nature of nucleotide binding to the minimal complex and γ complex. ATPγS is essentially non-hydrolyzable by γ complex (k_{cat} 1 X 10^{-4} at 37 °C) (Hingorani and O'Donnell, 1998). However, it does bind γ complex with a similar K_d, cause conformational change, and allow binding of β (Bertram et al., 1998). To determine if the ATPs bound to the minimal complex can readily exchange with ATPγS, pre-steady-state MDCC-PBP ATPase assays were performed where the hydrolysis of ATP bound to the clamp loader was chased upon the addition of excess ATPγS. This ATPγS-chase assay was performed in the presence and absence of β to determine if β could in fact bind to and "trap" the minimal complex in an active conformational state, "locking in" nucleotides.

To chase the ATP hydrolysis activity with ATPγS, the sequential-mix "three-syringe" stopped-flow setup was used. A syringe containing a solution of the minimal complex and β (when present) was mixed with a second syringe containing a solution of ATP. This mixture was preincubated for 1 s, and then mixed with the contents of the third syringe loaded with pt DNA, MDCC-PBP and ATPγS (i.e. at a 10-fold higher concentration than ATP). Figure 5-7 shows the results of ATPγS-chase assays in the presence and absence of β for the minimal complex. For the chase in the presence of β (Figure 5-7A, red trace), a single rapid phase of ATP hydrolysis occurred which ended at an amplitude approximately equal to the original MDCC-PBP ATPase assay.

Figure 5-7. Pre-steady-state kinetics of ATP hydrolysis by the minimal complex when chased with non-hydrolyzable ATPγS. ATPγS-chase assays were performed A) for the minimal complex in the presence of β, and B) in the absence of β. ATPγS-chase assays (+ ATPγS, red) show that ATP originally bound to the minimal complex was hydrolyzed faster than it dissociates. Using a sequential-mix "three-syringe" stopped-flow assay, one syringe loaded with a solution containing the minimal complex and β (when present) was mixed with the contents of a second syringe loaded with ATP. The resulting mixture was preincubated for a period of 1 s, and then mixed with the contents of a third syringe containing a solution of pt DNA, MDCC-PBP and ATPγS. All solutions were prepared in P_i-mopped assay buffer: 20 mM Tris-HCl, 50 mM NaCl, 5 mM DTT, 40 μg/mL BSA, and 8 mM $MgCl_2$. Final reactant concentrations were 270 nM minimal complex, 1.0 μM β (when present), 100 μM ATP, 1.0 μM pt DNA, 2.7 μM MDCC-PBP, and 1 mM ATPγS. The flat trace (gray) at the bottom of each figure showing no change is a negative control assay performed in the absence of ATP. Raw fluorescence data were transformed into the concentration of P_i-bound MDCC-PBP using the equation shown in figure 5-5.

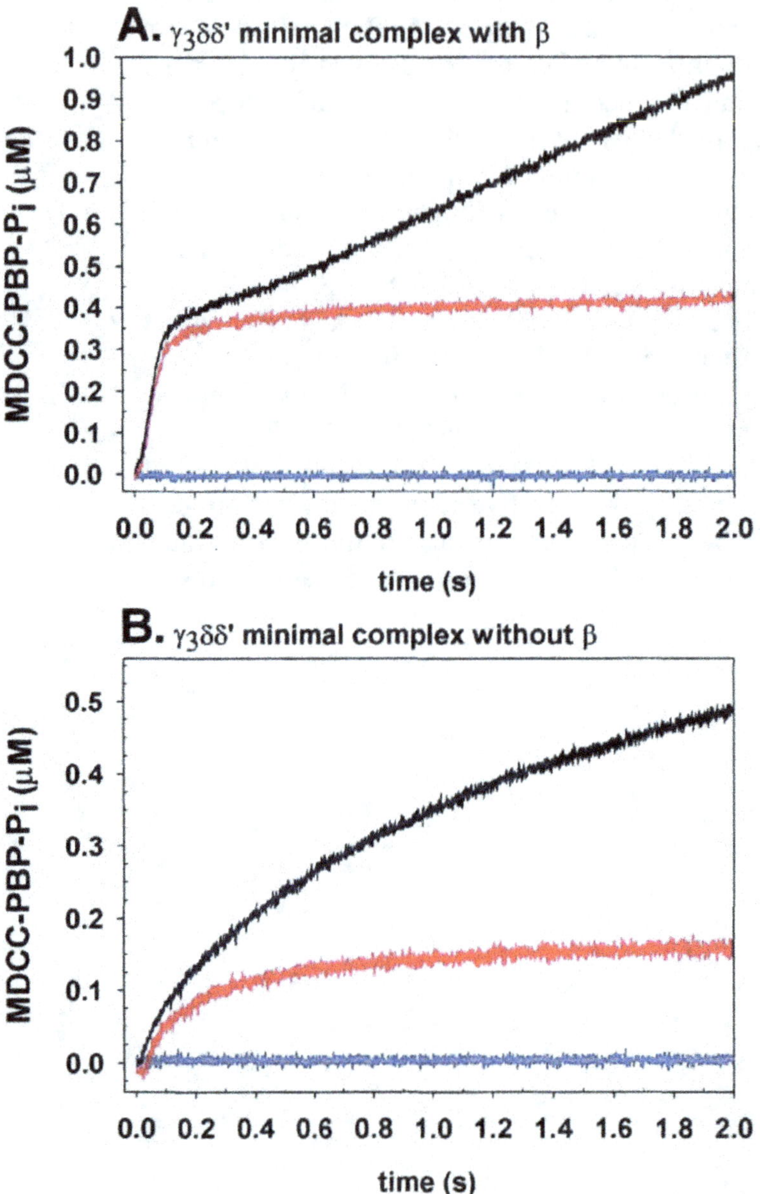

This is essentially an isolation of the first turnover of ATP hydrolysis in the clamp loading reaction, and there was no steady-state ATP hydrolysis activity. Increasing the preincubation period to 2 s in this assay resulted in no difference in activity (not shown). These results are similar to those observed for γ complex in the presence of β (Figure 4-5A), and indicate that all ATP molecules bound to the minimal complex active conformation are "locked in", and hydrolyzed faster than they dissociate.

In the chase assay without β (Figure 5-7B, red trace), the addition ATPγS reduced the amount of ATP hydrolyzed by the minimal complex to a short biphasic increase in ATP hydrolysis preceded by a short lag. The duration of the lag was similar to that observed in the original MDCC-PBP ATPase assay (Figure 5-7B, black trace). The total level of amplitude of these two pre-steady-state phases was roughly equal to the amplitude of the small initial rapid phase observed in the original assay. This result shows that ATP hydrolyzed in these two pre-steady-state phases by the minimal complex was hydrolyzed more rapidly than it dissociates, and that the ATP hydrolyzed in the original assay slow phase, or steady-state reaction, readily exchanged with ATPγS.

ATPγS-Chase of Steady-State ATP Hydrolysis Activity by the $\gamma_3\delta\delta'$ Minimal Complex or γ Complex

The effect of addition of ATPγS in the steady-state MDCC-PBP ATPase assay for the minimal complex and γ complex was studied along side the pre-steady-state ATPγS-chase assays. In the steady-state clamp loading cycle, there must be a nucleotide exchange step following hydrolysis of ATP. The ATPγS-chase of steady-state ATP hydrolysis activity was measured in assays containing the minimal complex or γ complex and pt DNA in the presence and absence of β using concentrations of 50 nM minimal complex or γ complex, 150 nM pt DNA, and 150 nM β (when present). ATP hydrolysis activity was initiated by addition of a solution of ATP giving a final concentration of 77 μM. After approximately 10-s of ATP hydrolysis activity, solutions containing varying concentrations of ATPγS were added in separate reactions providing a range of ATPγS to ATP ratios (Scheme 5-1). The ATPγS:ATP ratios for assays with β were, 0.25:1, 0.5:1, 1:1, 2:1, and 10:1. The ATPγS:ATP ratios for assays without β were 0.25:1, 1:1, 2:1, and

10:1. The assays where a 10:1 ATPγS:ATP ratio was present mirrored the ratio of ATPγS:ATP used in the pre-steady-state ATPγS-chase assays (Table 5-2).

In the steady-state assay with β, there ATP hydrolysis activity was completely poisoned after addition of ATPγS at all ATPγS:ATP ratios tested with both the minimal complex and γ complex.

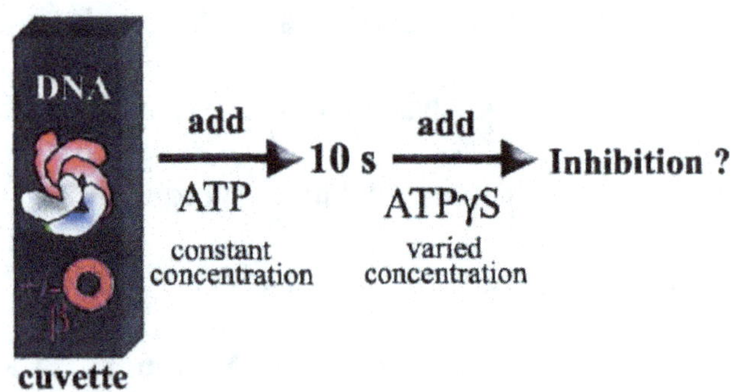

Scheme 5-1. Steady-state ATPγS-chase assay.

Table 5-2. Steady-state ATPγS-chase assay results: comparison of the minimal complex and γ complex in the presence or absence of β

ATPγS:ATP	minimal complex + β	minimal complex − β	γ complex + β	γ complex − β
0.25:1	−[a]	n.[b]	−	+
0.5:1	−	n.	−	n.
1:1	−	+[c]	−	+
2:1	−	−	−	+
10:1[d]	−	−	−	−

[a] Inhibition of ATP hydrolysis activity
[b] Assay not performed at this ratio
[c] ATP hydrolysis activity persists
[d] Pre-steady-state "mock-ratio"

When ATPγS was added to steady-state MDCC-PBP ATPase assays in the absence of β, a different effect was observed. Over the range of ATPγS:ATP ratios from 0.25:1 to 2:1, ATP hydrolysis activity persisted for γ complex, although a ratio of 2:1 was enough to inhibit ATP hydrolysis activity of the minimal complex. At a ratio of 10:1, which mimics the pre-steady-state ATPγS-chase condition, all ATP hydrolysis activity was abrogated for γ complex, and of course, the minimal complex as well. It appears that the

pt DNA-stimulated hydrolysis of ATP continued until there was a large enough excess of

ATPγS to compete with ATP for all of the nucleotide binding sites in the clamp loader to

stop the reaction. These steady-state MDCC-PBP ATPase assays chased with ATPγS

also show that the 10-fold excess ATPγS used in pre-steady-state ATPγS-chase assays

was sufficient to completely inhibit the steady-state ATP hydrolysis activity of both

clamp loaders.

Clamp Loading Activity of the Minimal Complex is More Sensitive to ADP than γ Complex.

A significant result of the ATPγS-chase analyses was the difference in sensitivity

for nucleotide binding of the minimal complex compared to γ complex. To determine in

more detail the nature of nucleotide binding to the minimal complex and γ complex, the

effect of ADP on the steady-state clamp loading reaction was studied using fluorescence-

based anisotropy binding assays to measure clamp loading with RhX-labeled pt DNA.

Previously it was shown that ADP causes a change in conformation of γ complex similar

to that caused by ATP in proteolytic digest assays (Hingorani and O'Donnell, 1998).

This suggested that ADP could stably bind to the clamp loader. However, the

conformation due to ADP binding must be different from that due to ATP, because ADP

cannot promote the clamp loading reaction.

To determine if the ATP-bound clamp loader-β clamp complex remained committed to

clamp loading in the presence of excess amounts of ADP, steady-state DNA

binding/clamp loading assays were performed in the presence of increasing

concentrations of ADP. The steady-state clamp loading activity of the minimal complex

or γ complex was measured in several assays with RhX-pt DNA.

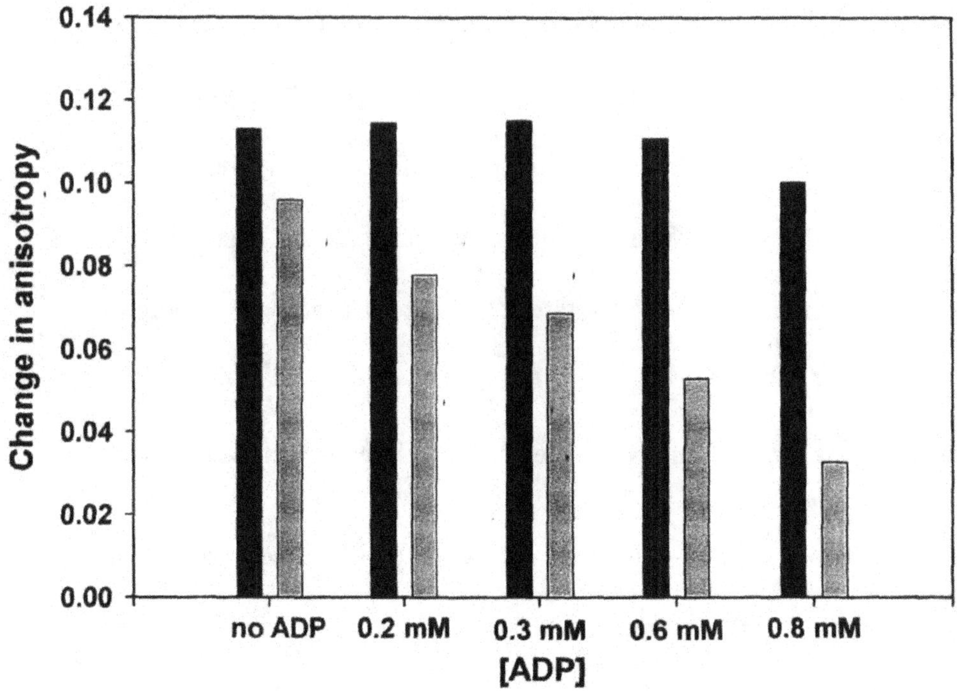

Figure 5-8. Effect of increasing ADP concentration on the steady-state β clamp loading
reaction for the minimal complex and γ complex. In separate steady-state
assays the polarized emission intensities of RhX-pt DNA were measured, for
use in calculation of anisotropy, upon the addition of solutions containing
either the minimal complex (500 nM) (gray bars) or γ complex (500 nM)
(black bars) to cuvettes containing RhX-pt DNA (50 nM), β (500 nM), ATP
(0.3 mM), and a range of concentrations of ADP (0.0, 0.2, 0.3, 0.6, 0.8 mM)
in assay buffer: 20 mM Tris-HCl pH 7.5, 50 mM NaCl, 5 mM DTT, 40μg/mL
BSA, and 8 mM $MgCl_2$. Final concentrations are shown in parentheses. The
anisotropy value for free RhX-pt DNA (r_{free}) in assay buffer with ATP and
ADP was subtracted from the anisotropy for RhX-pt DNA in the presence of
protein (r_{bound}) to give the value of change in anisotropy plotted.

The polarized emission intensities of the RhX-probe were measured for calculation of

anisotropy upon the addition of a solution containing either the minimal complex or γ

complex to a solution of RhX-pt DNA, β, ATP, and varied concentrations of ADP in

assay buffer. The value of anisotropy reflects the steady-state β clamp loading activity on

RhX-DNA, and the constant concentration of ATP in these assays was high enough to

promote the loading reaction for at least 200 s (see Figure 4-1, black trace).

The calculated values of the change in anisotropy shown in Figure 5-8 were averaged for separate 30 s clamp loading reactions. Increasing ADP concentration hindered the minimal complex-catalyzed clamp loading activity. The minimal complex clamp loading reaction was even inhibited at a concentration of ADP (0.2 mM), slightly lower than the constant concentration of ATP (0.3 mM). The γ complex-catalyzed clamp loading reaction remained unaffected until a large excess of ADP (0.8 mM) was present.

ADP does not completely stop the clamp loading reaction like ATPγS, because, as a product of ATP hydrolysis, ADP is part of the steady-state reaction cycle. ADP does not become "locked-in" to a clamp loader-β complex or even become part of a stable inactive complex with clamp loader because it is constantly exchanging with ATP. Only when ADP was at a concentration above 0.6 mM did it begin to inhibit the clamp loading reaction by γ complex. This indicates that a product of ATP hydrolysis (ADP) inhibits clamp loading by the minimal complex to a greater extent than by γ complex, and that the steady-state stages (i.e., conformational changes?) of the clamp loading reaction cycle for the minimal complex must differ from γ complex in some way.

Pre-Steady-State Kinetics of Clamp Loading by the Minimal Complex Initiated at Different Steps of the Reaction Cycle

The kinetics of β clamp loading were studied in real time to determine if similar populations of the minimal complex observed in pre-steady-state ATP hydrolysis assays could be observed in different stages of the loading reaction. Fluorescence-based anisotropy binding assays were used to measure polarized emission intensities of RhX-pt DNA in real time using a stopped-flow apparatus. The clamp loading reaction was initiated at different steps by changing the equilibration of, and order of mixing the clamp loader with ATP and β, prior to rapidly mixing with RhX-pt DNA.

Kinetics of Clamp Loading when the Minimal Complex is Equilibrated with ATP and β

In the ATP hydrolysis assays, there was only a single rapid phase observed when the minimal complex was preincubated with ATP and β (Figure 5-5A). The monophasic reaction suggested that the minimal complex was completely in the "activated" state. Will similar rapid monophasic clamp loading kinetics be observed for the clamp loading reaction in the anisotropy binding assay? A reaction was initiated after the ATP-dependent conformational change and assembly of the minimal complex with β (Figure 5-9A, scheme-1). A solution containing the minimal complex, ATP, and β in one stopped-flow syringe was rapidly mixed with a solution containing RhX-pt DNA and ATP from a second syringe to initiate the reaction.

The scheme-1 results show a rapid single-exponential rise to an anisotropy value of roughly 0.31 in about 100 ms, followed by a decay to steady-state anisotropy (~ 0.25) over an additional ~ 400 ms. The rapid rise in anisotropy most likely was the binding of the β-bound minimal complex, to RhX-pt DNA (see the correlation assays below). After β was loaded on DNA, the anisotropy decayed as the clamp loader was released from DNA, ending the first turnover of the clamp loading cycle. The rate of the rapid rise in clamp loading activity for the minimal complex was similar to that observed for γ complex, but the decay into the steady-state activity was more rapid and more pronounced or "deeper". The steady-state anisotropy of RhX-pt DNA in the clamp loading reaction never decayed totally to the level of free RhX-pt DNA in all assays since the clamp loader and β were in large molar excess over RhX-DNA. This also shows that the end-point of the clamp loading reactions did not change with respect to when the reactions were initiated.

Figure 5-9. Pre-steady-state kinetics of the clamp loading reaction initiated at different steps. The real-time change in polarized emission intensities of RhX-pt DNA was measured and used for calculation of anisotropy for clamp loading reactions by the minimal complex (red), compared to γ complex (black) initiated at different steps by using the mixing schemes shown above the plots. A) (Scheme 1), A solution from one syringe containing the minimal complex or γ complex, β, and ATP was mixed with a solution from a second syringe containing RhX-pt DNA and ATP. B) (Scheme-2), One syringe, loaded with a solution containing the minimal complex or γ complex and ATP was mixed with a solution from a second syringe containing RhX-pt DNA, β, and ATP. C) (Scheme-3), One syringe was loaded with a solution containing the minimal complex or γ complex only, then mixed with the contents of a second syringe loaded with a solution of RhX-pt DNA, β and ATP. In all cases, final reaction concentrations were 250 nM minimal complex or γ complex, 600 nM β, 0.5 mM ATP, and 50 nM RhX-pt DNA in assay buffer containing 20 mM Tris-HCl pH 7.5, 50 mM NaCl, 5 mM DTT, 40 μg/mL BSA, and 8 mM $MgCl_2$. The gray plots showing no change in anisotropy (r_{free}) was a DNA-only control where assay buffer was added to RhX-pt DNA. All steady-state clamp loading activity had matching anisotropy values respective to the clamp loader used, and the total intensities did not change with time indicating that initiating the clamp loading reaction at different steps had no effect on this steady-state activity.

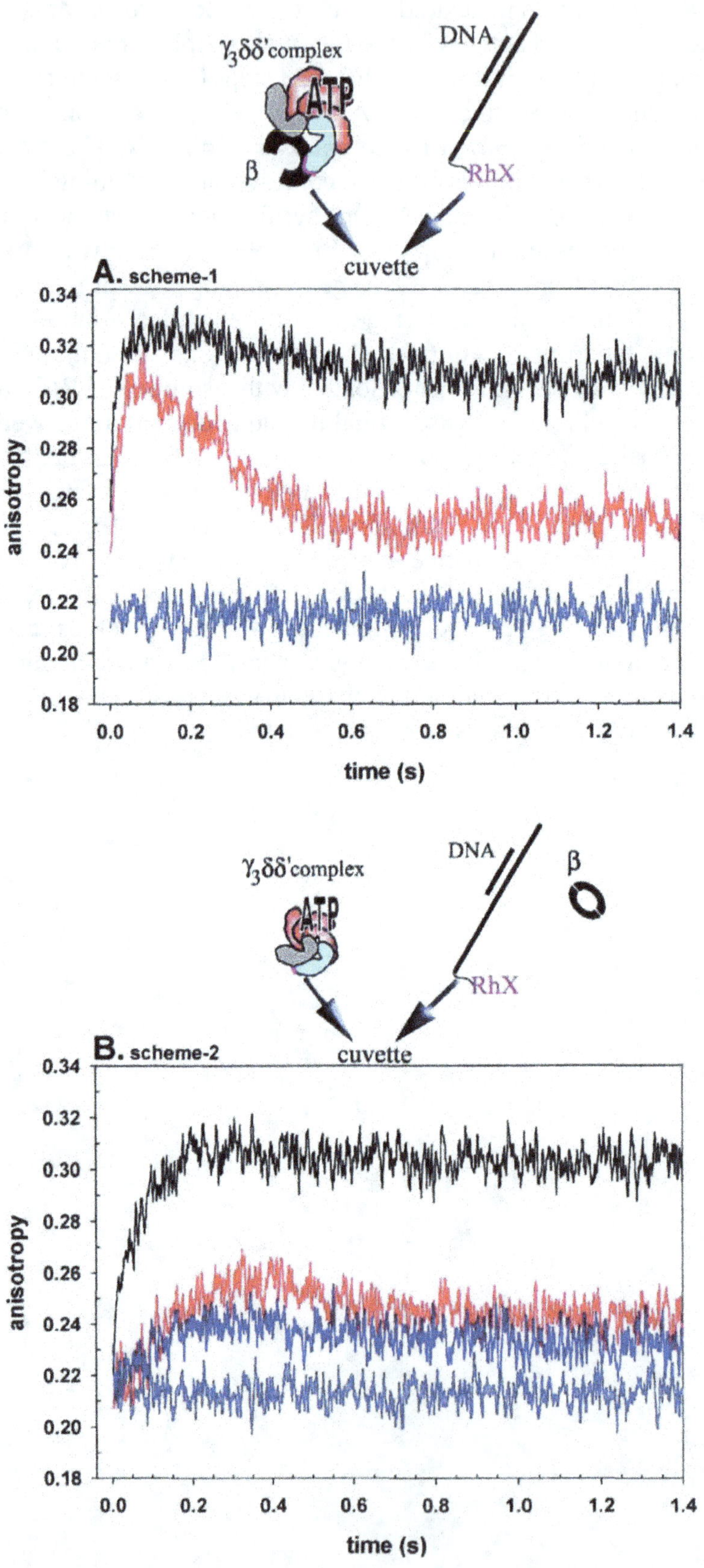
γ₃δδ'complex
DNA
ATP
β
RhX
cuvette
A. scheme-1
0.34
0.32
0.30
0.28
0.26
0.24
0.22
0.20
0.18
anisotropy
0.0 0.2 0.4 0.6 0.8 1.0 1.2 1.4
time (s)
γ₃δδ'complex
DNA
β
ATP
RhX
cuvette
B. scheme-2
0.34
0.32
0.30
0.28
0.26
0.24
0.22
0.20
0.18
anisotropy
0.0 0.2 0.4 0.6 0.8 1.0 1.2 1.4
time (s)

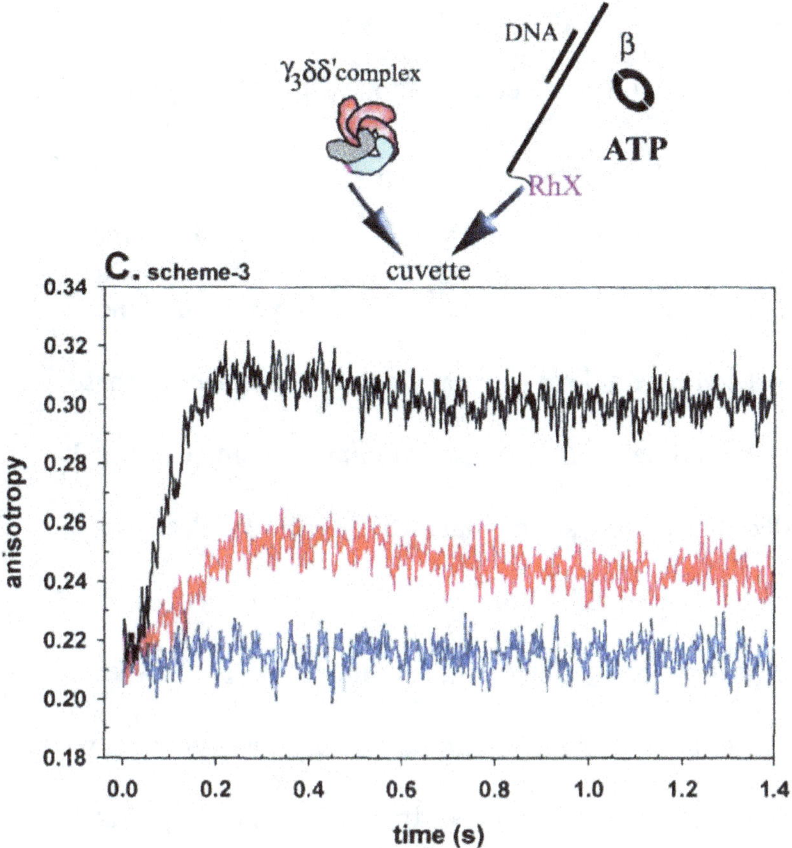

Figure 5-9. Continued.

Kinetics of Clamp Loading when the Minimal Complex is Equilibrated with ATP

In a second assay (Figure 5-9B: scheme-2), the clamp loading reaction was initiated

by adding an equilibrated solution of the minimal complex and ATP to a solution

containing β, RhX-pt DNA, and ATP. When the minimal complex was equilibrated with

ATP in the corresponding pre-steady-state MDCC-PBP ATPase assay, two kinetic phases

were observed (Figure 5-5A), which suggested that a small "activated" population of the

minimal complex was present and rapidly reacted with DNA followed by a slower

hydrolysis reaction. The amount of the "activated" population of the minimal complex

was estimated at the low value of ~25 % of total enzyme.

Unlike that observed for γ complex, biphasic clamp loading kinetics were not seen for the minimal complex in this assay following equilibration with ATP (Figure 5-9B, red trace). In fact, a lag of ~70-80 ms in duration was present, preceding a slow monophasic rise to an anisotropy value of roughly 0.26 that peaked at ~ 400 ms. The anisotropy then decayed into the same steady-state clamp loading reaction (~0.25) observed in the scheme-1 assay. The minimal complex did not show any rapid clamp loading phase, suggesting that there was little or no "activated" species of minimal complex initially present. This result contrasted with the pre-steady-state MDCC-PBP ATPase assay, where a small initial rapid phase after equilibration with ATP was observed. It is possible that the predicted concentration ($\leq$63 nM) of the minimal complex present in the rapidly ATP hydrolyzing population was unable to produce a rapid DNA binding phase in this clamp loading assay. In comparison to the DNA binding kinetics displayed by γ complex (black trace), the minimal complex may be slower to form a DNA binding surface when it binds β. Taken together, these results could indicate that the minimal complex undergoes a significant slow step after equilibration with ATP before it can bind and load β.

Kinetics of Clamp Loading When the Minimal Complex is Mixed Directly with a Solution of ATP, β, and DNA

To examine the kinetics of the slow phase, the clamp loading reaction was initiated prior to the ATP binding, conformational change(s), and β binding steps (Figure 5-9C, scheme-3). Anisotropy binding assays were performed where a solution containing the minimal complex alone in one stopped-flow syringe was mixed with a solution from a second syringe containing ATP, β, and RhX-pt DNA. If the minimal complex has a slower ATP-dependent conformational change compared to γ complex, then will the observed clamp loading kinetics by the minimal complex also be slower in this assay?

In this scheme-3 assay (Figure 5-9C, red trace), the minimal complex showed slow clamp loading kinetics rising to an anisotropy value of about 0.26, preceded by a 70-80 ms lag consistent with a reaction where a slow step forms prior to a fast phase. This result was nearly identical to that observed in the scheme 2 assay (Figure 5-9B, red trace), although in scheme-3 there was no equilibration of the minimal complex with ATP. Similar sigmoidal kinetics were observed for γ complex in the scheme 3 assay (black trace). The slower kinetics observed for the minimal complex in the scheme-3 assay do suggest that the minimal complex undergoes a slower rate-limiting transition than γ complex does.

Direct Real Time Correlation of the Minimal Complex DNA Binding and ATP Hydrolysis Kinetics in the Presence and Absence of β Clamp

To directly examine the timing of formation of the minimal complex clamp loader binding DNA with the kinetics of ATP hydrolysis, in the presence and absence of β, pre-steady-state fluorescence anisotropy binding and MDCC-PBP ATPase assays were preformed under identical conditions. These combined assays allow direct comparison of the kinetics of ATP hydrolysis activity (Figure 5-5A), and DNA binding activity (Figure 5-9A) by the minimal complex in the presence of β. These correlation assays were performed in a stopped-flow either outfitted for simple fluorescence emission detection of MDCC-PBP for measurement of ATPase kinetics, or for polarized emission detection for measurement of the RhX-pt DNA binding kinetics. A sequential-mix "three-syringe" setup was utilized where one syringe was loaded with the minimal complex and β (when present), and a second syringe was loaded with ATP. The contents of these two syringes were mixed and preincubated for 1 s, then mixed with the contents of a third syringe

containing RhX-pt DNA for RhX-anisotropy binding assays, or unlabeled-pt DNA and

MDCC-PBP for ATPase assays. The concentrations of all reactants and buffer were the

same in both assays.

Previously, similar correlated analyses were performed with γ complex for

investigation of mutant β clamps (Bertram et al., 2000), and recently for study of clamp

loading activity on elongation proficient pt DNA substrates (Ason et al., 2003). Key

results from these correlation assays with γ complex showed that a complex of the clamp

loader, β (when present), and DNA rapidly formed just prior to DNA-stimulated ATP

hydrolysis activity, which was then followed by release of the clamp loader from the

complex as the first turnover of clamp loading was completed.

Figure 5-10 shows the results of correlation assays performed with the minimal

complex. When equilibrated with ATP in the presence of β (Figure 5-10A), the

anisotropy (DNA binding/clamp loading) data showed that a ternary complex rapidly

peaked at an anisotropy value of about 0.25 in ~ 50-60 ms just prior to a "burst" of ATP

hydrolysis ending at approximately 100 ms. The anisotropy of the rapidly formed ternary

complex quickly decayed to the level of free RhX-pt DNA upon a single rapid phase of

ATP hydrolysis. In these assays, the concentrations of the minimal complex and β were

not saturating with respect to the concentration of RhX-pt DNA, and therefore the

anisotropy decayed as expected to a level near that of free RhX-DNA. During this

anisotropy decay phase, the minimal complex was released from β loaded on DNA, and

the ATP hydrolysis activity showed a corresponding pause before the steady-state.

Figure 5-10. Direct correlation of the kinetics of DNA binding and ATP hydrolysis by the $\gamma_3\delta\delta'$ minimal complex in the absence and presence of β clamp. A) Correlated RhX-pt DNA binding kinetics (red) and ATP hydrolysis kinetics (black) for the minimal complex in the presence of β. B) Correlated RhX-DNA binding kinetics (red) and ATP hydrolysis kinetics (black) for the minimal complex in the absence of β. Using a sequential-mix "three-syringe" stopped-flow assay, one syringe loaded with a solution containing the minimal complex and β (when present) was mixed with the contents of a second syringe loaded with ATP. The resulting mixture was preincubated for a period of 1 s, and then mixed with the contents of a third syringe loaded with a solution-containing RhX-pt DNA for RhX-anisotropy binding assays, or unlabeled-pt DNA and MDCC-PBP for ATPase assays. All reactions were performed in assay buffer: 20 mM Tris-HCl, 50 mM NaCl, 5 mM DTT, 40 μg/mL BSA, and 8 mM MgCl$_2$. Final reactant concentrations were 225 nM minimal complex, 500 nM β (when present), 200 μM ATP, 450 nM pt DNA, and 4.4 μM MDCC-PBP. The real-time change in polarized emission intensities of RhX-pt DNA was measured and used for calculation of anisotropy by the stopped-flow software. Anisotropy binding data were fitted to an empirical double exponential function and plotted as solid curves (blue). Raw MDCC-PBP fluorescence data were transformed into the concentration of P$_i$-bound MDCC-PBP plotted using the equation shown in figure 5-5.

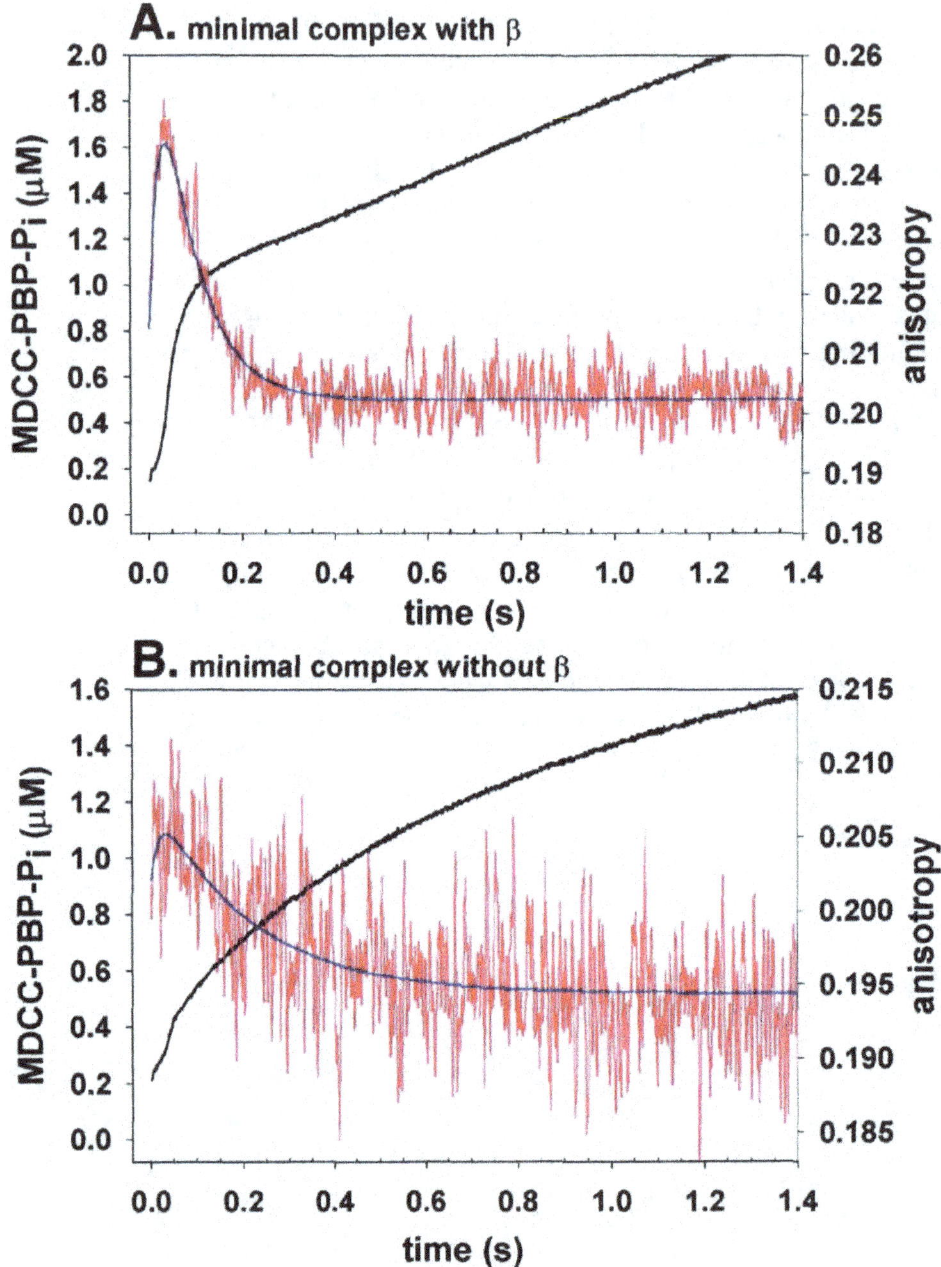

It is during this stage of the reaction where the nucleotide exchange steps are thought to occur, ending the first turnover. The results of these correlated assays are consistent with the pre-steady-state kinetics of DNA binding/clamp loading when the minimal complex was equilibrated with ATP and β (Figure 5-9A, scheme-1), and with the pre-steady-state kinetics of ATP hydrolysis by the minimal complex when equilibrated with ATP and β (Figure 5-5-A).

In the absence of β (Figure 5-10B), the anisotropy (DNA binding) data showed that a binary complex of the minimal complex and DNA rapidly formed to an anisotropy value of roughly 0.21 in ~70 ms just prior to stimulation of biphasic ATP hydrolysis activity. Unfortunately, there was considerable noise in the calculated anisotropy in the DNA binding assay for the minimal complex in the absence of β under these correlation assay conditions. The results, albeit noisy, do show that the minimal complex, after equilibration with ATP, bound DNA just prior to the initial rapid phase of ATP hydrolysis activity. The minimal complex then appeared to be releasing the DNA during the second phase of ATP hydrolysis before entering the steady-state reaction. The MDCC-PBP ATPase assay (Figure 5-5A) for the minimal complex was performed under more optimal conditions for examination of ATP hydrolysis kinetics (i.e., when the concentration of pt DNA was saturating). The previous assay showed much clearer results, to which the correlated assay appeared to be consistent.

Discussion

Understanding γ Complex Kinetics by Characterization of and Comparison with $\gamma_3\delta\delta'$ Minimal Complex

The results described in this chapter address the function of the clamp loading machine by comprehensive study of a minimal complex clamp loader ($\gamma_3\delta\delta'$), and direct comparison to the activity of γ complex. In the clamp loading reaction, the ring-shaped β clamp must be opened, placed at the correct location on DNA, and released to allow attachment to DNA polymerase for processive synthesis. To accomplish these well ordered tasks, the clamp loader acts in a switch-like manner, where ATP binding and hydrolysis modulate its affinities for β and DNA. The status of nucleotide binding to the three-γ motor subunits of the clamp loader coordinates intramolecular communication between all subunits through a structural "language" of conformational transitions, "switching" between inactive and active species. ATP binding induces conformational changes revealing sites for binding DNA and β in the clamp loader. DNA-stimulated ATP hydrolysis causes inactivation and release of the clamp loader from loaded β, and ultimately, during nucleotide exchange, conformational changes occur that reset the clamp loader for continued activity. The combination of $\gamma_3\delta_1\delta'_1$ subunits has been known to be the minimal subcomplex with clamp loading ability suitable for processive DNA synthesis in *in vitro* reactions (Onrust et al., 1991; Stukenberg et al., 1991). The $\gamma_3\delta\delta'$ minimal clamp loader structure has been solved, is currently the most complete structure of any clamp loading machine (Jeruzalmi et al., 2001a). By AAA+ structural homology to the eukaryotic and archaeal clamp loaders, the $\gamma_3\delta\delta'$ structure has provided investigators with many fundamental deductions for conserved clamp loading functions (Davey et al., 2002; Neuwald et al., 1999; O'Donnell et al., 2001).

Here, steady-state and pre-steady-state ATP hydrolysis and DNA binding/clamp loading analytical methodologies were utilized with fluorescence-based techniques, revealing the importance of the χ and ψ subunits for optimum ATP hydrolysis and clamp loading activity by γ complex. This work extends the previous analysis of γ complex by mirroring and comparing results in identical biochemical assays, and addresses the theoretical model presented in chapter 4 for the clamp loader ATP-dependent conformational dynamics by providing additional insight into these conformational transitions that drive the *E. coli* clamp loader.

$\gamma_3\delta\delta'$ is the Minimal Complex with DNA Binding Ability, and Binds ATP and β with Affinity Similar to γ Complex

Examination of the γ complex individual subunits and different combinations of these subunits initially revealed that only the combination of $(\gamma/\tau)_3,\delta,\delta',\chi,\psi$ ("γ complex") and $(\gamma/\tau)_3\delta,\delta'$("minimal complex"), had ATP-dependent DNA binding ability (Figure 5-1). Although the exact nature of the DNA binding site within the clamp loader is currently undefined, this study shows that a DNA binding "site" (or surface) was formed in the presence of ATP. A tetramer of the γ subunit, for example, cannot obtain an ATP-dependent conformation for formation of the putative DNA binding surface. This indicates that the δ and δ' subunits are required and may physically contribute to formation of this DNA binding surface.

The ATP- and β clamp-binding constants, estimated using the β^{pyrene} anisotropy binding assays, were both nearly identical for γ complex and the minimal complex. Both bind ATP with an apparent dissociation constant of 2 μM (Figure 5-4). The ATP-dependent affinity for β^{pyrene} was estimated at 3-4 nM for both clamp loaders (Figure 5-3). Thus under the assay conditions presented in this work, means that nearly all clamp loaders would be bound to β when it is present. The similar magnitude of these affinities for each clamp loader to ATP and β suggest that the γ-subunits' nucleotide binding site accessibility was similar in both clamp loaders, and the subsequent ATP-dependent changes inducing exposure of the δ-subunit β-interaction element were also similar.

Pre-Steady-State ATP Hydrolysis and DNA Binding Kinetics: Analyses of the Active and Inactive Clamp Loader States

The kinetics of ATP hydrolysis by the minimal complex are discussed in terms of the model presented in chapter 4, which was based in the computerized fitting of the pre-steady-state ATP hydrolysis kinetics of γ complex (Kuzmic, 1996).

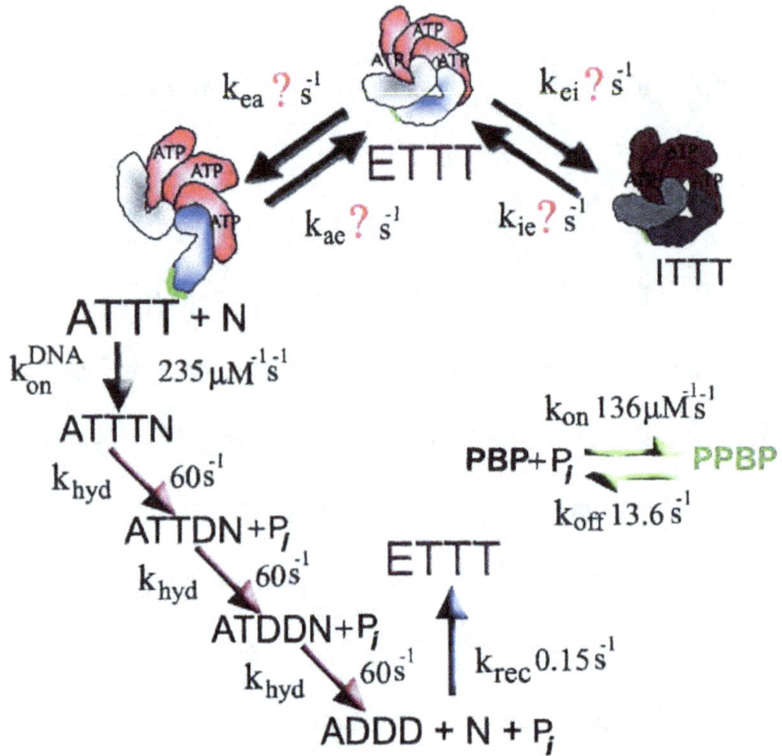

Figure 5-11. Kinetic modeling of ATP hydrolysis reactions. This model is reproduced from chapter 4 (Figure 4-10A). Three states of the clamp loader, ETTT, ATTT and ITTT are shown (cartoons are based on the structure of the clamp loader, see chapter2, figure 2-3). The ATTT form binds DNA (N) rapidly and hydrolyzes its three molecules of ATP (T) in succession producing ADP (D) and inorganic phosphate (P_i). The clamp loader dissociates from DNA on hydrolyzing its third molecule of ATP and is recycled back to the ETTT state. Experimentally determined rates of γ complex binding DNA (Ason et al., 2000; Bertram et al., 2000), k_{onN}, and MDCC-PBP (PBP) binding phosphate (Brune et al., 1994), k_{on} and k_{off}, are presented. All DynaFit program scripts and analysis for these models, and KinTekSim simulations are presented in the appendix.

This model demonstrated that at least three different species of γ complex were required to produce biphasic kinetics of ATP hydrolysis and that these three species could not all exist on the linear pathway to products. The minimal model that best fit the data has been reproduced here as Figure 5-11, and is presented without the rates determined for the γ complex conformational change, because the assays performed to reliably estimate those rates have not been repeated for the minimal complex (see computer modeling methods in chapter 3 for details).

In this model, the clamp loader is initially present in a state, ETTT, that can convert to two different species, one that is "activated" for DNA binding, ATTT, one that is inactive, ITTT, and does not directly form products. In this minimal model, it is assumed that each of these species is bound to three molecules of ATP because ATP hydrolysis rates were independent of the ATP concentration at this saturating concentration (400 μM ATP). However, it is possible that the E and/or I states have fewer molecules of ATP bound and that very rapid ATP binding steps and additional conformational changes exist between the three states.

In this model, the activated state, ATTT, rapidly binds DNA. This DNA-clamp loader complex then hydrolyzes each of its three molecules of ATP sequentially at the same rate. The ATP hydrolysis steps are modeled using single forward rate constants because the presence of MDCC-PBP limits the reverse reaction such that it cannot accurately be determined. Previous work has shown that the DNA dissociation step is likely to be very rapid relative to the steady-state recycling of the clamp loader, and because these data do not directly measure this step, DNA dissociation was set to be concurrent with hydrolysis of the third molecule of ATP. Finally, the steady-state portion of the reaction was modeled as a pseudo-first order conversion of the ADP-bound clamp loader (ADDD) back to the initial ATP-bound state (ETTT). This pseudo-first order approximation is reasonable because only a small fraction of the ATP substrate is consumed on the time scale of the reaction.

Here, examination of the pre-steady-state ATP hydrolysis and clamp loading activities reveal that the minimal complex also exists in a mixture of conformational states in equilibrium with ATP. However, several notable differences were observed for the minimal complex mechanism, suggesting a slower conformational transition and less stable "activated" ATTT conformational state in the minimal complex. The pre-steady-state biphasic kinetics of ATP hydrolysis, observed when the minimal complex alone, was equilibrated with ATP and mixed with DNA (Figure 5-5A, blue trace), or not equilibrated with ATP prior to mixing with DNA and ATP (Figure 5-6) suggested that two species of the minimal complex were present, but the estimated population of the rapidly acting ATTT species was significantly smaller (~25 % of total enzyme) in comparison to γ complex (Figure 4-3) (~40 % of total enzyme) present under similar conditions.

The two kinetic phases observed in the minimal complex assay may not represent the same biphasic kinetics described for γ complex in chapter 4. For the minimal complex, the short pre-steady-state rapid phase is followed by a second pre-steady-state phase that is likely indistinguishable from steady-state activity, or the pre-steady-state rapid phase is directly followed by steady-state hydrolysis kinetics. In assays with the minimal complex alone (figure 5-5A, blue trace), several possible mechanisms could produce the two kinetic phases. 1) A single population of the clamp loader was present in which the individual γ subunits hydrolyze ATP a different rates, 2) like γ complex, two distinct species of the clamp loader were present that differ in DNA-stimulated ATP hydrolysis activity, or 3) two different species of clamp loader interact with pt DNA in separate parallel reactions. For example, pt DNA binding may not be coupled with the pre-steady-state rapid hydrolysis phase, due to a 1.15 molar excess of ss-primer DNA over pt DNA, likely to be present in these assays. After the small population of "activated" ATTT species reacted with ss DNA, then the steady-state formation of ATTT species from ETTT would be the separate reaction with pt DNA. In the later two cases, the two species could differ by the ATP-dependent conformational change(s) required for binding DNA described by the model. One conformational species, ATTT, has high affinity for DNA, and therefore binds pt DNA (or ss DNA) and hydrolyzes ATP at a faster rate than the second. The second phase of ATP hydrolysis could arise from the ETTT species slowly changing conformation forming the ATTT species, and therefore slowly gaining affinity for DNA. The equilibrium mixture of the minimal complex species with ATP resulting in a small "activated" ATTT population (~25 %) could provide a mechanistic explanation for the slower steady-state ATP hydrolysis activity observed without β.

In the pre-steady-state DNA binding/clamp loading assay (Figure 5-9B scheme-2), when the minimal complex was equilibrated with ATP prior to mixing with a solution containing β, RhX-pt DNA, and ATP, biphasic clamp loading kinetics were not exhibited. This was in contrast to the expected biphasic activity based on the model that predicts the mixture of species in equilibration with ATP dominated by the ETTT / ITTT and ATTT states. A lag of ~70-80 ms, followed by slow clamp loading kinetics were observed. It is interesting to note that the lag phase directly related to the duration of the rapid phase of the biphasic kinetics displayed by γ complex (Figure 5-9B), showing that there is some significant disparity in the clamp loading reaction when χ and ψ subunits are missing from the clamp loader.

The differences between the DNA binding/clamp loading kinetics by the minimal clamp loader from γ complex could have been due to a difference in equilibrium between activated (ATTT) and inactive states (ETTT ⇆ ITTT) predicted modeling the pre-steady-state ATP hydrolysis analysis. Another likely possibility is that the model based on the ATP hydrolysis kinetics does not adequately describe the kinetics of DNA binding/clamp loading. Although similar populations of activated and inactive clamp loader species could be expected during equilibration with ATP based on the model, when mixed with DNA and β, a different reaction may have been measured in the DNA binding/clamp loading assays. For γ complex, the pre-steady-state biphasic DNA binding/clamp loading kinetics did show that distinct species were present, but it is possible that they were not the same as those described in the model.

It is currently unknown which of the ATP-dependent conformational states β can bind to. The data shown in figure 5-3 even shows that β binds the clamp loader in the absence of ATP, albeit with 50-fold less affinity than in the presence of ATP. The model based on the pre-steady-state ATP hydrolysis experiments satisfactorily describes the mixture of clamp loader ATTT, ETTT, and ITTT conformational states in equilibrium with ATP, but whether β binds any one or all of these species is unclear. For example (with γ complex), the activated ATTT clamp loader species may rapidly bind DNA before binding β, and the ETTT or ITTT states could bind β, and then convert into an activated "ATTTβ"-complex that loads the clamp in the second slower pre-steady-state phase.

For the minimal complex equilibrated with ATP, the pre-steady-state ATP hydrolysis kinetics predicted that a small population of activated ATTT species was present that hydrolyzed ATP in a rapid pre-steady-state phase. Why then, was there an initial lag observed in the DNA binding/clamp loading assay when the minimal complex was equilibrated with ATP? It is possible that the small-activated population did load β on DNA initially, but the resulting change in anisotropy was too low to observe. The minimal complex DNA binding/clamp loading kinetics suggested that the β clamp loading reaction was preceded by a slow phase. Following the above example (for γ complex), β may have bound to any one or all of the minimal complex species in equilibrium with ATP (ATTT, ETTT, ITTT), and a slow conversion of ETTT and/or ITTT to an activated "ATTTβ"-complex occurred while the small original activated ATTT population rapidly bound DNA and hydrolyzed ATP. When DNA binding kinetics (in the absence of β) were correlated with ATP hydrolysis for the minimal complex equilibrated with ATP (Figure 5-10B), a small amount of DNA binding activity just prior to ATP hydrolysis was observed, further suggesting that only a small activated population of the minimal complex was present. Although this experiment was not a clamp loading assay, it did show that a small amount of the active minimal complex reacted rapidly with DNA, instead of displaying an initial lag phase.

The minimal model (Figure 5-11) describes intramolecular conformational transitions that are likely to separate the mixture of states in equilibrium with ATP for γ complex and probably the minimal complex as well. However, without knowing to which ATP-dependent conformational state or states β binds, it is not yet possible to reliably model the DNA binding/clamp loading kinetics displayed in these assays when either clamp loader was initially equilibrated with ATP.

The monophasic ATP hydrolysis kinetics for the minimal complex in the presence of β indicate that there was only a single species of minimal complex present that was active for DNA binding, and therefore clamp loading. These results clearly show that the χ and ψ subunits, missing from the minimal complex, are required by γ complex to maintain a higher population of ATTT activated conformational species in equilibrium with ATP. For example, they may facilitate the conformational changes separating the ATTT and ETTT species of the clamp loader. The β clamp appears to help the minimal complex increase its pre-steady-state activity perhaps by bypassing a slower reaction step (i.e., the [ITTT ⇋ ETTT] ⇋ ATTT transition) that occurs in the first turnover of ATP.

In pre-steady-state ATP hydrolysis experiments in the presence of β, reactions were initiated following a preincubation period allowing equilibration of the minimal clamp loader with ATP and β. The minimal complex and γ complex displayed similar rapid monophasic kinetics in the first turnover (Figure 5-5A and B). Pre-steady-state ATPγS chase assays performed in the presence of β (Figure 5-7A) were also similar to γ complex (Figure 4-5B), showing rapid monophasic kinetics of a single turnover of ATP hydrolysis. These results indicate that equilibration with ATP and β converted or "trapped" all species of the minimal complex into an "activated" ATTT-β bound conformational state, effectively increasing its population before being mixed with DNA to initiate measurement of ATP hydrolysis activity for the clamp loading reactions.

The corresponding pre-steady-state DNA binding/clamp loading assays where the minimal complex was equilibrated with ATP in the presence of β to initiate the reaction after the ATP-dependent conformational change also showed a single rapid phase of activity (Figure 5-9A, scheme-1). The decay in anisotropy as the clamp loader released from DNA-β occurred at a faster rate and was "deeper" for the minimal complex compared to γ complex as the reaction entered the steady-state. The decay to a lower anisotropy value for steady-state activity indicates that there was little minimal complex rebinding to DNA during this phase. A possibility is that the steady-state reaction was impeded as a consequence of slower conformational change(s) (i.e., [ITTT $\leftrightarrows$ ETTT] $\leftrightarrows$ ATTT) required for conversion to the reactivated "ATTTβ" state. Slower conversion of the minimal complex, and less clamp loading activity could explain the quicker rate of the decay compared to γ complex, as well as the reduced steady-state anisotropy observed in this and other assays with the minimal complex. Further, the correlation of pre-steady-state ATP hydrolysis and DNA binding/clamp loading activity (Figure 5-10A) demonstrated that the rapid monophasic DNA binding/clamp loading reaction (analogous to scheme-1 in Figure 5-9A) directly preceded a monophasic "burst" of ATP hydrolysis by the minimal complex. Together, the pre-steady-state ATP hydrolysis and DNA binding assays in the presence of β clearly show that the clamp has the ability to convert the mixture of ATP-dependent conformational states of the clamp loader into a single activated population "ATTTβ" capable of rapid loading of β on DNA, followed by rapid ATP hydrolysis in the first turnover. Again, whether β binds the ATTT state only and "traps" it, or binds to the ETTT or ITTT states and increases the rate of their conversion to the activated "ATTTβ" species is unclear.

The χ and ψ Subunits, Missing from the Minimal Complex, May Facilitate the Conformational Dynamics of γ Complex

The only difference between γ complex and the minimal complex in these experiments is the presence of the χ and ψ subunits in γ complex. Based on the model for equilibration with ATP and pre-steady-state ATP hydrolysis activity (Figure 5-11), the absence of the χ and ψ subunits from the minimal complex may cause conformational

instability or some other deficiency in ATP-dependent conformational communication

between subunits within this clamp loader, resulting in a low quantity of activated ATTT

versus inactive [ETTT $\leftrightarrows$ ITTT] conformational states. A possibility is that the overall

forward rate of conformational changes (...ETTT $\rightarrow$ ATTT...) producing the minimal

complex activated species may be slower than the corresponding rate for γ complex.

Also, an increased reverse rate of conformational transitions (...ETTT $\leftarrow$ ATTT...) could

potentially result in an elevated population of the "inactive" minimal complex species.

The opposite conformational dynamics could exist for γ complex, resulting in the

difference in estimated activated species population observed between γ complex (~40

%) and the minimal complex (~25 %) when equilibrated with ATP. For example, γ

complex could be optimized for an overall faster forward rate of conformational changes

(i.e., facilitated by the χ and ψ subunits) towards an activated species, and a slower

reverse rate towards an inactive population.

β Clamp Enhances the Switch from Inactive to Active Clamp Loader Populations

It has been known for some time that the sliding clamp enhances the ATP

hydrolysis activity of the clamp loader, and therefore the clamp loading activity of γ

complex (Onrust et al., 1991; Stukenberg et al., 1991), and also clamp loaders of other

organisms (Ellison and Stillman, 2001). Here, steady-state and pre-steady-state ATP

hydrolysis assays show that the activity of the minimal complex is nearly identical to γ

complex in the presence of β, consistent with previous analyses. β enhanced the steady-

state ATP hydrolysis turnover rate approximately 2-fold for the minimal complex, and

characteristically, 3-fold for γ complex (Table 5-1). An even more significant result was

the enhancement of an apparent second-order binding constant ($k_{cat}^{complex}/K_m$) in the presence of β. This steady-state parameter is roughly 1000-fold slower than would be expected for diffusion-controlled binding of a small nucleotide cofactor such as ATP to an enzyme, and thus indicates that binding of ATP is followed by a slow step leading to hydrolysis such as the proposed conformational changes modulating β clamp and DNA binding affinities. The increase in the specificity constant for γ complex by the presence of β was about 2-fold, whereas, the minimal complex showed a much higher increase in the specificity constant (4-fold), demonstrating that β did more "work" in conversion of this putative slow phase following ATP binding.

Based on these steady-state data, along with the similarities observed for the minimal complex and γ complex in pre-steady-state ATP hydrolysis data, it is concluded that β clamp allows the clamp loader to overcome the slow phase, most likely by conversion of the population of inactive clamp loader into a population of all completely activated conformational state "ATTTβ". Further, the enhancement of the minimal complex specificity constant supports that β clamp had a much greater effect in

$$\text{ETTT} \rightleftharpoons \text{ATTT} + \beta \rightleftharpoons \text{ATTT}\beta$$
$$\text{ETTT} + \beta \rightleftharpoons \text{ATTT}\beta$$

Scheme 5-2. β clamp enhancement of "activated" clamp loader.

conversion of the large estimated population of minimal complex "inactive" species ($\leq$75 % of total enzyme). Enhancement by β effectively increases the concentration of the active species most likely by selectively binding to ATTT and "trapping" it into an active ATTTβ-complex poised for clamp loading. Alternately, β could conceivably bind to an "inactive" species such as ETTT and increase the rate of conversion to ATTTβ (see scheme 5-2).

Experiments when the Minimal Complex was not Equilibrated with ATP Reveal Slower Conformational Change Kinetics than γ Complex

In the pre-steady-state DNA binding/clamp loading assay where the minimal complex was not equilibrated with ATP, slow sigmoid-like clamp loading kinetics were seen in the first turnover (Figure 5-9C, scheme-3). Interestingly, this initial lag was of the same duration (~70-80 ms) observed when the minimal complex was equilibrated with ATP before mixing with DNA and β (compare with Figure 5-9B). These kinetics were similar to with those observed with γ complex in the same assay conditions, and show that the DNA binding/clamp loading kinetics were preceded by a slow step. The same step responsible for the lag phase most likely limited the slow clamp loading kinetics as well, and it is hypothesized here that this slow step was an intramolecular conformational change(s) following ATP binding. Previously, for γ complex it was observed that the ATP-dependent conformational change was limiting the clamp loading reaction progress.

Pre-steady-state kinetics of ATP hydrolysis were also measured when the minimal complex (Figure 5-6) (or γ complex, figure 4-8) was not equilibrated with ATP. The observed sigmoid-like kinetics were analogous to those observed in the DNA binding/clamp loading assay (Figure 5-9C, scheme-3). The minimal complex, when directly mixed with DNA and ATP, displayed a lag of ~60 ms, that was then followed by slow hydrolysis kinetics. These analogous clamp loading, and ATP hydrolysis kinetics indicate that a rate limiting complex-conformational transition switching an inactive species into an active species occurred when the clamp loader was mixed with ATP. However, it remains possible that the kinetics represent a mechanism different than when the clamp loader is equilibrated with ATP before mixing with DNA (+/-β). It is entirely possible, and likely, that the clamp loader is fluctuating between conformational states in the absence of any nucleotide that are different than the conformational states observed in equilibrium with ATP. Biphasic kinetics in either case are still consistent with the domination of two states overall including a "branch" pathway towards no product formation, modulating activity of the clamp loaders.

Computer modeling of the data shown in Figure 5-6 was done using the model presented in chapter 4 (Figure 4-10, here, figure 5-11). The computer modeling results for the minimal complex estimate that the forward conformational change towards an active complex is ~2-fold slower than for γ complex, and the reverse conformational change is roughly 3-fold faster than for γ complex. Fitting these single data sets for the minimal complex does not provide enough data for a completely reliable fit, however, the fit shown in figure 5-6, as well as simulations shown in the appendix (Figures A-3-sim, A-4-fit) are consistent with slower forward conformational changes producing a small active population of the minimal complex species. The small change in MDCC-PBP fluorescence due to the small active population limited the ability to further examine the ATPase kinetics as a function of equilibration time with ATP.

The Nature of Nucleotide Binding to the Minimal Complex and γ Complex

The pre-steady-state ATP hydrolysis activity of the minimal complex in the presence of β was chased by the addition of a 10-fold excess of non-hydrolyzable ATPγS to examine the nature of the nucleotide binding. The minimal complex was preincubated with ATP and β before mixing with DNA to trigger ATP hydrolysis in the presence of ATPγS (Figure 5-7A, red trace). It is known that β still binds the clamp loader in the presence of ATPγS, but cannot load β onto DNA without ATP hydrolysis (Bertram et al., 1998). Experiments for the minimal complex were performed under conditions identical to assays with γ complex (Figure 4-5B). For both clamp loaders in the presence of β, a single turnover of ATP hydrolysis occurred, and there was no further steady-state activity. The results demonstrated that in the presence of β, the ATPs bound to the clamp loader were hydrolyzed faster than they exchanged with ATPγS in the pre-steady-state, suggesting that they were stabilized or "trapped" in an activated conformation clamp loader-β complex "ATTTβ". In the steady-state, the excess amount of ATPγS easily out competed ATP for binding to the minimal complex or γ complex, poisoning further hydrolysis activity.

For steady-state ATPγS-chase of the minimal complex or γ complex activity in the presence of β, ATPγS was added directly to the steady-state ATP hydrolysis reaction approximately 10-s after reaction initiation (see Scheme 5-1). Several ATPγS:ATP ratios were tested in separate assays, including a 10:1 pre-steady-state mock-ratio. Under all ATPγS:ATP ratios tested, there was no further hydrolysis activity after chasing the reaction with ATPγS (Table 5-2). Due to the presence of β, the "activated" conformational state resulting from ATPγS binding was trapped in a state that could not load β on DNA, and therefore there was no further ATP hydrolysis or nucleotide exchange. Even a substoichiometric ATPγS:ATP ratio of 0.25:1 was sufficient to stop the reaction in these assays indicating the possibility that ATPγS may share one or two of the three nucleotide binding sites with ATP in the clamp loader, and still prevent ATP hydrolysis-dependent loading of β. These results support the idea that β clamp traps the clamp loader in an active conformation where the nucleotides (ATP or ATPγS) are stably bound. This provides additional evidence for a mechanism of clamp loader enhancement by the sliding-clamp, showing that an activated clamp loader-β complex "ATTTβ" is stable and committed to clamp loading upon pt DNA triggered ATP hydrolysis.

Pre-steady-state ATPγS-chase assays in the absence of β were also performed. The minimal complex was preincubated with ATP, and then mixed with DNA in the presence of a 10-fold excess of ATPγS (Figure 5-7B, red trace). A small biphasic level of ATP hydrolysis activity resulted that was roughly equal to the level of the small rapid phase seen in original assay (Figure 5-7B, black trace). This result could indicate that there was considerable conformational instability leading to more competition between ATP and ATPγS, even in the active species population. Another possibility is that these two pre-steady-state phases observed in the ATPγS-chase assay represent the same two kinetic phases seen in the original assay, however, they are significantly reduced in amplitude due to competition between ATP and ATPγS in binding the "unstable" minimal complex. When the preincubation time was increased to 2 s for this assay there was no difference in the resulting kinetics of ATP hydrolysis activity (not shown). This ATPγS-chase assay did isolate the small population of the activated conformational species (ATTT) of the minimal complex that was present when equilibrated with ATP. Under the same assay conditions with γ complex (Figure 4-5A), a single rapid phase of ATP hydrolysis followed by a second low-level slow phase was observed prior to complete loss of steady-state activity. Both assays show that ATP hydrolyzed in the rapid pre-steady-state phase is hydrolyzed more rapidly than it dissociates, and that the ATP hydrolyzed in the second-slow phase of the original assay readily exchanged with ATPγS.

When the steady-state ATP hydrolysis activity of the minimal complex in the absence of β clamp was chased with ATPγS, it was inhibited at a lower ATPγS:ATP ratio than γ complex was (Table 5-2). Since the ATPγS-chase was initiated during a steady-state reaction, the nucleotide exchange step following DNA-stimulated ATP hydrolysis was vulnerable. During this step, the clamp loader would be in equilibrium with ATP, if only for a short period of time, (i.e., the ATP concentration was saturating in these assays), and therefore only a small population of the minimal complex, compared to γ complex, would in be in the ATP-dependent conformational activated state. A possible reason for the elevated sensitivity of the minimal complex to ATPγS in these steady-state chase assays is that the estimated high population of "inactive" minimal complex conformational states would easily bind ATPγS and become unavailable for further ATP binding and hydrolysis, stopping the reaction.

In contrast, at similar ATPγS:ATP ratios, less γ complex "inactive" species is present during nucleotide exchange, and maintained continued ATP hydrolysis activity at nearly a normal steady-state rate. When chased with stoichiometric (or even slightly higher, i.e. 2:1) ATPγS:ATP ratios γ complex may either be able to maintain a population of an all ATP-bound active state, or sustain the ability to bind DNA and hydrolyze ATP even when there is mixed ATPγS and ATP binding during the steady-state reaction. It is known that ATPγS can induce conformational changes in γ complex allowing DNA binding. If the nucleotide binding sites were mixed with ATP and ATPγS, the results show that DNA binding is sufficient to trigger hydrolysis of a minimum of one ATP in γ complex. This indicates that when γ complex is in equilibrium with both ATP and ATPγS, additional conformational states could be identified in pre-steady-state ATP hydrolysis reactions that represent species where only one or two hydrolyzable ATPs are present.

Additional analysis of the nature of nucleotide binding during the steady-state clamp loading reaction by the minimal complex and γ complex was performed. It was determined that increasing relative concentrations of ADP could inhibit the DNA binding/clamp loading activity of the minimal complex, but not γ complex (Figure 5-8, gray bars). ADP alone is insufficient for DNA binding and clamp loading activity. In steady-state DNA binding/clamp loading assays, the concentration of ATP was held constant at a level that could sustain steady-state clamp loading activity for at least 200 s (see chapter 4, figure 4-1, black curve). When ADP was added at increasing concentrations, γ complex continued loading β, cycling through the steady-state reaction. Only when the ADP concentration was raised to a level ~2.7-fold above ATP, was the clamp loading cycle of γ complex somewhat hindered. The clamp loading activity of the minimal complex, however, was diminished beginning at a substoichiometric concentration of ADP relative to ATP, and decreased further with increasing ADP. It was previously shown that ADP could stably bind γ complex, and cause a conformational change (Hingorani and O'Donnell, 1998; Naktinis et al., 1995). The β clamp must do more work to convert and "trap" the minimal complex in an activated ATP-dependent conformational state. Therefore in these steady-state assays, there is predicted to be a significant level of "inactive-state" minimal complex during nucleotide exchange after ATP hydrolysis. It is believed that ADP can readily compete with the "inactive' conformational states of the minimal complex and inhibit steady-state clamp loading activity. These results add to the ATPγS-chase analyses, and support a mechanism where the sliding clamp enhances its clamp loader by affecting the conformational dynamics that modulate DNA binding, by trapping the clamp loader in an activated clamp loader-β

clamp complex with ATPs secured "ATTTβ".

Investigation of the minimal clamp loader complex has proven to be invaluable for

understanding the activity of γ complex in clamp loading. This comprehensive analysis

of the minimal complex shows that it does have activity similar to γ complex, however,

in the absence of the β clamp, it is less stable and has reduced DNA-stimulated ATP

hydrolysis activity. Together, these kinetic analyses show that in the presence of ATP,

both the minimal complex and γ complex clamp loaders exist in some dynamic

conformational equilibrium dominated by two states with distinct DNA binding abilities.

The ATP-dependent conformational dynamics of the minimal complex appear to be

dominated by a species inactive for DNA binding that must go through a "forward"

conformational change to become activated which may occur at a slower rate compared

to the related conformational change for γ complex. A key to the clamp loading

mechanism is that when either of the clamp loaders is equilibrated with ATP in the

presence of β, a slow step is bypassed, and their activities achieve considerable similarity

under identical experimental conditions. β clamp binding enhances the clamp loader

specificity for ATP, converts the mixture of ATP-dependent conformational species into

an activated conformational state, essentially increasing its concentration, and affects the

stability of the nucleotide binding sites. In comparison to γ complex, the minimal

complex is missing the χ and ψ subunits. These subunits are unique to the *E. coli* clamp

loader and probably other gram-negative prokaryotic clamp loaders as well (Xiao et al.,

1993a; Xiao et al., 1993b). Therefore, it is proposed that the missing χ and ψ subunits are

responsible for the differences in the kinetics observed here. This chapter encompasses the most detailed study of the $\gamma_3\delta\delta'$ minimal complex to date, and therefore provides additional significant biochemical data for understanding the mechanism of the clamp loading reaction catalyzed by the γ complex molecular machine.

CHAPTER 6
CONCLUSIONS AND RECOMMENDATIONS

Introduction

Within the DNA polymerase III holoenzyme, the principal enzyme responsible for

E. coli chromosomal replication, are processivity proteins. Processivity proteins confer

the ability of DNA polymerase III to replicate stretches of DNA thousands of nucleotides

long at a rate approaching 750 nucleotides per second without dissociation. These speed

and processivity characteristics are required by the polymerase for synthesis of the

leading and lagging strands of the entire *E. coli* chromosome in a period of about 40-60

minutes (Kelman and O'Donnell, 1995; Kornberg and Baker, 1992).

The processivity proteins form a ring-shaped DNA sliding-clamp and a complex

molecular machine, the clamp loader. These processivity proteins have been functionally

conserved in all branches of life, and are involved in all forms of DNA metabolism

including recombination and repair, underscoring their importance in the cell and the

importance of investigating their functions. The *E. coli* sliding-clamp is a dimer of

crescent-shaped β subunits that topologically links DNA polymerase to the template

DNA (Kong et al., 1992). The clamp loader is an energase enzyme consisting of seven

subunits (τ_2, γ, δ, δ', χ, and ψ). ATP binding and hydrolysis by the τ/γ-AAA+ ATPase

motor subunits drives conformational changes in the clamp loader that modulate affinities

of the clamp loader for the sliding clamp and DNA (Davey et al., 2002). ATP-dependent

conformational changes cause exposure of the δ subunit from occlusion by δ', and cause

formation of a DNA binding surface on the clamp loader. The δ subunit binds the β

clamp and induces conformational changes within the clamp that open it at a single dimer interface allowing threading onto DNA (Jeruzalmi et al., 2001b). As a final step in clamp loading, DNA stimulates ATP hydrolysis by the clamp loader that leads to its detachment from the loaded clamp (Ason et al., 2000; Bertram et al., 2000). DNA polymerase then binds the clamp for processive replication.

Although the basic clamp loading reaction mechanism is understood, several key questions remain to be answered. The kinetics of ATP-dependent conformational changes driving the clamp loader DNA and β binding interactions are unclear. X-ray crystal structures of the clamp loader and a truncated γ subunit have exposed details of the nucleotide binding sites within the clamp loader, and steady-state ATP binding constants have been estimated (Hingorani and O'Donnell, 1998; Jeruzalmi et al., 2001a; Podobnik et al., 2003). However, the nature of nucleotide binding to the clamp loader during the real-time clamp loading reaction has not yet been explored. Sliding clamps of all organisms are known to enhance the activity of their respective clamp loader, but the mechanism of enhancement remains poorly characterized (Ellison and Stillman, 1998; Onrust et al., 1991). A heteropentameric minimal clamp loader complex ($\gamma_3\delta\delta'$) can be reconstituted, which is missing the χ and ψ subunits. This "minimal" clamp loader is fully functional in clamp loading promoting processive DNA synthesis, and consists of the evolutionarily conserved AAA+ motor and clamp interaction subunits (Neuwald et al., 1999; Onrust and O'Donnell, 1993). The χ and ψ subunits are unique to many prokaryotic clamp loaders; therefore, what distinctive properties do they contribute to the clamp loading mechanism?

Each of these poorly understood facets have been addressed in this dissertation by two fluorescence-based methodologies designed for investigation of steady-state and pre-steady-state ATP hydrolysis and DNA binding kinetic analyses of the clamp loader. The kinetics of DNA and β clamp binding were addressed directly in solution-based anisotropy binding assays with fluorescent-labeled β and DNA substrates. The calculated changes in anisotropy in these assays directly reflect protein-protein or protein-nucleic acid binding by the clamp loader with labeled-β or DNA, respectively (Ason et al., 2003). The ATP hydrolysis activity of the clamp loader in the presence or absence of β was measured with fluorescent-labeled *E. coli* phosphate binding protein assays (Ason et al., 2003; Brune et al., 1994). These ATP hydrolysis assays were also used to address the nature of nucleotide binding to, and the kinetics of the conformational changes within the clamp loader. By comparison of the activities of γ complex ($\gamma_3\delta\delta'\chi\psi$) to the minimal complex ($\gamma_3\delta\delta'$) using both methodologies, the functions of the χ and ψ proteins in clamp loading were addressed.

The studies outlined in this dissertation show that the clamp loader exists in a mixture of conformations in equilibrium with ATP. There were two dominant populations of conformational states, one of species activated for DNA binding and clamp loading, and one of "inactive" species. The possibility of two major populations of the clamp loader was originally proposed based on observation of pre-steady-state biphasic ATPase kinetics in the absence of β, then confirmed based on relating pre-steady-state biphasic clamp loading kinetics observed in DNA binding assays. The kinetics of the conformational changes separating these species have been determined by investigating their evolution in increasing equilibration times with ATP. Computer modeling of the experimental data from these ATP hydrolysis experiments gave reliable estimates of the rates of the conformational changes occurring between a mixture of ATP-dependent conformational states of the clamp loader. Further ATPase and DNA binding assays when the clamp loader was not equilibrated with ATP showed that the conformational changes were likely to be rate limiting in the clamp loading reaction.

Together, ATP hydrolysis and DNA binding/clamp loading kinetics describe a possible mechanism of enhancement by the β clamp where the clamp essentially increases the concentration of the activated clamp loader species, "trapping" ATP, and committing this ATP-bound clamp loader-β clamp complex to the loading reaction upon binding primed-template DNA. Comparison of the minimal complex ($\gamma_3\delta\delta'$) to γ complex ($\gamma_3\delta\delta'\chi\psi$) revealed that the χ and ψ subunits most likely facilitate the conformational changes towards the active clamp loader species. As a result, an inactive population considerably dominated the mixture of conformational states of the minimal complex in equilibrium with ATP. The β clamp, through the enhancement mechanism, converted the "unstable" minimal complex into a nearly complete active species population with ATP hydrolysis and clamp loading properties similar to γ complex. A non-hypothesized aspect of this comparison of clamp loaders is the idea that χ and ψ are AAA+ adaptor proteins that provide the clamp loader with increased stability of the activated conformation by facilitation of conformational dynamics and communication between the AAA+ subunits of the clamp loader.

Steady-State Kinetics of the Clamp Loader in the Absence or Presence of β

Originally under study in this research project was how the β clamp affected the kinetics of DNA binding and ATP hydrolysis by γ complex. Among the aims of the research was to determine how β increases the DNA-stimulated ATP hydrolysis activity of γ complex and how β affected the stability of nucleotide binding in γ complex. The goal was to gain a better understanding of these aspects of the clamp loading reaction. The early findings showed that β clamp conferred upon the clamp loader an apparent higher affinity for DNA that could be sustained for hundreds of seconds depending on the concentration of ATP in steady-state DNA binding/clamp loading assays. The apparent higher affinity for pt DNA in the presence of β was due to an increase in γ complex affinity for ATP. A series of steady-state clamp loading reactions were performed at several different concentrations of ATP. Analysis of the resulting reactions gave an estimate for an apparent dissociation constant for ATP of ~5 μM. Along with the steady-state β^{pyrene} binding assays, it is concluded that γ complex and the minimal complex have a similar apparent K_d for ATP of ~2 μM (see Figure 5-4). The DNA binding assays for γ complex in the presence or absence of β indicated that β decreases the dissociation constant for ATP binding giving the clamp loader more affinity for ATP. Additional analysis of the steady-state clamp loading reaction where the relative starting concentration of ADP was increased in separate DNA binding assays showed that excess amounts ADP could not inhibit the clamp loading reaction. This result is consistent with the conclusion that β increases the binding constant for ATP specifically.

Steady-state ATP hydrolysis assays were performed to directly assess how the β clamp affects the binding and hydrolysis of ATP by γ complex. The "Michaelis-Menten" kinetic parameters were determined for γ complex in the presence and absence of β. The maximum velocity (V_{max}) and ATP specificity constant ($k_{cat}^{complex}/K_m$) for γ complex increased in assays with β, and thus V_{max}, which is dependent on enzyme concentration, is increased about three-fold over γ complex in the absence of β to ~0.07 $\mu M^{-1}s^{-1}$.

Steady-state analysis of ATP hydrolysis also showed that β increased both the V_{max} and $k_{cat}^{complex}/K_m$ values determined for the minimal complex to levels similar to γ complex (see Table 5-1). These results are consistent with the early studies that showed the minimal complex had ATP hydrolysis and clamp loading activity similar to γ complex. Concluded here, however, is that this similar activity by the minimal complex is produced only in the presence of β clamp.

The $k_{cat}^{complex}/K_m$ values in the absence or presence of β are relatively slow, about 0.8 to 2.0 x 10^5 $M^{-1}s^{-1}$, too slow to represent a bimolecular binding reaction occurring at diffusion controlled limit between a small nucleotide cofactor (ATP) and the clamp loader. The magnitude of $k_{cat}^{complex}/K_m$ is likely to be reduced relative to the theoretical upper limit (i.e. a diffusion controlled binding rate, 1 x 10^9 $M^{-1}s^{-1}$) by a two-step binding reaction consisting of relatively rapid ATP binding followed by slow step induced by ATP binding. The slow second step of this putative two-step binding reaction is concluded to be ATP-dependent conformational changes within the clamp loader.

Kinetics of ATP-Dependent Conformational Changes within the Clamp Loader

Binding and hydrolysis of ATP molecules by the γ subunits of the clamp loader modulate the activity of the clamp loader. The energy derived from ATP binding and hydrolysis is transduced mechanically into conformational changes that give the clamp loader differential (switch-able) affinities for β and DNA. In the clamp loading mechanism, binding of ATP to this "energase" enzyme drives all of the steps necessary for placement of the clamp at the correct position on primed DNA (Turner et al., 1999). DNA stimulated ATP hydrolysis drives further conformational transitions within the clamp loader that causes it to release β on DNA. Nucleotide exchange following hydrolysis and clamp loader release allows another round of ATP binding and continued steady-state clamp loading activity. There exists both biochemical (Hingorani and O'Donnell, 1998) and structural (Jeruzalmi et al., 2001a; Podobnik et al., 2003) evidence for a conformational change in the clamp loader upon nucleotide binding to the three γ subunits. The nucleotide binding sites lie within interfaces between the δ'-γ_1, γ_1-γ_2, and γ_2-γ_3 subunits of the clamp loader. At least one and possibly two of these interfacial nucleotide binding sites is thought to be constitutively open whereas the third is deeply buried in an interface. This lead to a model describing a sequential series of individual γ-subunit conformational changes whereupon ATP binding to an open interface caused a conformational change opening the adjacent interface and so on (Jeruzalmi et al., 2001a). These conclusions drawn respective to ATP-dependent conformational changes in this study of the clamp loader structure were strongly limited by the simple fact that the structure was solved in the absence of nucleotide, and its conformational state was thought to be largely due to a crystal packing artifact. The structure does predict that ATP binding to any one of the γ subunits will directly affect adjacent subunits by undergoing a conformational change. Along this line, it could easily be imagined that the γ subunits are not static, and could constantly be undergoing conformational transitions in the absence of ATP that result in random opening and closing of the interfacial binding sites allowing molecules of ATP to bind and stabilize the clamp loader in several different forms with varying affinities for the clamp and DNA. All together, these subjective remarks point out that the clamp loader most likely exists as a mixture of conformational states both in the absence and presence of ATP.

The previously undefined kinetics of ATP-dependent conformational changes separating these states were measured in ATP hydrolysis assays and DNA binding/clamp loading assays in this research project. The overall conclusion regarding ATP-dependent conformational dynamics is that the clamp loader exists in a mixture of conformational states in equilibrium with ATP, dominated by two populations, one activated for DNA binding, and one incapable of DNA binding.

The rate of the conformational change separating the two major species of γ complex was directly addressed in pre-steady-state ATP hydrolysis reactions when γ complex was equilibrated with ATP for defined periods of time prior to mixing with DNA to initiate biphasic ATP hydrolysis reactions. Simultaneous global fitting of the experimental data for all equilibration times using the kinetic model (Figure 4-10A, reproduced here as Figure 6-1) (see also, appendix) was performed with the computer program DynaFit (Kuzmic, 1996), and simulations of equilibration were performed with the program KinTekSim (Barshop et al., 1983; Dang and Frieden, 1997). This minimal model of three conformational species adequately fitted the experimental biphasic ATP hydrolysis kinetics. The model showed that the mixture of conformational species in equilibrium with ATP is dominated by two of these species, ATTT and ETTT, where ATTT is the activated conformational state, and ETTT is an inactive conformational state defined by the slow branch transition to an inactive ITTT state that does not produce product. The physiological relevance of the non-productive ITTT species is unknown, however a "branch-pathway" producing a slow step in the reaction was necessary for modeling of the slow pre-steady-state phase of the experimentally observed biphasic kinetics.

A significant limitation of this kinetic model is that it is not sufficient for describing the DNA binding/clamp loading kinetics from the pre-steady-state anisotropy assays in the presence of β. When γ complex was equilibrated with ATP, then mixed with DNA, β and ATP, biphasic DNA binding/clamp loading kinetics were observed. These biphasic kinetics did show that two species of the clamp loader were present after equilibration with ATP, but the exact nature of these two species cannot be determined without understanding how β affects the pre-steady-state kinetics in this assay.

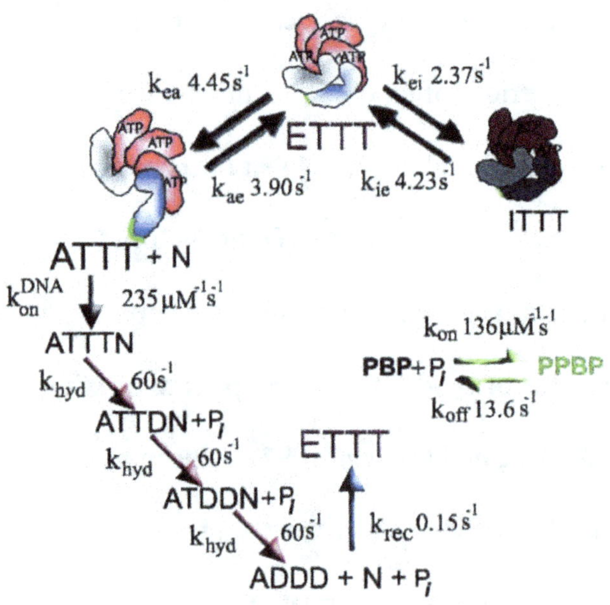

Figure 6-1. Kinetic modeling of ATP hydrolysis reactions. The minimal kinetic model is reproduced from Figure 4-10 used to fit the ATP hydrolysis data in Figure 4-9. Three states of the γ complex, ETTT, ATTT and ITTT are related by the kinetic constants shown (cartoons are based on the structure of the clamp loader, see chapter2, figure 2-3). The ATTT form binds DNA (N) rapidly and hydrolyzes its three molecules of ATP (T) in succession producing ADP (D) and inorganic phosphate (P$_i$). The γ complex dissociates from DNA on hydrolyzing its third molecule of ATP and is recycled back to the ETTT state. Experimentally determined rates of γ complex binding DNA (Ason et al., 2000; Bertram et al., 2000), k_{onN}, and MDCC-PBP (PBP) binding phosphate (Brune et al., 1994), k_{on} and k_{off}, were set constant in the fit. All other rate constants were determined by fitting the data to the model using DynaFit (Kuzmic, 1996). This model does not describe the only possible way to explain the experimental kinetics of ATP hydrolysis, and other possible mechanisms can and should be considered in the future. As an example, an alternate minimal model that contains a slow step following hydrolysis of two molecules of ATP and release of DNA fits the data well, and is presented in the appendix.

Although the above model was satisfactory in describing the conformational changes separating the mixture of clamp loader states in equilibrium with ATP, it remains an open question to which of these states β clamp binds. Therefore the biphasic DNA binding/clamp loading kinetics observed after equilibration with ATP may represent a different mechanism of clamp loader activity in the presence of β. Consistent with this idea is the result of a pre-steady-state ATP hydrolysis assay where γ complex was equilibrated with ATP prior to rapid mixing with DNA, β and ATP (see Figure 4-7). In that assay, only a single pre-steady-state rapid phase followed by steady-state ATP hydrolysis kinetics were observed. This assay is analogous to the DNA binding/clamp loading assay were pre-steady-state biphasic kinetics were observed (above, Figure 4-6A), and indicated that some amount of activated γ complex alone binds DNA and rapidly hydrolyzes ATP, then loads the clamp in a slower pre-steady-state step that occurs at a rate which limits the continuing steady-state activity.

Based on comparison of these two analogous ATP hydrolysis and DNA binding/clamp loading assays, it is concluded that the clamp loader entered the pre-steady-state reaction in the mixture of ATP-dependent conformational species described by the computer model, and then β clamp converted this mixture by binding to one or more of the conformational species. Conversion of the mixture of conformational species by β to a single new species activated for clamp loading (ATTTβ) would then relate to the rate-determining step of the clamp loading mechanism. In the pre-steady-state ATP hydrolysis and DNA binding/clamp loading assays where γ complex was equilibrated with ATP and β prior to rapid mixing with DNA, only a single pre-steady-state rapid phase of activity was observed, followed by transition into the steady-state reaction.

From these assays it was concluded that β converted the equilibrium of ATP-dependent conformational states of the clamp loader into a single activated species before reacting with DNA.

Modeling the ATP hydrolysis kinetics for the reaction of γ complex with DNA gave an estimate of the rate of the conformational change forming ATTT from ETTT (4.45 s^{-1}). This estimated "forward" conformational change rate is consistent with the complex conformational changes this multi-subunit clamp loader must make to become activated, perhaps by β binding the ATTT species or binding and converting the ETTT species (scheme 6-1). This estimated conformational change rate is in consistent with the ~2 s^{-1} turnover rate ($k_{cat}^{complex}$) calculated in the steady-state ATP hydrolysis analysis.

$$ETTT \rightleftharpoons ATTT + \beta \rightleftharpoons ATTT\beta$$
$$ETTT + \beta \rightleftharpoons ATTT\beta$$

Scheme 6-1.

The turnover rate measured by steady-state analysis is generally smaller than the actual rate constant limiting the formation of product, therefore, the rate of the conformational change estimated from pre-steady-state assays and computer modeling is likely to represent a closer approximation of the rate-limiting step in the clamp loading reaction.

The contribution of ATP binding to the rate of the conformational change was addressed in assays when γ complex was not equilibrated with ATP, then mixed with DNA and ATP. The assay was repeated at several concentrations of ATP in an attempt to determine the ATP-binding rate to γ complex. It was concluded that when the ATP concentration was ≤ 200 μM in the assay conditions, ATP binding contributed to the rate of the conformational changes, but the complexity in modeling three possibly independent ATP-binding steps hampered estimation of the ATP binding rate. The experimental data from an assay where the ATP was above 200 mM was fitted to the minimal model shown in Figure 6-1. An additional species of γ complex (PTTT) that converts relatively rapidly ($53\ s^{-1}$) to the ETTT state is required to fit to the lag phase in the experimental data. When this single initial step was included, the minimal model using the rate constants shown in Figure 6-1 gave an adequate fit to the data (see Figure 4-10C). It was concluded from this fitting that γ complex went through the same reaction pathway producing biphasic kinetics when not equilibrated with ATP as when it was equilibrated with ATP. A simulation of the equilibration with ATP for γ complex including the PTTT $\rightarrow$ ETTT step is shown in the appendix (A-1).

For the minimal complex, the above assays wherein equilibration times with ATP were changed to determine the rate of the conformational change for γ complex were not plausible due to the reduced amplitudes of biphasic ATP hydrolysis by the minimal complex in the absence of β, and therefore the reduced fluorescence change during the assay. Although the research was limited by this experimental factor, the minimal model was applied to a single data set where the minimal complex was equilibrated with ATP for a period of 1 second (Figure 5-7B, appendix). The estimated the rate of the minimal complex forward conformational change from this reasonable fitting was ~2.6 s^{-1}. Although this approximation is limited by the absence of additional data sets for increasing equilibration times, it is in agreement with the lower calculated steady-state turnover rate for ATP hydrolysis limiting the clamp loading reaction compared to γ complex. A large reverse conformational change (ATTT $\rightarrow$ ETTT) rate (12.6 s^{-1}) was also estimated by this fitting analysis, and is consistent with the greatly reduced activated species population of the minimal complex. Simulation of the minimal complex equilibration with ATP using rates determined from the fitting of the single data set from Figure 5-7B gave estimate of ~10 % ATTT species at 1 second (appendix, A-2).

A Possible Mechanism for β Clamp Enhancement of Clamp Loader Activity

The majority of experiments presented in this dissertation were performed in the absence and presence of β to understand how this clamp affects γ complex. A comprehensive comparison of the kinetics of ATP hydrolysis, and DNA binding/clamp loading has produced a possible answer to the question: How does sliding-clamp enhance the activity of its clamp loader? Due to the functional conservation between clamps and clamp loaders of all forms of life, the possibilities presented here may be applicable across evolution as well.

The steady-state ATP hydrolysis kinetics of the clamp loader are affected by the β clamp (Hingorani and O'Donnell, 1998; Onrust et al., 1991). The Michaelis-Menten analysis presented in this dissertation produced parameters consistent with the previous studies, and additionally provided novel analysis of the minimal complex. An important conclusion is that the increase of the ATP specificity constant for the minimal complex was of greater magnitude than for γ complex. The predicted active population of the minimal complex in equilibrium with ATP in the absence of β is considerably small in comparison to γ complex, and the presence of β, confers upon this population enhanced activity similar to γ complex. This reflects on the overall effect the clamp has on the clamp loader. The clamp appears to partition the clamp loader to an activated form kinetically more robust than in the absence of the clamp.

The combined results of the kinetic assays presented here point to a possible mechanism where β clamp selectively binds the activated species of the clamp loader during equilibration with ATP and β, converting all clamp loaders into an activated ATP-clamp loader-β complex (ATTT + $\beta \leftrightarrows$ ATTTβ) poised for rapid clamp loading. Both the minimal complex and γ complex have similar affinity for ATP, $K_d \sim 2$ μM (Figure 5-4) (Hingorani and O'Donnell, 1998), and the ATP-bound clamp loaders have low nanomolar affinity for β, $K_d^{apparent} \sim 3\text{-}4$ nM (Figure 5-3) (Naktinis et al., 1995), therefore during the equilibration period in the assays β would be able to pull the equilibrium between the mixture of ATP-dependent conformational clamp loader states from the inactive species towards a completely activated species. An alternative to this mechanism is a possible situation where β could bind to an inactive species as well as the activated species, promoting conversion to the activated species (see scheme 6-1). This possibility would require a change in the conformational change kinetics rather than β simply changing the equilibrium between the states by selectively binding to the activated species. Future work will be required to differentiate between these two possible mechanisms. Pre-steady-state ATP hydrolysis assays should be repeated when the clamp loader is equilibrated with ATP and β for several different periods of time before mixing with DNA to initiate the reaction. Analysis of the resulting kinetics from these assays could be done in a manner analogous to the analysis of the series of assays when γ complex was equilibrated with ATP for different periods of time. If β selectively binds to the activated species only, then a decrease in rapid hydrolysis pre-steady-state kinetics would be observed with decreasing equilibration times. The resulting pre-steady-state kinetics may also be biphasic as the inactive species is intrinsically converted to the activated species similar to the pre-steady-state ATP hydrolysis kinetics observed in assays without β clamp (ETTT $\leftrightarrows$ ATTT + $\beta \leftrightarrows$ ATTTβ).

Preliminary pre-steady-state experiments when the equilibration time for the clamp loader, ATP, and β was decreased to 250 ms have already been performed in this dissertation project (not shown). The resulting pre-steady-state hydrolysis kinetics were biphasic in nature. Previous pre-steady-state analysis using radio-labeled ADP release as a measure of ATP hydrolysis kinetics was performed when γ complex was incubated with ATP and β for a duration of 80 ms. The published results showed that the pre-steady-state formation of ADP was in fact biphasic (Hingorani et al., 1999). Although the possible mechanism explained by the equilibrium shift by selective binding to the activated clamp loader is supported by these data, further analysis like that proposed above is required, perhaps along with computer modeling, to fully understand how β converts the clamp loader into a completely activated form within a ≥ 1-second incubation period. These experiments would also help the future investigator acquire more kinetic data concerning which clamp loader conformational species β binds.

Conclusions from the ATPγS-chase assays shed additional light on the putative clamp enhancement mechanism. An essentially non-hydrolyzable ATP analogue, ATPγS, was used in pre-steady-state and steady-state ATP hydrolysis assays to determine the nature of nucleotide binding by the γ complex or minimal complex clamp loaders. Pre-steady-state ATPγS-chase assays in the presence of β revealed that the molecules of ATP bound to either clamp loader were hydrolyzed faster than they dissociate (see γ complex, Figure 4-5, and the minimal complex, Figure 5-7). Pre-steady-state ATPγS-chase assays were also performed in the absence of β, and showed that only the molecules of ATP bound to the activated clamp loader species were hydrolyzed faster than they dissociate, and that ATP molecules bound to any inactive species exchange more rapidly with ATPγS than they were hydrolyzed. It is thus concluded that molecules of ATP bound to the activated clamp loader are more secure or stably bound than ATPs bound to any inactive species. In the presence of β, all of the clamp loader is in the ATP-bound active state with the molecules of ATP essentially "trapped" in the activated complex with β. The ATPs remain securely bound until the activated clamp loader-β complex forms a ternary complex with DNA. DNA then triggers hydrolysis of the bound ATP molecules causing the clamp loader to lose affinity for the clamp and DNA, finishing the clamp loading reaction.

It is concluded that the β clamp enhances the clamp loader by a mechanism wherein it selectively binds to the activated species of the clamp loader. Selective binding to the activated clamp loader effectively increases its concentration by driving conformational equilibrium between inactive and active species towards the activated state (ETTT → ATTTβ). The sliding-clamp secures the molecules of ATP bound to this active state, and commits it to a productive clamp loading reaction on primed-template DNA for processive replication.

The Missing χ and ψ Subunits are Responsible for the Kinetic Differences Between the Minimal Complex and γ Complex

As stated above, the χ and ψ subunits are missing from the minimal complex, and are the likely causes of the differences described here for the DNA binding, clamp loading and ATP hydrolysis activities by the minimal complex and γ complex. The most notable dissimilarities between the minimal complex and γ complex appeared in assays without the β clamp. Without β, the clamp loader is dynamically changing between a range of conformations that acquire some new equilibrium mixture when molecules of ATP bind, essentially defined or separated in majority, by inactive and activated species populations. The ATP-dependent conformational dynamics in the clamp loader are the driving force of this energase molecular machine. Without the χ and ψ proteins, the minimal complex conformational dynamics appear to be hampered in such a way that this sub-complex clamp loading machine had diminished intersubunit communication.

The χ and ψ Subunits are *E. coli* Clamp Loader AAA+ Adaptor Proteins

The minimal complex has activities similar to γ complex in the presence of β. This should be expected given that the five-subunit assembly of AAA+ proteins in the minimal complex appears to be conserved across evolution (Davey et al., 2002; Ellison and Stillman, 2001). It should be noted that the *E. coli* clamp loader differs from clamp loaders of other organisms. There is the need for the additional χ and ψ clamp loader subunits at the replication fork for direct interaction with single-stranded DNA binding protein (SSB), and for indirect interaction with primase. The χ and ψ subunits, based on the work presented in this dissertation, are also required for optimum β clamp loading activity. They confer additional conformational stability to the clamp loader when in equilibrium with ATP, essentially "tightening" it up. This is accomplished possibly by facilitation of the ATP-dependent conformational changes towards the activated clamp loading species, or by slowing the rate of conformational change towards the inactive population. It is likely that χ and ψ confer both of these functions to the γ complex clamp loader, giving it a true advantage over the evolutionarily conserved five-subunit clamp loader architecture found elsewhere.

It is proposed in this dissertation that the χ and ψ subunits are members of a growing group of AAA+ adaptor proteins (Dougan et al., 2002). Among this group are the small subunits of Clp proteases (Zeth et al., 2002) in *E. coli*, and the general function of these adaptor proteins appears to be in conferring substrate specificity for the AAA+

motor proteins they assist. This group of proteins has only recently been identified along with the continued identification of new AAA+ proteins. The χ and ψ are small structurally unrelated protein subunits of a AAA+ machine. The χ subunit, by its interaction with SSB at the replication fork and sites of DNA repair where ss DNA would be present, gives the clamp loader additional substrate specificity by drawing it to the sites where β must be loaded for DNA synthesis. By these characteristics, it is proposed that the χ and ψ subunits are AAA+ adaptor proteins at the replication fork.

Additionally, they also perform "adaptor" functions within the clamp loader as outlined in this dissertation. The χ and ψ subunits are directly required for optimal conformational dynamics and intramolecular communication in the clamp loader for the β clamp loading reaction. The possible facilitation of conformational change producing an active AAA+ motor, could be a novel function for AAA+ adaptor proteins. It will be interesting to learn in the future if other AAA+ adaptor proteins help modulate the "switching", or conformational dynamics of their respective AAA+ machines. Interestingly, the four small subunits of the eukaryotic RFC (2-5) clamp loader form a functional core, and are adapted for other cellular processes. The large RFC-1 subunit swaps with another protein, specific for functions such as replication termination (Kouprina et al., 1994), and cell cycle checkpoint control (Green et al., 2000). Differing the nature of ATP binding affinities and putative conformational changes in the RFC-2-5 core may provide differential reaction mechanisms for these diverse functions brought on by the "adaptor" proteins, analogous to that presented here for the *E. coli* clamp loader.

Programmatic Recommendations

The recommendations are presented in order of their relevance to this work.

Pre-Steady-State Kinetics of β Clamp Binding

One of the most important aspects of continued exploration of the mechanism of clamp loading is the determination of the kinetics of β clamp binding to the clamp loader. In this work, the steady-state dissociation constant was identified, but in order to fully understand the clamp loading mechanism, the pre-steady-state kinetics of β clamp binding must be determined. For example, the kinetics of ATP hydrolysis and DNA binding/clamp loading in this work could not be reliably modeled without knowing the rate constants for binding and dissociation of β. Modeling and simulation of experimental data is important for not only understanding the mechanism under study, but also for prediction of experimental results. The ability to predict experimental results by simulation based on a working (developing) model can be a significant time and money saving process.

During the course of this dissertation study, the measurement of pre-steady-state β clamp binding kinetics was attempted in stopped-flow experiments using the fluorescent-labeled β^{pyrene} substrate described for the steady-state analyses. The measurement of pre-steady-state β^{pyrene} binding kinetics was hampered due to instrumental limitations of the two stopped-flow setups. The fluorescence emission characteristics of β^{pyrene} are complex in nature, and this along with the disperse nature of the emitted fluorescence (i.e., emission bandwidth broadening) due to the optical setup of the stopped-flow instruments caused the inability to measure polarized emissions for calculation of real-time anisotropy of β^{pyrene}. The steady-state emission intensities measured for the results presented here, were recorded in a steady-state fluorimeter setup that used monochromators for well-defined selection of the β^{pyrene} emission. For use of β^{pyrene} in pre-steady-state experiments, the emission optics of the stopped-flow should be directed to the same monochromators. Another perhaps easier approach would be the use of β covalently labeled with a different fluorophore (i.e., tetramethyl-rhodamine (TMR). The use of β^{TMR} in anisotropy binding assays may provide the needed data for determination of the bimolecular reaction rates with the clamp loader. Also possible would be the determination of which clamp loader species β binds by observation of pre-steady-state fluorescent-β binding kinetics in assays when the clamp loader is either equilibrated with ATP or not equilibrated with ATP prior to rapid mixing with fluorescent-β. If selective binding to the activated clamp loader species occurs, as proposed here, then the investigator may expect to observe biphasic β binding kinetics followed by steady-state equilibrium binding. Upon observation of pre-steady-state biphasic β binding kinetics it would be in the investigator's best interest to repeat the experiment at several equilibration times to determine if the kinetics defining the evolution of the activated clamp loader species binding β are similar to those presented here as a function of DNA-stimulated ATP hydrolysis. If β can bind to any species of the clamp loader when the clamp loader is equilibrated with ATP, then a single pre-steady-state rapid β binding phase may be observed. When the clamp loader is directly added to fluorescent-β and ATP, slow binding kinetics may be observed, at which time the concentration dependence of ATP in the β binding kinetics should be explored. Just as the pre-steady-state kinetics of DNA binding/clamp loading and ATP hydrolysis were directly correlated in this work (see Figure 5-10), the pre-steady-state analysis of fluorescent-β binding should be correlated with both as well. Pre-steady-state fluorescent-β binding analysis should prove to be a powerful tool for the continued development of the kinetic model presented here, as well as for modeling of the clamp loading mechanism.

Fluorescence Lifetime Measurement of MDCC-PBP Titrated with Inorganic Phosphate

In the development and characterization of the MDCC-bound PBP the investigators found that the inorganic phosphate (P_i)-responsive form of MDCC-PBP consisted of equal amounts of two diastereoisomers by tryptic digest (Brune et al., 1998). It was postulated that the equal populations of diastereoisomers affected the fluorescence response in P_i-titrations. The populations of diastereoisomers could not be distinguished kinetically or spectroscopically in their work. A 50:50 mixture was modeled where one form was responsible for the fluorescence increase to approximately 50% saturation, and the other form was responsible for saturation of fluorescence to 100%. In our P_i-titrations we propose that the point of intersection of the linear rise of fluorescence and saturated fluorescence represents the accurate population of labeled-MDCC-PBP in our assays. Thus we currently model the change in fluorescence in the P_i-titrations as due to mostly a single isomer population. To determine if the fluorescence change observed in our P_i-titrations (for characterization of our preparations of MDCC-PBP) are due to one or more populations of isomers, fluorescence lifetime measurements should be done as a function of increasing P_i concentration, mimicking the P_i-titrations. In these experiments, number of fluorescence lifetimes calculated at each concentration of P_i should be proportional to the number of isomers present. The longest relative lifetime(s) at each concentration of P_i will relate to the isomer with the highest quantum yield, thus the isomer that is mostly responsible for the observed change in fluorescence. The fluorescence decays measured from the time-resolved data for each concentration of P_i should be integrated to calculate total emission for regeneration of a P_i-titration from the lifetime data. In their work, (Brune et al., 1998), MDCC-PBP in the absence or presence of P_i showed a change in fluorescence lifetime of 0.3 ns and 2.4 ns respectively, consistent with a single population of responsive MDCC-PBP with an increase in quantum yield when bound to P_i. The proposed fluorescence lifetime measurements should be performed under "magic-angle" conditions measuring emission polarization at $54.7°$ to correct for polarization artifacts (Brune et al., 1998; Lakowicz, 1999).

Further Investigation of ATP Binding and Nucleotide Exchange by the Clamp Loader

The pre-steady-state rate constants for ATP binding the clamp loader were unable to be estimated as a function of hydrolysis kinetics in this work. In the future, pre-steady-state assays utilizing fluorescent-labeled ATP molecules could be performed to determine the kinetics of ATP binding to the clamp loader. It is recommended that ribose-modified ATP analogues containing the N-methylanthraniloyl (MANT) fluorophore be used as they are well characterized (Hiratsuka, 1983), and have been successfully used with other Walker-A motif containing energases, suggesting that the MANT fluorophore does not hinder nucleotide binding to these enzymes (Moore and Lohman, 1994; Moyer et al., 1998). A fluorescent-ATP analogue (Bodipy-FL-ATP) was used during the course of this study, and it was concluded that this fluorescent analogue most likely did not bind to the nucleotide binding sites in γ complex due the relatively large Bodipy-FL derivative. Based on the analyses and computer modeling described in this work, it is predicted that the rate of ATP binding and clamp loader conformational dynamics will be intimately related at low ATP concentrations. The complexity of determining pre-steady-state binding rates to an enzyme with three nucleotide binding sites may still impede this work, however, mutations in γ complex which change the nature of one or more of the nucleotide binding sites exist, and could help in these proposed studies.

These assays performed in this work for investigation of the nature of nucleotide binding to the clamp loader probed nucleotide exchange, a largely unexplored facet of the clamp loading reaction cycle. Future work will be required to understand the nucleotide exchange steps and conformational changes that occur during this stage of the reaction cycle. Ribose-modified ADP analogues containing the *N*-methylanthraniloyl (MANT) fluorophore can be used to explore the kinetics of ADP binding and exchange by the clamp loader (Hiratsuka, 1983; Moyer et al., 1998). Pre-steady-state nucleotide exchange assays using either mant-ADP or mant-ATP could be imagined in two complementary assays. In the first, mant-ADP, equilibrated with γ complex would be rapidly mixed with DNA, +/-β, and non-fluorescent ATP. These assays would be expected to produce a decay of mant-ADP fluorescence as it is exchanged for ATP in the pre-steady-state reactions. The observed rate of this decay will be dependent at least on the exchange rate of mant-ADP for non-fluorescent ATP. In the complementary assay, γ complex would be equilibrated with non-fluorescent ADP, and then rapidly mixed with DNA, +/-β clamp, and mant-ATP. After rapid mixing into the reaction cuvette, an increase in mant-ATP fluorescence could be expected, which will represent rate(s) of non-fluorescent ADP exchange followed by mant-ATP binding in the pre-steady-state reaction. Understanding of the kinetics of nucleotide exchange steps of the clamp loading reaction will significantly aid in computer modeling. The aspects of computer modeling are eased as rates of additional steps of the clamp loading reaction become defined, and fewer assumptions are made about reaction steps lacking in experimental data. The steady-state

CD spectroscopy analyses described below are likely to provide information on the conformational change rates during nucleotide exchange, and may show that they are similar to the rate-determining conformational change determined from pre-steady-state ATP hydrolysis and DNA binding/clamp loading assays in this dissertation.

Analysis of Clamp Loader Conformational Dynamics by Circular Dichroism Spectroscopy

For continued analysis of the conformational dynamics of the clamp loader and identification of the conformational change rates with another assay, it is recommended that Circular Dichroism (CD) spectroscopy be utilized. The γ complex clamp loader is readily applicable to CD spectroscopy, as are most proteins (Fersht, 1999; Woody, 1995). The peptide bonds, as well as aromatic side chains are optically active centers from which CD arises. For example, aromatic side chains absorb circularly polarized light in two components. It is the difference between the two absorptions that is represented by the CD spectra. While probing the electronic transitions of the aromatic groups, a change in CD reveals a change in the protein tertiary structure. The amides in the peptide bond group also present strong characteristic CD and therefore are exploited as probes of protein secondary structure. CD easily reveals fractions of α-helical and β-sheet content. Each of the five subunits of γ complex has aromatic side chains. This would allow a global CD analysis of the clamp loader complex upon induction of the ATP-dependent conformational change. It would also be informative to examine isolated γ subunits and the minimal complex in the presence of different nucleotides (ATP, ADP, ATPγS) to gain further insight into other possible conformational intermediates (i.e., due to mixed nucleotide binding). CD, although not as spatially sensitive as other methods of protein structural analysis (X-ray diffraction, magnetic resonance), is a much simpler method both instrumentally and experimentally. CD also allows for the use of large complexes, and rapid measurement making it suitable for the determination of the molecular and kinetic details of the γ complex ATP-dependent conformational changes. Once γ complex conformational transitions are well characterized by steady-state CD, time-resolved pre-steady-state CD analysis should be completed for measurement of the conformational change rates. These pre-steady-state experiments should be performed under conditions similar to those presented in this dissertation project to allow closer comparison to the data presented here, and correlated computer modeling.

Identification of the Putative Clamp Loader DNA Binding Surface

The pre-steady-state ATP hydrolysis assays in the presence of β indicated that each γ subunit of the clamp loader hydrolyzed one molecule of ATP for a total of three per clamp loading reaction (see Figure 4-4, inset). Also, the combined biphasic pre-steady-state ATP hydrolysis amplitudes from assays of γ complex in the absence of β gave a value of about three molecules hydrolyzed. Therefore, DNA stimulates hydrolysis of any ATP bound to each γ subunit, only when the clamp loader has acquired the activated conformation that can bind DNA. That DNA stimulates hydrolysis by each γ subunit in the clamp loader implies that all three γ subunits contact the DNA and form part of the putative DNA binding surface on the clamp loader. This provisional conclusion will require further investigation of the exact nature of the DNA binding properties of the clamp loader. Four subunits of the clamp loader, including all three γ subunits, contain cysteine-coordinated zinc modules (see Figure 2-3). These zinc modules are located the N-terminal domains of the clamp loader; the same surface where the δ subunit β-interaction element is exposed upon ATP binding (Jeruzalmi et al., 2001a; Jeruzalmi et al., 2001b). For future determination of whether these zinc modules are part of the putative DNA binding surface, and to explore if they are linked to DNA-stimulated ATP hydrolysis, it is recommended that the cysteine-coordination be disrupted by site-directed mutagenesis such that the zinc-binding module is disrupted. This action should be first taken on the γ subunits, as they most likely form the larger portion of the putative DNA binding surface. Also, the clamp loader crystal structure should be utilized as a guide for additional site-directed mutagenesis to obtain information about the DNA binding site. The structure did show some focused positive electrostatic potential, usually observed for protein-DNA binding sites, on the N-terminal domains of the clamp loader (Jeruzalmi et al., 2001a).

The χ and ψ AAA+ adaptor hypothesis

The hypothesis that the χ and ψ subunits could be addressed initially by a bioinformatics approach, for comparison to the other identified AAA+ adaptors (Dougan et al., 2002) and a search for other AAA+ protein partners. The goal in terms of continuation of the work in this dissertation would be to identify other AAA+ adaptor proteins that confer stability to multi-subunit complexes, as well as those adaptors that may facilitate the conformational dynamics and intersubunit communication between AAA+ subunits. It will be interesting to learn in the future if other AAA+ adaptor proteins help modulate the "switching", or conformational dynamics of their respective AAA+ machines.

Concluding Statement

The details uncovered in this dissertation reveal the dynamic interconnections within a molecular machine, essentially switching it between "on" and "off" states. The ATP-dependent conformational changes within the clamp loader modulate the affinities for DNA and β clamp that drive the clamp loading reaction. Understanding of how protein quaternary (intersubunit) dynamics and substrate-dependent communications

drive multi-protein complexes such as the clamp loader is a fundamental aspect of

understanding the function of many complex biological processes or other enzyme

machine consisting of interactions between several protein subunits. These results

heighten the importance of solution-based exploration of protein structural and functional

dynamics for determination of their mechanistic modes of activity using fluorescence-

based assays like those presented here, as well as NMR spectroscopy, circular dichroism

spectroscopy, EPR measurements etc. Familiarity of the inner workings of multi-

component biological machines will greatly improve understanding of their functions,

and will aid in the understanding and treatment of disease caused by their malfunction.

Knowledge of the dynamic molecular communication within these machines will also aid

in research and development of novel molecular devices with the goal of providing a

contribution to the quality of human life.

COMPUTER MODELING OF EXPERIMENTAL KINETIC DATA

DynaFit Script For Fitting Shown in Figure 4-9

See output index below

```
;----------------------------------------------------------
[task]
  task = fit
  data = progress

[mechanism]
; isomerizations of Gamma.ATP complexes:
  ETTT <==> ATTT      : kea    kae
  ETTT <==> ITTT      : kei    kie

; DNA binding and hydrolysis:
  ATTT + N ---> ATTTN    : konN
  ATTTN ---> ATTDN + P   : khyd
  ATTDN ---> ATDDN + P   : khyd
  ATDDN ---> ADDD + N + P   : khyd
  ADDD ----> ETTT   : krec

; phosphate probe:
  P + mPBP <==> PmPBP      : kpon    kpoff

[constants]
  kea   = 4.453 ?
  kae   = 3.604 ?
  kei   = 2.37 ?
  kie   = 4.376 ?
  konN  = 235
  khyd  = 60
  krec  = 0.15
  kpon = 136
  kpoff= 13.6

[concentrations]
  N = 1.0
  mPBP = 2.7
```

```
[responses]
 PmPBP = 1 ?


[progress]
 directory  DATAmastr/atpase
 extension  txt
 offset     auto ? local
 file       XDdt2_15Pi_2s
 incubate   ETTT = 0.54 , dilute 1:2, time 0.015
 file       XDdt2_48Pi_2s
 incubate   ETTT = 0.54 , dilute 1:2, time 0.048
 file       XDdt2_98Pi_2s
 incubate   ETTT = 0.54 , dilute 1:2, time 0.098
 file       XDdt2_247Pi_2s
 incubate   ETTT = 0.54 , dilute 1:2, time 0.247
 file       XDdt2_498Pi_2s
 incubate   ETTT = 0.54 , dilute 1:2, time 0.498
 file       XDdt2_1000Pi_2s
 incubate   ETTT = 0.54 , dilute 1:2, time 1.0


[output]
 directory  users/CRWDISS/gcXDdt/outfit1


[settings]
<Marquardt>
  Interrupt = 100
<Constraints>
  ConcError = 10


[end]
```

DynaFit Script For Fitting in Shown in Figure 4-10C

See output index below

```
;------------------------------------------------------------
[task]
  task = fit
  data = progress

[mechanism]
 ; isomerizations of Gamma.ATP complexes:
 PTTT --> ETTT        : kpe
 ETTT <==> ATTT        : kea   kae
 ETTT <==> ITTT        : kei   kie
```

```
; DNA binding and hydrolysis:
ATTT + N ---> ATTTN    : konN
ATTTN ---> ATTDN + P   : khyd
ATTDN ---> ATDDN + P   : khyd
ATDDN ---> ADDD + N + P   : khyd
ADDD ----> ETTT   : krec
; phosphate probe:
P + mPBP <==> PmPBP      : kpon    kpoff


[constants]
 kpe  = 61 ?
 kea  = 4.9
 kae  = 3.8
 kei  = 3.1
 kie  = 4.8


 konN  = 235
 khyd   = 60
 krec  = 0.15


 kpon = 136
 kpoff= 13.6


[concentrations]
 N = 1.0
 mPBP = 2.7


[responses]
 PmPBP = 1 ?


[progress]
 directory   DATAmastr/atpase
 extension   txt
 offset     auto ?
 file       Zdt500Pi_2s
 conc PTTT = 0.27


[output]
 directory   users/CRWDISS/gcXZdt/outfit1


[settings]
<Marquardt>
 Interrupt = 100
<Constraints>
 ConcError = 10
```

[end]

DynaFit Script For Fitting in Shown in Figure 5-6

Note low [γδδ'] for "dead" enz.
see output index below
;---+--
[task]
  task = fit
  data = progress

[mechanism]
  ; isomerizations of Gamma.ATP complexes:
  PTTT --> ETTT   : kpe
  ETTT <==> ATTT   : kea   kae
  ETTT <==> ITTT   : kei   kie

  ; DNA binding and hydrolysis:
  ATTT + N ---> ATTTN   : konN
  ATTTN ---> ATTDN + P : khyd
  ATTDN ---> ATDDN + P : khyd
  ATDDN ---> ADDD + N + P : khyd
  ADDD ----> ETTT : krec

  ; phosphate probe:
  P + mPBP <==> PmPBP   : kpon   kpoff

[constants]
  kpe = 53 ?
  kea = 1.72
  kae = 8.8
  kei = 2.5
  kie = 4.99

  konN = 235
  khyd   = 60
  krec = 0.15

  kpon = 136
  kpoff= 13.6

[concentrations]
  N = 1.0
  mPBP = 2.7

[responses]

```
  PmPBP = 1 ?

[progress]
   directory   DATAmastr/atpase
  extension   txt
  offset      auto ?

  file      gddXZdtA02_Pi_20
  conc PTTT = 0.135 ?

[output]
  directory   users/CRWDISS/gdd/outfitlowgdd1

[settings]
<Marquardt>
  Interrupt = 100
<Constraints>
  ConcError = 10
[end]
```

Simulation Mechanisms for Equilibration Steps: KinTekSim Program

See Figure 4-10B

```
"gcatpEQ1.mec"
$: γ complex or minimal complex equilibration with ATP

ETTT == ATTT
ETTT == ITTT

*OUTPUTS:
F1*ETTT
F2*ATTT
F3*ITTT


-----------------------------------
-----------------------------------
"gcatpNOEQ.mec"
$: γ complex or minimal complex, no equilibration with ATP

PTTT == ETTT
ETTT == ATTT
ETTT == ITTT

*OUTPUTS:
```

F1*ETTT
F2*ATTT
F3*ITTT
F4*PTTT

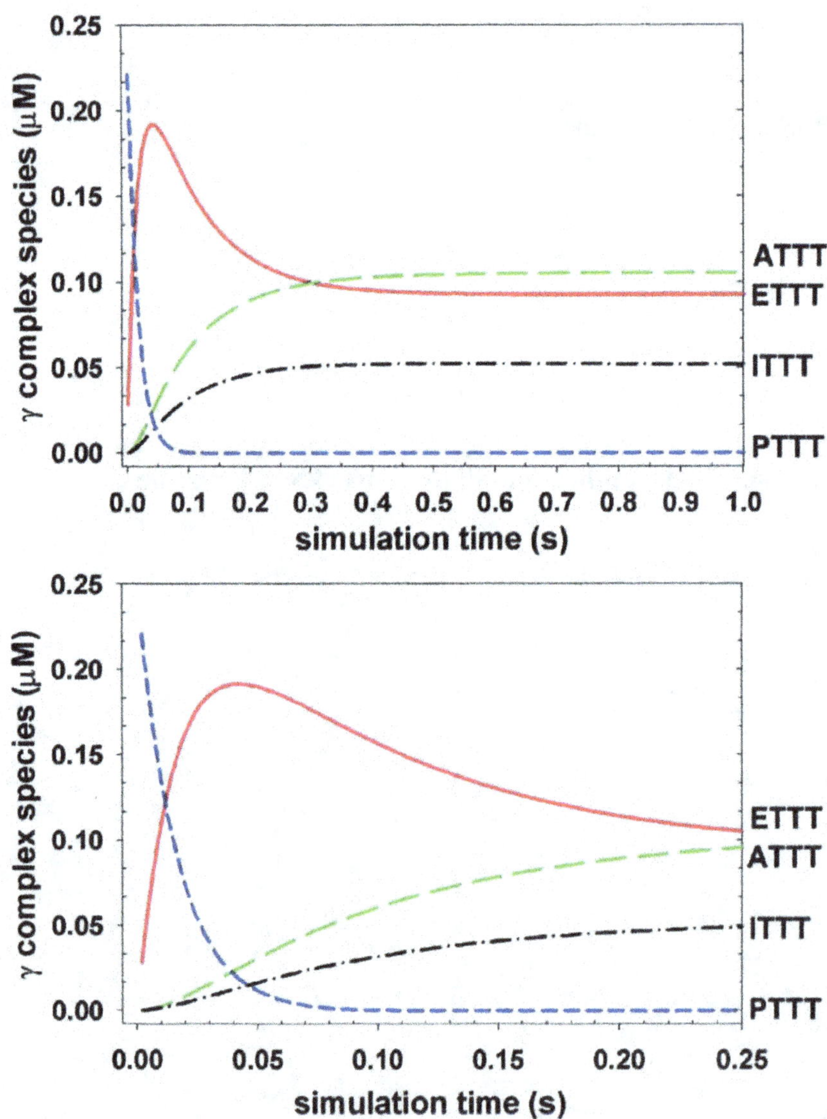

A-1. KinTekSim simulation of γ complex, no equilibration with ATP. An expanded time scale is shown in the bottom panel. This simulation uses the rates determined by DynaFit fitting (figure 4-10C). See output index below.

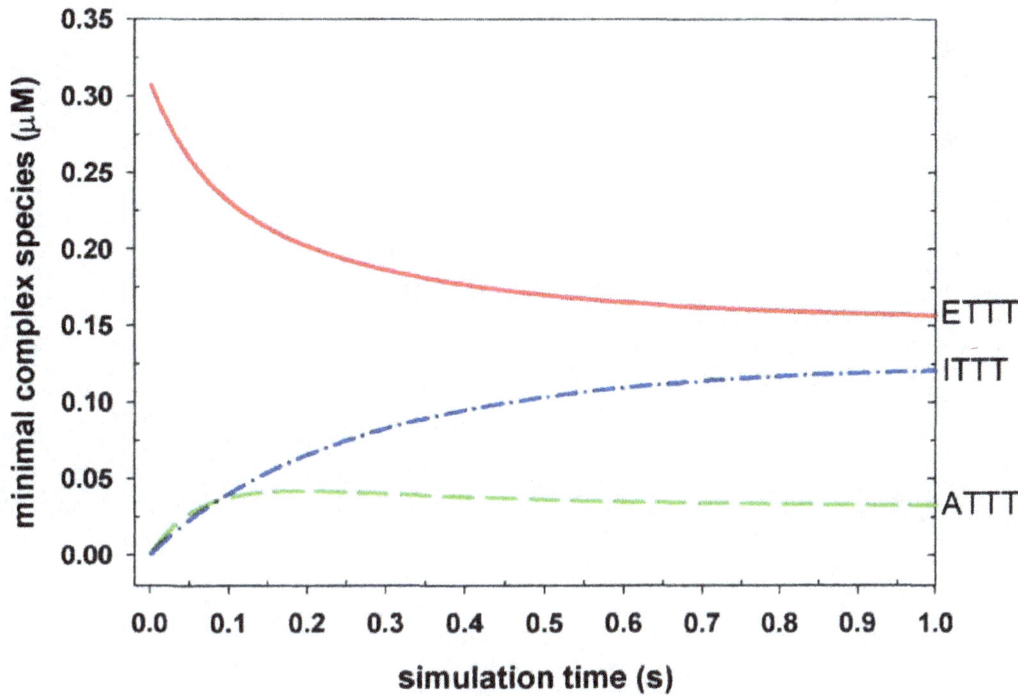

A-2. KinTekSim simulation of the minimal complex equilibrated with ATP. This
simulation uses the rates determined by DynaFit fitting a single 1000 ms preincubation time data set shown in figure 5-7B, black trace. [kea = 2.6. kae =
12.6, kei = 1.7, kie = 2.1 (s^{-1}) Conc ETTT = 0.31 μM] See output index below

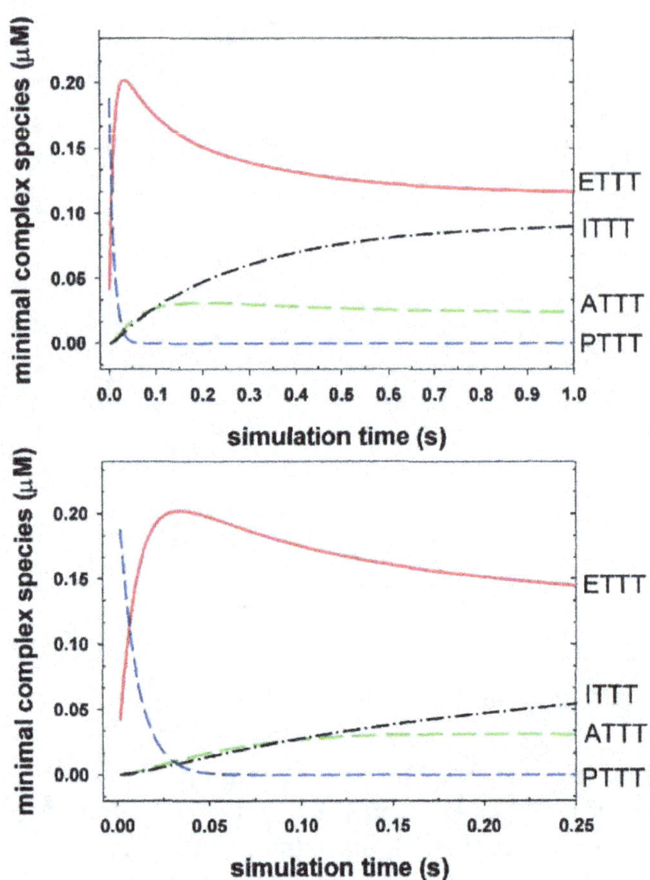

A-3. KinTekSim simulation of the minimal complex, no equilibration with ATP. An
     expanded time scale is shown in the bottom panel. This simulation uses the
     rates determined by DynaFit fitting (figure 5-6). [kpe = 102, kea = 2.6. kae =
     12.6, kei = 1.7, kie = 2.1 (s^{-1}) Conc PTTT = 0.23 µM] See output index
     below.

DynaFit Output Indices

Fitting for γ Complex Data in Figure 4-9

SCRIPT FILE: "XDdtfit1_2s.txt"

TASK
Fit of progress curves
DATA
file DATAmastr\atpase\XDdt2_15Pi_2s.txt
file DATAmastr\atpase\XDdt2_48Pi_2s.txt
file DATAmastr\atpase\XDdt2_98Pi_2s.txt
file DATAmastr\atpase\XDdt2_247Pi_2s.txt

file DATAmastr\atpase\XDdt2_498Pi_2s.txt
file DATAmastr\atpase\XDdt2_1000Pi_2s.txt

REACTION MECHANISM
ETTT <===> ATTT : kea kae
ETTT <===> ITTT : kei kie
ATTT + N ---> ATTTN : konN
ATTTN ---> ATTDN + P : khyd
ATTDN ---> P + ATDDN : khyd
ATDDN ---> N + P + ADDD : khyd
ADDD ---> ETTT : krec
P + mPBP <===> PmPBP : kpon kpoff

DIFERENTIAL EQUATIONS
d[ETTT]/dt = -kea[ETTT]+kae[ATTT]-kei[ETTT]+kie[ITTT]+krec[ADDD]
d[ATTT]/dt = +kea[ETTT]-kae[ATTT]-konN[ATTT][N]
d[ITTT]/dt = +kei[ETTT]-kie[ITTT]
d[N]/dt = -konN[ATTT][N]+khyd[ATDDN]
d[ATTTN]/dt = +konN[ATTT][N]-khyd[ATTTN]
d[ATTDN]/dt = +khyd[ATTTN]-khyd[ATTDN]
d[P]/dt = +khyd[ATTTN]+khyd[ATTDN]+khyd[ATDDN]
        kpon[P][mPBP]+kpoff[PmPBP]
d[ATDDN]/dt = +khyd[ATTDN]-khyd[ATDDN]
d[ADDD]/dt = +khyd[ATDDN]-krec[ADDD]
d[mPBP]/dt = -kpon[P][mPBP]+kpoff[PmPBP]
d[PmPBP]/dt = +kpon[P][mPBP]-kpoff[PmPBP]

OUTPUT

LEAST-SQUARES FIT
mean square   5.01823e-005
standard deviation   0.00708395
log(determinant)   -0.993
log(condition number) -3.66
Marquardt parameter 8
execution time (min) 0.06033
datapoints   9000
parameters   11
iterations   20
subiterations   29
function evaluations 54
error status   0

PARAMETERS & STANDARD ERRORS

Set	Parameter	Initial	Fitted	Error	%Error
	kea	4.453	4.853	0.034	0.7

kae	3.604	3.781	0.019	0.51
kei	2.37	3.142	0.072	2.3
kie	4.376	4.752	0.054	1.1

	r:PmPBP	1	0.9265	0.00068	0.073
1	offset	0	-0.008696	0.00066	7.6
2	offset	0	-0.013	0.00066	5.1
3	offset	0	-0.009641	0.00065	6.7
4	offset	0	-0.006036	0.00065	11
5	offset	0	7.357e-005	0.00068	920
6	offset	0	-0.02251	0.00071	3.1

COVARIANCE MATRIX
Set Parameter Covariances

```
        kea a   a
        kae b -10   b
        kei c  94 -17   c
        kie d  74  -2  89   d
    r:PmPBP e  35 -49  17 -12   e
1   offset f -64  22 -43 -22 -81   f
2   offset g -62  23 -41 -20 -82  92   g
3   offset h -59  27 -38 -18 -83  92  92   h
4   offset i -55  37 -37 -19 -84  90  90  91   i
5   offset j -59  45 -45 -28 -82  88  89  89  91   j
6   offset k -61  48 -49 -33 -80  87  87  88  91  93
```

EIGENVECTORS AND EIGENVALUES

```
            Eigenvectors
 Eigenvalues   0.00 0.01 0.05 0.09 0.36 1.00 1.00 1.00 1.00 1.26
     log(C)    3.54 2.75 2.02 1.79 1.17 0.72 0.72 0.72 0.72 0.62
Set Parameter     1   2   3   4   5   6   7   8   9  10
        kea 1  52 -22 -55 -34  23   0   0   0   0 -12
        kae 2  -1   7 -18 -45 -65   0   0   0   0 -46
        kei 3  75  18  -3  43 -12   0   0   0  -1   6
        kie 4  34  37  65 -32 -13   0   0   0   0
    r:PmPBP 5   6 -58  14  43 -50   0   0   0   0 -15
1   offset 6  -7  26  -6  24  11 -23  51 -36 -42 -43
2   offset 7  -7  27  -7  23   9  65 -44   8 -17 -39
3   offset 8  -6  27 -10  21   4 -61 -28  46  27 -29
4   offset 9  -6  27 -18  13 -11  31  48   0  70   2
5   offset 10 -7  26 -25   7 -26 -17 -43 -64   7  31
6   offset 11 -8  26 -29   3 -34   5  18  46 -46  45
```

```
            Eigenvectors (contd.)
 Eigenvalues  5.23
     log(C)   0.00
```

```
Set   Parameter    11
          kea  1   42
          kae  2  -32
          kei  3  -43
          kie  4   42
      r:PmPBP  5   40
1     offset  6   14
2 '   offset  7   15
3     offset  8   15
4     offset  9   17
5     offset 10   19
6     offset 11   20
```

CONDITION INDICES
```
          Index  4  3  2  2  1  1  1  1  1  1  0.
Set   Parameter
          kea  27  4 30 12  5  0  0  0  0  1 17
          kae   0  0  3 20 43  0  0  0  0 21 10
          kei  56  3  0 19  1  0  0  0  0  0 18
          kie  12 13 43 10  1  0  0  0  0  0 18
      r:PmPBP   0 34  2 18 25  0  0  0  0  2 16
1     offset   0  7  0  5  1  5 26 13 18 18  2
2     offset   0  7  0  5  0 43 20  0  2 15  2
3     offset   0  7  1  4  0 37  8 21  7  8  2
4     offset   0  7  3  1  1  9 23  0 49  0  3
5     offset   0  6  6  0  7  3 18 42  0  9  3
6     offset   0  6  8  0 11  0  3 22 21 21  4
```

Fitting for γ Complex Data in Figure 4-10C

SCRIPT FILE: "gcXZdtfit1_2s.txt"

TASK
Fit of progress curves

DATA
file DATAmastr\atpase\Zdt500Pi_2s.txt

REACTION MECHANISM

PTTT ---> ETTT : kpe
ETTT <===> ATTT : kea kae
ETTT <===> ITTT : kei kie

ATTT + N ---> ATTTN : konN
ATTTN ---> ATTDN + P : khyd
ATTDN ---> P + ATDDN : khyd
ATDDN ---> N + P + ADDD : khyd
ADDD ---> ETTT : krec
P + mPBP <===> PmPBP : kpon kpoff

DIFERENTIAL EQUATIONS

d[PTTT]/dt = -kpe[PTTT]
d[ETTT]/dt = +kpe[PTTT]-kea[ETTT]+kae[ATTT]-
kei[ETTT]+kie[ITTT]+krec[ADDD]
d[ATTT]/dt = +kea[ETTT]-kae[ATTT]-konN[ATTT][N]
d[ITTT]/dt = +kei[ETTT]-kie[ITTT]
d[N]/dt = -konN[ATTT][N]+khyd[ATDDN]
d[ATTTN]/dt = +konN[ATTT][N]-khyd[ATTTN]
d[ATTDN]/dt = +khyd[ATTTN]-khyd[ATTDN]
d[P]/dt = +khyd[ATTTN]+khyd[ATTDN]+khyd[ATDDN]
      kpon[P][mPBP]+kpoff[PmPBP]
d[ATDDN]/dt = +khyd[ATTDN]-khyd[ATDDN]
d[ADDD]/dt = +khyd[ATDDN]-krec[ADDD]
d[mPBP]/dt = -kpon[P][mPBP]+kpoff[PmPBP]
d[PmPBP]/dt = +kpon[P][mPBP]-kpoff[PmPBP]

OUTPUT

LEAST-SQUARES FIT
mean square   5.11789e-005
standard deviation   0.00715394
log(determinant)   -0.645
log(condition number) -2.25
Marquardt parameter 0.000488
execution time (min) 0.003167
datapoints   1500
parameters   3
iterations   14
function evaluations 17
error status   0

PARAMETERS & STANDARD ERRORS

Set	Parameter	Initial	Fitted	Error	%Error
	kpe	**61**	**52.99**	**1.7**	**3.3**
	r:PmPBP	1	0.9628	0.0014	0.14
1	offset	0	-0.006878	0.0011	17

COVARIANCE MATRIX
Set Parameter Covariances
        kpe a   a
        r:PmPBP b 79   b
1   offset c -86 -98

EIGENVECTORS AND EIGENVALUES
    '   Eigenvectors
    Eigenvalues   0.01 0.60 2.39
        log(C)   2.21 0.60 0.00
Set Parameter   1   2   3
        kpe 1 22 -84 49
        r:PmPBP 2 62 51 59
1   offset 3 -74 17 63

CONDITION INDICES
        Index 2 1 0
Set Parameter
        kpe 5 70 24
        r:PmPBP 38 26 35
1   offset 56 2 40

Fitting for Minimal Complex Data in Figure 5-6

SCRIPT FILE "gddXZdfit1_lowgdd_2s.txt"

TASK
Fit of progress curves

DATA
file DATAmastr\atpase\gddXZdtA02_Pi_20.txt

REACTION MECHANISM
PTTT ---> ETTT : kpe
ETTT <===> ATTT : kea kae
ETTT <===> ITTT : kei kie
ATTT + N ---> ATTTN : konN
ATTTN ---> ATTDN + P : khyd
ATTDN ---> P + ATDDN : khyd
ATDDN ---> N + P + ADDD : khyd
ADDD ---> ETTT : krec
P + mPBP <===> PmPBP : kpon kpoff

DIFERENTIAL EQUATIONS
d[PTTT]/dt = -kpe[PTTT]

$d[ETTT]/dt = +kpe[PTTT]-kea[ETTT]+kae[ATTT]-kei[ETTT]+kie[ITTT]+krec[ADDD]$

$d[ATTT]/dt = +kea[ETTT]-kae[ATTT]-konN[ATTT][N]$

$d[ITTT]/dt = +kei[ETTT]-kie[ITTT]$

$d[N]/dt = -konN[ATTT][N]+khyd[ATDDN]$

$d[ATTTN]/dt = +konN[ATTT][N]-khyd[ATTTN]$

$d[ATTDN]/dt = +khyd[ATTTN]-khyd[ATTDN]$

$d[P]/dt = +khyd[ATTTN]+khyd[ATTDN]+khyd[ATDDN]-kpon[P][mPBP]+kpoff[PmPBP]$

$d[ATDDN]/dt = +khyd[ATTDN]-khyd[ATDDN]$

$d[ADDD]/dt = +khyd[ATDDN]-krec[ADDD]$

$d[mPBP]/dt = -kpon[P][mPBP]+kpoff[PmPBP]$

$d[PmPBP]/dt = +kpon[P][mPBP]-kpoff[PmPBP]$

OUTPUT

LEAST-SQUARES FIT

mean square	3.04854e-005
standard deviation	0.00552136
log(determinant)	-4.85
log(condition number)	-6.34
execution time (min)	0.02467
datapoints	1500
parameters	4
iterations	100
subiterations	29
function evaluations	130
error status	1

PARAMETERS & STANDARD ERRORS

Set	Parameter	Initial	Fitted	Error	%Error
	kpe	53	102.4	27	27
	r:PmPBP	1	0.5588	0.18	32
1	c:PTTT	0.135	0.2292	0.074	32
1	offset	0	0.002729	0.0011	39

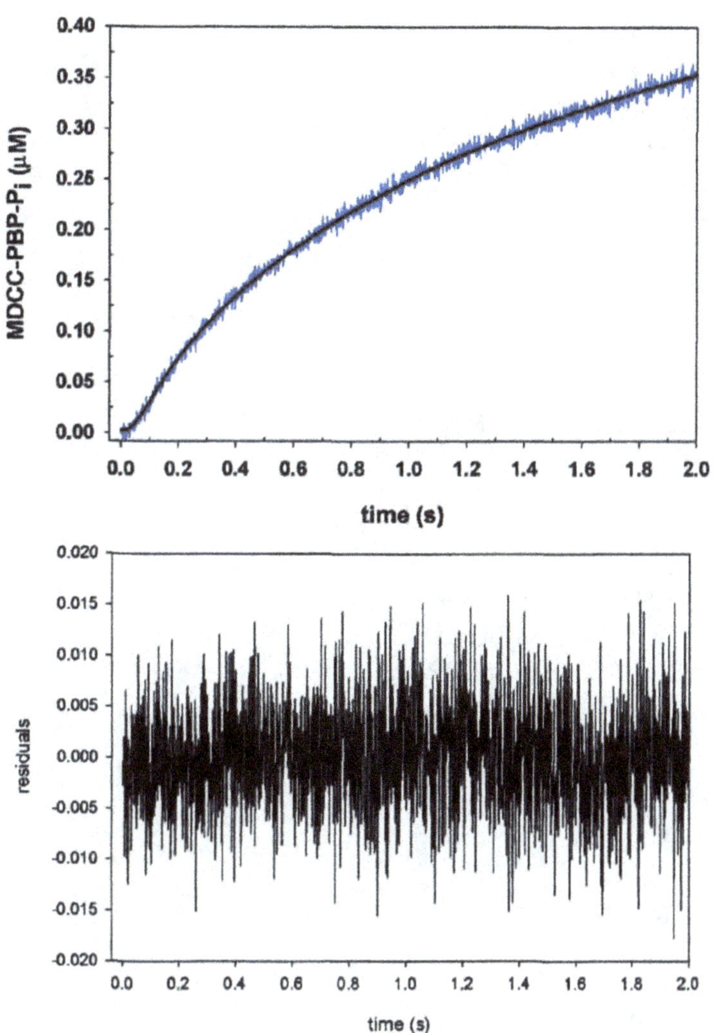

A-4. DynaFit Fitting of the minimal complex (Figure 5-6), no equilibration time with ATP.

Fitting for Minimal Complex Data For Estimation of Conformational Rate Constants From a Single Data set

SCRIPT FILE "gddptX3Pi_fit1_low.txt"

TASK
Fit of progress curves
DATA
file DATAmastr\atpase\gddptX3_14.txt

REACTION MECHANISM
ETTT <===> ATTT : kea kae
ETTT <===> ITTT : kei kie

ATTT + N ---> ATTTN : konN
ATTTN ---> ATTDN + P : khyd
ATTDN ---> P + ATDDN : khyd
ATDDN ---> N + P + ADDD : khyd
ADDD ---> ETTT : krec
P + mPBP <===> PmPBP : kpon kpoff

DIFERENTIAL EQUATIONS
d[ETTT]/dt = -kea[ETTT]+kae[ATTT]-kei[ETTT]+kie[ITTT]+krec[ADDD]
d[ATTT]/dt = +kea[ETTT]-kae[ATTT]-konN[ATTT][N]
d[ITTT]/dt = +kei[ETTT]-kie[ITTT]
d[N]/dt = -konN[ATTT][N]+khyd[ATDDN]
d[ATTTN]/dt = +konN[ATTT][N]-khyd[ATTTN]
d[ATTDN]/dt = +khyd[ATTTN]-khyd[ATTDN]
d[P]/dt = +khyd[ATTTN]+khyd[ATTDN]+khyd[ATDDN]-
kpon[P][mPBP]+kpoff[PmPBP]
d[ATDDN]/dt = +khyd[ATTDN]-khyd[ATDDN]
d[ADDD]/dt = +khyd[ATDDN]-krec[ADDD]
d[mPBP]/dt = -kpon[P][mPBP]+kpoff[PmPBP]
d[PmPBP]/dt = +kpon[P][mPBP]-kpoff[PmPBP]

OUTPUT

LEAST-SQUARES FIT
mean square   2.90901e-005
standard deviation   0.00539352
log(determinant)   -20.7
log(condition number) -11.8
Marquardt parameter 3.81e-006
execution time (min) 0.01317
datapoints   1399
parameters   7
iterations   32
subiterations   4
function evaluations 39
error status   0

PARAMETERS & STANDARD ERRORS

Set	Parameter	Initial	Fitted	Error	%Error
	kea	4.453	2.579	2.8	110
	kae	3.604	12.59	2.4	19
	kei	2.37	1.71	0.81	47
	kie	4.376	2.07	3.1	150
	r:PmPBP	1	1.136	130	11000
1	ce:ETTT	0.27	0.3096	35	11000
1	offset	0	-0.003348	0.0016	48

COVARIANCE MATRIX
Set Parameter Covariances
         kea a   a
         kae b 95   b
         kei c 72 80   c
         kie d 100 92 67   d
      r:PmPBP e 99 90 59 99   e
  1   ce:ETTT f -99 -90 -59 -99 -100   f
  1   offset g -56 -33 -18 -60 -62 62

EIGENVECTORS AND EIGENVALUES
           Eigenvectors
   Eigenvalues   0.00 0.00 0.00 0.01 0.02 0.25 6.72
        log(C) 11.84 5.79 4.29 3.04 2.55 1.42 0.00
Set Parameter   1   2   3   4   5   6   7
         kea 1   0 -59 54 -22 39   1 38
         kae 2   0 -4 16 -52 -21 -72 -35
         kei 3   0 -71 -3 39 -42   6 -38
         kie 4   0 -34 -79 -14 11 -26 38
      r:PmPBP 5 70   9 16 37 -22 -36 37
  1   ce:ETTT 6 -70   9 16 37 -22 -36 37
  1   offset 7   0   0   4 -46 -70 36 37

CONDITION INDICES
        Index 12 6 4 3 3 1 0
Set Parameter
         kea 0 34 29 5 15 0 14
         kae 0 0 2 27 4 52 12
         kei 0 51 0 15 17 0 14
         kie 0 12 62 2 1 7 14
      r:PmPBP 49 0 2 14 5 13 14
  1   ce:ETTT 50 0 2 13 4 13 14
  1   offset 0 0 0 21 50 13 14

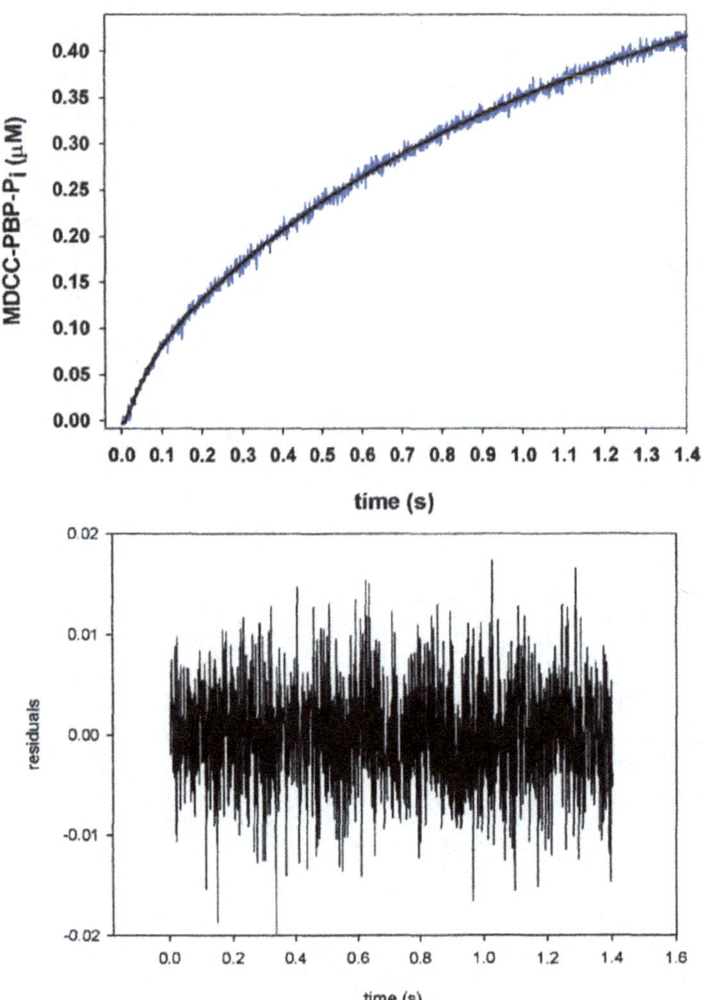

A-5. DynaFit fitting of the minimal complex (Figure 5-7B, black trace), Single
      experimental data set, 1000 ms equilibration with ATP.

Alternate Dynafit Model with a "Branch" Step at After Hydrolysis of Second ATP

There is no branch to the "inactive" ITTT species in this model.

Alternate Dynafit Model Applied to γ Complex in Figure 4-9

Script File: "XDdtfitALT1_2s.txt"

[task]
 task = fit
 data = progress

[mechanism]

```
; isomerizations of Gamma.ATP complexes:
ETTT <==> ATTT        : kea    kae
; DNA binding and hydrolysis:
ATTT + N ---> ATTTN    : konN
ATTTN ---> ATTDN + P   : khyd
ATTDN ---> ATDD + N + P  : khyd
ATDD ---> ADDD + P    : khydS
ADDD ----> ETTT   : krecy
; phosphate probe:
P + mPBP <==> PmPBP      : kpon   kpoff

[constants]
 kea   = 4.453 ?
 kae   = 3.604 ?
 konN  = 235
 khyd   = 66
 khydS  = 1.7
 krecy  = 0.17
  kpon = 136
 kpoff= 13.6

[concentrations]
 N = 1.0
 mPBP = 2.7

[responses]
 PmPBP = 1 ?

[progress]
  directory  DATAmastr/atpase
 extension  txt
 offset     auto ? local

 file      XDdt2_15Pi_2s
 incubate   ETTT = 0.54 , dilute 1:2, time 0.015

 file      XDdt2_48Pi_2s
 incubate   ETTT = 0.54 , dilute 1:2, time 0.048

 file      XDdt2_98Pi_2s
 incubate   ETTT = 0.54 , dilute 1:2, time 0.098

 file      XDdt2_247Pi_2s
 incubate   ETTT = 0.54 , dilute 1:2, time 0.247

 file      XDdt2_498Pi_2s
```

incubate   ETTT = 0.54 , dilute 1:2, time 0.498

file   XDdt2_1000Pi_2s
incubate   ETTT = 0.54 , dilute 1:2, time 1.0

[output]
 directory   users/CRWDISS/gcXDdt/outfitAltBranch1

[settings]
<Marquardt>
 Interrupt = 100
<Constraints>
 ConcError = 10

[end]

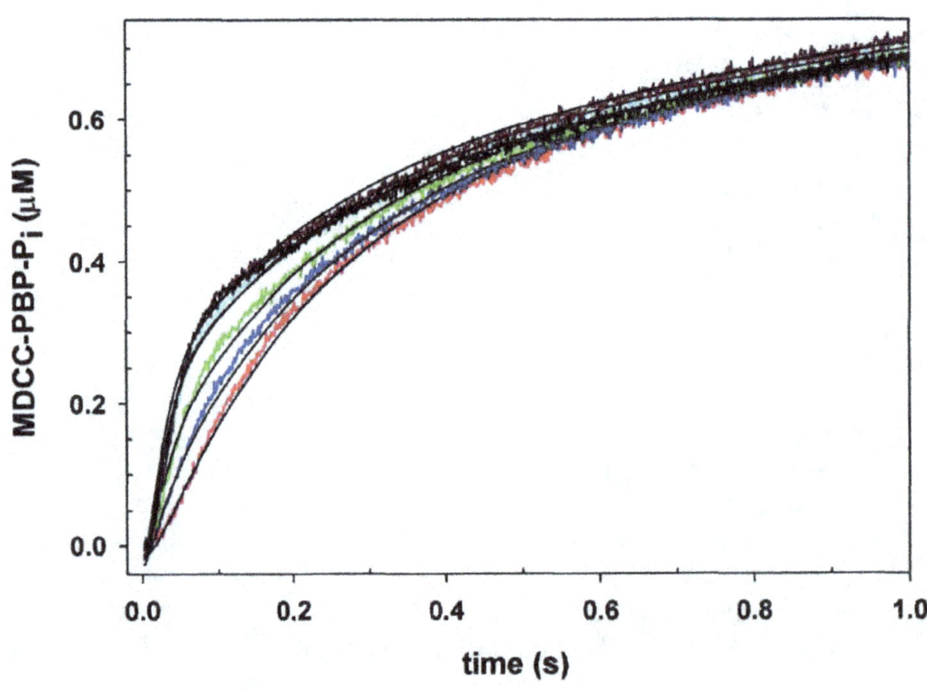

A-6. Figure 4-9 data for γ complex fit to alternate model with a "branch" step after
      hydrolysis of two ATPs

Alternate Dynafit Model Applied to γ Complex in Figure 4-10C

Script File: "XZDtfitALT1_2s"
;--
[task]

```
task = fit
data = progress

[mechanism]
 ; isomerizations of Gamma.ATP complexes:
 PTTT ---> ETTT        : kpe
 ETTT <==> ATTT        : kea    kae
 ; DNA binding and hydrolysis:
 ATTT + N ---> ATTTN    : konN
 ATTTN ---> ATTDN + P   : khyd
 ATTDN ---> ATDD + N + P   : khyd
 ATDD ---> ADDD + P    : khydS
 ADDD ----> ETTT   : krecy
 ; phosphate probe:
 P + mPBP <==> PmPBP       : kpon    kpoff

[constants]
 kpe   = 74 ?
 kea   = 5.34
 kae   = 4.633
 konN  = 235
 khyd   = 66
 khydS   = 1.7
 krecy  = 0.17
  kpon = 136
 kpoff= 13.6

[concentrations]
 N = 1.0
 mPBP = 2.7

[responses]
 PmPBP = 1 ?

[progress]
 directory  DATAmastr/atpase
 extension  txt
 offset     auto ?

 file     Zdt500Pi_2s
 conc PTTT = 0.27

[output]
 directory  users/CRWDISS/gcXZdt/outfitAltBranchZdt1

[settings]
```

```
<Marquardt>
  Interrupt = 100
<Constraints>
  ConcError = 10

[end]
```

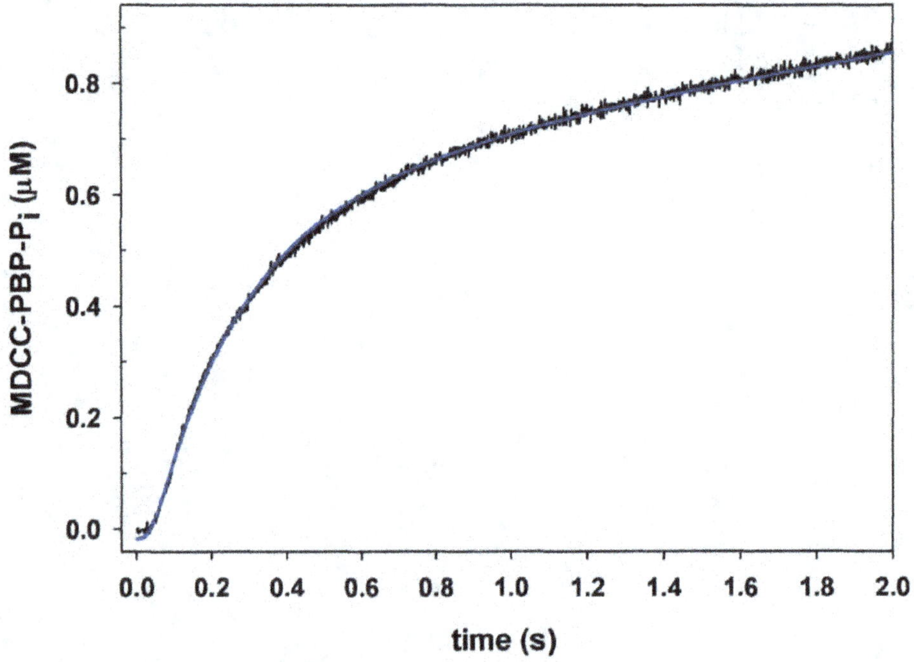

A-7. Figure 4-10C data for γ complex fit to alternate model with a "branch" step after hydrolysis of two ATPs

Alternate Dynafit Model Applied to the Minimal Complex in Figure 5-7

Script File: "gddptX3Pi_ALT1.txt"

```
;--------------------------------------------------------
[task]
  task = fit
  data = progress

[mechanism]
  ; isomerizations of Gamma.ATP complexes:
  ETTT <==> ATTT        : kea    kae
  ; DNA binding and hydrolysis:
  ATTT + N ---> ATTTN    : konN
  ATTTN ---> ATTDN + P   : khyd
```

```
ATTDN ---> ATDD + N + P   : khyd
ATDD ---> ADDD + P   : khyds
ADDD ----> ETTT   : krec
; phosphate probe:
P + mPBP <==> PmPBP       : kpon   kpoff

[constants]
 kea  = 4.453 ?
 kae  = 3.604 ?
 konN  = 235
 khyd  = 60
 khyds  = 1.0
 krec  = 0.15
 kpon = 136
 kpoff= 13.6

[concentrations]
 N = 1.0
 mPBP = 2.7

[responses]
 PmPBP = 1 ?

[progress]
 directory  DATAmastr/atpase
 extension  txt
 offset     auto ?
 file       gddptX3_14
 incubate   ETTT = 0.54  , dilute 1:2, time 1.0

[output]
 directory  users/CRWDISS/gdd/outfitALT1

[settings]
<Marquardt>
 Interrupt = 100
<Constraints>
 ConcError = 10

[end]
```

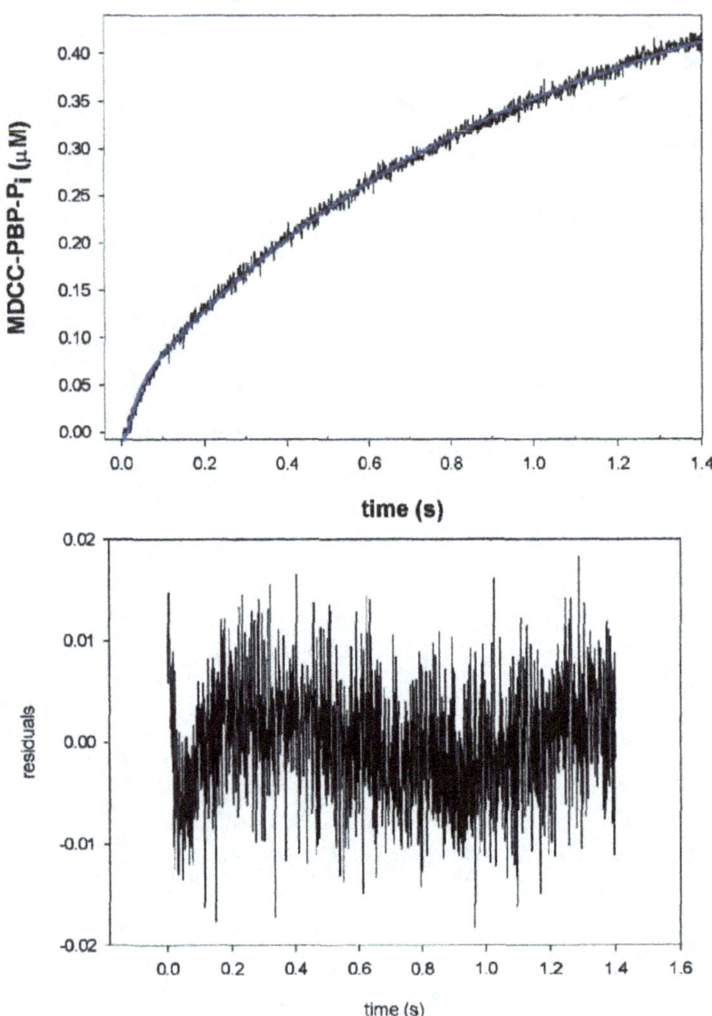

A-8. Figure 5-7, black trace data for the minimal complex fit to alternate model with a "branch" step after hydrolysis of two ATPs

Alternate Dynafit Model Applied to γ Complex in a DNA Binding Assay in the Absence of β Clamp

Script File: "gcptP2XBfitALT1_2s.txt"

```
;------------------------------------------------------------
[task]
  task = fit
  data = progress

[mechanism]
  ; isomerizations of Gamma.ATP complexes:
  ETTT <==> ATTT        : kea    kae
  ; DNA binding and hydrolysis:
  ATTT + N ---> ATTTN    : konN
```

```
ATTTN ---> ATTDN + P   : khyd
ATTDN ---> ATDD + N + P   : khyd
ATDD ---> ADDD + P   : khydS
ADDD ----> ETTT   : krec
; phosphate probe:
P + mPBP <==> PmPBP       : kpon   kpoff

[constants]
 kea   = 4.453 ?
 kae   = 3.604 ?
 konN  = 235
 khyd  = 66
 khydS = 1.7
 krec  = 0.15
 kpon = 136
 kpoff= 13.6

[concentrations]
 N = 0.050
 mPBP = 2.7

[responses]
; PmPBP = 1 ?
ATTTN = 1.7 ?
ATTDN = 1.7 ?
N = 0.1187 ?

[progress]
 directory   DATAmastr/rbind
 extension   txt
 offset     auto ?
 file       gcptP2XBaniso_20
 incubate   ETTT = 0.50  , dilute 1:2, time 2000

[output]
 directory   users/CRWDISS/gcptP2XB/outfitrALT

[settings]
<Marquardt>
  Interrupt = 100
<Constraints>
  ConcError = 10

[end]
```

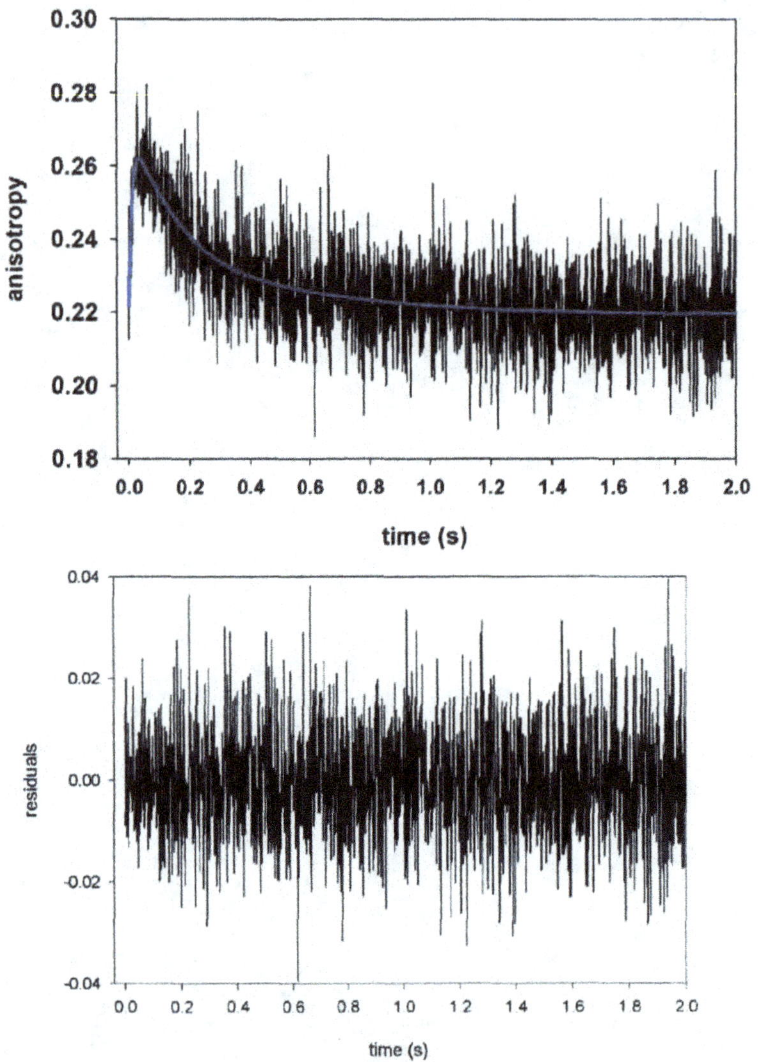

A-9. Example of the alternate model applied to DNA binding (anisotropy) data for a reaction of γ complex and Rhx-pt DNA.

LIST OF REFERENCES

Ahmadian, M. R., Stege, P., Scheffzek, K., and Wittinghofer, A. (1997). Confirmation of the arginine-finger hypothesis for the GAP-stimulated GTP-hydrolysis reaction of Ras. Nat Struct Biol *4*, 686-689.

Alley, S. C., Abel-Santos, E., and Benkovic, S. J. (2000). Tracking sliding clamp opening and closing during bacteriophage T4 DNA polymerase holoenzyme assembly. Biochemistry *39*, 3076-3090.

Alley, S. C., Shier, V. K., Abel-Santos, E., Sexton, D. J., Soumillion, P., and Benkovic, S. J. (1999). Sliding clamp of the bacteriophage T4 polymerase has open and closed subunit interfaces in solution. Biochemistry *38*, 7696-7709.

Alley, S. C., Trakselis, M. A., Mayer, M. U., Ishmael, F. T., Jones, A. D., and Benkovic, S. J. (2001). Building a replisome solution structure by elucidation of protein-protein interactions in the bacteriophage T4 DNA polymerase holoenzyme. J Biol Chem *276*, 39340-39349.

Ason, B., Bertram, J. G., Hingorani, M. M., Beechem, J. M., O'Donnell, M., Goodman, M. F., and Bloom, L. B. (2000). A model for Escherichia coli DNA polymerase III holoenzyme assembly at primer/template ends. DNA triggers a change in binding specificity of the gamma complex clamp loader. J Biol Chem *275*, 3006-3015.

Ason, B., Handayani, R., Williams, C. R., Bertram, J. G., Hingorani, M. M., O'Donnell, M., Goodman, M. F., and Bloom, L. B. (2003). Mechanism of loading the Escherichia coli DNA polymerase III beta sliding clamp on DNA. Bona fide primer/templates preferentially trigger the gamma complex to hydrolyze ATP and load the clamp. J Biol Chem *278*, 10033-10040.

Baker, T. A., and Bell, S. P. (1998). Polymerases and the replisome: machines within machines. Cell *92*, 295-305.

Barshop, B. A., Wrenn, R. F., and Frieden, C. (1983). Analysis of numerical methods for computer simulation of kinetic processes: Development of KINSIM--a flexible, portable system. Anal Biochem *130*, 134-145.

Bertram, J. G., Bloom, L. B., Hingorani, M. M., Beechem, J. M., O'Donnell, M., and Goodman, M. F. (2000). Molecular mechanism and energetics of clamp assembly in Escherichia coli. The role of ATP hydrolysis when gamma complex loads beta on DNA. J Biol Chem *275*, 28413-28420.

Bertram, J. G., Bloom, L. B., Turner, J., O'Donnell, M., Beechem, J. M., and Goodman, M. F. (1998). Pre-steady state analysis of the assembly of wild type and mutant circular clamps of Escherichia coli DNA polymerase III onto DNA. J Biol Chem *273*, 24564-24574.

Biswas, S. B., and Kornberg, A. (1984). Nucleoside triphosphate binding to DNA polymerase III holoenzyme of Escherichia coli. A direct photoaffinity labeling study. J Biol Chem *259*, 7990-7993.

Bloom, L. B., Turner, J., Kelman, Z., Beechem, J. M., O'Donnell, M., and Goodman, M. F. (1996). Dynamics of loading the beta sliding clamp of DNA polymerase III onto DNA. J Biol Chem *271*, 30699-30708.

Bonner, C. A., Stukenberg, P. T., Rajagopalan, M., Eritja, R., O'Donnell, M., McEntee, K., Echols, H., and Goodman, M. F. (1992). Processive DNA synthesis by DNA polymerase II mediated by DNA polymerase III accessory proteins. J Biol Chem *267*, 11431-11438.

Brune, M., Hunter, J. L., Corrie, J. E., and Webb, M. R. (1994). Direct, real-time measurement of rapid inorganic phosphate release using a novel fluorescent probe and its application to actomyosin subfragment 1 ATPase. Biochemistry *33*, 8262-8271.

Brune, M., Hunter, J. L., Howell, S. A., Martin, S. R., Hazlett, T. L., Corrie, J. E., and Webb, M. R. (1998). Mechanism of inorganic phosphate interaction with phosphate binding protein from Escherichia coli. Biochemistry *37*, 10370-10380.

Burgers, P. M., and Kornberg, A. (1982). ATP activation of DNA polymerase III holoenzyme of Escherichia coli. I. ATP-dependent formation of an initiation complex with a primed template. J Biol Chem *257*, 11468-11473.

Cai, J., Yao, N., Gibbs, E., Finkelstein, J., Phillips, B., O'Donnell, M., and Hurwitz, J. (1998). ATP hydrolysis catalyzed by human replication factor C requires participation of multiple subunits. Proc Natl Acad Sci U S A *95*, 11607-11612.

Cann, I. K., and Ishino, Y. (1999). Archaeal DNA replication: Identifying the pieces to solve a puzzle. Genetics *152*, 1249-1267.

Coontz, R., Fahrenkamp-Uppenbrink, J., and Szuromi, P. (2003). Speeding up chemistry: Special section on catalysis. Science *299*, 1683-1706.

Cullmann, G., Fien, K., Kobayashi, R., and Stillman, B. (1995). Characterization of the five replication factor C genes of Saccharomyces cerevisiae. Mol Cell Biol *15*, 4661-4671.

Dallmann, H. G., Kim, S., Pritchard, A. E., Marians, K. J., and McHenry, C. S. (2000). Characterization of the unique C terminus of the Escherichia coli tau DnaX protein. Monomeric C-tau binds alpha AND DnaB and can partially replace tau in reconstituted replication forks. J Biol Chem *275*, 15512-15519.

Dang, Q., and Frieden, C. (1997). New PC versions of the kinetic-simulation and fitting programs, KINSIM and FITSIM. Trends Biochem Sci *22*, 317.

Davey, M. J., Jeruzalmi, D., Kuriyan, J., and O'Donnell, M. (2002). Motors and switches: AAA+ machines within the replisome. Nat Rev Mol Cell Biol *3*, 826-835.

De Felice, M., Sensen, C. W., Charlebios, R. L., Rossi, M., and Pisani, F. M. (1999). Two DNA polymerase sliding clamps from the thermophilic archaeon *Solfolobus solfataricus*. J Mol Biol *291*, 47-57.

Dillingham, M. S., Wigley, D. B., and Webb, M. R. (2000). Demonstration of unidirectional single-stranded DNA translocation by PcrA helicase: Measurement of step size and translocation speed. Biochemistry *39*, 205-212.

Dong, Z., Onrust, R., Skangalis, M., and O'Donnell, M. (1993). DNA polymerase III accessory proteins. I. holA and holB encoding delta and delta'. J Biol Chem *268*, 11758-11765.

Dougan, D. A., Mogk, A., Zeth, K., Turgay, K., and Bukau, B. (2002). AAA+ proteins and substrate recognition, it all depends on their partner in crime. FEBS Lett *529*, 6-10.

Ellison, V., and Stillman, B. (1998). Reconstitution of recombinant human replication factor C (RFC) and identification of an RFC subcomplex possessing DNA-dependent ATPase activity. J Biol Chem *273*, 5979-5987.

Ellison, V., and Stillman, B. (2001). Opening of the clamp: An intimate view of an ATP-driven biological machine. Cell *106*, 655-660.

Erzberger, J. P., Pirruccello, M. M., and Berger, J. M. (2002). The structure of bacterial DnaA: Implications for general mechanisms underlying DNA replication. Embo J *21*, 4763-4773.

Fay, P. J., Johanson, K. O., McHenry, C. S., and Bambara, R. A. (1981). Size classes of products synthesized processively by DNA polymerase III and DNA polymerase III holoenzyme of Escherichia coli. J Biol Chem *256*, 976-983.

Fay, P. J., Johanson, K. O., McHenry, C. S., and Bambara, R. A. (1982). Size classes of products synthesized processively by two subassemblies of Escherichia coli DNA polymerase III holoenzyme. J Biol Chem *257*, 5692-5699.

Fersht, A. R. (1999). Structure and mechanism in protein science (New York, W.H. Freeman and Co.).

Fersht, A. R., and Shakhnovich, E. I. (1998). Protein folding: Think globally, (inter)act locally. Curr Biol *8*, R478-479.

Flower, A. M., and McHenry, C. S. (1990). The gamma subunit of DNA polymerase III holoenzyme of Escherichia coli is produced by ribosomal frameshifting. Proc Natl Acad Sci U S A *87*, 3713-3717.

Franco, R., Canals, M., Marcellino, D., Ferre, S., Agnati, L., Mallol, J., Casado, V., Ciruela, F., Fuxe, K., Lluis, C., and Canela, E. I. (2003). Regulation of heptaspanning-membrane-receptor function by dimerization and clustering. Trends Biochem Sci *28*, 238-243.

Gao, D., and McHenry, C. S. (2001). tau binds and organizes Escherichia coli replication through distinct domains. Partial proteolysis of terminally tagged tau to determine candidate domains and to assign domain V as the alpha binding domain. J Biol Chem *276*, 4433-4440.

Gefter, M. L., Hirota, Y., Kornberg, T., Wechsler, J. A., and Barnoux, C. (1971). Analysis of DNA polymerases II and 3 in mutants of Escherichia coli thermosensitive for DNA synthesis. Proc Natl Acad Sci U S A *68*, 3150-3153.

Glover, B. P., and McHenry, C. S. (1998). The chi psi subunits of DNA polymerase III holoenzyme bind to single-stranded DNA-binding protein (SSB) and facilitate replication of an SSB-coated template. J Biol Chem *273*, 23476-23484.

Glover, B. P., and McHenry, C. S. (2000). The DnaX-binding subunits delta' and psi are bound to gamma and not tau in the DNA polymerase III holoenzyme. J Biol Chem *275*, 3017-3020.

Glover, B. P., and McHenry, C. S. (2001). The DNA polymerase III holoenzyme: An asymmetric dimeric replicative complex with leading and lagging strand polymerases. Cell *105*, 925-934.

Gomes, X. V., and Burgers, P. M. (2001). ATP utilization by yeast replication factor C. I. ATP-mediated interaction with DNA and with proliferating cell nuclear antigen. J Biol Chem *276*, 34768-34775.

Gomes, X. V., Schmidt, S. L., and Burgers, P. M. (2001). ATP utilization by yeast replication factor C. II. Multiple stepwise ATP binding events are required to load proliferating cell nuclear antigen onto primed DNA. J Biol Chem *276*, 34776-34783.

Green, C. M., Erdjument-Bromage, H., Tempst, P., and Lowndes, N. F. (2000). A novel Rad24 checkpoint protein complex closely related to replication factor C. Curr Biol *10*, 39-42.

Guenther, B., Onrust, R., Sali, A., O'Donnell, M., and Kuriyan, J. (1997). Crystal structure of the delta' subunit of the clamp-loader complex of E. coli DNA polymerase III. Cell *91*, 335-345.

Hakimi, M. A., Bochar, D. A., Schmiesing, J. A., Dong, Y., Barak, O. G., Speicher, D. W., Yokomori, K., and Shiekhattar, R. (2002). A chromatin remodelling complex that loads cohesin onto human chromosomes. Nature *418*, 994-998.

Hingorani, M. M., Bloom, L. B., Goodman, M. F., and O'Donnell, M. (1999). Division of labor--sequential ATP hydrolysis drives assembly of a DNA polymerase sliding clamp around DNA. Embo J *18*, 5131-5144.

Hingorani, M. M., and O'Donnell, M. (1998). ATP binding to the Escherichia coli clamp loader powers opening of the ring-shaped clamp of DNA polymerase III holoenzyme. J Biol Chem *273*, 24550-24563.

Hingorani, M. M., and O'Donnell, M. (2000). Sliding clamps: A (tail)ored fit. Curr Biol *10*, R25-29.

Hiratsuka, T. (1983). New ribose-modified fluorescent analogs of adenine and guanine nucleotides available as substrates for various enzymes. Biochim Biophys Acta *742*, 496-508.

Hirota, Y., Gefter, M., and Mindich, L. (1972). A mutant of Escherichia coli defective in DNA polymerase II activity. Proc Natl Acad Sci U S A *69*, 3238-3242.

Jarvis, T. C., Paul, L. S., and von Hippel, P. H. (1989). Structural and enzymatic studies of the T4 DNA replication system. I. Physical characterization of the polymerase accessory protein complex. J Biol Chem *264*, 12709-12716.

Jeruzalmi, D., O'Donnell, M., and Kuriyan, J. (2001a). Crystal structure of the processivity clamp loader gamma (gamma) complex of E. coli DNA polymerase III. Cell *106*, 429-441.

Jeruzalmi, D., O'Donnell, M., and Kuriyan, J. (2002). Clamp loaders and sliding clamps. Curr Opin Struct Biol *12*, 217-224.

Jeruzalmi, D., Yurieva, O., Zhao, Y., Young, M., Stewart, J., Hingorani, M., O'Donnell, M., and Kuriyan, J. (2001b). Mechanism of processivity clamp opening by the delta subunit wrench of the clamp loader complex of E. coli DNA polymerase III. Cell *106*, 417-428.

Johanson, K. O., Haynes, T. E., and McHenry, C. S. (1986). Chemical characterization and purification of the beta subunit of the DNA polymerase III holoenzyme from an overproducing strain. J Biol Chem *261*, 11460-11465.

Johnson, A., and O'Donnell, M. (2003). Ordered ATP hydrolysis in the gamma complex clamp loader AAA+ machine. J Biol Chem *278*, 14406-14413.

Katayama, T. (2001). Feedback controls restrain the initiation of *Escherichia coli* chromosomal replication. Mol Microbiol *41*, 9-17.

Kelman, Z., and Hurwitz, J. (2000). A unique organization of the protein subunits of the DNA polymerase clamp loader in the archaeon *Methanobacterium thermoautotrophicum* (delta)H. J Biol Chem *275*, 7327-7336.

Kelman, Z., and O'Donnell, M. (1995). DNA polymerase III holoenzyme: Structure and function of a chromosomal replicating machine. Annu Rev Biochem *64*, 171-200.

Kelman, Z., Yuzhakov, A., Andjelkovic, J., and O'Donnell, M. (1998). Devoted to the lagging strand-the subunit of DNA polymerase III holoenzyme contacts SSB to promote processive elongation and sliding clamp assembly. Embo J *17*, 2436-2449.

Kim, S., Dallmann, H. G., McHenry, C. S., and Marians, K. J. (1996a). Coupling of a replicative polymerase and helicase: A tau-DnaB interaction mediates rapid replication fork movement. Cell *84*, 643-650.

Kim, S., Dallmann, H. G., McHenry, C. S., and Marians, K. J. (1996b). tau couples the leading- and lagging-strand polymerases at the Escherichia coli DNA replication fork. J Biol Chem *271*, 21406-21412.

Kim, S., Dallmann, H. G., McHenry, C. S., and Marians, K. J. (1996c). tau protects beta in the leading-strand polymerase complex at the replication fork. J Biol Chem *271*, 4315-4318.

Kitani, T., Yoda, K., Ogawa, T., and Okazaki, T. (1985). Evidence that discontinuous DNA replication in Escherichia coli is primed by approximately 10 to 12 residues of RNA starting with a purine. J Mol Biol *184*, 45-52.

Kong, X. P., Onrust, R., O'Donnell, M., and Kuriyan, J. (1992). Three-dimensional structure of the beta subunit of E. coli DNA polymerase III holoenzyme: A sliding DNA clamp. Cell *69*, 425-437.

Kornberg, A., and Baker, T. A. (1992). DNA Replication, 2nd edn (New York, W.H. Freeman).

Kouprina, N., Kroll, E., Kirillov, A., Bannikov, V., Zakharyev, V., and Larionov, V. (1994). CHL12, a gene essential for the fidelity of chromosome transmission in the yeast Saccharomyces cerevisiae. Genetics *138*, 1067-1079.

Krishna, T. S., Kong, X. P., Gary, S., Burgers, P. M., and Kuriyan, J. (1994). Crystal structure of the eukaryotic DNA polymerase processivity factor PCNA. Cell *79*, 1233-1243.

Kuzmic, P. (1996). Program DYNAFIT for the Analysis of Enzyme Kinetic Data: Application to HIV Protease. Anal Biochem *237*, 260-273.

LaDuca, R. J., Fay, P. J., Chuang, C., McHenry, C. S., and Bambara, R. A. (1983). Site-specific pausing of deoxyribonucleic acid synthesis catalyzed by four forms of Escherichia coli DNA polymerase III. Biochemistry *22*, 5177-5188.

Lakowicz, J. R. (1999). Principles of fluorescence spectroscopy (New York, Plenum Publishers).

Lee, S. H., Pan, Z. Q., Kwong, A. D., Burgers, P. M., and Hurwitz, J. (1991). Synthesis of DNA by DNA polymerase epsilon in vitro. J Biol Chem *266*, 22707-22717.

Leu, F. P., Georgescu, R., and O'Donnell, M. (2003). Mechanism of the E. coli tau processivity switch during lagging-strand synthesis. Mol Cell *11*, 315-327.

Leu, F. P., Hingorani, M. M., Turner, J., and O'Donnell, M. (2000). The delta subunit of DNA polymerase III holoenzyme serves as a sliding clamp unloader in Escherichia coli. J Biol Chem *275*, 34609-34618.

Levine, C., and Marians, K. J. (1998). Identification of *dna*X as a high-copy supressor of the conditional-lethal and partition phenotypes of the *parE10* allele. J Bacteriol *180*, 1232-1240.

Li, X., and Marians, K. J. (2000). Two distinct triggers for cycling of the lagging strand polymerase at the replication fork. J Biol Chem *275*, 34757-34765.

Maki, H., Horiuchi, T., and Kornberg, A. (1985). The polymerase subunit of DNA polymerase III of Escherichia coli. I. Amplification of the dnaE gene product and polymerase activity of the alpha subunit. J Biol Chem *260*, 12982-12986.

Maki, H., and Kornberg, A. (1987). Proofreading by DNA polymerase III of Escherichia coli depends on cooperative interaction of the polymerase and exonuclease subunits. Proc Natl Acad Sci U S A *84*, 4389-4392.

Maki, H., Maki, S., and Kornberg, A. (1988). DNA Polymerase III holoenzyme of Escherichia coli. IV. The holoenzyme is an asymmetric dimer with twin active sites. J Biol Chem *263*, 6570-6578.

Maki, S., and Kornberg, A. (1988). DNA polymerase III holoenzyme of Escherichia coli. II. A novel complex including the gamma subunit essential for processive synthesis. J Biol Chem *263*, 6555-6560.

Mazumder, B., Seshadri, V., and Fox, P. L. (2003). Translational control by the 3'-UTR: The ends specify the means. Trends Biochem Sci *28*, 91-98.

McHenry, C. S. (1982). Purification and characterization of DNA polymerase III'. Identification of tau as a subunit of the DNA polymerase III holoenzyme. J Biol Chem *257*, 2657-2663.

McHenry, C. S., and Crow, W. (1979). DNA polymerase III of Escherichia coli. Purification and identification of subunits. J Biol Chem *254*, 1748-1753.

Mehta, A. D., Rock, R. S., Rief, M., Spudich, J. A., Mooseker, M. S., and Cheney, R. E. (1999). Myosin-V is a processive actin-based motor. Nature *400*, 590-593.

Moarefi, I., Jeruzalmi, D., Turner, J., O'Donnell, M., and Kuriyan, J. (2000). Crystal structure of the DNA polymerase processivity factor of T4 bacteriophage. J Mol Biol *296*, 1215-1223.

Moore, K. J., and Lohman, T. M. (1994). Kinetic mechanism of adenine nucleotide binding to and hydrolysis by the Escherichia coli Rep monomer. 1. Use of fluorescent nucleotide analogues. Biochemistry *33*, 14550-14564.

Mossi, R., Keller, R. C., Ferrari, E., and Hubscher, U. (2000). DNA polymerase switching: II. Replication factor C abrogates primer synthesis by DNA polymerase alpha at a critical length. J Mol Biol *295*, 803-814.

Moyer, M. L., Gilbert, S. P., and Johnson, K. A. (1998). Pathway of ATP hydrolysis by monomeric and dimeric kinesin. Biochemistry *37*, 800-813.

Naktinis, V., Onrust, R., Fang, L., and O'Donnell, M. (1995). Assembly of a chromosomal replication machine: Two DNA polymerases, a clamp loader, and sliding clamps in one holoenzyme particle. II. Intermediate complex between the clamp loader and its clamp. J Biol Chem *270*, 13358-13365.

Naktinis, V., Turner, J., and O'Donnell, M. (1996). A molecular switch in a replication machine defined by an internal competition for protein rings. Cell *84*, 137-145.

Neuwald, A. F., Aravind, L., Spouge, J. L., and Koonin, E. V. (1999). AAA+: A class of chaperone-like ATPases associated with the assembly, operation, and disassembly of protein complexes. Genome Res *9*, 27-43.

Nixon, A. E., Hunter, J. L., Bonifacio, G., Eccleston, J. F., and Webb, M. R. (1998). Purine nucleoside phosphorylase: its use in a spectroscopic assay for inorganic phosphate and for removing inorganic phosphate with the aid of phosphodeoxyribomutase. Anal Biochem *265*, 299-307.

O'Donnell, M., Jeruzalmi, D., and Kuriyan, J. (2001). Clamp loader structure predicts the architecture of DNA polymerase III holoenzyme and RFC. Curr Biol *11*, R935-946.

O'Donnell, M., and Lopez de Saro, F. J. (2001). Interaction of the beta sliding clamp with MutS, ligase, DNA polymerase I. Proc Natl Acad Sci U S A *98*, 8376-8380.

O'Donnell, M., Onrust, R., Dean, F. B., Chen, M., and Hurwitz, J. (1993). Homology in accessory proteins of replicative polymerases--E. coli to humans. Nucleic Acids Res *21*, 1-3.

O'Donnell, M. E. (1987). Accessory proteins bind a primed template and mediate rapid cycling of DNA polymerase III holoenzyme from Escherichia coli. J Biol Chem *262*, 16558-16565.

Olson, M. W., Dallmann, H. G., and McHenry, C. S. (1995). DnaX complex of *Escherichia coli* DNA polymerase III holoenyzme: The chi-psi complex functions by increasing the affinity of tau and gamma for delta-delta' to a physiologically relevant range. J Biol Chem *270*, 29570-29577.

Onrust, R., Finkelstein, J., Naktinis, V., Turner, J., Fang, L., and O'Donnell, M. (1995). Assembly of a chromosomal replication machine: Two DNA polymerases, a clamp loader, and sliding clamps in one holoenzyme particle. I. Organization of the clamp loader. J Biol Chem *270*, 13348-13357.

Onrust, R., and O'Donnell, M. (1993). DNA polymerase III accessory proteins. II. Characterization of delta and delta'. J Biol Chem *268*, 11766-11772.

Onrust, R., Stukenberg, P. T., and O'Donnell, M. (1991). Analysis of the ATPase subassembly which initiates processive DNA synthesis by DNA polymerase III holoenzyme. J Biol Chem *266*, 21681-21686.

Otto, M. R., Lillo, P., and Beechem, J. M. (1994). Resolution of Multiphasic Reactions by the Combination of Fluorescence Total-Intensity and Anisotropy Stopped-Flow Kinetic Experiments. Biophys J *67*, 2511-2521.

Oyama, T., Ishino, Y., Cann, I. K., Ishino, S., and Morikawa, K. (2001). Atomic structure of the clamp loader small subunit from Pyrococcus furiosus. Mol Cell *8*, 455-463.

Peterman, B. F. (1979). Measurement of the dead time of a fluorescence stopped-flow instrument. Anal Biochem *93*, 442-444.

Phillips, R. A., Hunter, J. L., Eccleston, J. F., and Webb, M. R. (2003). The mechanism of Ras GTPase activation by neurofibromin. Biochemistry *42*, 3956-3965.

Pietroni, P., Young, M. C., Latham, G. J., and von Hippel, P. H. (2001). Dissection of the ATP-driven reaction cycle of the bacteriophage T4 DNA replication processivity clamp loading system. J Mol Biol *309*, 869-891.

Pisani, F. M., De Felice, M., Carpentieri, F., and Rossi, M. (2000). Biochemical characterization of a clamp-loader complex homologous to eukaryotic replication factor C from the hyperthermophilic archaeon *Solfolobus solfataricus*. J Mol Biol *301*, 61-73.

Podobnik, M., Weitze, T. F., O'Donnell, M., and Kuriyan, J. (2003). Nucleotide-Induced Conformational Changes in an Isolated Escherichia coli DNA Polymerase III Clamp Loader Subunit. Structure (Camb) *11*, 253-263.

Pritchard, A. E., Dallmann, H. G., Glover, B. P., and McHenry, C. S. (2000). A novel assembly mechanism for the DNA polymerase III holoenzyme DnaX complex: Association of deltadelta' with DnaX(4) forms DnaX(3)deltadelta'. Embo J *19*, 6536-6545.

Purich, D. L. (2001). Enzyme catalysis: a new definition accounting for noncovalent substrate- and product-like states. Trends in Biochemical Sciences *26*, 417-421.

Salinas, F., and Benkovic, S. J. (2000). Characterization of bacteriophage T4-coordinated leading- and lagging-strand synthesis on a minicircle substrate. Proc Natl Acad Sci U S A *97*, 7196-7201.

Scheuermann, R. H., and Echols, H. (1984). A separate editing exonuclease for DNA replication: the epsilon subunit of Escherichia coli DNA polymerase III holoenzyme. Proc Natl Acad Sci U S A *81*, 7747-7751.

Schmidt, S. L., Gomes, X. V., and Burgers, P. M. (2001a). ATP utilization by yeast replication factor C. III. The ATP-binding domains of Rfc2, Rfc3, and Rfc4 are essential for DNA recognition and clamp loading. J Biol Chem *276*, 34784-34791.

Schmidt, S. L., Pautz, A. L., and Burgers, P. M. (2001b). ATP utilization by yeast replication factor C. IV. RFC ATP-binding mutants show defects in DNA replication, DNA repair, and checkpoint regulation. J Biol Chem *276*, 34792-34800.

Sexton, D. J., Kaboord, B. F., Berdis, A. J., Carver, T. E., and Benkovic, S. J. (1998). Dissecting the order of bacteriophage T4 DNA polymerase holoenzyme assembly. Biochemistry *37*, 7749-7756.

Seybert, A., Scott, D. J., Scaife, S., Singleton, M. R., and Wigley, D. B. (2002). Biochemical characterization of the clamp/clamp loader proteins from the euryarchaeon *Archaeoglobus fulgidus*. Nucleic Acids Res *30*, 4329-4338.

Shamoo, Y., and Steitz, T. A. (1999). Building a replisome from interacting pieces: Sliding clamp complexed to a peptide from DNA polymerase and a polymerase editing complex. Cell *99*, 155-166.

Shiomi, Y., Usukura, J., Masamura, Y., Takeyasu, K., Nakayama, Y., Obuse, C., Yoshikawa, H., and Tsurimoto, T. (2000). ATP-dependent structural change of the eukaryotic clamp-loader protein, replication factor C. Proc Natl Acad Sci U S A *97*, 14127-14132.

Song, M. S., Pham, P. T., Olson, M., Carter, J. R., Franden, M. A., Schaaper, R. M., and McHenry, C. S. (2001). The delta and delta ' subunits of the DNA polymerase III holoenzyme are essential for initiation complex formation and processive elongation. J Biol Chem *276*, 35165-35175.

Spanos, A., Sedgwick, S. G., Yarranton, G. T., Hubscher, U., and Banks, G. R. (1981). Detection of the catalytic activities of DNA polymerases and their associated exonucleases following SDS-polyacrylamide gel electrophoresis. Nucleic Acids Res *9*, 1825-1839.

Stewart, J., Hingorani, M. M., Kelman, Z., and O'Donnell, M. (2001). Mechanism of beta clamp opening by the delta subunit of Escherichia coli DNA polymerase III holoenzyme. J Biol Chem *276*, 19182-19189.

Studwell-Vaughan, P. S., and O'Donnell, M. (1991). Constitution of the twin polymerase of DNA polymerase III holoenzyme. J Biol Chem *266*, 19833-19841.

Studwell-Vaughan, P. S., and O'Donnell, M. (1993). DNA polymerase III accessory proteins. V. Theta encoded by holE. J Biol Chem *268*, 11785-11791.
Stukenberg, P. T., and O'Donnell, M. (1995). Assembly of a chromosomal replication machine: two DNA polymerases, a clamp loader, and sliding clamps in one holoenzyme particle. V. Four different polymerase-clamp complexes on DNA. J Biol Chem *270*, 13384-13391.

Stukenberg, P. T., Studwell-Vaughan, P. S., and O'Donnell, M. (1991). Mechanism of the sliding beta-clamp of DNA polymerase III holoenzyme. J Biol Chem *266*, 11328-11334.

Tang, M., Shen, X., Frank, E. G., O'Donnell, M., Woodgate, R., and Goodman, M. F. (1999). UmuD'(2)C is an error-prone DNA polymerase, Escherichia coli pol V. Proc Natl Acad Sci U S A *96*, 8919-8924.

Tougu, K., and Marians, K. J. (1996). The Interaction between helicase and primase sets the replication fork clock. J Biol Chem *271*, 21398-21405.

Tsuchihashi, Z., and Kornberg, A. (1989). ATP interactions of the tau and gamma subunits of DNA polymerase III holoenzyme of Escherichia coli. J Biol Chem *264*, 17790-17795.

Tsuchihashi, Z., and Kornberg, A. (1990). Translational frameshifting generates the gamma subunit of DNA polymerase III holoenzyme. Proc Natl Acad Sci U S A *87*, 2516-2520.

Tsurimoto, T., and Stillman, B. (1991). Replication factors required for SV40 DNA replication in vitro. II. Switching of DNA polymerase alpha and delta during initiation of leading and lagging strand synthesis. J Biol Chem *266*, 1961-1968.

Turner, J., Hingorani, M. M., Kelman, Z., and O'Donnell, M. (1999). The internal workings of a DNA polymerase clamp-loading machine. Embo J *18*, 771-783.

Uhlmann, F., Cai, J., Flores-Rozas, H., Dean, F. B., Finkelstein, J., O'Donnell, M., and Hurwitz, J. (1996). In vitro reconstitution of human replication factor C from its five subunits. Proc Natl Acad Sci U S A *93*, 6521-6526.

Uhlmann, F., Cai, J., Gibbs, E., O'Donnell, M., and Hurwitz, J. (1997). Deletion analysis of the large subunit p140 in human replication factor C reveals regions required for complex formation and replication activities. J Biol Chem *272*, 10058-10064.

Vale, R. D. (2003). The molecular motor toolbox for intracellular transport. Cell *112*, 467-480.

Vale, R. D., and Milligan, R. A. (2000). The way things move: Looking under the hood of molecular motor proteins. Science *288*, 88-95.

Waga, S., and Stillman, B. (1994). Anatomy of a DNA replication fork revealed by reconstitution of SV40 DNA replication in vitro. Nature *369*, 207-212.

Waga, S., and Stillman, B. (1998). The DNA replication fork in eukaryotic cells. Annu Rev Biochem *67*, 721-751.

Walker, J. E., Saraste, M. J., Runswick, M. J., and Gay, N. J. (1982). Distantly related sequences an the alpha- and beta-subunits of ATP synthase, myosin, kinases and other ATP-requiring enzymes and a common nucleotide binding fold. Embo J *1*, 945-951.

Wickner, S. (1976). Mechanism of DNA elongation catalyzed by Escherichia coli DNA polymerase III, dnaZ protein, and DNA elongation factors I and III. Proc Natl Acad Sci U S A *73*, 3511-3515.

Willsky, G. R., and Malamy, M. H. (1976). Control of the Synthesis of Alkaline Phosphatase and the Phosphate-Binding Protein in *Escherichia coli*. Journal of Bacteriology *127*, 595-609.

Woody, R. W. (1995). Circular dichroism. Methods Enzymol *246*, 34-71.

Wu, C. A., Zechner, E. L., Hughes, A. J., Jr., Franden, M. A., McHenry, C. S., and Marians, K. J. (1992). Coordinated leading- and lagging-strand synthesis at the Escherichia coli DNA replication fork. IV. Reconstitution of an asymmetric, dimeric DNA polymerase III holoenzyme. J Biol Chem *267*, 4064-4073.

Xiao, H., Crombie, R., Dong, Z., Onrust, R., and O'Donnell, M. (1993a). DNA polymerase III accessory proteins. III. holC and holD encoding chi and psi. J Biol Chem *268*, 11773-11778.

Xiao, H., Dong, Z., and O'Donnell, M. (1993b). DNA polymerase III accessory proteins. IV. Characterization of chi and psi. J Biol Chem *268*, 11779-11784.

Xiao, H., Naktinis, V., and O'Donnell, M. (1995). Assembly of a chromosomal replication machine: Two DNA polymerases, a clamp loader, and sliding clamps in one holoenzyme particle. IV. ATP-binding site mutants identify the clamp loader. J Biol Chem *270*, 13378-13383.

Yao, N., Leu, F. P., Anjelkovic, J., Turner, J., and O'Donnell, M. (2000). DNA structure requirements for the Escherichia coli gamma complex clamp loader and DNA polymerase III holoenzyme. J Biol Chem *275*, 11440-11450.

Yao, N., Turner, J., Kelman, Z., Stukenberg, P. T., Dean, F., Shechter, D., Pan, Z. Q., Hurwitz, J., and O'Donnell, M. (1996). Clamp loading, unloading and intrinsic stability of the PCNA, beta and gp45 sliding clamps of human, E. coli and T4 replicases. Genes Cells *1*, 101-113.

Yoder, B. L., and Burgers, P. M. (1991). Saccharomyces cerevisiae replication factor C. I. Purification and characterization of its ATPase activity. J Biol Chem *266*, 22689-22697.

Yuzhakov, A., Kelman, Z., and O'Donnell, M. (1999). Trading places on DNA--a three-point switch underlies primer handoff from primase to the replicative DNA polymerase. Cell *96*, 153-163.

Zeth, K., Ravelli, R. B., Paal, K., Cusack, S., Bukau, B., and Dougan, D. A. (2002). Structural analysis of the adaptor protein ClpS in complex with the N-terminal domain of ClpA. Nat Struct Biol *9*, 906-911.

Christopher R. Williams was born in the state of New Jersey in 1974. In 1997, he received a Bachelor of Science degree in Biotechnology from William Paterson College, Wayne, NJ (now known as William Paterson University). During his undergraduate experience, Chris performed an independent study under the guidance of Drs. Robert Chesney and Edith Gardner titled: "The search for an intermediate filament gene in *Saccharomyces cerevisiae*". In 1996 Chris was inducted as a member of the Beta Beta Beta Biological Honor Society. In August of 1997 Chris entered the Molecular and Cellular Biology Ph.D. program at Arizona State University (ASU), Tempe, AZ, with a Ph.D. Fellowship under the Research Training Grant in Optical Biomolecular Devices (National Science Foundation). In the Department of Chemistry, Chris began his dissertation research project under the direction of Linda B. Bloom, Ph.D. Chris was admitted to candidacy at ASU in May 1999. In August 1999, the laboratory moved to the University of Florida (UF), Gainesville, FL. Chris was admitted to candidacy at UF in June 2000. During his graduate school experience Chris was a teaching assistant for Biochemistry each fall semester (2000, 2001, 2002). Chris was awarded a Keystone Symposia Scholarship for meeting A1 in January 2002, "Mechanisms of DNA Replication and Recombination". In March 2003, Chris was a recipient of the Boyce Award for Graduate Research by the Department of Biochemistry and Molecular Biology, and was chosen to represent the department in the College of Medicine Medical Guild Graduate Research Competition were he placed fourth. After receiving his Ph.D.

Chris will be a postdoctoral associate on a DARPA (Department of Defense) funded project at the University of Florida, in conjunction with the Whitney Laboratory (UF), and the Max Planck Institute for Polymer Research, Mainz, Germany. This work will be for the development of a novel biosensor device based on addressable immobilized ion channels.

I certify that I have read this study and that in my opinion it conforms to acceptable standards of scholarly presentation and is fully adequate, in scope and quality, as a dissertation for the degree of Doctor of Philosophy.

Linda B. Bloom, Chair
Assistant Professor of Biochemistry and
Molecular Biology

I certify that I have read this study and that in my opinion it conforms to acceptable standards of scholarly presentation and is fully adequate, in scope and quality, as a dissertation for the degree of Doctor of Philosophy.

Daniel L. Purich
Professor of Biochemistry and Molecular
Biology

I certify that I have read this study and that in my opinion it conforms to acceptable standards of scholarly presentation and is fully adequate, in scope and quality, as a dissertation for the degree of Doctor of Philosophy.

Arthur S. Edison
Associate Professor of Biochemistry and
Molecular Biology

I certify that I have read this study and that in my opinion it conforms to acceptable standards of scholarly presentation and is fully adequate, in scope and quality, as a dissertation for the degree of Doctor of Philosophy.

James B. Flanegan
Professor of Biochemistry and Molecular
Biology

I certify that I have read this study and that in my opinion it conforms to acceptable standards of scholarly presentation and is fully adequate, in scope and quality, as a dissertation for the degree of Doctor of Philosophy.

Alfred S. Lewin
Professor of Molecular Genetics and
Microbiology

This dissertation was submitted to the Graduate Faculty of the College of Medicine and to the Graduate School and was accepted as partial fulfillment of the requirements for the degree of Doctor of Philosophy.

December 2003

____
Dean, College of Medicine

____
Dean, Graduate School

FINDING THE GOOD LIFE IN THE FAMILY AND SOCIETY:
THE TSWANA AGED OF BOTSWANA

By

ELIZABETH A. GUILLETTE

A DISSERTATION PRESENTED TO THE GRADUATE SCHOOL
OF THE UNIVERSITY OF FLORIDA IN PARTIAL FULFILLMENT
OF THE REQUIREMENTS FOR THE DEGREE OF
DOCTOR OF PHILOSOPHY

UNIVERSITY OF FLORIDA

1992

Copyright 1992

by

Elizabeth A. Guillette

PREFACE

Becoming old. It is something we all think about, including our own personal
securities and insecurities that accompany the individual process of aging. People see
themselves as being old in a setting similar to the one they know, without marked
alterations in the way the family incorporates its aged members or radical change in the
social and political environment. We have been socialized to expectations of how
individuals and groups will accept us as old people, who will provide for our needs, what
our roles will be, and the strategies we will use. These lessons provide a passageway to
a satisfactory old age, and cast a light over this final phase of life.

But what happens to the passageway when rapid social change associated with
the modernization of developing countries means that perceptions of life are topsy-turvy
and life's expectations are remodeled? Which elements from the past can be utilized?
Which are broken or lost? Can the resulting combination of new social and materialistic
advances combine with traditional ideology to create a secure life for the aged? Or do
the old face a barricade that prevents the finding of security and happiness.

This dissertation deals with social change and cultural continuity and how these
factors block or encourage the aged of Botswana in their quest for a good life. Like all
ethnographies, a specific cultural group is involved. In this case, the people are the
Malete, a Tswana cultural group of southern Africa. The setting is an old, established
Botswana village experiencing rapid development as the country becomes modernized.
Both the "old" and the "new" are embodied in the decorum of life. The scene involves
riches for a few and poverty for many. One also finds the repeated calamities of
drought, as created by nature.

The actors are the aged involved with two underlying social processes. Foremost is the interplay between traditional beliefs and behaviors and a new world view generated from modern technology, education, medicine, and mass media. Second is the resulting social conflict as individuals, in their attempt to gain status and acceptance, rely on abstract and material resources that have unequal value between generations.

A common proposition is that modernization promotes youth and is deleterious to the old. One must not to be too quick in assuming that development automatically reduces the quality of life for the aged. Modernization can provide goods, services, and new avenues of opportunity, which improve life for the old as well as the young. Neither must one assume that traditional culture automatically provided a high quality of life for the aged. A better understanding of the traditional ways in which communities defined and responded to the aging process is necessary before reasons for fault can be assigned to moderizations and avenues for betterment in the future can be identified.

Old age is multi-dimensional, involving a multitude of factors reflecting the past, present and future. Too often we think only of immediate and the negative: the cessation of work, alterations in the family constellation, illness and progressive decrepitude, as well as the increased realization that possibly one may be disliked solely because of advanced years. Therefore, the study of aging is often thought of as synonymous with investigations into crises that adversely modify or even inhibit the ability to function in the social realm. Concentration on the impact of lack of productive activities, the disappearance of roles and increasing senescence does a disservice to the aged and society alike. It is equally important to recognize cultural and individual strengths. Unification of the calamities and joyful provides new perspectives for determining whether events are defined as problems or challenges, social burdens or conquerable responsibilities. Thus, this dissertation addresses two aspects of aging: the liabilities and the advantages with growing old.

A standard approach in gerontological studies is to emphasize diversity within the group to provide ideas about how problems are defined and addressed. Equally relevant are the similarities that stimulate group identity, which can then provide previously unrecognized depth and breadth to social policy. The same is true across groups and across cultures. Within the diversity spectrum, concepts too often include only overt behaviors and responses. One must go beyond this to identify that which is covertly universal to all aged, and distinguish those factors generated by specific cultural and linguistic responses. I have accepted this challenge, as world aging is gaining in import.

By the turn of the century, developing nations will contain 61% of the world's aged (Hoover and Siegel, 1986). The largest increase in number of old will occur in Africa, where individuals 60 years and over are already the fastest growing age group (United Nations, 1985:98). In southern Africa, an increase of 76% in the older population is predicted between 1980 and 2000 (United Nations, 1985:99). Declining fertility rates and the loss of adults through AIDS will have some impact on the relative increase in the numbers of aged. More importantly, increased survival rates from childhood infections, and longer life expectancy among the healthy, will account for the generalized aging of the population (Hoover and Siegel, 1986). Therefore, it is to be expected that not only will there be more aged, but the old will become older.

In 1980, the United Nations stressed ignorance of potential problems and the lack of factual data as the main deficits in planning for the aged. Ten years later it is safe to say that consciousness of the imminence of an aging population has become widespread, although the present level of knowledge regarding the intricacies of African aging is no more than minimal. In part, this may be due to the emphasis placed on maternal child health by social scientists, as birth rates continue to far exceed death rates in developing countries. The needs of the African aged should carry equal weight, as it is the aged who provide much of the supportive care and teaching to the young

(LeVine and LeVine, 1985; Guillette, 1990; Biesele and Howell, 1981). Care-giving, as well as care-receiving, is an integral part of aging in Africa.

For some years, the main models for gerontological social programs available to developing countries have been provided by established, economically secure nations. These established programs for the aged include guaranteed income, government supported group housing, and long term health care facilities. While social security and long-term care facilities may seem easy solutions to increasing problems with the increased numbers of aged, these solutions may not be best on the national, community or individual level. As a rule, developing nations do not have the monetary resources to invest in such problem-solving approaches. On the community level, the transference of such programs tends to discourage rather than encourage already existing beneficial moral and social ties that provide support to the aged. For the individual, other salient solutions available within the geo-political and sociocultural settings are overlooked in favor of what is known.

I want to stress, that while the theoretical physical and social aspects of aging in a modern society may have catholic implications, the approaches and solutions to the aging process should not be universal. Local factors that support the aged, as shaped by each unique sociocultural system, need to be incorporated into specific program development. Thus, qualitative identification of local systems that influence the socioeconomic and humanitarian aspects of aging warrant as much consideration as qualitative evaluation of the aged as old people.

The contents consist of four parts. The first three chapters deal with the research study itself, explaining how gerontological thought in combination with anthropological theory and methods were used to investigate human actions and reactions regarding the process of aging in a dynamic environment. Chapters 4 and 5 explain the evolution of the village and the determinants of who is considered old.

Chapters 6 through 8 present what it is like to be old in the village: living conditions, family, and community life. The last three chapters (9, 10 and 11) delve into what the aged desire and need in the present context, with suggestions on meeting old age goals within the community and nation. The realisms involved with aging, social change, and cultural continuity in the one village are also considered on a national and international level.

None of this work would have been possible without the help of many people. It all began with my parents, Caroline and R. C. Arnold, who taught me the value of adaptability and creativity. Dr. Gordon Streib encouraged this creativity and provided much direction for gerontological thought. Drs. Allen Burns and Tony Oliver-Smith opened windows for thinking beyond the obvious and for challenging old assumptions. Great appreciation is extended to Dr. Leslie Sue Lieberman, who went beyond academic guidance to encourage a healthy mix of motherhood and personhood with professional duties. Dr. Art Hansen deserves special and warm merit for extending my view beyond the aged to include a "back-door" approach toward culture and society, and for always encouraging me to take advantage of the serendipitous and unknown.

The people of Ramotswa, both young and old, deserve more than a simple thank you, especially those who became Mma (mother) and Rra (father), grannies, and sisters. I have changed all names, but all are very real people. There are also unknown people whose financial assistance is appreciated. The Social Science Research Council provided a Predissertation Fellowship. The University of Florida Foundation provided post-research funds.

Above all is my appreciation for my own nuclear family: Kaiya, Tammy, and John, who provided constant encouragement and expression of faith, Matt, who experienced and explained Africa with the wise eyes of a ten year old, and my husband, Lou. It was his willingness to participate in African life and his warm understanding of the trials and tribulations associated with research that allowed for my own growth and

understanding.

TABLE OF CONTENTS

Abstract of Dissertation Presented to the Graduate School
of the University of Florida in Partial Fulfillment of the
Requirements for the Degree of Doctor of Philosophy

FINDING THE GOOD LIFE IN THE TSWANA FAMILY AND SOCIETY:
THE AGED OF BOTSWANA

By

Elizabeth A. Guillette

August 1992

Chairman: Dr. Art Hansen
Major Department: Anthropology

This quantitative and ethnographic study involved the Tswana aged of Botswana.
The setting was an established rural village experiencing rapid development. A variety
of interviewing tools were used to promote abstract thought and obtain statistical data
for 105 individuals, age sixty and above.

Traditionally, gerontocratic principles placed the elders of Botswana in a favored
position. The indigenous life cycle separated the elder from the helpless aged. The
helpless aged were regarded as children, incapable of meaningful thought and
judgement. Contemporary life has advanced the assignment of childhood to encompass
the active old, resulting in premature social disenfranchisement and denial of worth.

The aged attempt to maintain elder status with the use of personal/physiological,
social/familial, and fiduciary assets. Many experience an additional age-induced poverty
in conjunction with generalized village poverty. Cultural continuity and social change
create generational conflicts in the value and use of remaining assets. Those without
valued assets frequently fail to obtain care, security, and respect as an elder. Some
aged, with adequate resources, could not manipulate the setting to their benefit.

Traditional ideology continues to influence expressions of thought and behavior. Social change alters the directions flow of benefits from the aged to the younger members of the village. Families, which continue to be thought of as insurance against old age, usually provide basic physical care but most aged are excluded from roles of authority and self-expression. A community disregard of the social contributions made by the aged adds to their disenfranchisement from a meaningful life.

CHAPTER 1
INTRODUCTION

One must always "be doing" as the process of achieving gives meaning to
life. (79 year old man)

How can I do things? The door to my house is broken and my son does
not obey the laws to fix it. To toss aside the Tswana laws is to toss aside
the value of living. (60 year old woman)

This is a study of broken doors. A door controls a passageway. It has two

functions. It lets people into our homes and lives, and keeps the unwanted and

undesirable out. We tend to think of the door to our house as something that should

be closed and locked. We use it selectively at our own discretion, opening it to others

when sound reasoning and knowledge provide good reason.

In the Botswana village of Ramotswa the door to the Tswana home is usually

ajar. The open door is a signal that one is home, and the passerby is welcome into their

homes and life. It also means that the person inside wants to be included in the large

and small events happening outside.

There were no true doors when the present day aged were younger. A stick

placed across the entrance indicated no one was at home to provide company or assist

you if you were in need (Schapera, 1944; 1953). Actual doors were introduced during

British colonization (Schapera, 1953:26). In contemporary rural Botswana, all homes

have wooden doors. Many are imperfect. The people say it is because of "nature,"

external to the way the door has been used. The people overlook small cracks, as these

are considered normal. Other doors have larger gaps and deficient hinges. They do not

swing properly. This is more serious as the occupant and visitor have to pause and expend unnecessary work to open the door wider. At times the passerby will keep on walking, or the occupant will stay indoors. Both know the door is broken and question if the extra work is worth the visit. Occasionally one sees a door so damaged that it has been sealed shut. The house has nothing to offer, and the past occupant is said to have moved or "disappeared." No one exits and no one visits.

To Americans, the closed door provides a safe haven. To the Tswana, an open door provides access to a safe haven. It shows that people are part of life and have a respected place in it. The open door is more than a passageway. It is a giver-of-life in itself. The Tswana use many doors in their movement through old age.

I use the Tswana meaning of door as an analogy for a continuing access to opportunities to maximize potential as a person. The small cracks are those that affect the process of "being someone." They may be disliked but are an integral part of life. As the cracks enlarge and the door refuses to swing properly, access to personhood and the good things in life becomes more difficult. Additional breakage creates barriers in bringing others into one's life, and in entering the lives of others. Some obstacles are more difficult to overcome than others, depending on the type of damage. Extra effort is always needed to use the door. The door so damaged that it is permanently sealed shut severs access to a full and meaningful life.

Two things will be highlighted in this study. First are the many doors that the Tswana aged use for a rewarding life, and the fact that many no longer work properly. Cultural continuity provides both strength and imperfections in their structure. Social change can provide new doors. Usually, social change misshapes the old doors so that additional effort is needed for use. Sometimes the old cracks and new imperfections combine to form a deadlock. The door is inoperable from either side. No one can pass through it. Both the Tswana aged and society must deal with these various damaged and broken doors in finding access to the things they hold dear.

Secondly, just as there are cultural differences in the meaning of a door's use, there are cultural differences in interpretation of the meaning of aging. I present contrasts between aging in Botswana and developed nations, and contrasts between concepts of what old age should be like. This is the sort of thing that may provide additional information for opening new approaches to cross-cultural gerontology.

In this introduction, I will present the generalized salient elements of the study. Following this overview, I will discuss the aged and their physical and ideological setting. The chapter ends with an interpretation of success as delineated by sociocultural values. This information provides the foundation for understanding how the Tswana doors have been broken, and why our own should be unlocked in the interpretation of cross-cultural gerontology.

<u>Overview of the Study</u>

The study involved 105 aged (age 60 or greater) of the Malete tribe of the Tswana ethnic group of southern Africa. The setting was an established, moderate sized, rural village in Botswana. The village presented a typical mix of traditional and modern thought, social rites, and technology. The investigation was done during two research periods, each involving separate steps. The first was a two month period during 1988, looking at society as a whole to extract specifics for gerontological research. The second was a seven month period extending from the end of 1989 through March 1990. At this time an in-depth study of the aged was undertaken. At both times research methodologies of formal and ethnographic interviewing, oral histories and participant observation were used. Visual aids were added as interviewing tools with the intensive research on the lives of 30 individuals. These tools encouraged abstract thought, which is difficult for the Tswana to express and promoted discussion to include that which others were hesitant to mention or considered irrelevant.

The study was based on a gerontological adoption of social exchange theory, in which the aged trade personal/physiological, social/familial and fiduciary resources in exchange for the services and support needed for a good life (Gubrium, 1973). It was my belief that the elements of a good life are present in the transitional village environment, and that those who control goods and labor wanted by others find satisfaction. It was anticipated that aspects of tradition and modernization both promote and hinder the process of exchange.

Little was previously known about the definition, let alone the acceptance, of the Tswana aged as a unique population within traditional Tswana society. The commonly held assumption that high status and power were automatic for all aged was investigated. Central goals were to learn how definitions of old age traditionally affected application of the eldership principle, and if supportive behaviors were withdrawn with advanced aging. Cultural change and continuity in these social processes were seen as affecting the present as much as the social and material changes associated with modernization. The intent was to evaluate the assumption that the process of modernization is the sole agent in the devaluation and degradation of old people.

The main unit of analysis was the individual. Cultural continuity and social change set the stage for the use of their resources. Access to the good life was hypothesized as difficult for those who are poor in resources. I found that most people lived with broken doors. The poverty of old age superimposed upon widespread economic deprivation, and the fact that traditional attributes of childhood with a lack of social value were no longer limited to the extremely decrepit, were barriers too strong to overcome.

My research includes the similarities and differences found between the aging process in a developing country and the developed nations. Much of the theory of social gerontology comes from the modern world. I apply Western-based theoretical concepts to aging in a transitional peasant agricultural setting. This necessitates looking at aging from two perspectives: the outsider's theoretical and experiential view of

gerontology and the Tswana view of ideals and reality. Identification and understanding

of the grass root desires and yearnings of the Tswana aged, as based on their approach

to life, is necessary. I assume that hidden behind the overt behaviors and values one

can find universals in the goals of the aged and the processes used to obtain them.

Research design purposely reflects the inclusion of possible universal components.

<u>The Setting</u>

Botswana, slightly larger than France, is a land-locked country in southern Africa.

(See Figure 1.1.) The Kalahari desert occupies two-thirds of the country but sand is

found everywhere. Sand is a main component of the soil in the flood plains of the

Okavango Delta to the north and the low rocky hills in the east. Only the east is

hospitable for agriculture, concentrating the population within this area. Even there one

finds environmental risks with crops and livestock. Droughts and erratic rainfall, varying

from year to year and site to site within a given area, have caused repeated withdrawal

and resettlement (Campbell, 1982:20-22).

The eastern area of Botswana and the South African Transvaal were sparsely

populated by Sotho-Tswana groups by 1200 AD (Tlou and Campbell, 1984:32).

Pastoralism and peasant agriculture provided subsistence (Campbell, 1982:20-22).

Gradually the Tswana culture emerged, encompassing various tribal groups and resulting

in the seven Tswana nations found in the country today. The Malete were first

identified as a tribal nation near the turn of the eighteenth century (Ngcongco, 1982:25).

Botswana, known as Bechuanaland, was claimed by the British in 1885. The

Protectorate served as a "labor reserve"' for South Africa, providing workers for mines,

farms, and later, industry. At the turn of the twentieth century, a railway was built in

the eastern corridor to interconnect the wealth of South Africa and Rhodesia (Tlou and

Campbell, 1984:156, 161). Other development was directed toward supporting the cost

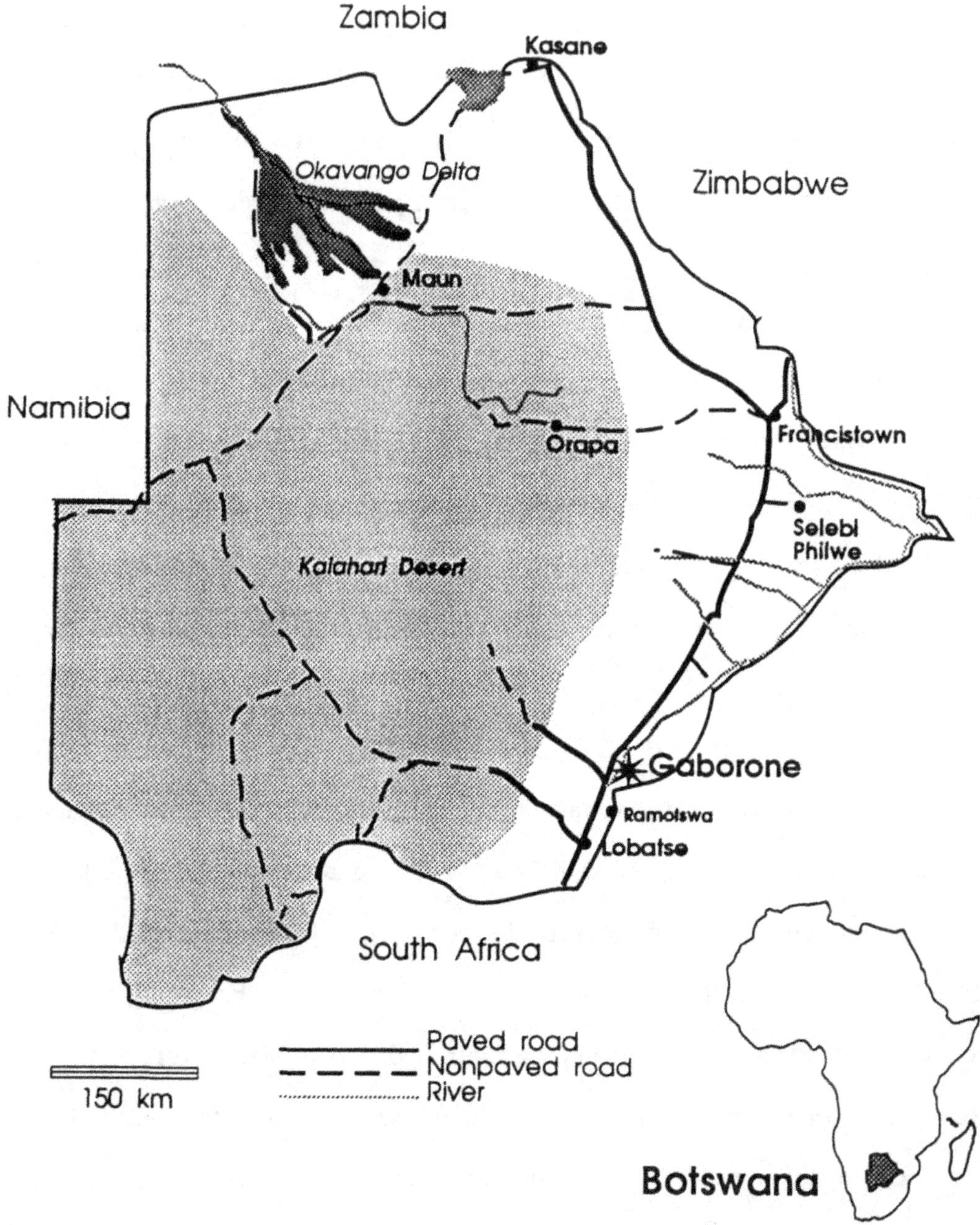

Figure 1.1: Map of Botswana

of Protectorate administration (Parson, 1984).

National independence came on September 30, 1966. Seretse Khama, a Bamangwala tribal chieftain, became the first president. He and his Anglo-European wife served Botswana until his death in 1980 (Tlou and Campbell, 1984:213-228). Quett Masire was elevated from vice president to president at that time, and has been constantly re-elected in the multiparty elections, held every five years.

The national flag was designed to represent the interracial policy of Botswana. A thick black band in the center represents the African majority. Thin white stripes on either side stand for the non-Africans. A blue background represents *Pula* (rainwater), which is the source of all life and blessings (Tlou and Campbell, 1984:228). A non-racial policy that permeates central and local government, and urban and village life, has existed since the country's beginning. In addition there is freedom of speech and open access to the media (Parson, 1984).

At Independence, Botswana was one of the poorest countries in the world. Today it is among the richest of the developing countries. The discovery of diamonds at Orapa and the development of copper-nickle mining at Selegi Phikwa, an increase in beef exports, and successful negotiations with South Africa for more balanced trade contributed to national economic growth. This allowed Botswana to develop social and economic infrastructures that now bring health care, education and employment to many of its people. The traditional peasant subsistence system has become a "Peasantariat" subsistence system, incorporating a peasant type of agriculture and society with the proletarian attributes resulting from wage labor (Parson, 1984:123). There is still much to be done in raising the standard of living, as poverty and unemployment dominate in rural areas (Tlou and Campbell, 1984:238-255).

In 1981, 941,027 people lived in Botswana (Republic of Botswana, 1981). The population is a young one, growing at a rate of 3.48% a year (Knudsen, 1988). Slightly less than 80% of the people live rurally, with others living in the relatively new urban

centers (Picard, 1987:4). There is only one true city, as defined by a population of

100,000. That is Gaborone, the capital.

The village of Ramotswa, located near the country's southeastern juncture with

South Africa, had an early history of intermittent settlement. It became a permanent

village near the turn of the twentieth century, when entered by the Ba-Ga-Malete. The

tribal name was shortened to Ba-Malete during the period of colonization (Ellenberger,

1937). Today, the people prefer to be known as Malete, the most modern name of all.

Most village inhabitants are Malete, with the others having a Tswana heritage.

Approximately 4- to 5,000 of the 13,000 area inhabitants do not live in the village itself.

Some are migrant workers returning home for weekends, either frequently or irregularly.

Others reside on a more permanent basis at their agricultural lands, returning briefly for

a change of clothes or purchased foods. There are almost no technological advantages

pressing for a return home. Almost every household is without running water, electricity

and telephones. Weekend entertainment is limited to funerals, weddings and trips to

the bars.

Social hierarchy is now based on economic elitism rather than kinship. About

10% of the population can be considered upper class (Parson, 1984:34). The majority

live in rural poverty, relying on agriculture, livestock, and remittances from employed

emigrated family members for household support. The scarcity of rural employment and

harsh environment contribute to the lack of a rural middle class (Hudson, 1977;

Hitchcock, 1989).

Poverty is widespread in the village. What do I mean by poverty? Absolute

poverty is a comparison of each household's income with a Poverty Datum line (PDL), a

line below which a household is unable to meet the minimum needs for life. A 1977

survey, which included Ramotswa, found that about 45% of the households had incomes below the PDL (Hudson, 1977). Since that time, rural poverty has been increasing, with estimates of households in absolute poverty always ranging well over 50% (Knudsen, 1988:17). The number of impoverished households reflects the fact that the richest 10% of Botswana families control 42% of national household income. That leaves 58% of household income to support 90% of the people (Botswana Central Statistics Office, 1974)

Rural household income, whatever the amount, must be considered in conjunction with dependency ratios, or the numbers of economically dependent. The 1981 national census for Ramotswa reported a population of 13,000 with 2,845 actively employed (Republic of Botswana, 1981). Thus, 20% of the village population supports the others. The employed include those in agriculture, but most are migratory workers who claim Ramotswa as their home. The majority of households are without a locally employed adult. Remittances from migratory family members are a must in order to survive. Many grow the majority of their own food to ease expenditures. (These individuals are included in the employed if any food is sold.) Even this attempt at survival does not always guarantee food, as rainfall is erratic and drought is common.

Deprivation is rampant, according to the outsider's view. There is a lack of food and lack of what we consider necessities for normal living. To the Tswana, the situation is reality. It is real not to be sated after a meal, to be cold in the winter and to have the roof leak during rain. Life could be better, just as yours and mine. Like us, they strive to improve life, only their expectations are placed in village reality.

Members of the household work to maintain a status quo, as gain is improbable and loss pushes people further into poverty. Families who wants to survive cannot afford to be lazy or stupid, for then they will lose what they have (Chambers, 1983:107). Chambers (1983) calls this situation a "deprivation trap." Poverty unites with a loss of power to increase physical weakness, vulnerability, and isolation that cannot be overcome. Any additional loss serves to pull the household into deeper poverty (Chambers, 1983:108-113).

The extensive poverty does not deter the villagers from seeking modernization. The progressive modernization of the tribal name parallels progression from "the days before civilization" to "being modern." Modernization is an important quest in village development. The inhabitants are proud of the one paved road. Many approve of the installation of a village sewer system but do not understand water pollution and ecological degradation. The village boasts of its hospital and two churches. The South-East District Headquarters, the federal building for local district government, is on the outskirts of town. Ramotswa is also the site of the Malete Tribal Administration, for the Malete chief has always lived in the village. These are unique structures for any medium sized village. They affect the process of growing-up and of aging.

<u>Traditional African Elders</u>

A common assumption is that the characteristics of agrarian African societies, with their interpersonal relationships reflecting a positive value for increasing age, promote status and power for all aged. The loyalty to the seniority principle by children and kin provides respected personal and social old age identities and functions (Schapera, 1953; LeVine and LeVine, 1985; Cohen, 1986; Rosenmayr, 1989). The longer one lives, the more highly one is valued (LeVine and LeVine, 1985:30-33). These characteristics, as presented in ethnographies, imply that old age is the pinnacle of life.

A difficulty lies in the fact that ethnographies do not separate "elders" from the "aged." The same is true with the extensive writings on the Tswana by Isaac Schapera. Only once does he refer to the aged, who are those who must be supported (Schapera, 1955:179). He discusses elders freely. Deference, honor and respect are given to an elder. The elder has authority and controls the power in relationships (Schapera, 1944, 1953, 1955). One must be careful in delineating who are the elders, as an elder is anyone who is older than the speaker (Schapera, 1953:38). Such a definition does not limit elderhood to the chronologically aged. There has been a tendency to unite the two, resulting in a mix of truth and myth (Diouf, 1985).

Schapera only hints at a division between the socially defined aged and the elder. This division can be very strong, with a removal of status and supportive care when decrepitude occurs (Glascock and Feinman, 1981).

The Aged

In contrast to developed nations, old age occurs earlier in developing countries. Harsh environments and working conditions, limited health care and malnutrition result in a physiologically aged body long before age 65 (Diouf, 1984; Tout, 1989). The United Nations (1985:98), in recognition of premature aging, demarcates old age as beginning at age 60. Although chronological age alone does not present an undisputable representation of biological and social age, it is used in this research to provide consistency, and for cross cultural comparisons.

Chronological aging is distinct from social aging. Some of the Malete are socially old, or designated as "children," before the chronologically assigned date. Other chronologically old are perceived as modern "adults." Imposed social labeling does not remove the cherished traditional values and behaviors held by the aged, which previously promoted them as elders. Some are treated as elders in the true spirit of the word, leaders with meaningful family and social roles. The majority of aged are regarded as children.

I questioned the aged about their own parents. Once an old person could no longer function as an active social elder in society, status and power were taken away and replaced with the care and supervision needed by a child. They were treated as a child incapable of meaningful thought and action. The same is true today, except the timing of childhood occurs much earlier. Any contributions are seen as valueless. Their advisement is seen as based on uncomprehending thought. Even the chronologically old who are recognized as active and valuable social elders, with multiple resources and vigor, are constantly and actively working against the label of "child."

<u>Tradition and Cultural Continuity</u>

Modernization of village structure has not supplanted tradition and cultural beliefs in the varied aspects of daily life. Even the most modern of people exhibit actions and thoughts reflecting cultural continuity. Traditional ideology still dominates, although frequently in a modified manner. Underlying all behaviors are two cultural traits: the continuous "process of achieving" and a following of what the Tswana refer to as "the laws." These will be defined and discussed separately.

<u>The Process of Achieving</u>

Achieving is an on-going, perpetual process. One never achieves, as there is always a need for continuing participation in the "'Great Works" of supplying food, building a family and keeping community commitments. The person is not judged by past achievements, but the continuation of them (Alverson, 1978). The focus is on the present. The past is completed and to be forgotten. The future is unpredictable and should not be the foundation for worry. There is no such concept as self-caused personal failure with achieving. It is external circumstances that get the upper hand and prevent progress within the on-going process . It is the failure to strive for betterment under hardship makes one useless, for it is the "doing" that provides social value, plus contentment and fulfillment in life.

In today's world, the process of achieving continues to emphasize family corporacy, generational symbiosis and social solidarity. Actions are important in that they not only accentuate one's own welfare but help others. Family roles and a meaningful place in society are paramount. These attributes direct the group-centered approach to achieving. The process should involve interdependency between family members and with society. Great importance continues to be placed on the extended family as a social and economic unit. Achieving for women is in the realm of raising children and producing food, either through agriculture or employment. Men are the family administrators, controlling the income of self and others, and overseeing family activity and welfare. As a rule, family interdependence reigns over individual independence. Individual independence usually involves "the breaking of the law."

<u>The Laws</u>

The "doing" in the process of achieving is regulated by a traditional system of cultural rules for behavior, which the Tswana call "the laws." The laws are more than norms for behavior as they define and regulate moral obligations to others and society. They are a system to insure rights, safety and cooperation. All further references to the laws refer to these unwritten codes of behaviors unless explicitly stated otherwise.

The laws cover all aspects of daily life, regulating what can and cannot be said and done, including proper voice and steps for action. Laws are explicitly taught to children. "One must always obey an elder." "The son must fix the parent's roof." "The daughter must cook for her parents." The laws are numerous and varied: one must not raise one's voice in anger; to loan is to give; one should seek the advice of others before making a decision. As times have changed, some of the lesser laws have been dropped by the old and young alike, such as prohibitions against washing during darkness. Other laws regulating personal behavior are still followed by the aged, but the young have permission to discontinue their observance, such as the ban on drawing water during darkness. Most laws concern interpersonal behavior. These laws should never be broken by any age at any time.

There is almost no diversity in the interpretation of the law. Obedience is through social and family sanctions. The desire to comply is propelled by outside forces. Neither Alverson (1978) nor I found any indication of self-guilt with misbehavior. One follows the law because of possible punishment. The punishment is infliction of shame. Beating, as a painful method of shame, is no longer permissible under federal law. Today, shame is placed on the disobedient individual either verbally or by avoidance. Shame is made public when there is chronic or severe breaking of the law. Public shame is the worst shame possible as it also brings shame to the family. The defiant individual is teased, belittled or ignored. The family is pitied as they could not influence kin behavior.

I observed great heterogeneity among individuals in the effectiveness with which shame controls obedience. There is more variation found between generations than within them. Attempts by elders, either parents or grandparents, to place shame on the young tend to be ignored. In contrast, the youth are quick to notice deviation from the laws by elders and encourage the placement of shame on them.

The old carry out the law as well as direct others in its use. Law enforcement, by right of seniority, is part of the process of achieving. Age seniority brings the privilege of being treated as an elder, being obeyed and respected according to the law. The enhanced authority of the Tswana elder should not be equated with Western style autonomy in decision making. By law, decision making involves consideration of others with group input.

Seniority laws dictate that the very old will receive care-provisioning and economic support. According to the laws, no old person should be deprived, always having an adequate, safe environment. These laws, like others, are specific. "The very old must be washed and kept clean." "The very old must be given food." Thus, through laws and the process of achieving, the individual Tswana abstracts qualities of positiveness that make for a good life throughout the phases of old age.

The Good Life

The "good life" concept was initially used by Lawton (1983:349) to incorporate both the process and the goals of success that all individuals strive to meet. It is essential to understand the ways in which the Tswana conceive success in life. This differs considerably from Western concepts.

Theoretical Components of The Universal Good Life

The components of the good life are assumed to be legitimate goals. Behavioral competency, psychological well-being, and life satisfaction are based on individual behaviors within the physical, personal and social environment. Lawton (1983) insists that each individual defines the idiosyncratic goals that have personal importance in the quest for the good life. This implies that the good life ceases with an inability to make rational independent decisions or with a lack of orientation to the future.

In contrast, Keith, Fry and Ikels (1990) place success or life satisfaction in a community context. The environment rather than the individual defines the determinants of success. The nature of the local society directs the individual in making choices and determining strategies for the meeting of goals that create a good life. Both the individual and society are united, giving the theoretical possibility that the good life can continue in some way with loss of control over the environment, or the good life can be an impossibility in a negative environment regardless of the strengths of the aged.

I will use the term "a good life" to reflect a positive quality in life that can theoretically continue until death. The individual has personalized goals based on idiosyncratic definitions of what is desired. The environment is the framework for determining what the goals should be, their timing, and rules for fulfillment. The degree of satisfaction with old age is dependent on the present attainment of a rewarding blend between individual goals and social standards.

The aged Tswana have expectations and desires of things to come, but believe that very limited control over the future is possible. Conceptually, the future is blurry at best. "Nature" (the interrelationship between self and outside forces) determines the future and the potential meeting of goals. It is the present, with involvement in the process of achieving, which is the central issue, not the long range outcome or the future. Thus, old age goals are synonymous with everyday behaviors involving self, family and community.

One must ponder whether there is a theoretically universal uniformity regarding the goals of all aged, or whether the cross-cultural context differs too radically to be bridged. Differing cultural value systems and socialization processes produce varying roles and behaviors expected of the aged (Sokolovsky, 1990). It is logical that the positive social outcomes of these expectations produce culturally specific old age goals. Do culturally specific variables transcend differences to produce a higher level of uniformity? It is vital to answer this question, using data from a variety of cultures from both developed and developing nations to produce sound theory and to plan prophylactically for the increase in world aging.

The goals of old age in America are usually described abstractly as independence, autonomy, good morale, health and peace of mind. To use leisure time effectively and have sufficient guaranteed income are vital inclusions (Gubrium, 1973:ix; Jacobs, 1974; Myerhoff, 1978). Such goals reflect the social horizons of the world in which we live.

Presentations of goals tend to be more concrete when one looks at aging on a universal scale. They include the desire to have and control resources acquired during life and the guarantee of food, shelter and care. They also include features of retaining status and respect, being useful participants in family and group affairs while safeguarding health and energies, and the right to withdraw from life honorably when the advantages of death outweigh the pleasures of physical existence (Simmons, 1945; Brathwaite, 1986; Jacome, 1988). These broader goals closely parallel the culturally specific Tswana goals of seniority rights, interdependence and usefulness, as gained through continual achieving and the application of laws.

These broader theoretical goals can also be superimposed on the intents of the Western aged. In America, goals with aging are defined by the middle class. Their more affluent life style allows them to take the basics of the above for granted. This is not true with the aged poor who view success similarly to the aged in developing countries (Stack, 1974; Chatters, 1988; Cohen and Sokolovsky, 1989).

The Specifics of the Good Life in Ramotswa

The intent of this section is to present the specific concepts of the good life as defined by the Tswana. I will develop the major thoughts but will not consider degrees of individual variation. General strategies for the operation of the concepts are brought out, mainly for reasons of clarification. The constructs of successful aging emerge from the generalized operational definition of success in adulthood.

I investigated the question of who is a successful person in the present society. These conclusions are based on conversations with individuals of all ages during my stays in the village. The successful person has stable employment, owns a house as compared to a hut, and has basic furniture. Customary laws and traditions are still followed in the home and village. The home will be a setting of harmony with respect for eldership and cooperation between generations. There is no doubt that whatever money enters the household will be used for the basic maintenance of its economically dependent family members, although benefits need not be divided equally. The spouse will perform all assigned roles without question. The children, who represent support for old age, obey all orders to help and serve, before and after school. Socially, the successful person willingly accepts the responsibility to give to others, however small the gifts may be.

Frequent visits to kin about the village are interspersed with paying respects to bereaved families and attending social and political functions.

These norms may clash with introduced ideas of individualism, self-reasoning and self-concern that are promoting new values concerning property, less division of labor between sexes, and more emphasis on the economic expressions of social status. Following new norms, without piety to traditional responsibilities, often negates social success in this rural area, especially when judged by the aged. To the young, economic success is often equated with social success.

For the aged, the good life is not the same as the meaning of success to a younger adult. With the aged, success is not based on conquering poverty or having a better life than before. Success is determined by the temper of current family affairs and being included in acts of on-going family and social solidarity. Status comes from continual striving, in opposition to previous achievements or past events. Involvement in the process is more important than the ownership and control over multiple possessions. Material goods reflect economic achievements as a working adult, not success in old age, and are not considered in successful aging. Deprivation should be absent, but plenty is not a requisite. An old person finds more joy in living in the hut built as a young adult than in moving into a more modern square house provided by son or daughter.

The successful aged person also contributes to the economic welfare of the family, either directly with money-making schemes or indirectly with agriculture and/or household management, including child care. This old person also has ultimate authority within the household, even if major responsibilities have been delegated to others.

A caring family is an overt marker of success: a daughter to tend house and prepare sufficient food, a grandchild to sleep with and to fetch a cup of water, and a son to do heavy chores or repairs. The old man should have a wife although a husband is not necessary to a woman's success. The more intensive and comprehensive family care is, the better. But such aspects of family life alone do not represent complete success, for with service must come respect. Household members should not show disagreement with the wishes of the old concerning proper behavior or action. Even migratory children must show respect with visits and with the remittance of money. The showing of respect does not represent power, but the value of the old person to members of the family.

The successful aged continue meeting the moral obligations encompassed in the laws. Like the successful adult, visitations and socializing are mandatory. This includes maintaining proper relationships with multiple kin and contributions to the welfare of the village. For the man, this may be participation in the *Kgotla* (tribal meetings). For the woman, it is participation in village organizations, such as a modern burial society or committees for community improvement. No true division exists between leisure and work. The successful aged are always at work. To sit idle is to be worthless.

In summary, the aged of Ramotswa base their goals on a continuing but changing society. Their expectations for life come from tradition and from life experiences with the new and old ways. Doors should be open to keep a constant flow among self, others, and the world about them. The following chapters present the doors and degrees of success the aged have in obtaining the good life.

CHAPTER 2
THEORY

> Modernization has been a bad thing for the old people. They have no
> control over their children. No one seeks my advice or asks about the
> past. (68 year old woman)

> The old are now poor because of drought and old age. People don't
> treat us well. What I want is someone who cares. I'm lucky if people do
> things for me and I don't have to pay them. (91 year old woman)

Some Tswana aged are efficacious in old age, retaining meaningful activities and status. Their doors are open and work well. Others fail to meet the goals of successful aging and lack life satisfaction. They cannot overcome the barriers of broken doors and are unable to move forward or have others enter their lives in a meaningful way. This chapter will explore the theoretical reasons behind the possibility of obtaining success in aging, as an individual and as a social phenomenon. This includes the social processes that can either open or block doors. Gerontology is presented as a specific field within the realm of the social sciences. Selected theories on social aging are discussed. The study's hypotheses are presented, as extracted from the presented theory.

Early Gerontological Theories

Early gerontological theories focused on the declines and losses experienced with aging. In 1961, Cummings and Henry presented the Disengagement Theory, stating that aging people, with an awareness of diminishing capabilities and a desire to promote self over society in the limited time before death, reduce their amount and intensity of social interaction. A process of disengagement takes place, beginning with the relinquishment of given roles, such as those surrendered at retirement. This act leads to withdrawal from other roles and activities. With increasing age and diminution of energy, social

20

participation may be narrowed to only those role relationships and activities that are most necessary or rewarding. This withdrawal accompanies a preoccupation with self, leading to a new equilibrium characterized by a greater distance from social relationships. In this manner an optimum level of personal gratification may be found (Cummings and Henry, 1961).

Society also retreats away from the aged person. Society retracts because of the need to fit younger people into the positions occupied by older people, who are thought to be no longer as useful or dependable as they once were. Difficulties may occur when either society or the individual is not yet ready to begin the disengagement process. It is this lack of synchronization that leads to adjustment problems on the part of the individual. High morale is reestablished as the gaps close and a new orientation to life occurs (Cummings and Henry, 1961).

This model sought clues to both personal and social stability with the process of social aging (Atchley, 1987a:186). Adjustment, leading to the attainment of the good life, was based on the ability of the individual to separate self from the social environment. The roots of generational and social conflict were perceived as stemming mainly from the individual's failure to withdraw interest and commitment to others (Cummings and Henry, 1961). The social contributions of the aged to themselves, the family, and society were disregarded. The presumed inevitability and inherent nature of the process left no room for active participation in life. Self-centered activity and social idleness become the "normal and inevitable" rewards of aging (Cummings and Henry, 1961).

Disengagement Theory challenged the conventional wisdom that activity and contribution were the best way to adjust to aging. This challenge led to an opposing Activity Theory. Activity Theory held that older people are the same as middle-aged people with each group having the same psychological and social needs. Activity provided need fulfillment to compensate for the withdrawal of society from the aging person (Havighurst, Neugarten and Tobin, 1963:309).

Again, it was posited that it was the sole responsibility of the aged to adapt to the social situation. The aged had to find substitutes as previous roles and activities were withdrawn. Replacement was mandatory for adjustment, regardless of the availability or desirability of new substitutes (Atchley, 1987b:4). Little attention was directed to differences among types of activity, abilities to perform activities, differential rewards, or control over interactions (Hendricks and Hendricks, 1986:90). Forsaken was the idea that meaningful interaction with others, not merely activity, created satisfaction. As with disengagement, conceptual threads continued. The aged were people of decreasing social value and were to be shunned and displaced.

These two simplistic theories of social aging, developed through research on aged Americans, were outgrown by the field of gerontology. Practitioners and the aged rebelled against the thought that maladjustment during old age was due solely to the inability of the aged to reorient attitudes and to restrict activity to fit new and negative circumstances. Theory had to be developed on a broader footing, as both theories contradicted knowledge about aging in other parts of the world where active aged are incorporated into society and have continuing roles (Hampson, 1982; Shostak, 1983; Cohen, 1986). In addition, it was realized that other factors affected social aging, such as the use of technology, and political and economic systems. As people became more interested in aging beyond America, the framework for study had to include cultural, historical, and contemporary impacts.

In spite of drawbacks, these early theories do contain lessons that should not be forgotten. The first is that meaningful interactions are important to the aged, whether the aged individual selects to be socially active or reclusive. While older people may be more selective and self-serving in their activities and relationships, it is the quality, not quantity, of activity that determines if satisfaction exists (Hendricks and Hendricks, 1986). Second is that the availability and performance of meaningful interactions are determined through the offerings of society. It is now clear that psychological disengagement, or continuous meaningless behavior, is neither natural or inevitable, and that most cases of poor activity selection result from a lack of opportunities for continued involvement (Atchley, 1987a:186).

Modernization Theory

Modernization Theory began with the social sciences as a means to analyze social change. The theory assumed unilinear evolution, with universal patterns in structural growth that developed along set lines, regardless of context (Parsons, 1964). As technology was introduced, all aspects of a society were thought to move away from the diffused activities based on the closed ascriptive status systems that were associated with extended kin networks . Eventually, industrial-intensive societies, based on achieved status, would evolve. The need for integration of modern bureaucracies and money market economies would create convergence of modern structural characteristics in all societies. New cultural values to justify the emergent dimensions of human society would evolve from changing language, religion, and other belief systems (Hendricks and Hendricks, 1986:100). These values would then be incorporated into human relationships and reflect secular and instrumental rationality.

Gerontologists perceived within Modernization Theory a useful framework for describing and explaining continuities and changes in the social position of the aged across space and over time. The theory had drawbacks because of its generalized, political nature and its emphasis on the reliance of developing nations on established countries for socioeconomic direction. Instead, communities became the focus, rather than nations. The interdependencies in the local socioeconomic-political arenas directed the aging process. Social change was the cause of change in relationships among various age groups. All ages would be involved with this change as critical shifts in attitudes, ways of life, and social structure occurred. In turn, this change influenced the experience and interpretation of aging (Achenbaum, 1987:453). It was believed that economic and technological innovation transformed the roles of the elderly. The aged lost ascriptive status and could not compete with others in the new environment. The assumption was that modernization marked the beginning of the end for old people.

Leo Simmons presented this assumption in *The Role of the Aged in Primitive Societies* (1945) and later argued that the role of the aged in a given society was inversely related to the level of technological development and occupational specialization (Simmons, 1960). The premise was that the structured framework of primitive societies favored the elderly as advisors, carriers of culture and controllers of civic and political powers. With introduced technological change, seniority rights disappeared as new power-producing roles were allocated to the young (Simmons, 1960).

Cowgill and Holmes, the leaders in the application of Modernization Theory to the aged, conclude that all factors of modernization lead to social change that is inimical to the status of the aged (Cowgill and Holmes, 1972). Society moves away from giving beneficial ascriptive status for the old and institutes achieved status based on a money market economy (Parsons, 1964). Whatever seniority rights and authority the aged possess are undermined and eventually eliminated as the young usurp power (Cowgill and Holmes, 1972). The speed or intensity of change, either in the present or future, cannot be predicted (Achenbaum, 1987).

Modern health technology, economic technology, urbanization and rising levels of education were identified as the prime movers in the devaluation of old age. These factors unite in a functional manner to deprive the aged of their central importance in the affairs of daily life. In any event, status for the aged inverts as the old are delegated to function at the periphery of society (Cowgill, 1979). The aged are thus placed in an isolated position at the undesirable end of the life cycle.

Contrasting with Disengagement and Activity Theories, Modernization Theory, as applied to gerontology, totally attributes failure of the aged to reach goals on social change. The value of Modernization Theory is that it has stimulated scholars to construct an historical record of aging in various settings as modernization progressed. The theory has a problem however, as it implies that an historical *before/after* bifurcation occurs with the onset of industrialization and modernization. The "before" carries the uncritically accepted assumption that the aged were inevitably treated as honored members of the family and community as long as they lived. The "after" stresses that, with the introduction of social change, the aged have no valued functions at all. At this new point in time, it is the aged's responsibility to reorient attitudes and activities to mesh with their unwelcome place in society (Cowgill and Holmes, 1972). (The antithetical position of assuming that a changing society's goal should be to facilitate social efficiency and personal adjustment of aged members is just as naive.)

The assumption that all old people always held a high position in society is quickly becoming recognized as myth. The aged, whether contributing members or living liabilities, were not treated the same in all societies prior to the introduction of modernization. Glascock and Feinman (1984), in a Human Relation Area Files cross-cultural survey, found practices of non-supportive and death hastening behaviors directed towards the aged in 84% of traditional societies.

In many places of the modern world the aged are regarded as lacking ability to direct their lives and become known as the undesirable (Hendricks, 1982). They are cast out of main-stream society (Cowgill and Holmes, 1972). In many ways the situation of the aged is similar to that of other dislocated people. Instead of physical displacement from war or natural disaster, the aged are uprooted-in-place. Displaced refugees are no longer thought of as powerless individuals dependent on the receptiveness of the host society. Refugees recognize loss, develop strategies, make decisions and act on them, and have the ability to be incorporated into a new social world (Oliver-Smith and Hansen, 1982:6-8). The same has been shown with some groups of aged who are displaced-in-place with the process of modernization (Hendricks, 1982; Jacome, 1988; Coles, 1990).

Participation in life and society is demonstrated in a wide variety of sociocultural contexts and socio-political levels. The aged actively participate to maintain status and roles in some newly changing societies (Biesele and Howell, 1981; Colson and Scudder, 1981; Rosenmayr, 1989; Rosenberg, 1990). More important, the aged, as a result of their actions, continue to have respect and power in instances where change has been on-going for decades (Streib, 1972; Amoss, 1981; Simic, 1990). Other aged purposely take advantage of social change to promote their position in society (Amoss, 1981; Peterson, 1990). In each of these instances, the aged are active actors in directing markers of status. As a group, their refusal to be passive agents challenges the assumption that industrialization and modernization are irresistible agents driving the lack of social reward with aging.

Theories continue to assume there is a consistency in passive reactions accompanying personal change with aging. The focus remains on how old people should react to be accepted within the social situation. This concept leaves little room for a positive or creative "modernization of old age" with valued contemporary roles within the new socioeconomic-political arena. No consideration is given to the range of successful adaptive behaviors the individual is capable of doing with a continuation of self.

The concept that the aged function as passive agents needs to be replaced with a concept that the aged are participating individuals within a social system. The aged should be viewed as intentional directors of their lives, accepting the fluid relationship between people and social contexts and actively integrating role definitions with opportunities experienced and to be had. Like other people, they seek need fulfillment using what is available from the environment, from culture, and from others. Their efforts involve both active and reactive responses. Individuals strive to meet ideals, which can, and do, express a variance from social ideals and from reality. Context can vary, as with the differing reality and ideals between the African village and American city. The positiveness of aging, with a recognition of continuing yet varying capacities for interactions with and within the environment, must be incorporated into theory.

<u>Social Environmental Theory</u>

The social sciences have provided a framework for more comprehensive gerontological theory. Social Exchange Theory explains how individuals exchange goods and nonmaterial resources in society in order to meet their needs. Gerontology has adapted this theory to the special needs of the aged.

Social Exchange theory arose from Goodenough's (1963) interpretation of culture as emerging from shared knowledge among individuals with each person having his or her own mental template for behavior. Behavior is interpreted as voluntary actions on the part of the individuals, who are motivated by the returns they expect to receive from others. The unit of analysis is always the individual who is seen as an independent, self-directing person involved with a series of trade transactions (Cook, 1987). Each individual negotiates for a favorable exchange. Trade is possible only when both partners perceive a reward that is advantageous to themselves (Dowd, 1980:58).

A person must have resources in order to make trades (Blau, 1973). A social exchange resource is any labor, service or commodity exchanged during social interaction. Resources include tangible and abstract goods, including the use of power and subordinate behaviors. An individual may have direct control of a resource, such as one's own money or labor. In other cases the control is indirect, such as the use of family labor or influence (Sen, 1981).

Society sets expectations or norms for the give and take of social exchange. Ideally, a reciprocally satisfying balance is achieved during transactions, as the assumption is made that people are constrained when, over time, equal exchanges are not maintained. A profit making exchange results in power. James Dowd (1975:589) defines power as "an abstract entity creating privilege," and as such is used to demand, rather than command, control over exchange. As such, power can be used when resources are lacking for equal exchange. Power is not always abstract as it can arise from direct or indirect control over multiple resources. Age seniority plays a role in many societies as elders have control over the labor of others (Sen, 1981). Power promotes situations where equal exchanges are not mandatory, usually to the benefit of the power holder (Dowd, 1975).

An exchange resulting in a deficit leads to dependency. Dependency, as used here, means the actor is unable to reciprocate in an equivalent manner (Dowd, 1980:128). Dependency does not necessarily mean the actor lacks power with others beyond the particular exchange partner. Power in any situation leads to deference, or a ritualistic dramatizing of an individual's priority. Deference contrasts with respect, which is an expression of honest concern based on esteem (Dowd, 1975, 1980).

Gubrium (1973) adapted social exchange theory to social gerontology, taking into account the unique personal and social needs of the aged and their behavior. His resulting Social Environmental Theory of Aging provides a framework wherein the actors, who are the aged, function as active participants in their environment. This allows for the needed separation of intrinsic (individual and cultural variations) and extrinsic factors (the accompanists of modernization) that contribute to diversity. Thus, it becomes possible to evaluate the individual and the aging process independently of the degree of modernization or by standardizing the sociocultural environment.

This separation does not mean the specific social context, including the people, social structures, and institutions, can be ignored. Historical, political and economic features, as well as cultural values and the ongoing social construction of everyday living experiences, are important in the process of aging. These factors can be approached objectively in that they are external to, and partially independent of, the dynamics of the mind (Gubrium, 1973:36).

The theory states that three overlapping dimensions of resources under the direct control of the aged individual are necessary in order to maximize satisfaction and to maintain the individual in a viable position as a creative agent in the exchange process (Gubrium, 1973). The broad outline of resources cuts across all social and cultural situations. The *personal/physiological* resources are of a physical, mental and psychological nature, and are essential for integrity and active participation in the exchange process within the environment. The *social/familial* dimension refers to the people and support facilities whose contextual presences can provide positive reinforcement for personhood. The *fiduciary* dimension consists of the "coin of the realm" in a particular context. Money, goods and other barterable items, such as access to land, are included (Gubrium, 1973).

A major assumption of Social Environmental Theory is that resources decrease with aging (Gubrium, 1973). Maladjustment increases and social acceptance decreases as holdings in the dimensions become increasingly limited (Lawton, 1983:665; Dowd, 1984). Resources may be used or "spent" without replacement, disappear with failing body integrity, or may be simply lost to the environment, such as cattle death with drought.

It is possible for the aged to continue satisfying trade with a lessening of resources. This can be done by using processes that call into play any cultural norm that allocates power based on age or status alone (Gubrium, 1973). Power can also arise through the creation of fear, such as assignment of supernatural abilities to the aged. This power can be used in the process of finding the good life.

The resource dimensions in Social Environmental Theory vary from the usual social exchange resources in that power and control over others are not included. Such "abstract" entities are regarded as part of the process of exchange rather than actual trade items (Dowd, 1975). The presence of family, friends and social support agencies only indicates the degree to which trade partners are available. The service and promotion of positive values and self-concepts that this group provides occur through equal trade relationships.

The drawback to relying on power alone during the process is that others see it as a demand (Dowd, 1975:590). As trade is unequal, the person upon whom the demand is made must act out of deference rather than exchange (Dowd, 1980). Over time, continuous unequal exchanges, and deference to the aged, will create bias toward old people and change attitudes (Dowd, 1984).

Interaction is required to create attitudes, thereby making attitudes the outcome of interactions. Attitudes may be positive in one situation and negative in another, depending on context. Gubrium (1973:31,34) believes attitudes are stable and have no general or direct relationship in determining trade relationships. This assumption may be true in his American research setting. In societies experiencing rapid social change, attitudes can change quickly as sources for power are questioned (Cowgill, 1979). Roles change, and so does the accompanying power.

All societies produce prescribed normative behavioral expectations and roles. Norms are not systematically fixed but reflect variation between individuals and groups of individuals within the social context. The context places limitations on the choices an individual can make in terms of activity, and on which roles can be assumed. These limitations produce a social definition of acceptable age-related behavior. The range of possible actions and roles a person may take usually remains relatively stable (Gubrium, 1973). The definition of acceptable behavior can change over time, as individuals alter their interpretation of what is correct and group consensus occurs.

Action is behavior that takes into account the individual self, the environment, and the resources of an individual in relationship to those of others. Conceptually, people can act independently of norms and expected roles, as they impinge on but do not determine the course of action (Blau, 1973). Action can be verbal or behavioral. In theory, action is always rational and reflects the nature of dispositions or attitudes towards self and others (Gubrium, 1973). People weigh their strengths as an individual against the expectation of others in evaluating options for behavior and adapt accordingly. (The degree that this holds true in real life is questionable.)

Adaptability is a duel-edged process of adjusting to and influencing one's environment. The process of adaptation is both active and reactive. With aging, the intent is to master a changing situation while extracting from it what is needed (Keith et al., 1990). Expectations change and resources can dwindle. New wants arise. These changing circumstances call for a revision in exchange relationships, yet the people involved still want to maximize the rewards while minimizing the costs (Gubrium, 1973). Adjustment is contingent on maintaining effective and rewarding exchanges within this new setting. Behavior continues to include the existence of normative expectations and the weighing of options.

Command in social exchange is said to exist when a want is fulfilled through trade. Command is possible through the use of resources and power. The use of power during the exchange becomes increasingly important as resources decrease, as power provides strength to limited resources. The ideal is to have both sufficient resources and more power than others. (Any power held by the aged should be seen as relative to the power held by other age strata.) When either power or resources is absent, command may or may not be possible, as either resources or power alone may control one particular exchange but not another.

Older people in industrialized societies have minimal age-related power to control daily social interaction (Gubrium, 1973). They are required to rely on the dimensions of their resources. The debasing of exchange commodities is recognized by both parties. When resources are insufficient, and power is absent, the social exchange process ceases and is replaced with demand (Gubrium, 1973).

Demand is the claim that services and goods are due without anything given in exchange. There is no ability to enforce this claim. Others are requested to give and have nothing given in return. Fulfillment of a demand depends on the relationship between the donor and the receiver. The aged use compliance to improve relationships. Complying to the wishes or family and friends is the only means for survival (Dowd, 1984).

Compliance does not stimulate resource regeneration. Any remaining power disappears (Dowd, 1975:590-593). Aged in this situation obtain what they need by relying on others, as beneficence has replaced reciprocity (Dowd, 1984). Support, through beneficence, is always available unless generalized prejudice against the aged becomes great. Then social breakdown occurs (Dowd, 1984). At this point, no aged are successful in social transactions. Old people are regarded as incompetent and unwanted trade partners, and undeserving of beneficence. Once individuals are labeled as socially aged, they experience this negative feed-back, which reinforces their social exclusion (Hendricks and Hendricks, 1986:106). With social breakdown, the position of an aged individual is less a function of age than one of social values. Continuing adherence to middle-age values with visible productivity is the only means for acceptance (Hendricks, 1982).

Social Environmental Theory, as an adaptation of social exchange, provides the basic framework and methodology for this research. Both the aged individual and society play an active role in the process of aging. Exchange behaviors, reflecting resources, action and attitudes, affect the process and individual outcome. This provides the essential structure in evaluating the aged as individuals, and as a functioning group within a society.

Social Exchange Theory, involving resources and success in trade, has been used in famine research (Sen, 1981). As far as I know, the gerontological Social Environmental Theory has never been applied in total, with the delineation of actual resources, the reception of others in accepting resources in trade, and gaining the good life. I attempt to apply the complete theory to the Tswana aged as they experience life in Botswana, a contextual setting very different from America. Therefore, I must take a critical look at the concepts, including the asking of some questions without answers.

Many of the references cited for the actual Social Environmental Theory reflect thoughts on its application to political issues and the generalized reaction of society to the aged. Much of the background material for theoretical claims came from relatively stable Western societies. Gubrium (1973) provides much data for the reasoning behind concepts but provides only brief categories of resources in the resource dimensions. I must take these fragmentary categories, adjust them for the Tswana and make them complete.

The theory, as a reflection of Western thought, definitely presents challenges when applied across international settings. Both social exchange and social environmental theory are set in a money market context, stressing that the basics of provisioning are obtained through the trade of resources having use-value. The assumption is that individual welfare is paramount, leading to reciprocal exchange with equal balance between individuals. This assumption perhaps presents the greatest challenges in cross-cultural use. The question arises of what constitutes equal balance in reciprocal exchanges when the family or society is paramount and group good is placed above individual good. In traditional Africa, individual outcome is frequently secondary to group welfare (LeVine and LeVine, 1985; Cohen, 1986). In this village of Botswana, the process of achieving still includes emphasis on the group. To what extent does family and social welfare override individual equality during exchange in such a setting? Here there is no known answer.

The focus of the theory is on aging and the aged, not society as a whole. This demands that a high level of awareness be directed towards the Tswana social structure and functions both inside and outside of the direct aging process. One factor is communication patterns, which affect all social exchanges. Other factors are as diverse as kinship lines and social schisms, both of which may place any individual either at the core or periphery of society. Of course, the rapidity of social change needs to be considered along with the traditional patterns and practices with social aging. It is in this context, which I regard as the setting in which social exchange occurs, that I draw upon Modernization Theory. This theory explains the changes in rules for exchange, generational differences in perceptions of equal trade, and how incomplete knowledge and limited options intertwine to direct the process of decision-making.

The assumption of any theory involving rational choice is that the knowledge for choice is complete and that unlimited options exist. Needed knowledge is not always complete when decisions must be made. Knowledge is very often bound by culture and life experiences. The individuals participating in exchange may be drawing upon different data bases involving disparate information. In developing countries cross-generational exchanges increase in complexity. The aged are very apt to base decisions on traditional knowledge without having the more formal education and informal knowledge about Westernization, which has been obtained by the young.

Limited knowledge can also produce limited options for possible behavior. In addition, the life situation in itself can limit options, as I pointed out earlier in regards to extreme poverty. With limited resources, individuals cannot consider all options: a small risk can make potential reward too threatening. Emotions, hopes, and fears detract from the assumed rationality in decision-making.

The essence of the theoretical outcomes of aging are said to be problems with decreasing power sources and trade resources (Dowd, 1984). The decrease in age-related ascriptive power of the Tswana aged is well known (Schapera, 1953; Ingstad and Saugestad, 1987; Suggs, 1987). The aged are also known to be economically poor (Tlou, 1986). This has created conflicting ground rules between generations for the essence of trade.

Ability to participate in social exchange also decreases in other ways as old age progresses. The number of kin and friends with whom trade occurs shrinks with death of friends and relatives. Loss of significant others gains additional momentum in rural Africa as emigration of the young is the norm. This situation is confounded with sensual and ambulatory losses. These common circumstances are not deterrents in using exchange theory as these losses can be incorporated in the dissolution of resources and a shrinking of command and power. It is the presence and use of existing resources that has import, especially in regard to meeting old age goals. By regarding bodily and personal loss as lost assets for trade allows for the testing of the commonly held assumption that such loss is a major factor in personal disintegration.

The past use of the theory in situations of famine assumes control over labor is an asset. The presence of family and friends equates with success (Sen, 1981). I see the inclusion of family and friends in the social/familial resource dimension as an indication of the amount of access to trade partners. Control is not to be considered automatic. It is the access to trade partners that determines if social exchange for labor is possible.

A major theoretical assumption is that all resources diminish with age. There has been no past attempt to determine if a threshold point exists, a point where remaining resource holdings no longer work effectively. It seems very possible that resources in one dimension could diminish, such as in the fiduciary area, yet health and family remain strong. Do resources actually diminish because they are used or lost, or does the value of held resources alter with time although the amount may remain more or less the same? It is here that questions from early theories can be asked. Do aged people limit activity, either by choice or necessity, and thus lose opportunity to regenerate resources, or does the sociocultural and/or physical environment prevent such rejuvenation? Again, one must include comprehensive social processes, this time adding cultural continuity in conjunction with modernization. What traditional opportunities for resource generation and power and control can still be utilized? How does the inheritance factor, which in Botswana is usually cattle and/or land, contribute to resource accumulation?

Modernization theory provides clues for the movement away from traditional care-giving and support systems. The aged were, and are, involved in social evolution of the village. The aged individual's degree of modernization or traditionalism cannot be overlooked. The most salient attitudes involve differences between generations in resource use and value. (Goldstein, Schuler and Ross, 1983; Gubrium, 1973). Resource use and value involve the process of exchange. How the aged deal with generational conflicts can be explored if exchange is viewed as existing in the on-going social process of change. This social process involves cultural continuity and social change, as both direct who will exchange with whom, and act as stimulants for equality or inequality.

The decision-making and related behaviors for social exchange is not limited to the sociocultural arena. The geographical, ecological and physical aspects of the environment have direct impact. Such factors as terrain, climate and natural disasters demand on-going adaptations on the individual level (Sen, 1981). In Sub-Sahara Africa, where agriculture and cattle play an important role in subsistence, changing responses to drought and the impact of drought relief programs affect daily living. Harsh climate and terrain influence the emergent behaviors during modernization, plus the ability of the aged to engage in these behaviors. Unfortunately, the impact of physical environment, beyond that of architectural barriers, is often lacking in Western gerontological thought. We assume buffers for the aged during natural disasters are built into local and natural policy (Wolensky and Wolensky, 1990). Actually, the American aged find no buffers with environmental calamities and frequently fail to recover (Bell, Kara and Batterson, 1978; Parr, 1987).

One must also consider the long-term effects of widespread poverty and under-development. This places the majority of rural Africans in a situation where everyone is resource-poor. Does this increase the value of the resources of the aged as they use physical abilities to contribute to the household and/or are the owners of land and livestock? In contrast, maybe the aged, in spite of real or potential contributions, cannot use their resources to compete with the demands of social environment.

<u>Research Hypotheses</u>

In brief, many evaluations of aging in Africa follow a set pattern. It is usually assumed the African aged once had a large degree of command generated from resources and power generated from resource use in customary roles, norms and social behaviors, all of which potentiated the position of the aged (LeVine and LeVine, 1985; Cohen, 1986). Social change involving loss of tradition, the erosion of the extended family and education of youth is preventing the aged from obtaining power and is

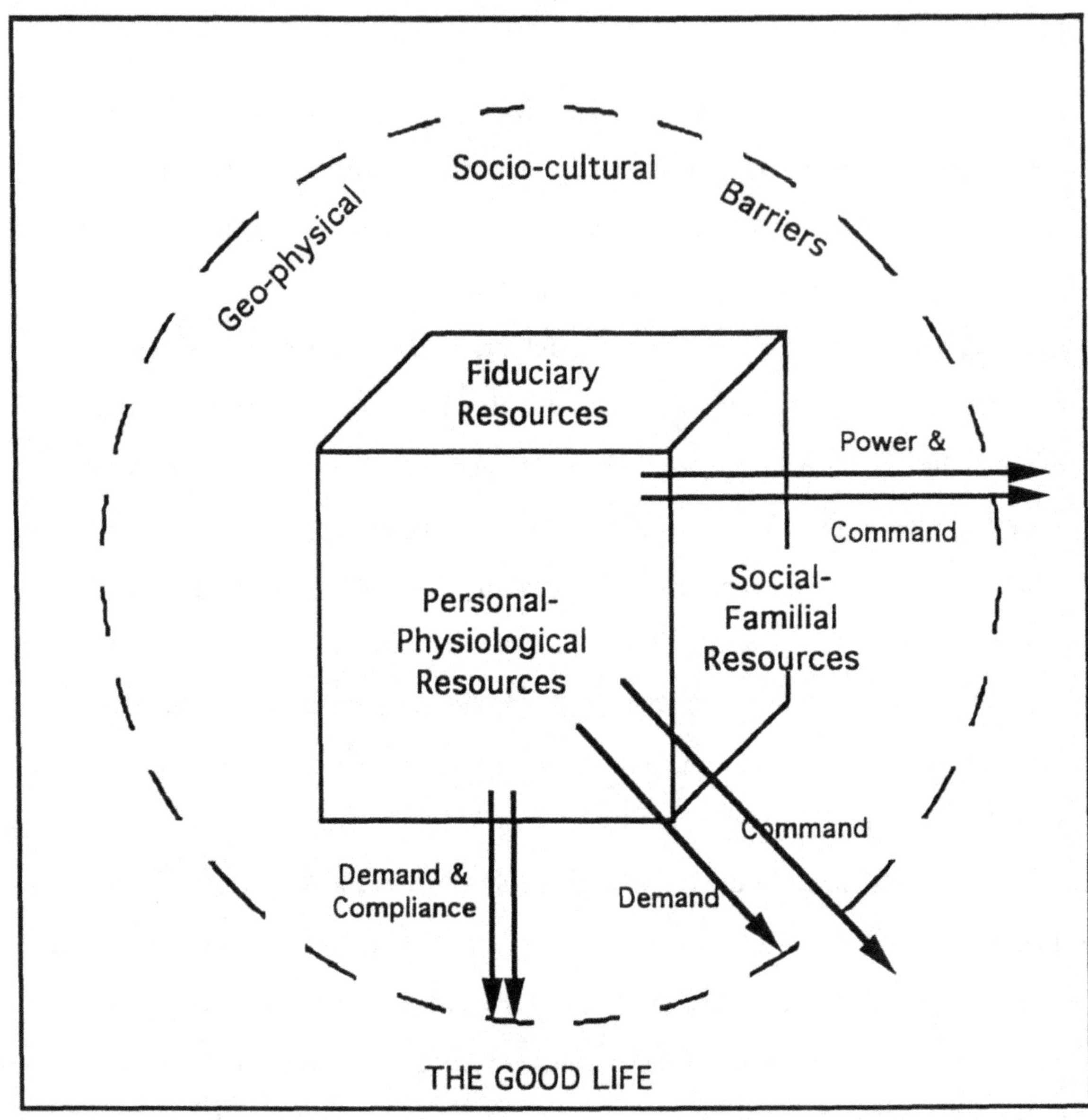

Figure 2.1: Theoretical Model of Access to the Good Life

diluting the resources used for command (Diouf, 1984). In addition, the desire of the aged to return to their place of origin under conditions of widespread poverty has placed unwanted and insufferable demands on village populations (Diouf, 1985; Tout, 1989:22-23). The result is a social milieu where the aged live in a situation where comprehensive provisioning is not available with either command or demand. Families and society lack the ability to provide care and economic support to the aged (Hampson, 1982; Diouf, 1985; Hay, Burke and Dako, 1985; Ingstad et al., 1991; Thomas, 1992).

I am building on the opposite approach: that provisioning factors are present in the environment. This care is available to aged individuals who control resources that provide them with power and command in social exchange for meeting the goals associated with the good life. Each person has various amounts of the three resource dimensions (personal/physiological, social-familial and fiduciary). Modernization has effected the process of exchange but has not eliminated the role of resources in finding the good life. The strength of resources and the ability to use them effectively to command reciprocal exchange and to generate power allows one to overcome the sociocultural barriers of bias and the geo-physical barriers produced in the environment. Those with minimal resources, or those having to rely on a mix of command and demand, can overcome some barriers but not others. If resources become insufficient, power is lost and the individual must rely on demand and compliance. This may or may not create benevolence from others. Aged without resources cannot overcome the sociocultural and geo-physical barriers and fail to find the good life, as demonstrated in Figure 2.1.

Amartya Sen (1981) applied this concept with famine, assuming food was always available but not accessible. Families with desired resources, such as money or control over labor, had an effective command for obtaining food. Those without resources were powerless. Ineffective demand would not provide food, although other families were well fed. Thus, family resources, not the situational locale, determined access to food during famine (Sen, 1981). Just as food during famine is accessible by the resource rich, so is old age care and support during scarcity stemming from social change in developing countries. With command, middle-age values and behaviors, or old age expectations of compliance and reliance, can be replaced with active eldership roles, security and status, which constitute the good life in the eyes of the aged.

A direct correlation exists between the strength of resources at a given time and the meeting of old age goals. The needed weight in each area is unknown, but is assumed to be relative to the general resources of others in the community. There is a level, or threshold, when the amount of controlled resources is deemed to be without value in exchange relationships. This threshold can occur either because an individual enters old age with insufficient resources, or because the individual has since lost them. These people have no control in directing their lives. At the same time the reliance on beneficence produces a passive reaction on the part of society. Society may provide the minimal basics of food, housing and care, but provisioning is insufficient for life satisfaction.

The historic change found in Botswana, which encompasses personal and social change, as well as environmental calamities, has altered not only the value of resources but also individual's abilities to accumulate, rejuvenate, and retain them. The oldest old are expected to have fewer valued resources under their direct command than the younger aged. I anticipate that the oldest old, as a group, would have less life satisfaction as a result. As individuals, they would be equal to others having minimal resources. Resources, not age, make the difference in obtaining the good life.

New and on-going trends can and must be analyzed, as they reflect what people desire. It is people, of all ages, who constitute the essence of society and direct its action (Blau, 1973). The aged experience friction between past acceptable behavior and emerging trends. This friction extends beyond moral conflict, as trends set new rules for the values associated with specific resources, and changes what is regarded as an acceptable reciprocal exchange. Cultural continuity, in contrast and in relationship to change, must also be analyzed. On-going social customs and beliefs also affect present reality. Such customs and beliefs can either be transferred directly to the new society or altered to fit the new society. In either case, they affect the value of resources and the exchange process, both in the present and in the future. My goal is to make clear the strong interrelationship between the individual and society in establishing the good life during old age. This includes the behaviors and values of the aged and the responses of others towards the aged, for flow with process is not one-way.

My approach assumes the aged must have open doors for flow between themselves and the contextual setting. Old people enter into the lives of others as well as having others enter their lives. I do not assume the doors were originally perfect. Large and small cracks have always existed. The process of social change can create more damage to some and repair others. The swinging of the door is as important as its actual condition. Resources determined how well the aged individual can use the assorted doors.

CHAPTER 3
ORGANIZATION OF THE STUDY

> In Botswana, families give good care to their old parents. That is one
> tradition that everyone believes in. There is no need to study old people,
> as they are taken care of by their family. (a middle-aged government
> employee)

> I am old. Tradition has changed. You should learn about the old people
> and tell the world how badly our family forgets us. It is alright that you
> are white on the outside, but to learn about the hidden parts of being
> old, you must become a Tswana on the inside. (67 year old woman)

The democracy, social freedoms, and racial equality of Botswana provided a

research locale where I could investigate the impact of modernization on aging while

blending into village life for research depth and breadth. This chapter presents the

formal steps taken for such work, beginning with permission for research and village

selection. Methodology and analysis reflect the incorporation of anticipated and actual

difficulties. The chapter ends with a synopsis of village population demographics. The

need for adaptability in, and during, research is stressed, as I had to unlock my

American door in order to find the Tswana doors and develop open and meaningful

communications with understanding.

<u>Formal Permission for Research</u>

The myriad of necessary steps required for formal research were undertaken

during my initial visit. Rigid regulations dictated the order of these steps. The first

action was to gain affiliation with a ministry or governmental research institute.

There is no specific governmental unit that deals directly with aging nor a

specific government program targeted to the old. Policy emphasis was placed on such

matters as maternal child health, increasing rural employment and decreasing the public expenditure. A series of appointments with various agency heads met with disappointment. I was informed that there was no need to study the aged, as tradition dictated the family should take care of and support their elderly members. In other words, as I was told, "the aged have no problems."

No matter who I talked with, the high official or the person on the street, the custom of providing family support for aging parents was stressed. However, outside of government buildings, individuals admitted that this was a custom often given allegiance in abstract principle but not in reality. They pointed to social change as creating inroads on the continuation of status and welfare of the aged. Therefore, I had no doubt that the situation on which I based my study was present in the country.

Now was the time for self-critique for failure to obtain permission for research. Analysis indicated my problem-centered approach for seeking approval was in error. No agency had said the study was faulty; it just did not fit their needs. Maybe a fresh approach was needed, something that did not threaten the concept of traditional family support. I had used the problem-solving approach, stressing identification of social change and areas of need. Why not have a study that reflected strengths?

With only one possibility of approval left, I applied to the Ministry of Local Government and Lands. This Ministry administers village development programs including the destitute funding program. This time I changed my opening format and said, "Since the tradition of family support for the aged is working so well in this country, I would like to investigate why, with the hope of finding strengths that could be incorporated by other countries." Approval was immediate with a signature on the needed formal Presidential Permission for Research application form. At the same time, I was given the necessary written permission and letters of introduction for visits to various villages.

<u>Village Visitations</u>

I now had to leave the familiar safety of the capital city of Gaborone and venture out into the unknown. The first unknown was the bus station. In spite of dire warnings of personal danger and the dictum of "Don't go near the bus station" from European and American workers, I had no other transport. Eventually the bus station became one of the highlights of my trips to Gaborone, meeting new friends or sitting on the curb and sharing an ear of boiled maize with old friends. Masses of travelers, street venders, trash and aromas generated a sense of accomplishment when boarding the correct unmarked bus.

I must admit that ambiguous thoughts and anxiety dominated that first bus ride. I was meeting the unknown and anticipating differences and similarities with the known. I planned to base the wording of my study on the local reflections toward the aged. This seemed very logical for meetings with local government officials. My key concern was how I would approach the chief, as he was the protector of his people and culture. In Ramotswa, this person was the "paramount chief of the Malete," in contrast to a village chief. His approval was mandatory! The stereotypical image of a tribal authoritative figure, gained from old pictures and ethnography, was firmly locked in my mind. I knew the image was not reality but I could not shake my expectation of required submissiveness under his power. My reverie was cut short with numerous nudges and shouts that I had arrived in Ramotswa, and the South-East District Council was on my right.

Contrasting with the earlier central government presentation of nonchalance toward the aged, local district government was open and frank regarding the necessity to address social problems relating to old age. Some officials placed the problems on the shoulders of the aged, claiming that laziness and adherence to the old ways were the sources of their difficulty. Others blamed social change with the refusal of adult children to maintain traditional support patterns. Overall, the welfare of the aged was placed on economic terms with governmental short-fall of money for program development. A study was definitely needed and acceptable, if approved by the tribal administration.

My walk on the village roads toward the tribal administration building, known as the *kgotla*, alternated between a fast pace and a slow crawl. A mishmash of the modern and the traditional was everywhere: fancy houses next to *rondavals* (round mud huts with a thatched roof), cars passing donkey carts, a woman carrying a bag of refined flour talking to a woman pounding grain. Children greeted me with "Good morning, teacher" while adults addressed me with the customary *Dumella Mma* (Good morning, mother).

I arrived at the *kgotla*, feeling the final moment of judgment had come. Two men were raking the central yard for public shame, as they had broken laws concerning family relationships. The Tribal secretary greeted me. Alas, Chief Kelemogile Mokgosi was at lunch. His assistant, Deputy Chief Ikanena Mokgosi, was available and would be glad to meet with me. This tall, well dressed, late middle-aged man quickly put me at ease with a liberal dose of honest laughter mixed with questions and answers. He highlighted the similarities between myself and other village women. There were many Elizabeths, all with children the ages of mine, and who relied on the bus for transport.

Slowly, I explained my interest in old people and seeing how they lived. I leaned back in my chair as his expressions indicated a vehement reaction was forthcoming. He began.

> Good! You must come here! I am growing old myself and I do not like what I see in store for me. People do not like the old. You should come here to find out how to make life better for our old people. All the old people of the village need help! You must use this village!

My need to sell the study was displaced with his need of explaining why I must select his village. His insights paralleled the theoretical constructs of modernization and aging, yet he had never heard of gerontological paradigms. He explained that:

> the aged are being pushed out of the home and society as the village is modernized. Values have changed and old people are no longer liked. Children migrate and do not send money for the old people, so there is poverty and hunger.

On that note, he suggested I talk to his people and look around the village, returning later to meet Chief Mokgosi, who "was much older" than himself.

My return visit to the *kgotla* went equally as well. Chief Kelemogile Mokgosi was evidently primed by his assistant, as formal permission was immediate. His caring concern for individual welfare shone beneath his reserved, yet truthful, personality. "How soon could the study begin, as the aged are suffering?" "It would be good to bring your family with you. The water is safe to drink and you should be able to rent a mud house." The tables had been turned with the tribal administration, local government and people beseeching me for the study. Suddenly I was in the position of not being able to say yes. Other villages still needed to be visited and evaluated.

The seven other villages I visited were equally friendly with political awareness of decreased economic and social welfare with aging. Family requirements, which included a setting for my husband's zoological research and schooling for my 10 year old son, were no problem. My own research requirements were an old, traditionally established Tswana village experiencing the impact of modernization and westernization. It needed to be large enough to have a representative sprinkling of social structures and services, yet small enough for community solidarity. I decided Ramotswa was most representative of a village in transition, although it was unique in that it contained a hospital and two, instead of usual one, churches. These disadvantages later turned into assets for understanding the expanding interweaving of cultural continuity and thought with exposure to increased technology. The cooperation among villagers, local government, and tribal administration provided additional assets within the research setting. Now began the reverse movement up the chain of authority to give notification that Ramotswa was selected.

Why was it that the major welfare planners in Gaborone were hesitant to verbalize what the urbanite employee and village leaders were adamant about? Botswana is the sole multi-party democracy in Africa, whose high GNP prevented any famine in the recent drought. The three-pronged government of presidency, parliament and tribal administration promotes equality in all regions without tribal or racial prejudice. The admission of a fault in the social structure resulting from modernization of the infrastructure is a major step. To admit this fault to an outsider whose actual status or position is unknown requires commitment and trust by all involved parties. This trust could not be established overnight. Happily, by the end of my initial visit, many of the same individuals, who initially stressed the universality of family support for aged parents, freely discussed the breakdown of the custom and admitted it was time to start considering effective government intervention. By the completion of the study the government was as open and receptive as Chief I. Mokgosi.

<u>Early Insights into Culture and Society</u>

Within Ramotswa, grape-vine communication was rapid. There is no word in *Setswana*, the language of the Tswana, encompassing the concept of research or formal analytical study. The people knew I would return to learn how old people lived. The reasons why I would do so were perplexing to them. On subsequent organizational visits I was constantly interrupted with questions of why I wanted to learn about their life as an old person. "If you are not with the government, what will you do with what I tell you? Only government workers write names down, ask questions, but then they proceed to do nothing to improve my life." The tribal administration and local council also had a valid grievance: "people come to study the area and we never find out what they found."

Should the local people benefit from research? The nature of my work would involve more than time commitments from the respondents as they would have to give of themselves, opening "closed windows" (a phrase used in psychology for bringing hidden thoughts to the surface). The aged men and women I encountered in the village included the pain of remembering with responses to my questions. Yet, they insisted they be included in the study when I returned. Is contribution to theoretical growth, sharing new knowledge with professionals, and sending publications to the National Library of Botswana enough to justify the stirring of thoughts and emotions with the congruent raising and dashing of hopes of the populations studied? Thus began my strong commitment to share findings and thoughts, in both verbal and written forms, on the local level which included government, tribal administration, and the people themselves.

It was also during these early visits that I began discovering my own suppressed thoughts regarding cultural differences. These windows needed to be opened in order to be successful in my goals. Even though behavior may have been correct in American society, it did not mean it was correct in Botswana.

During an early visit, I was seeking the health clinic. Everyone I approached with a simple request of directions said, "I do not know." In complete frustration I settled myself on a roadside rock to recuperate. Two old men ambled along and we exchanged names and asked of family and health. Afterwards, we began talking in depth. They talked of past and present living conditions. When I asked, "May I visit your house?" Both old men doubled over in laughter, pointing to each other and to me. Their broken English and my limited Setswana was insufficient to clarify what was so hilarious. A bystander finally explained that my selection of words intoned I was offering myself as a prostitute! Luckily, the faux pas was taken humorously.

When I finally chanced to ask the men where the clinic was located, they pointed to the building directly behind me. My earlier attempts to gain directions failed because of another social faux pas. Thus, I learned my first lesson in the Laws: always begin any conversation, however minor, with a greeting and introductions. The discovery and application of laws were relatively easy in regards to asking questions. Others laws, as I found later, forced me to open myself to the existence of a new value system, and temporarily disregard some tenets of my own.

Insight into potential research questions began with service providers, such as the clinic and village leaders. The most valuable insights into the differences between research design for aging in developed and developing nations came from the aged. Participant observation in their household activities, long talks on the meaning of life and aging, and the simple joining with aged in various activities allowed me to see differences in life style and world view. Potential key informants were identified, including John, age 75, and Monate, age 73.

John and I first met when he was walking home from the *kgotla* after attending a planning meeting for the annual "United Nations Children's Day." John was jubilant about his old age. He thought that maybe some old people had problems, although he was not sure as, "such things are not discussed with other people." He considered himself exempt from difficulty, as he had his involvement with *kgotla*, a caring wife and family, and good health. I offered to walk with him, being careful not to mention the destination of "house."

Monate, his wife, was waiting for him on the porch. We shared a delightful afternoon. They answered my questions. They asked their own. "I have often wondered, where does the sun go at night?" "I heard America sent a man to the moon. If the moon is up, why didn't he fall off?" "How long does it take to get to America? Only 29 hours! Why I can get to Johannesburg on the train in 23 hours so it must not be that far."

These questions were not a mark of stupidity but indicated two reflective minds at work, searching for answers within their own realm of knowledge. In this way the couple was typical of many of the aged: intelligent and exhibiting the reasoning patterns taught through culture, yet bound by limited education . I found I was similarly culturally bound with limited Tswana education. Their answers to my "Hows" and "Whys" were elusive and confusing. I left the village, and Botswana, recognizing the research was possible but in a very different way from the usual American approach to gerontology.

Research Design

Social environmental theory, proposing that the aged use personal-physiological, social-familial and fiduciary assets in a process of social exchange to provide a good life, formed the basic constructs of research design. These basics became the foundation of a Gerontological Assessment Form, a formal survey instrument to provide quantitative data on the aged. The actual questions are presented in Appendix A. From this form, specifics could be evaluated singly or grouped for comparisons.

The categorical data for the resource dimensions, such as possessions or the use of social services, could not be equally equated in meaning or content with Western society. Questions regarding the assets held by the aged were devised by modifying the proposed contents, listed by Gubrium (1973), to the local environment and level of village development. Adaptations were based on observations and statements accrued during the initial visit. For instance, the distance to water had great meaning. The number of indoor bathrooms did not.

Originally, the inclusion of proven life-satisfaction questions from known valid instruments was considered for determining the good life, as this would provide comparisons with aged in other countries. Even though such questions were previously judged to be applicable in situations of cultural differences among Western ethnic groups, many were found irrelevant to Tswana reality. For example, the standard question of "How do you see yourself five years from now?" actually brought forth anger, as the Tswana refuse to predict the future. Questions regarding individual decision-making and personal independent functioning were not applicable at all, in light of the marked norm for interdependent behaviors for household functioning.

Tswana world view does not include comparisons and ranking. One knows only about him/herself. There are no gradations with emotions or opinion. One believes or does not believe. One feels emotions totally or not at all. Open-ended questions with ethnographic interviewing provided the best measurements of life satisfaction, but these too met with limitations. Questioning about perceptions of life as it is today (rewarding, dull, interesting, etc.?) was limited, as adjectives describing the quality of life are absent in the Setswana vocabulary. Life is either happy or sad. Discussions of the reasons for happiness or sadness did little to illuminate the degree of feelings, as learned thought processes constrict perceptions to the categorical black-or-white. However, responses did shed much light for the development of constructs for the process of achieving, including perceptions of utility, providing direction to others and the value of self in social interactions.

Functional utility, both in the home and community, was expounded upon by altering the often used question of "How often does your health, and changes with age, stand in the way of your doing the things you want to do?" When the unmodified question was used, all respondents answered "All the time," regardless of the degree of observable physical limitations. In the same question, aged always stated if they were "useful" or "useless." This again reflected the dichotomy of their world view. While reasons for the response would be easily stated, the individual could offer no variations on a continuum, i.e., could not see oneself as having varying degrees or types of functional value. When the initial question was changed to a request for specific activities that could or could not be done and why this was do, data concerning health and activity emerged.

As I have demonstrated, many facets of the research could not be approached with standard instruments or Western concepts of personhood and aging, without comprising validity. Some conventional questions were used with revision. Other question were devised for this particular ethnic group. I sought additional validity with pre-testing and evaluation of instruments with 20 aged villagers before use. This pre-testing stimulated adaptation of some questions and the elimination or addition of others. Throughout the study, the aged continued to identify the more elusive segments of culturally specific domains while answering ubiquitous questions regarding life in general. Areas that were identified early in the study were added to the assessment form. Other areas became topics for conversation. The final constructs should not be considered valid for all Tswana age groups, nor should they be used to say the aged are more or less happy than other generations.

Linguistic patterns and the absence of ranking as part of the thought process made it necessary to capitalize on innovative research techniques. The use of the "ladder technique" (Hansen, 1990) allowed for placement on a continuum. While individuals cannot perceptually rate themselves in an abstract manner, they can see themselves as individuals in a particular setting. Two opposing situational occurrences involving people were placed at opposite ends of a small, five-rung ladder. (The situational questions are described in Appendix B.) The individual was then asked to place him/herself on the ladder, according to self-perceptions. (Holding the ladder on its side eliminated ideas of worst, bad, good, better and best or hierarchy among people.) Discussion revolved around the reasons the individual selected a particular rung. This brought forth thought on how the individual perceived similarities and differences between self and others and what they thought was important and/or acceptable and what was disliked.

At other times, one historical statement was assigned to the middle rung, and the respondent selected the rung representing the present truth and explain why this was so. The ladder technique, was also used during interviews with primary care-givers to obtain data on perceptions of self and the aged. Knowing that each person interpreted the rungs differently, and that statistical analysis on rung selections would not be done, I encouraged response on reasons for rung selection. Similarities between respondents between directional flow in present truths, such as position of the aged in modern life, could were noted, but the reasons for choice became the important data.

I devised another interviewing technique, based on the concept of a felt board for story telling, to obtain data relating to family and village interactions. Instead of a story, the respondent created a picture of the ideal. Game pieces were drawn from photographs and painted to stimulate reality. They represented a wide variety of traditional and modern items and buildings. Multiple people of all ages were available for selection. Two pictures were made. One was the ideal house, and the second the ideal village.

I began by asking what type of house they would like to live in, and letting the respondent place the selected house on a felt board. Respondents then selected the items they wanted in their house and yard. Much positive feed-back was required, with emphasis on the fact that there was no correct way to make the picture. I asked many questions during the construction of the pictures, including the reasons they selected (or did not select) various goods and people. I then asked questions about the chosen selection in comparison to real life and what they would like for the future. After completing of the picture, I asked about the inter-relationships between selected items, and how the picture compared to past experiences and future expectations.

This procedure eliminated the boundaries of the reality of poverty in descriptions of wants and needs. It also allowed for some prediction of the future with questions about what would happen if the items and people were real. Probably, most important, the respondents saw their picture as a pretend situation where discourse regulations and taboos did not apply. Conflicts between the ideal and actuality could be openly discussed. When asked what the people in the picture were doing compared to what they actually did, participants would critique life and the individuals. These critiques often contrasted with earlier descriptions of family relationship, as respondents could openly say what they did want, and did not want. They forgot social facts and opinions and the way things should be according to law. I believe some of my most accurate and complete data resulted from this exercise. The hour or more required for completion was well worth it.

The last technique involved the use of ethnographic interviewing to obtain folk taxonomies on the meaning of aging and the intricacy of disease (Spradley, 1979). The respondents were encouraged to enlarge upon their thoughts after describing themselves. Although the initial questions remained constant, conversations took many directions. For instance, responses to the statement, "Tell me about your heart," included mystical, emotional and physiological answers. Ethnographic interviewing content data were constantly evaluated and re-directed as previously unrecognized concepts emerged. For instance, from the taxonomy on the meaning of life, card sorts (Fry, 1986) were developed in the field to clarify the emics of life-cycle progression.

My key to obtaining data with all these techniques was to present the instrument as a means of teaching me about old age. The exercises were not presented as games. This was very important, as games are for children! The aged were proud to accept the role of teacher. I noticed this role helped them overcome observed hesitancy to answer questions. (Several aged told me that much of their hesitancy to provide me with data was because others considered old people to be childlike in thought.) The serious approach did not mean that laughter and fun were avoided with discussion.

The hour spent in completing each type of data-gathering exercise was enjoyed by the aged although they said the questions were hard. The difficulty was understandable, as participants were exposed to new methods of thought and reasoning. Constant reassurance that there were no right or wrong answers helped eliminate trepidation. Verbal recognition of interesting and good thoughts, and positive feedback for their ideas, prompted openness in their replies.

<u>Identification and Selection of the Aged</u>

In Botswana, one does not ask, "How old are you?" It is not impolite, but only a question that people cannot answer. Age is not measured chronologically, as the number of years a person has is not important in itself. The date of birth is more important, as eldership is determined by who was born first.

Many knew the year of their birth, some of whom were able to verify birth year with yellowed, but well cared for, baptism certificates. All aged knew the name of their "church-initiation" (confirmation) or tribal-initiation age-set, both of which had a graduation age of 17 or 18. Using a incomplete published list of male and female class names and dates (Ellenberger, 1937), adding data obtained from those with verifiable birth dates, and double checking of birth order among related kin, I was able to determine the unknown year of birth for others. Ages were felt to be accurate within two years.

In Botswana, none of the aged live segmented lives, apart from others. There are no retirement centers or old age communities within populated areas. For this reason, the demographic composition of their neighborhoods, and variation among various sections of the village, become important. Accurate representations depend on valid techniques of random selection. To accomplish this, a numbered grid was superimposed over a village map containing major roads, buildings and landmarks. Ten census tracts, out of an N of 42, were selected using random number identification.

These areas included sections at, or near, the major shopping area, the hospital, one

church, the *kgotla*, the recently developed areas on the village outskirts and the

immediate eastern side of the government council buildings, as shown in Figure 3.1.

Each census tract was approximately half a square kilometer. I omitted distances over

three kilometers from my residence, because of walking time. These were the hilly

section to the far east and the area containing the high-school and government

employee housing west of the council buildings. These two areas were used for pre-

testing, with findings comparable to the randomly selected areas.

Each house within the identified areas was visited to obtain a census of ages and

gender. (Only one household was not directly interviewed, as no one was ever home

during multiple return visits. Three other households provided identical data on that

household's residents) This census varied somewhat from the national census as only

household members residing in the house on a regular basis were included. Migratory

workers returning only on weekends or less frequently were excluded. Individuals who

were 60 years or over were interviewed, either at the time of census or with a return

appointment. Three attempts were made, during different periods of different days, to

contact the aged who were not at home.

The selection of 30 individuals for in-depth study was made with stratified

random sampling. These individuals were selected to represent one of each of three

categories reflecting level of activity. These levels were the active-old, the limited-active,

and the decrepit-old. Division was based on initial high, medium, and low scores of

physical functioning obtained from the Gerontological Assessment Form. One person

from each category was selected in each census area, being sure to include at least one

male and one female from the sample. As sometimes only one male and only one

decrepit-old resided in a census tract, absolute randomness was difficult to achieve.

(The only two decrepit males who resided in the areas were automatically included.) All

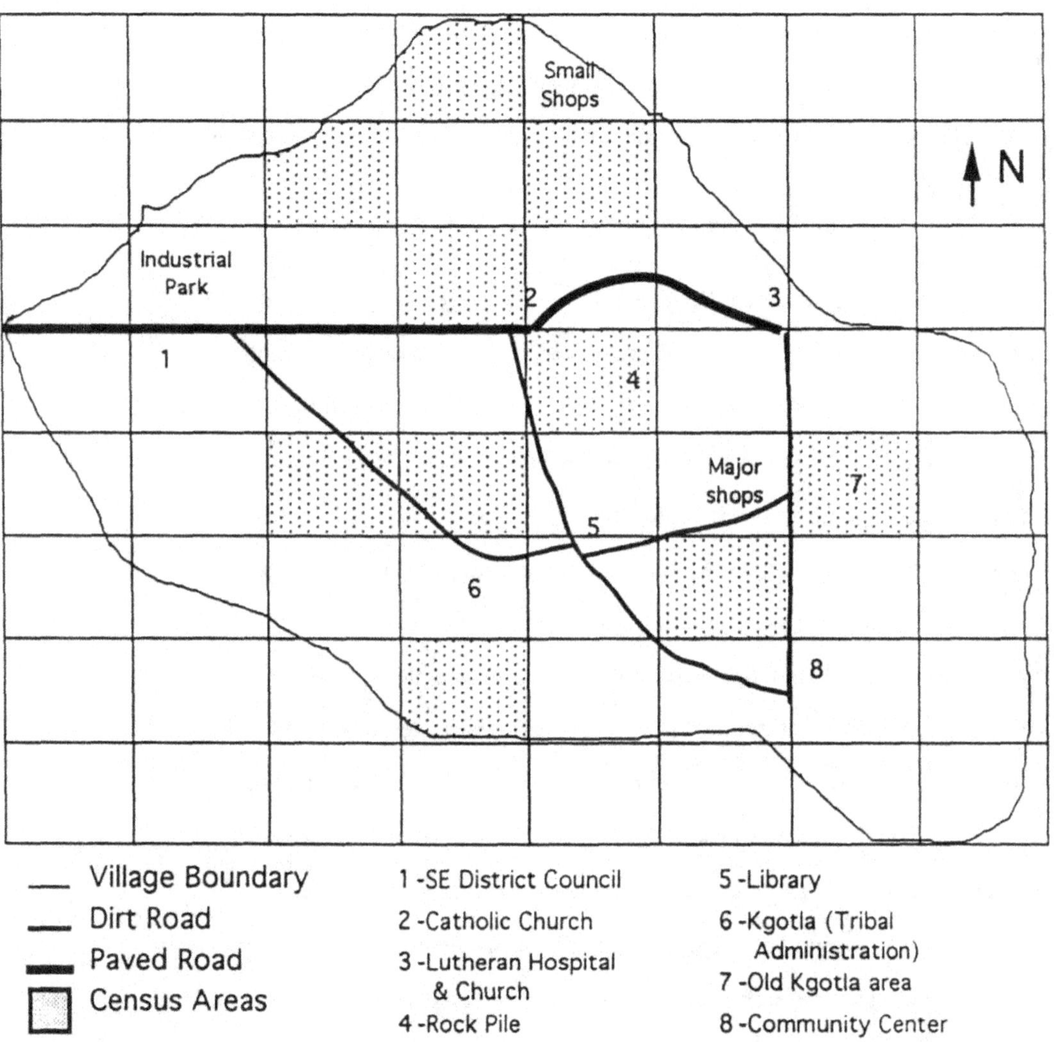

Figure 3.1: Map of Ramotswa indicating Census Areas (not drawn to scale)

agreed to participate in the three additional interviews involving the felt board, ladder and ethnographic interviewing. These individuals were also frequently visited throughout the study period to provide data on continuity and changes with time. An additional female was added in mid-study, as one woman was eliminated due to her extensive responsibilities in agriculture.

<u>Analysis</u>

The Geriatric Assessment Form was purposely designed to permit statistical analysis, serving as a bridge between social exchange theory and data on the individual level. Both the ordinal variables of assets held by the aged and the mix of ordinal and nominal variables for the good life were analyzed with the STAT-VIEW statistical program. Specific questions were asked, including means, ranges and frequencies of variables. My questions were many. "How are age and sex related to held resources, or assets?" "Which assets are most important to control the good life?" "Can people find the good life without assets?" Spearman correlations, and simple and stepwise regressions were used to determine relationships among variables. No doubt some questions were overlooked but the data are available for more detailed analysis.

Ethnographic interviewing associated with the in-depth studies provided much qualitative data of statistical value. Some lent itself to coding with frequency distributions, such as household items desired. The actual numbers on the ladder rungs had varied meanings to different individuals but could be used to show directional flow of relationships between self and others, and self and time. The majority of ethnographic data, like that from participant observation, was simply coded, with repetitive events, behaviors and thoughts singled out for use in quotes and descriptions. This method provides validity to observed family interactions and social processes (Bernard, 1988).

The description of formal routes for research approval, and the scientific facts of research design, method and analysis, are distinctly different from becoming an accepted village member. To become accepted I had to "become Tswana." This meant learning and obeying the laws while continually "achieving." I my case, achieving was directed toward the understanding of social aging.

Becoming a Village Member

Winter had just left Ramotswa when I returned with my family. Our first stop was at the tribal administration to acquire the socially-proper approval for village residency. Chief Mokgosi pulled my now-worn business card from his top desk drawer. With a grin, he said, "We have been waiting for you." Plans were made for a *kgotla* meeting in two days time when he would formally introduce me to the village, explain my reasons for being there, and ask for a community vote on acceptance. Meanwhile, he knew of people renting houses. It would take him a week or so to find us a place to live as people were without telephones. If we needed something faster, we could ask about town ourselves.

We proceeded to visit John and Monate. Monate saw us before we entered the gate and had her seven year old grandson running to the shops to get John before we got to the door. I followed the law for introducing my family and then asked the traditional, "How does the sun shine on you and your family?"

John told us about the United Nations Children's Day. It was a grand event with traditional dancing and singing. He is not working on the committee this year. "There are too many young people wanting modern entertainment and they vote against the elders who want traditional dancing." He now goes to the *kgotla* only for formal community meetings, and sometimes to talk informally with the old men of the village.

Our son, Matt, hands John our gift, a basic elementary school science book on the universe. John is ecstatic: the reading will give him something to do and answer his questions about the world. Monate says she has lots to do. Yesterday she had helped cook for a wedding. "But you should understand I don't do the cooking. I supervise, which means to sit and share talk with others."

With the preliminary exchanges now over, I introduce our need for rental housing. John is happy to accompany us to a place he knows of, the compound of Mr. and Mrs. Mosimi. The family had built a four-room square house on their plot of village land for an adult son who died unexpectedly. Moving in was easy, with the transfer of three pieces of luggage from a borrowed truck, each suitcase weighing the airline's 20 kilo limit. Household furnishings were simple: two beds, a table, six chairs, and a one burner kerosene stove. All we had to buy were candles for light, a couple of pots, and tableware. We carried in enough water to last until morning and made the latrine our last stop before going to bed.

The crack of dawn brought sounds of wood being chopped, radios, and voices. We were soon to learn there is no such thing as private space, either auricular or visual. Mrs. Mosimi was in our kitchen at six o'clock that first morning. "My master sent me to see if our new children slept well." Thus began our incorporation into an extended family of aged parents, two adult children and a grandchild. Family titles replaced given names, and new role-norms for extended family corporacy with intergenerational interdependency, replaced the American nuclear family independence.

Rra (Father) effectively, yet lovingly, demonstrated that he was the household elder and master, befitting his 76 years. Severe arthritis kept him from walking long distances but he remained the effective manager. *Mma* (Mother), born two years after Rra, guided me through the learning of female verbal and nonverbal behavior. Above all, I was to serve my own master (husband), show deference to Rra, and teach my son the law. Brother was to be my guardian when my master was not home. Sister was my equal. Her 6 year old child was also my child to discipline, teach and command. It was expected that "all my children" would be sent on errands and assist with simple chores. Matt was given all the responsibility for daily grocery shopping and carrying water. In doing so, he found new self-confidence and new friends.

It did not take long to discover the supposed equality between races was not quite true. Preferential treatment was always given to an "English woman," as all non-native females are labeled. I tried to discourage their efforts to elevate my being with a concentrated awareness on my overt behaviors. I immediately sat on a goat skin instead of looking for a chair, ate with my fingers before a fork could be found, and waited my turn instead of going to the front of a line. Having a Tswana mother and being a mother in my own right contributed to acceptance. Mma introduced me as her worthy daughter, but I became known as Mrs. Matt, a title implying respect for me as a Tswana woman.

Feedback on acceptance came through Mma when she related the village gossip. With hands on her hips, head high in the air and a swag in her walk, she said:

> this is an English woman in the village, here for her own good. She does
> not see the people or their life. The village women decided you were not
> like that. You see us as we are. You a person without color. We can
> tell you what we want to say, not what an English woman wants to hear.

Prior to this I had found a translator and together we were administering the pretesting. (Three other English-speakers verified accuracy in translations and interpretation of adages.) I had noticed a marked change in answers from the first to the third week, when the above occurred. It was now time to begin serious research on being old in a Tswana village. I began with the village census to obtain demographics, from which springs all further findings.

<u>Village Demographics</u>

The overall census involved 207 households. A household was defined as one or more individuals functioning as a family unit. They share one or more houses on family property. The mode for household size tied between 7 and 8 members. A high number households without children (25) and few with 11 to 15 (10) gave a distorted mean of 5.5 individuals, and a median of 5. In almost every household females outnumbered males in each age group. This disproportion is illustrated in Figure 3.2.

Aged lived in 111 of the 207 households. Of the 53.6% households containing aged, 34% of these had two or more old people. Aged lived by themselves or with another aged in 8% of all households and with young children only in another 3%. Thus 89% percent of the aged had an adult living in the household, usually in a three-generation, predominantly female family.

Children were everywhere. Children were defined according to the government definition: those 14 years and under (Knudsen, 1988). They numbered 501 in the sample population of 1146 (44%), a figure similar to national childhood population of 47% (Republic of Botswana, 1981). Most of the annual rate of village growth results from the natural increase with births (Republic of Botswana, 1981). Girl outnumbered boys, 57% to 43%, a fact supported by primary school statistics and the national census. School personnel feel more girls than boys are born, which is true nationally but not to this extent. The skewed proportion could also reflect the tradition of urban adult children sending daughters to villages to serve their grandparents. I found proportionally more daughters than sons of absentee adults in homes, regardless of overall household composition. Both factors are probably true, as nationally girls outnumber boys by 2% (Republic of Botswana, 1981).

Adults, spanning ages 15 through 59, comprised 43% of the total sampled population. Nationally, this percentage of adult population is much higher and the numbers of males and females are about equal (Republic of Botswana, 1981). In the village, the number of adult women dominates over the number of males, as females compose two/thirds of the adult population. This statistic reflects the importance of

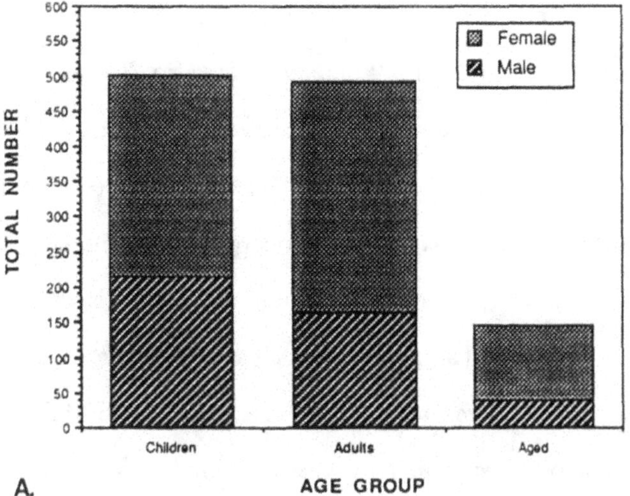

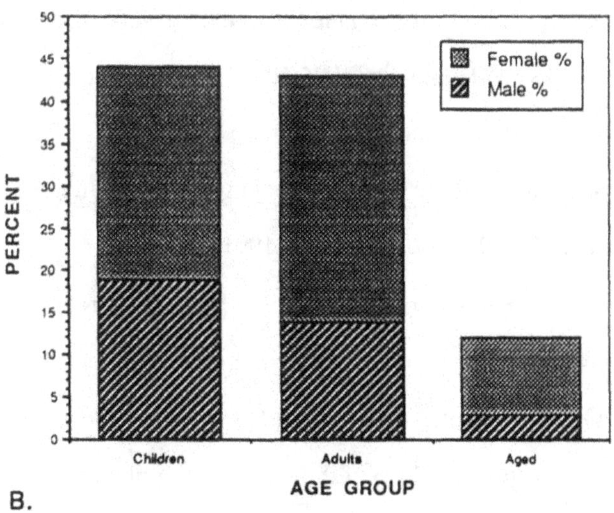

Figure 3.2: Population Breakdown in Census Areas by Age Groups and Sex

migration for men seeking wage labor, although the majority of households had one adult or aged male present. Women migrate also. The absence of daughters is the main reason aged live alone or serve as the sole adult caring for very young children.

In Botswana only 2%-5% of the population is 60 or more years old: this estimate varies with the source (Republic of Botswana, 1981; United Nations, 1985; Third World Guide, 1988). The percentage of aged in this country is similar to the rest of Africa. Aged men are almost as numerous as aged women throughout all developing countries, contrasting with figures from developed nations. Throughout Africa, it is not until the 80 year and over age group is examined do marked differences in sex ratios occur, with 100 very aged women for every 88 men of similar age (United Nations, 1985:31). This sex-ratio reflects extreme variation from the sex-ratios of the United States, where there are three women for every two men at age 65, and 66% more women than men by age 75 (Hendricks and Hendricks, 1986:68). Part of the ratio discrepancies between developed and developing nations can be accounted for through high maternal mortality with proportionally fewer females reaching old age (United Nations, 1985:31). Other reasons are difficult to isolate (Gage, 1991).

The total number of aged in the census areas represent 28% of the known 375 aged in Ramotswa, as reported in the national 1981 census (Republic of Botswana, 1981). Overall, there were 147 aged (41 males and 106 females) living in the 207 households, comprising 13% of the sampled population. The customary return to the village of birth with old age has caused governments to raise questions about old people being concentrated in the older villages and absent in cities and new towns (Hay et al., 1985; United Nations, 1985:105). Throughout Africa, 78% of all aged live in rural areas (United Nations, 1955:98). A concentration of aged appears to be true in Ramotswa, giving the village a proportion of old people that is similar to that found in the United States. The big difference between countries is that the aged must interact with proportionally more children and fewer adults. Another way of looking at the total village is that there are relatively few adults available to tend to the many aged and the very young.

The ratio of 5.3 village aged females to every 2.2 aged males could not be adequately explained, as this contrasts with national statistics of 100 females per 90 males, and the outside extreme sex ratio of 110 aged females per 90 males (Republic of Botswana, 1981; Tlou, 1986). The village health and social service providers supported my findings with reports of many more known aged females than males. Seasonal changes did not increase my chance of finding old men.

It is generally believed that men remain in a migratory work status until age forces retirement, or death occurs. I question if this is the actual reason for this setting. The national census reports 20 of the aged men, and 9 females, are employed (hence living) outside of the village (Central Statistical Office, MFDP, 1982). This number is not great enough to markedly influence the sex ratio. None of the sampled women report an absentee aged husband. It could be the past mates of divorced women are still employed or young wives excluded de facto aged husbands during my census. This latter cause does not seem reasonable for every case. Even though a young wife was mentioned as ideal, almost all of the interviewed married couples share no more than five years age difference. The question remains: Where are the aged men?

A large proportion (37%) of the aged males residing in the village households were reported to be at the lands, coming home several times a week, sometimes for only an hour, for clean clothes and a supply of food. (Does this hint at the possibility that aged males have a difficult time adjusting to village life with return from employment?) One thing that is known is that the simultaneous living at the lands and in the village by males distorted the sex-ratio of my interviews. I was able to interview 77% of eligible females compared to 56% of the males. Overall, 72% of the identified aged were interviewed. Many of the men unavailable for interviews were reported to be in their eighties. Absent women varied greatly in age, ranging from 60 to mid-eighties. Since absentee ages could not be verified, they are excluded from age-related statistics, but their reported ages, when averaged, were similar to the interviewed group.

All reported aged men, tended to fall between the aged of 70 and 85. The age-mode for the interviewed men fell in the 80-89 year old group. None were over the age of 88. For women, the mode was in the 60-69 age group. Six were over the age of 90, a proportion similar to a group studied by Ingstad, Brunn, Sandberg and Tlou (1991). Figure 3.3 demonstrates the found age grouping and sex. The noted age-grouping imbalance between sexes supports the concept that men remain employed as long as possible.

Despite the age group imbalance, average ages of the old were similar. The average age of the 82 interviewed females was 74.49. The average age of the interviewed males was similar at 75.82 (N = 23). Fifty-nine percent of the males were 75 or older, compared to 44% of the females (although no males were past 88 years). I use the age of 75 in this instance to make comparisons to the United States aged. In developed countries, the young-old (below 75) outnumber the old-old (75 and above) two to one, including the African-American population (Hendricks and Hendricks, 1986: 41, 380). In Ramotswa is a situation where nearly half (47%) of the aged are demographically classified as old-old.

Employment may affect the village demographics in regards to the ages of the aged population, but there is good reason to think there are other causes. Usually mortality curves are thought of as bath-tub shaped, with a falling of death rates after the neonatal period and a rise with senescence. In sub-Sahara Africa, a different pattern occurs. Life expectancy is lower, ranging in the 50 year old bracket for most countries (United Nations, 1955:101). In Botswana, the 1980-1985 life expectancy is 50.8,

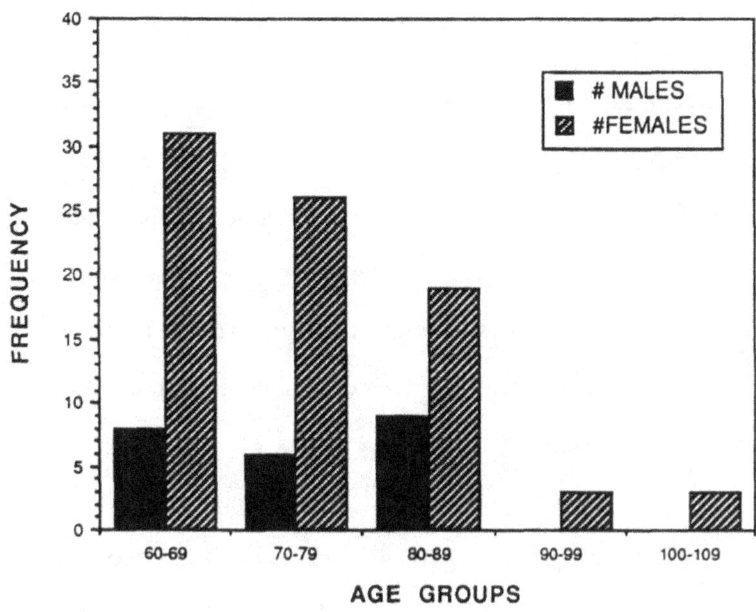

Figure 3.3: The Interviewed Aged by Age and Sex

according to the United Nations (1955:101). Botswana projects a life expectancy of 53.5 years for males and 60.6 for females for children born in 1986 (Knudsen, 1988). A high mortality rate from accidents, infectious disease, hypertension, and various other reasons occurs between the ages of 30 and 60. This results in a low life expectancy and extended range in age at senescent death. The extension can result in extreme old age (Gage, 1991). In Ramotswa, it is not only that the old are getting older but they are "the fittest of the fit" (Gage, 1991). Their bodies are strong with immunity to the common causes of death. Underlying reasons for differential mortality patterns between Africans and Anglo-saxons, including senescent mortality, have never been adequately studied (Gage, 1991). A similar extension of life, if the age of 60 is reached, exists for African Americans (Anderson, 1988).

The demographic facts, as determined by specified variables, are not the same as social facts regarding social age. The way in which people see themselves and each other is related to their cultural and social history. History, as it influences the doors that all people use, is presented next.

CHAPTER 4
THE EVOLUTION OF THE VILLAGE

> Ramotswa is better than ever before. We have schools, a tar road and
> firms to employ people. Botswana is ruling itself and the people are
> happy. (73 year old woman)

> It is heart-breaking to tell you all that has happened to Ramotswa. Our
> culture is becoming lost with development. Our richness is replaced with
> poverty. (78 year old woman)

The aged, with descriptions of their own lives, provided the organization for this
chapter on village history . It begins with the history of the Malete and their
introduction to colonization, and then explains the movement from the "days before
civilization" through "the days of becoming civilized" and ending with "being modern."
These time periods provide a framework for the discussion of social change and cultural
continuity. The goal is to examine in detail the historical and on-going sociocultural
facets that affect the meaning of life, especially for the aged. Oral histories are
compared to written history, with agreement and discrepancies discussed.

History of the Malete Prior to 1885

The eastern area of Botswana and the South African Transvaal were sparsely
populated by Sotho-Tswana groups by 1200 AD. The forerunners of the Malete tribe
lived in Transvaal, east of the Ngotwana River, which today separates South Africa from
Botswana. Tribal groups migrated, split and reformed, as droughts and population
growth occurred. Eventually, the parent groups of the seven Tswana tribal nations,
including the Malete, emerged (Tlou and Campbell, 1984: 66-67).

The Malete were first identified as a tribe of the Tswana nation near the turn of
the eighteenth century (Ngcongco, 1984:25). At this time in history, the whole of

southern Africa was affected with wide spread assassinations and conflict, known as the *Difaqane*. Tribe displaced tribe and nations were divided. The Malete, under the leadership of Chief Malete, left the Transvaal and settled in the iron-rich Tswapong Hills, straddling the present Botswana border with South Africa. Metal craftsmanship flourished. The trade of three iron hoes for an ox, or four for a cow, soon made the tribe rich in cattle (Ellenberger, 1937).

Boers moved into the area during the 1830s although the Malete's first battle with Europeans did not occur until 1852. It was then the tribe first encountered guns and defeat. Crossing the Notwani River where Ramotswa now stands, the tribe moved on to Ditheyane, a Bakwena tribal village to the west headed by Chief Sechele (Tlou and Campbell, 1984:115-116). The local residence of Dr. Livingstone, and his teaching of Christianity, was seen as added protection against further upheaval (Tlou and Campbell, 1984:131).

Again the Malete took advantage of their trading skills, this time bartering with early European traders in the area for guns. The price of a single barrel was four or five head of cattle; a double barrel was eight or nine head (Ellenberger, 1937). Without knowing herd sizes, who can say if guns were cheap or expensive? Guns were used to protect cattle from the numerous lions of the area and to provide an easier means of obtaining trade items during this time of peace (Tlou and Campbell, 1984:126-128). Ivory, skins and feathers were exchanged with traders for cooking pots, matches and cheap trinkets (Tlou and Campbell, 1984:126-128).

The stay in Ditheyane was disrupted after 10 years when Sechele, who was poor in cattle, demanded tribute from the visiting Malete. A short distance move to Mankgodi did not eliminate the pestering for cattle and guns by the former hosts. In 1875, Chief Mokgosi I, who had taken over from Chief Malete, took his people back to the Tswapong Hills, and established a new village named Ramotswa. (Ellenberger, 1937).

Accompanying the group was a new advisor to the chief, The Reverend Christopher

Schulenburg, of the Hermannsburg Missionary Society of Germany.

The Malete soon dominated the existing Babirwa, BaGananwa, and BaNgwaketsi

tribes of the area (Ngcongco, 1982). The dominance was not without bloodshed, as

conflict over land rights erupted into battle (Ellenberger, 1937). Today's old men of the

village recalled the stories they had been told of the battle. Sitting under a tree at the

kgotla on a quiet afternoon, tales of intellect and bravery emerged.

> The young and the women were sent to protect the cattle, with
> the men staying behind. Our fathers had never fought with guns before
> but had watched the ways the Boer's organized battle. For us, battles
> were fought in the open but the Boers hid behind trees. Our fathers and
> grandfathers copied the white man's way of hiding, with one man loading
> guns, another shooting. Only eight of our men were killed. The foes
> fought in the open and wore red turbans so they were easy targets.

The heavy fighting resulted in many casualties, with over a hundred dead among

the foe (Ellenberger, 1937). Growing pensive, the men continued. "Yes, our

forefathers were wise men, learning from others and not afraid of to fight for the land

that was ours." This battle coincided in time with the end of the pre-colonial era of

Botswana (Parson, 1984:15).

Early Colonization: 1885-1935

In 1885 the British declared a Protectorate over a wide area known as

Bechuanaland. Each Tswana tribe retained authority under the supervision of the new

European administration. In December of that year the English arranged a treaty

between the local warring tribes, giving the Malete rights to land in Ramotswa and a

limited surrounding area. Although the village belonged to the Malete, it was not until

1909 that the boundaries of the Bamalete Reserve were defined by proclamation

(Ellenberger, 1937). This same time period was also a time of drought. Rinderpest also killed many cattle and the remaining large native animals in the surrounding areas (Tlou and Campbell, 1984:126-128).

The Malete, like other Tswana at the beginning of colonization, were a self-subsistent agricultural and pastoral people. Agricultural practice was, and still is, limited more by resource availability, mainly water, than lack of creativity. Planting was mixed and eclectic, with sorghum, maize, beans, melons, pumpkins, and sweet cane (Schapera, 1967:16). Animal husbandry was valued equally with arable agriculture, representing an investment-portfolio with practical and symbolic implications (Alverson, 1978:10).

The Tswana based class divisions on kinship lines and rank within class on herd size (Campbell, 1971). Each person knew their status, and could move up in rank but not class. Wealth differences were outwardly minimized through the process of *mafisa*, the custom of loaning cattle and small livestock. The holders of such loan cattle benefited from the milk and the offspring, and could use the cattle in plowing. In turn, they were expected to provide service and political support (Schapera, 1953:28).

The peasant economy and lack of formally recognized religion were considered by the colonialists and missionaries as the "the heart of paganism." Christianity based on a money market economy was regarded as a necessity (Ellenberger, 1937; Thou and Campbell, 1984:183). Thus, early development was aimed at eliminating the traditional economic and religious practices in order to promote economic gain for the British. Lutherans built the first church in Ramotswa at the turn of the century. In the 1930s, the village was described as mostly Christianized. The claim was that the heathen practices for mystical control over the universe were rapidly disappearing, and that the people were becoming "civilized" (Ellenberger, 1937).

Economic change was instituted in 1899, with the imposition of a monetary hut tax. An additional "native tax" was placed on each hut in 1919. These taxes were used to pay British administrative costs, with extra money going to the Native Fund for village improvement (Picard, 1987:98). The penalty for non-payment was severe (Tlou and Campbell, 1984:181).

Access to money became a requirement for taxes and additional reasons. The Malete, because of the delineation of the size of their reserve by the protectorate, suffered from chronic agricultural and grazing land shortage (Schapera, 1953:24). In 1921, and again in 1926, Chief Seboko was forced to purchase additional lands from the British. Every man in the tribe had to contribute five English Pounds for the land purchases (Ellenberger, 1937). People also wanted to purchase the growing numbers of trade goods (Picard, 1987:31).

Money became an absolute necessity, forcing at least one male household member to become employed. Employment was available only through labor migration. The Malete did not relish the idea of working underground in the South African mines, but many went. Others chose farm labor in the region of Transvaal. Men, migrating for employment, were separated from wives and children, with remittances being sent home for the family's required expenses (Ellenberger, 1937).

At the end of the first third of the twentieth century, the English claimed success in meeting many of their goals for "rehabilitation" and "civilization" with only "relics of barbarism" remaining (Ellenberger, 1937). Schools had been established;, the church had over 3000 members; and tribal customs were being discarded. To the Malete, these were still the days before civilization.

Although changes were being made, the aged regard this time period as the days before civilization. Some were born before the 1899 hut tax, others as late as 1929. In either case, it was the time of their youth, the formative years when, as children, they were socialized to their world and expectations in later life. The following is based on their perceptions of the time period. Much of the data are supported by Ellenberger (1937) and Schapera (1944, 1955).

<u>The Village</u>

Land within the village was divided into seven wards, with each ward under the leadership of a hereditary headman. In turn, headmen were under the chief who resided near his headquarters, or *kgotla*. Ward members were grouped according to kinship lines. The village was small enough for everyone to know each other. Life involved much physical labor in a reported setting of peace and harmony.

Security for well-being dominated village life. A sense that others would provide, if the process of achieving failed, predominated in all aspects of life. Those who were without were always given by those who had, be it milk, grain, labor, or money. "No tally was kept, as some day the giver would be without and need to receive." Other comments support Schapera's (1944:46) observation that ward members associated together, shared work and supported each other in times of trouble. The chief also gave to individuals or assisted with ward needs, sharing his tribute with others.

The times of chatting and story telling were not lazy-hours as no division between work and leisure existed. All aspects of life were interpreted as work, including teaching of the tribal ways and maintaining interpersonal relationships between kin and friends. Prescriptions for proper interpersonal communication, especially with elders, were clearly taught. Tone of voice, body stance, and gestures, as exhibited by today's aged, were learned as part of the law. Children were specifically instructed in how to say hello to any elder and to ask others about their health and family before talking about oneself. The arts of communication were an integral part of the process of achieving.

Crimes such as robbery or rape were reportedly unknown. Serious disputes that threatened the extended family or ward, as with continuing adultery or failure to pay sufficient attention to the husband, were taken to the headman for trial. It he felt it was beyond his control, the issue went to court to be settled at the *kgotla*. Here the

Chief, with his advisors, would listen to the case. Such cases usually involved non-support and abandonment. Great attention was paid to the utterances of his advisors as they were the old men of the tribe. Although they did not necessarily belong to the highest classes or have royal blood, they were the repository of oral traditions on which justice was based.

The grapevine communications system quickly let the public know when and how such laws were broken, and by whom. Punishment was by public beating to inflict physical pain and the pain of shame. Social pressure, taking the form of ridicule, contempt, scorn, or ostracism, was dealt out accordingly. The greatest shame was to be banished from the village.

<u>The Family</u>

The basic unit for subsistence was the extended family, which lived together in a compound. The compound, consisting of multiple houses, was the central family residence. Each house was constructed with a circular wall of mixed dung and earth, surmounted by a conical roof made of poles covered with thatch. A nuclear family resided in each hut, with the male as head, or Master, of the house. Huts were built around a central family courtyard. The eldest male in the grouping was the authoritative figure, or Master of the compound. This elder served to maintain peace and order over his extended family.

The concept of shame with pain was applied in the home. The aged say, "it was perfectly acceptable to beat one's wife when she failed to perform duties to the husband's liking." Other men in the village reported returning their wife to her parents for proper beating. Children were to always act on the orders of elders, be it an older sibling, parent, relative or villager. Failure to obey resulted in physical punishment. "It was through beating that the children learned proper behavior and the laws of the tribe. They knew what they had to do, and to do it with demonstration of respect for the elder." Fear of beating gave strength to any elder's dominance, making disputes a rarity. By customary norms, anger was not shown and there was no back talk.

The austerity of obedience was tempered with security within the family. The aged claim food was shared equally; children were given to grandparents who would otherwise be alone; and unmarried men or women shared a compound with extended family. "We were as one, with each helping the other."

The village was the center for communal activity. All families maintained two additional seasonal homes. When the rains began women and children moved to the agricultural land, and men to the cattle post. Families reunited in the village during the dry, or winter, season.

<u>The Cattle Post</u>

The family cattle post was an outlying area in the open veld for the grazing of livestock. A small hut served as the temporary home. All boys received instruction and practice in the arts of cooking, sewing, laundry and cleaning from their mothers during the winter. They were well prepared to perform these strictly feminine chores at the cattle posts as women were generally absent.

Grazing could change sites on tribal lands, depending on the availability of water and quality of grass. Livestock herds consisted mainly of cattle, but could include goats, sheep, fowls and dogs. The men and their older sons, including boys over six or seven, tended animals. They returned to the village, with livestock, only when grass and water disappeared with the cessation of rain.

Donkeys and horses were introduced with early colonization. The donkey provided a new source of wealth and status for the man who amassed enough for plowing and wagon transport. The donkey gained special significance as this was the animal Christ rode. One man condensed the thoughts of many when he said:

the donkey is hard working animal, which devotes its life to the welfare of people. It is not to be eaten but paid respect until a natural death. Like the child and the dog, he must be trained and controlled through beating. We do this because we love him.

Another loved animal was the dog. Every man owned a dog for protection and hunting. Many of the men talked about their dog's value, including the two village men I met the first day. The eldest of the two said:

> a trained dog was the way I could get a rabbit in my pot. Spears were for wars or when attacked by a beast. They were not allowed in hunting. Instead, I would dig a trap in the ground or make a rope noose from plants like the men of the Kalahari {San or Bushman} did. My dog would chase an animal into the snare.
>
> I never went anywhere without my dog. A snake would always go after a dog before he went after a person. The most dangerous snakes were the black colored ones. Dogs saved my life several times, particularly when walking to Gaborone or Otsi.

The Lands

The "lands" were an assigned agricultural plot, usually some distance from the village. A simple shelter was home to the family women, girls, and very young boys during the growing season. Their men had plowed the field with oxen or donkeys in November after the early rain had softened the ground but before the grass was green at the cattle posts. Now it was the women's responsibility to plant, weed and harvest. According to Mma, all ages toiled together, chatting about life and the universe.

> I learned stories about baboons at the lands. Baboons were plentiful. They would come down from the hills, and rob our crops. The baboons are always frustrated. They want to be humans, but never develop into people. Therefore, baboons must be treated nicely and not thwarted, or they will use their people-like brains to become very mischievous and bring destruction to the lands. Night was the time you really had to be careful, as a thwarted baboon could run across the paths and cause crops to wilt.

Agriculture was one means by which a women could develop pride. A field kept free of weeds and birds was envied by others. Outside influences such as rain, and the mysteries of the baboon, determined the size of the harvest. One could expect crop failure about once in every three or four years. The fields were small, usually two to three hectares, but " 100 kilograms of Kafir corn {sorghum} could be grown in a good year."

Crop land, like land in the village, was assigned by the chief. This was not done haphazardly or without consideration of the potential user. The selection would involve rationality. John chose his land carefully.

> As a newly married man I had one piece of high ground. My growing family required more land for sufficient food production. I requested a second parcel about 20 kilometers from the village. I saw that this piece of land was in a gully that carried water. I chose it because it would produce in drought, when my other land would not. Of course, if there are heavy rains, I lose out with flooding. This land may seem far away but other people walked even further.

Daily Life

The above is the setting of the present day aged when they were children, or as children who had become young adults. I will now present the early life-histories of three individuals. Each had different family and social experiences.

Elizabeth. Elizabeth was born in 1888. Other than her cropped gray hair and slow gait, her body belies her present age of 101 years. Quick in wit and tongue, she relates the following experiences while breaking up wood and feeding the open cooking fire.

> When I was born I was given my Tswana name of Botlhale, meaning wisdom, as Elizabeth is my Christian name. My childhood was a time of pleasure, working with my family at the lands and being responsible as a very small girl for sweeping the hut and gathering fire wood. My father reared sheep, and we used the skins as blankets. These *mmaselekwana* were used only at night for they were very warm and soft to sleep with. My father and mother made us children clothes from the sheep skins. Girls wore fringed leather skirts that covered our lower front. Boys, like the men, covered their private parts with a animal skin. Everyone wore clothes to hide their nakedness.
> Sometimes my father would travel to trade with the men of the Kalahari. He would give them our home grown tobacco in exchange for tanned hides. The best ones were from the bucks of the antelopes. They were smoother than sheep skins, as the San had a special talent for making hides soft and strong. We used these skins for special clothes.

Elizabeth talked about how Chief Mokgosi I respected the church leadership, becoming one of the first Malete to convert to Christianity and attend the new Lutheran Church. She then talked of her family's conversion.

> I remember when we first went to the church. I was still a very young girl. There was a large room full of beautiful colored materials. Everyone wanted this cloth, but the only way to get it was to become a Christian. My mother joined the church, and she later had beautiful cloth to cover her entire body.
>
> My parents insisted I go to the church confirmation school instead of *Boyale* {female initiation training}. I was taught reading and writing. I was baptized at that time, and given the Christian name of Elizabeth. The girls of the school felt very different and separated from those of *Boyale* {tribal initiation} so we went to the chief and asked that he make us part of the age-regiment system. He did this by giving us a group name.
>
> It was easy for people to become a Christian. The Tswana have always believed in one god, *Modimo*. The Christian god and the Tswana god were really the same. With Christianity, you could pray directly to God for important things, like rain. We continued to seek guidance and knowledge during dreams from our *Badimo* {ancestors, mainly dead parents and grandparents}. It seemed to us that, as Christians, we could live a better life and still keep our customs. We drank tea with sugar. This was the food of the civilized people. The church looked down on the heathens, or people who did not join the church. The church said they were uncivilized. Actually, we all thought and acted the same.
>
> My parents had always used the *Ngaka ya Setswana* {traditional medical doctor} and the *Moprofiti* {an explainer of the past and predictor of the future, doctor of ill fortune}. The church said this was wrong, as they were heathen witch doctors. The people of the church gave us powerful new medicines to make us well. After that we never used the traditional doctors, mainly because the new medicine worked so well, not because we changed what we believed.

<u>Senatla</u>. Senatla, Elizabeth's nephew, is 88 years old. Today, he and Elizabeth live together with his 17 year old granddaughter and seven preschool great-grandchildren. They reside on a small piece of land near the original *kgotla*. This land was given to Senatla's father by the Chief Mokgosi I. According to custom, it was here that Senatla built his house. It is the only remaining living quarters on the original family compound.

Senatla, meaning Industrious-Person in Setswana, was not a church member in his youth. He spent much of his childhood away from the village. As a young boy he went with his mother to the lands. When about six years old, he began accompanying his father to the cattle post. His early experiences with Europeans and the acquisition

of imported material goods vary from Elizabeth's.

> I was at the cattle post with another lad my age. We saw a
> strange, large animal coming towards us. The upper part of it had many
> brilliant colors, like red and yellow. We were afraid and ran into the
> bushes to hide. Then we remembered what others had said, of people
> having white skin, dressed in colorful clothes and riding an animal called
> a horse. We stayed hidden until the man passed. After that I became
> accustomed to seeing the white man.
> I did not go to the Lutheran school. My parents were afraid of
> Christianity, as they made you give up the Tswana ways that were
> important for living. For instance, we buried people in the yard so their
> spirits would be with us. Children were buried in the house to be close
> to the mother. The church did not approve of this.
> As I grew older but before I was a man, I learned about money.
> I had heard about money as some of my friends' fathers worked on the
> South African farms during harvest to earn money for the hut tax. I did
> not know how to use money or what a sweet {candy} was until I was a
> big lad. A trader had opened a store in the village. I spent my first coin
> for my first sweet at the trader's store. Then I wanted to work for
> money, to buy a pair of trousers. My parents said 'No, it was the job of
> a lad to stay on the cattle post.' So I stayed.

Senatla broke off the conversation and went into his dwelling. A few minutes

later he emerged with a metal plaque about the size of a small index card. As he

handed it to me I saw the seal of England imprinted near the top. Around the seal

read "Bechuanaland Protectorate - Hut Tax Receipt 1902-1903, Gaborones." He

explained the plaque to me.

> You must understand these were the days before paper. This is
> the way we showed the British that the hut tax had been paid. Every
> house had one of these by the door. This plaque is from the year I was
> born, and it had to be paid for in money. If you didn't have money, you
> had to sell a goat or a sheep. To pay the tax we walked to Gaborones
> {Gaborone}. It took all day and was a hard trip for some people. There
> were no roads, only paths through the lands. If the rivers were not dry
> we could get water on the way.

<u>Sego</u>. Sego feels his name of "Lucky" is true, having now lived 76 years. He sees the

past clearly although his eyes are now dimmed with cataracts. One boyhood delight was

watching the 1917 installation of Chief Seboko Mokgosi II. An age-regiment had been

sent to kill a leopard and the skin was placed on Seboko's shoulders during the ceremony. The presence of The Resident Commissioner and District Magistrate made him stand in awe.

A second delight was the thrill of obtaining water from one of four communal water faucets that Chief Seboko had installed. Instead of walking many kilometers, he could get the household water in 30 minutes. These water lines were connected to bore holes that tapped into the large underground artisan springs, which continue to provide village water. He proudly explained that, although the water system has been enlarged, these same taps are present today.

Sego, sitting with other aged men at the *Kgotla*, shared memories of hard times. Hard times came from nature, usually in the form of drought or excessive damaging rain.

> According to custom, every household gave part of the crop to the chief as tribute. Grains, mainly maize and sorghum, were stored in silos, to be distributed to the hungry and poor. After the land purchases of 1926, a drought came. We had some food, but new crops were dying from lack of rain. Our elders questioned the status of Chief Seboko, as it appeared he was no longer a good provider for the tribe. The village was very discouraged. Our maize was wilting in the fields, and communion with God and our ancestors did little good.
>
> At that time, the chiefs were believed to have powers to make rainfall. Seboko, because he was a Christian like his father, had never been instructed in the secrets of rainmaking. Fhologang Peba was the one man in village who knew about rainmaking. He and another man began the ceremony. They grew afraid and quit, because the chief was a spectator. Seboko sent word to South Africa to have a famous rainmaker come to the village.

Laughingly, all the men described how the famous rainmaker spent three days in ceremony, praying and crying out for rains that did not come. They act out how they chased him out of the village as a hoax. Then, growing serious, they talked of the days that followed.

> As soon as the rainmaker left the village, the rains began to fall. They did not stop so our fields flooded and the maize turned black. We call 1926 "The Year of the Black Corn." Everything was ruined. People said trying to make rain was against the wishes of God. God acted

against us by making too much rain. This was the last time a rainmaking ceremony was performed in the village!

Sego pointed out how the later years of this time period contained manageable conflict between church and tribe. He gave the history of this evolving problem as it related to his *bogwera* (tribal initiation) in 1929. Sego was a member of the *Matsaakgang*, or "Those associated with a dispute."

> The National School of Ramotswa, established by the church, was operating at this time. Many older children went for two or three years of formal education, independent of church confirmation class. Girls were more apt to attend than boys, because of boy's duties with cattle. A Malete boy, who had been attending school in the Transvaal, came to Ramotswa with a letter of introduction for the missionary of the Lutheran Church. He was admitted to the church classes. The boy's new father removed him from school, as that family did not want him to become a Christian. They placed the lad in *bogwera* for his education. The church and the Magistrate tried to have the boy returned from *bogwera*, but the chief refused. A large dispute resulted. It was decided that both types of schooling was initiation-training and the chief then gave identical age-set names to the graduates. After that, we worked together as a unified age-regiment, clearing communal tribal land and building fences.
>
> It was silly for the church to have divided us according to whether we were a civilized Christian or a heathen . In reality, no one was really different. The heathens were not ruffian people, but good people. As a village, we were all Tswana. We all followed the Tswana laws and could understand each other.

Pre-1935: Social Change or Social Continuity

The traditional cultural system incorporated the natural and manmade environment with laws and customs providing control over the supernatural and natural ingredients of village life. Gerontocracy and principles of eldership penetrated all aspects of Malete daily life, directing interpersonal behaviors and social, judicial, and political structure and function. Extended kinship, based on principles of age seniority, served to integrate families and establish social cohesion within the community (Schapera 1944, 1953). The result was a communal system of solidarity and reciprocity, which probably occurred in varying extents between individuals.

Colonialists, aiming to establish Ramotswa as a Christian labor reserve, interpreted the traditional culture as heathenish and barbaric. The thrust of change was for total cultural conversion without regard for the impact on social and economic life (Tlou and Campbell, 1984:140). By the end of the 1930s, Europeans in Ramotswa claimed success in their goals although "a few relics of barbarism" remained (Ellenberger, 1937). There was some trepidation whether the acceptance of Christianity was a matter of course, as missionaries "labored amongst an unresponsive people" (Schapera, 1953:58).

There is no denial on the part of the aged Malete that the ingress of Europeans brought delightful foods, Western clothing, and money. But through the eyes of those who lived during this period of importation, new experiences were found within the bounds of the Tswana cultural setting. To them, it was still "the time before civilization."

In other words, this was still a time when the traditional reigned. Traditional here is defined as that quality of social life in which contact with the Western world did not play a significant part of the lives of the majority of the population (Cohen, 1986:136). It does not encompass the substitution of a metal pot for a clay one, or the covering of the entire body with cloth instead of a leather loin cloth. A traditional life is one where basic behavior, regardless of the implements used, continues to be built on the non-Western organizational principles of the society.

Social unification was stressed by many. The change that did occur was a series of adaptive responses. In part, change was to allow for continuation of the traditional through accommodations. On the individual level, money became another trade item, not the basic necessity for subsistence. Taxation and growing desire for luxury items encouraged, rather than forced, labor migration of a single family member (Parson, 1984;23). Employment became new way of achieving without altering the basic structure of the culture. The comprehensive process of achieving, embracing the concept of gerontocracy, remained the underlying ideology.

As a pragmatic group, the Tswana willingly gave up the external trappings of life and substituted sturdier and more time-efficient imported technology. These material objects, such as the metal plow, were incorporated into peasant farming with the retention of underlying structure and function. The same was true with the new water sources and extension of lands. Participation in the introduced activities of school and church was to "learn the European's ways in the hope of making life easier," not to make their life like his. The principles of eldership continued, with obedience to the laws perpetuating family interdependency and community solidarity.

On the structural-functional level, the traditional and new were intertwined. As in the case of conflicts for initiation into adulthood between the church and tribe, accommodation with unification of the two types of schooling under one name allowed for the continuation of horizontal social and political village interactions. Peers remained unified and continued to function under the rigid seniority principles.

The concept of English taxes was mitigated by the concept of tribute. Colonial government was a parallel to chieftainship, with both playing an active role in vertical social and political relations (Parson, 1984:17,23). The giving of allegiance to authoritative leaders and displaying decorous behavior towards social elders was unquestioned.

The English government and the Lutheran church both claimed to have influenced individual and tribal decisions to discontinue "heathen practices" (Schapera, 1970; Tlou and Campbell, 1984:134). The cessation of rain-making was frequently attributed to outside intervention (Ellenberger, 1937; Schapera, 1970). The disastrous results of the last Malete rainmaking ceremonies were probably quite influential in the decline of ceremonies to alter supernatural events. Other heathen customs were reported to "have disappeared because of conversion", such as placing a stone in a tree on arrival to the village and the purifying of the Army before war (Ellenberger, 1937).

There was no future war, hence no need for military cleansing. Even the very old did not remember their parents placing stones in trees, or had heard of the custom. The changes noted by colonialists may have resulted from selective perceptions, noting the natural demise of impractical customs during the on-going process of social evolution (Chambers, 1983:42,82-85).

The concept of modernization implies a condition in which outside leaders are dedicated to the goal of changing the nature of a society (Cohen, 1986:136). There is no doubt that at this time the colonizers were attempting to instigate change. Culturally adaptive responses of accommodation and resistance by the Malete mitigated the pressures and prevented the new from eradicating the old. Culture did change, but partly in order not to change. With the use of boundary-maintaining and self-correction mechanisms, life could continue within the traditional cultural setting (Social Science Research Council, 1953). Concurrently, the seeds of Westernization were planted, some more firmly that others, but all had the potential of growth.

"Becoming Civilized": 1936 - 1966

History presents this time period, extending to the time of national independence in 1966, as one of increasing and marked change. How does the broad, political view of change compare to the villagers' inside or "emic" view of village life during the period of "becoming civilized?" This section concentrates on oral histories, which diverge, and at times disagree, with the written. Such discrepancies are discussed, with a questioning if change was total acceptance of the new with rejection of the old, or if it was a modified expression of traditional social attributes.

Structural Growth

Village growth was evident by the late thirties. Assorted shops, numerous homes and church buildings began to crowd the lands surrounding the *kgotla*. More and more people were using the basic health clinic of the Lutheran mission. A second mission was established on the western outskirts of town by the Roman Catholics. Chief Seboko Mokgosi paved the way for its acceptance in the village by switching church affiliation.

At the same time, Seboko felt a need to move the tribal headquarters. A new *kgotla* was built, about a ten minute walk northeast of the new Catholic mission. Two round silos for storage of the annual tribute of grain were constructed to the side of the traditional round court building. The royal cattle corral sat across the large open courtyard. The corral could be used by anyone in need of a temporary livestock enclosure. According to custom, it was also to be the burial site of the present and future chiefs. The traditional design of the *kgotla* was distorted with the addition of a new type of building, a square one with glass windows and a corrugated iron roof. The present chief, Mr. Kelemogile Mokgosi, gives his version for the deviation.

> The people were walking to Gaborones quite often to pay taxes and do the increasing amount of government business. The elders complained to my father that the distance was to great for the aged and took too much time from the young. The British were approached and they agreed to build a office in the *kgotla*. The British District Magistrate collected taxes in the square building. There was no problem in paying taxes. If a person did not have money as he could bring a lamb or a goat and get the correct change. But, at the same time, local tax collection signified the end of giving grains to the chief for distribution to those in need. Our new silos were never used.

Such action was the beginning of political subordination of tribal authority, which was occurring throughout Botswana in the 1930s (Parson, 1984:22).

<u>Religion and Marriage</u>

Monate and John were one of the first couples to be married at the new *kgotla*. Monate, now 73 years old, sprawled herself out on a blue plaid synthetic blanket in her courtyard to recall the time of her wedding 55 years ago.

> I had finished initiation school at the church, which meant it was time for marriage. My parents arranged a marriage with a man I had never seen before. That was the way it was done in those days. The church could not take that away from us. Oh, I was so scared. What if I did not like him or if he treated me bad? Mother encouraged me to go nicely, as my name implies. My ancestors assured me all would go well.

The first day of the ceremony, the women of the village collected
me and we walked through the village. I cried out, 'I need a man to
cook and tend house for.' We looked and looked for the man I could
take care of. That afternoon, we gathered at Mother's house. She and
the other married women instructed me on the secrets of marriage. The
next day, John was taken through the village to cry out for a women he
could hunt and provide for. He was then led to the rocks, where he was
told the men's secrets of marriage by other married men. These are such
secrets that we do not tell each other what was said. They can only be
discussed among married Malete women or men.

Monate was hesitant to discuss her *Bogadi* or bride price. "Gifts were exchanged
so our future children could be claimed by the families in case of our death. Otherwise,
the bride price meant little to us as newlyweds." She stresses that this practice was
strongly condemned by the church, as well as her arranged marriage and the tribal
ceremony.

The chief led the ceremony. It took place at the *kgotla*. The
wedding was beautiful. We wore fine skins. Many of the others wore
real {western} clothes. Afterwards there were two days of feasting, one
day at my parents' house and the other at his parents' house. Cattle and
goats were slaughtered, with parents getting choice parts to save for later.
It was a long time until wedding ceremonies were held at one of the
churches with children selecting their own husband or wife.

<u>Home Life</u>

The aged are quite aware of the changes in daily life during this era, sometimes
approving and sometimes not. The following informal interplay mirrors the thoughts of
many aged. One morning I hear soft singing coming from inside the crumbling wall
surrounding the compound's courtyard. Inside sits Mogolokwane, or Happy Sound, as
she prefers she prefers to be called when speaking with me. At age 69, she radiates in
the freedom she now has to sit and play with her latest grandson.

I follow the customary greeting, asking how the sun shines upon her today.
There is never a simple answer to this mandatory question. Happy Sound bases her
answer on reflections of her young adulthood.

When I was a young adult, the village was so different. Everyone
plowed and grew food. The men earned money and controlled cattle.
The Malete have always been poor, but the meaning of poverty gradually

changed. Once, everyone had enough to eat. If crops failed, others gave
you food. The law was carried out. If a person was in need for
something, he could ask another and it was given. No tally was kept as
someday the person who gave would be in need and he would be given.
We always gave if we had what was needed. Unfortunately, it became so
that all you could ask for was a little bit of sugar or salt. Now people
ask for money. If I have money, I give part of it.

Once it was good to live in a round house. The children would
gather dung and bring water and I would always be mixing mud to keep
the house looking nice. In the late 1950s the chief gave permission for
people to make four-cornered {square} houses. My children were then
old enough to have their own house. They wanted the modern house
with the tin roof, even though it is very noisy when it rains and is hot as
the wind doesn't flow.

It wasn't just my children who wanted to be modern. I did too.
My mother used to make clay pots. I preferred the iron pots so I did not
learn how to make the clay ones. Now I wish I had clay pots for beer.
Beer was always cool to the tongue in clay pots in contrast to the metal
ones. There were many things my mother did not teach me as she also
preferred the modern new substitutes. Sometimes I must go without, as I
know nothing about wild foods or working with leather. The one thing
elders made sure I learned were the laws about what is right and wrong.
They would beat me, even when I was a young adult, if I disobeyed the
laws.

An older relative, passing by the house, stops to see what we are talking about.

She reconfirms the multiple beatings. "People seldom did wrong as they knew the

punishment of pain and shame. The entire village would know that you did wrong."

Wanting to know more, I ask if they ever purposefully went against parent's wishes.

The answer was a loud "NO!" Just for fun I tell them about one of my childhood

escapades. Happy Sound lives up to her name and laughs loudly. Not to be outdone,

she says:

we never went against elders but we had ways of having fun. My mother
wanted me to go to school every day. There were days I did not want to
go so I pretended to have a stomachache. She would take good care of
me in the morning and then I would play. She thought I was a sick child
as I did this many times.

The women giggle. It is the type of laugh that says, "you are right but I will not

say so." One woman becomes more bold, not wanting Happy Sound to become the only

star.

Oh, you think that is clever. When I was a big girl, my mother was very strict about being with boys without an elder present. I would wait until she went in the house and then sneak into the bushes where the boys were. We would have lots of fun. All the time Mother thought I went to visit other girls. No, I won't tell you what the boys and I did, but you did it too.

Happy Sound smiles with a reminiscent look in her eyes, but the baby begins to cry and the group breaks up. Never again would any of the aged relate tales of mischief, always insisting that respect for elders remained paramount.

<u>Village Life</u>

Maria, now in her late eighties, loves to talk about the growth of civilization.

Her 67 year old daughter is frequently at her side, for Maria can no longer walk.

Lameness does not keep Maria from acting out stories with her arms and face.

Maria beings bouncing on her stool, imitating a car on a dirt road. "Cars began to come to the village in the 1940s. There were three cars belonging to the British that drove the road regularly." She covers her ears and screams:

> varoom! varoom! I still can't figure out how a cart goes without an animal. I don't like them. Nearer Independence there were many cars in the village. Once my brother insisted I ride to church. I sat in the back seat and covered my eyes but couldn't shut out the varoom. I walked home from church and still prefer walking. Cars go too fast.

The daughter chuckles as she hears this new story about her mother's life. I ask what was the most important change prior to Independence in 1966. Without hesitation, Maria discusses sorcery. The daughter listens and nods in agreement as her mother proceeds.

> Witches can do many bad things like bring illness, bad luck, or even death. There is little recourse against witches, as you usually don't know you've been bewitched until it is too late. In the old days, it wasn't so bad as the *moprofiti* {prophet} charged us very little to throw his bones for interpretation and cure. Also, the village had very few witches. They were usually old spinsters. If we thought a witch was in the village, the chief would call everyone together. He passed out a drink. The drink would make the witch crazy. She would run around, and yell, and cry. The tribe knew who was the witch and all was well again.
>
> The British took away the powers of the chief. He could no longer perform this and other ceremonies. No one knew who were the

> witches, and they multiplied; some being old ladies; some being young
> mothers teaching their children the secrets of witchcraft. By
> Independence there were many witches of all ages. People no longer
> trusted people outside of relatives. Cures for being bewitched became
> very expensive, like one or two cows. We continue to live with the
> problem of disease and misfortune from witches because the white man
> denied it was so.

The daughter agrees that witchcraft is a daily threat that outsiders continue to overlook.

<u>Growth of the Money-Market Economy</u>

The primary activity of Bechuanaland administration was directed towards tax collection to maintain a modest number of officers for law and order, while developing Botswana as a labor reserve for South Africa (Parson, 1984:22). Opportunities for local entrepreneurship were restricted. In 1949, there were only ten African-owned stores in Bechuanaland as opposed to 155 European-owned ones (Parson, 1984). Small numbers of natives entered teaching or became store assistants, clergyman, or native clerks for the English administration. Even smaller numbers worked for themselves as dressmakers, thatchers, or carpenters (Schapera, 1955:33). The dominant avenue for economic relief was the road to the mines in South Africa. Low wages kept workers in a situation of poverty. Rules forced a return home after employment. Economic needs at home forced a return to the mines. Oscillating migratory labor became a necessity (Parson, 1984:24-25).

John, like so many of his age-set, left for the mines soon after his marriage to Monate. He says:

> there was a recruiting officer at Ramotswa Station. I just walked up to
> him. He asked which mine I wanted to go to. After I told him, he gave
> me papers and a ticket to get there. It was required that you stay eleven
> months, then you got a ticket to come home. After several years, I could
> have stayed longer for a bonus but never did. I did this again and again.
> I never thought much about it. That was the way things were done, as all
> people, not just the family, now needed money.

Monate joined her husband in South Africa a few years after his initial departure. She found work as a maid to an English family in the area. The couple openly talk of their approach to migratory labor.

> We knew our success was dependent on the way God wanted our life to go. People have no control over such outside forces, or can one make definite plans for the future as God brings that, but one must try to better themselves. All men are good and without evil, so we were not afraid of what would happen if we left Ramotswa. To better ourselves we learned English as we worked. When times changed, we knew enough English to find work in the cities.

In contrast, Elizabeth remained in the village to work the lands while her husband worked at the mines.

> Before he left, we made sure all his possessions were in place, where no one would touch them. Just to touch the belongings of someone working in the mines would cause his death. Even a shirt left on his bed must stay there until he returns. Our house was on my in-laws' compound. I had to obey my mother-in-law as she taught me to take over chores. That wasn't so bad, as all things come to people who wait. Eventually I was in charge of the house.

World War II brought new economic demands to the village. The colonial administration increased taxes for the war fund and worked through chiefs to increase migration to the Rand mines for gold production (Bhila, 1984). It was estimated that 40% of adult males and 5% of the females of Ramotswa were absent at labor centers in the Union of South Africa during the 1940s (Schapera, 1953:30). Other men were enlisted in labor battalions to serve the British Army in northern Africa and southern Europe (Bhila, 1984).

The Money Market Economy and Family Life

The effects of early massive labor migration on family welfare differ between British ethnographers and oral history. Writings emphasize the lack of economic support from absentee workers, including defection from family responsibility and complete desertion. Many aging parents left unprovided and had to "fend for themselves" (Ellenberger, 1937; Schapera, 1944; 1953).

The present aged refute these writings. "It has always been the duty of children to provide for parents. As a lad, I lived very meagerly in Kimberly, sending most of my money home." "We always respected our parents, and the aged were well cared for." "Money was always sent by working sons. Their wives, or the remaining siblings, gave good care to aging parents." "I myself took good care of my mother, giving her gifts of food and clothes, as well as sending money."

According to the aged, there have always been old people in the village, some living a very, very long time. Old people never lived alone during this time period. They shared their compound with married daughters and/or daughters-in-law, as very few women were not married by the age of twenty. Elizabeth says:

> the old people were always respected in the old days and never went hungry. When I married, our place was with my parents, as that was the law. When my husband left, he sent me money. I used it to pay the costs of his parents, my parents, and our children. If we did not take care of our parents, our children would not take care of us as we grew older!

It is impossible to determine the true historical situation at this time. One cannot deny the effects of time and social change on recall. The "Good Old Days" tendency is real, even in Botswana. No doubt, some parents were not supported monetarily by their absentee children. Their numbers cannot be told. Life histories repeatedly report that the husband sent his income to the wife. This money, under her control, had varying degrees of adequacy for household expenses. The claim is that the aged were supported economically, and given food and care equal to that of the rest of the family..

The aged's claim of past economic support is very different from the picture painted by Schapera (1953, 1966). There are two possible reasons for a misinterpretation by this English anthropologist. Western monetary economic principles, when applied to traditional subsistence systems, can easily overshadow the true picture.

Old age support emerging from the congruent peasant economy, with kin providing subsistence and care, could occur independently of the newly superimposed Western economic system. Secondly the Tswana, like the neighboring San, use complaint during discourse as a means of control over others (Rosenberg, 1990). Increasing household poverty may have prevented the aged of this era from receiving as much as they wanted. Complaint is a means of insuring fulfillment of as many wants as possible, and not to be taken at face value.

The Money Market Economy and Social Life

Social change resulting from a money market economy came with startling rapidity in all Tswana villages during the 1950s (Picard, 1987:118). People did not want to appear antiquated. Traditional crafts of pottery, leather work, woodwork and basketry were tossed aside as purchased items carried more social value (Campbell, 1971). Cash was needed on a daily basis to purchase the necessities and services of modern living, including school fees, clothing, chairs, beds, and metal roofing for square houses (Thema, 1972; Parson, 1984:25). At the same time, a permanent change in diet occurred. Tea and coffee replaced milk as the preferred beverages. White bread was now regarded as a necessity. Milk, meat and agricultural produce became items for sale, as items that could be sold were no longer given away (Thema, 1972).

Cattle, once used to emphasize important kin relationships and events and thus cementing position in society, became a direct economic asset. Variations in herd size reflected household economic status, with cattle purchased and sold as desired (Campbell, 1972). A survey of this area in the early 1950s showed 11% of the households had no cattle; 18% owned fewer than 10 head each; 19% had 10-20 cattle; and 40% owned 20-50. Chiefs and a few others had as many as 100-500 cattle each (Schapera, 1953:23).

Traditionally, labor was given as freely as food and other needed goods. Labor was regimented according to age-sets, as determined by tribal/church initiation. The group would work together to clear areas in the village. At other times, all the people of the village would work together, either directly for the chief or for village good. According to Senatla, the giving of labor provided reward with achieving.

> We used to have to work for the chief, helping him with his fields and village improvement. We would help him when we came home from South Africa. Sometimes it was inconvenient but it was good as everyone benefited. For example, about eight years after Chief Seboko built the new *kgotla*, he decided it needed a fence. Every man in the village had to cut a tree and place it on the fence. It is still a good fence. My log is one near the end.

The free giving of organized labor conflicted with the British concept of organized labor for material gain. The traditional labor system was abolished in 1965. The same year it was reintroduced in a new form - *Food for Work*. This was soon followed with specific employment programs that assisted internal development (Prah, 1979). Volunteer labor became paid labor, and money became the reward (Campbell, 1979). Sego recognized the impact of the change.

> In some ways it was good that we no longer had to give time and energy to the chief, as we could make money instead. In other ways it was bad, as no one really cared about the village, only the job. Today, no one gives labor for free.

<u>Pathways of Culture Change</u>

The process of becoming civilized was summarized by Sego. After leading me to the open area in front of his four-cornered house, he spread his arms in a wide circle.

> See all the houses here; they are the homes of all my family. The village was a good place years ago. In those days, the eldest male was in charge of the home. Wives knew their place was to cook and care for husbands and children. Children and old people both had roles and they knew what they were. Every family was involved with plowing and raising cattle.
>
> The village grew in size. We had children who had children. Change occurred, especially between the chief and new government. Some, like myself and my nephew, who lives in that black house, stressed education and good employment and became very modern. My sister's

daughter, in this *rondaval* beside me, stressed the old culture, insisting that the boys go to the cattle post and the girls plow. Life changed. New problems replaced old problems as we became civilized, but all these families here worked together. We learned to achieve and solve problems in different ways. Yet, everyone still followed the laws of achieving and showing respect to elders.

Sego was one of the few who talked about the two developing pathways of change: those who became submerged in the white man's methods of economic self-betterment through education and progression up the work ladder in contrast to those who sought accomplishment in a more traditional manner. There was no judgment involved, as both were considered a valid method of achieving.

<u>Sociocultural Change and Continuity</u>

Present household aged include the entrepreneurs like John and Monate and the agriculturalists like Elizabeth. Although differing in the main mode of providing subsistence during the working years, both groups continued to share the underlying values of personal and social achieving. Individual growth was fostered. Individual competition was not.

The use of external causality in explaining life and its successes was an physical, and cultural, environmental necessity (Schapera, 1953:34). All individuals based causality for their degree of success with achieving on external factors. Supernatural powers, in conjunction with political regulations, had ultimate control over the individual. The erosion of the chief's power resulted in a growing emphasis on supernatural power. The past generalized fear of the aged as the interveners with the supernatural became a fear of others, as anyone could now be a witch. Traditional religious and social behaviors, which once provided a degree of mastery over the supernatural, were gradually eliminated or placed out of reach.

Control over life depended more and more on application of the laws for proper daily living. The admitted pranks of trickery, lying and duplicity were not considered rebellion or going against the wishes of elders. These were acceptable ways to control the universe without stimulating guilt or shame (Alverson, 1978). Diligent adherence to tribal laws for harmonious solidarity continued to be rewarded with social approval and respect (Schapera, 1953:61).

The claim is that labor migration and economic change weakened parental control. Schapera (1953:50) stated that young people no longer looked to their parent for guidance, and tended to act more upon their own responsibility. Comments from the present aged indicate that they did not feel they were independent of elder authority. They say the efficacy of beatings and traditional sanctions continued to promote wide observance of social gerontocratic standards.

I believe that the present aged contributed to the power-shifts within households as the employed, not the aged, began to control expenses and payments. Yet, household family cohesiveness remained all important as the money market economy increased in intensity. Power-shifts were unrecognized by the younger income-producers. They did recognize that modern did not always mean better, as pointed out by Happy Sound. Economic deprivation mandated changes in the laws, such as restrictions on giving, and increased limitation on the use and exchange of items.

The structural environment created during this time was one of underdevelopment (Parson, 1984:33). Ramotswa was but part of an entire country of poverty. The mechanism was the new economic system. The outcomes of labor reserve economy were the creation of new social strata and new economic and political relationships. The subsistence system became a Peasantariat, a term embracing the persistence of a peasant type of agriculture and society with the proletariat attribute of working for wages (Parson, 1984:123). Professional and educational elites emerged in the transformed and redefined tribal and village hierarchy (Parson, 1984:34).

The structural shift of political power from the chief to colonial government was accompanied by a functional shift. In the public's eye, colonial governmental administration became the new source of provisioning and planning for village development. The customary shameless dependency on the chief for grain during times of economic hardship was transferred to the British administrators. With the Food for Work program, money replaced grain as the gift to those in need. Thus, the spirit of personal welfare through socio-political integration lived on.

<u>Being Modern: Post-1966</u>

At the time of independence in 1966, much of Botswana remained a rural country. Planning was directed to improve social service and employment opportunities in rural areas. (Chambers and Feldman, 1973). District Councils were formed, with Ramotswa becoming the center for the South-East District. This comparatively small district encompassed the reserve lands of the Malete. Other than the locally elected District Councillor, official representatives for council operations were, and still are, appointed from outside the district (Silitshena, 1979).

These political elites see themselves as modernizers with a responsibility to develop the area to the highest level possible with the given economic resources (Picard, 1987:173). The South-East District Council is charged with the overall running and development of the district. Such matters as primary education, provision of health facilities, construction and maintenance of roads, water supplies, and implementation of national programs for village development and rehabilitation fall under its jurisdiction.

Emphasis was placed on physical/technological developmental projects until mid-1980. Accelerated rural agricultural development has now become the major concern. (Picard, 1987) Increases in crime, desertion, single motherhood, unemployment, and alcoholism have only very recently been considered as potential threats to village welfare. Other more intangible aspects of village life remain overlooked, such as increases in latch-key children and the keeping of the handicapped and mentally ill behind closed doors.

With the formation of District Councils, tribal chiefs lost their remaining legislative and administrative functions, sweeping away any power to regulate the political and economic life of the tribe. The Tribal Land Acts of 1968-1973 and the creation of national land boards withdrew chiefly power over land allocation and use (Silitshena, 1979). These once powerful men could only hope they could preserve at least some residual authority over formal matters of traditional law and protocol by serving as government advisors in a upper house of Parliament known as The House of Chiefs (Picard, 1987:179).

Within Ramotswa today, the chief's functions are mainly concerned with non-criminal judicial matters and with serving as a vital link in communications between the federal government and the people (Silitshena, 1979). Although diminished in autocratic authority and power, the Malete regard the Chief Kelemogile Mokgosi and his son, Deputy Chief Ikanena Mokgosi, as symbols of tribal identity and cultural heritage.

The *kgotla* continues to serve as a place of guidance and rectification of perceived wrongs. On many mornings a family dispute, unresolved by the elder and headman, is solemnly judged by the chief and his advisors. The *kgotla* remains a place where people congregate, usually older men. Young men and women visit also, some seeking a answers to questions, others just to talk. Private affairs are discussed in either the old British Tax Collectors building, which today serves as chief's official office, or in the original *kgotla* building. Tucked in this building's walls are the impala horns, still used to call initiation classes back from the hills.

The large iron bell still notifies the people of impending important announcements and meetings. Its ringing signifies the necessity to stop activity and proceed immediately to the *kgotla*. The original *kgotla* building can no longer hold all the people. Instead they almost fill the new rectangular shaped, but traditionally thatched and painted meeting-hall. Directly next door is the district Police Station with uniformed officers.

The royal corral, once shared with villagers for the overnight pasturing of cattle, stands empty. The royal burial ground is still used for the chief and his immediate family. The grounds around the cemetery and *kgotla* building are kept clean by those judged guilty of crime. They work during daylight, advertising their shame to the village.

Not far from the *kgotla*, about a five minute westerly walk, is the new village library, a modern building with numerous books. The community center with a large meeting hall is anther 10 minute walk to the northwest. Various groups, such as the Village Development Committee, Red Cross, and burial societies, meet here during the day. Frequently large parties are held here by different organizations on Friday or Saturday nights.

The main shopping area is a ten minute walk south of the community center. The post office, a small bank, grocers, general produce stores and one of the two public pay phones line the dirt road. Bus service, northwest to Gaborone and southwest to Lobotse, originates here. There are no signs or benches; one just knows where to stand, which bus to ride and when it leaves. Early in the morning smartly dressed men and women going to work in the capital or Lobotse fill the bus. Later in the day the riders are mothers with babies and the aged, going shopping, visiting or to work the lands.

The first bus stop, directly south of the village center, is at the Ba-Malete Lutheran Hospital. This medical center offers out-patient and in-patient care, pharmaceutical service, plus surgical and maternity facilities. The physicians are European with Tswana nurses serving as translators. It is also here that the bus begins its journey on the one paved village road.

Traveling west, the bus passes several government housing units with electricity and running water, small grocers, the Catholic church with its elementary school and clinic, and what is known as the heathen cemetery. This cemetery is still in use for those who have never joined a Christian church. (The Catholic and Lutheran cemeteries are some distance from the paved road). Approaching the western outskirts of the village is the new industrial park with small family enterprises. The brigade, a carpentry training center, is across the street. Among the last buildings seen before leaving the village is the council administration, the major source of village employment.

If you ride the bus, the person sitting next to you will probably live in the village. He or she may know the village contains a national school for the deaf and one pre-school. Everyone can count the five primary schools and knows the location of the secondary school. You will be told the village has grown considerably in the last ten years, but they will not know the population is now over 8,000 with another 1,700 claiming residency although they live elsewhere and another 4,200 living outside of the main village in splinter settlements (Central Statistical Office, 1982).

The modern structures along the tarmac road support claims that social identity and value transformation that began with colonization has resulted in a new society (Comaroff, 1953; Parson, 1984). Others claim that the observed infrastructure and development has not penetrated into the self-identify, value system, and overall life of the rural dweller. The material trappings and government projects are but symbolic efforts primarily designed to create a veneer of progressiveness, with "beneficial change" occurring for only a small number of political, bureaucratic and land-owning elites (Campbell, 1971, Knudsen, 1988:17; Staugard, 1985:18-21; Picard, 1987:145).

As I traveled the tarmac road for the first time, my impression was of growth, development and modernization What I did not see, until I left the bus, was the woman stamping last year's harvest into meal, the smoke coming from the sides of a *rondaval* as the family tried to keep warm, the parent and child sitting on the ground to save their energy as there was no food today, and the look of deprivation as an aged couple stood by their broken fence. Unmarked homes often made it impossible to identify the places of practice of the reported 51 traditional doctors and faith healers or the residences of the eleven birth attendants (Knudsen, 1988).

Culture change can be documented but not precisely measured. Events can be dated but distinct social alterations in behavior cannot. Social facts are not necessarily objective facts, as already demonstrated by the conflicting historical interpretations. Yet, historical forces molded culture in the past and continue to mold it anew. Throughout these writings, I often try to look through the eyes of the Tswana to see the continuity and change. The purpose is to explain the present as influenced by their perceptions of the past. The current individual, family and social life, with its continuities and changes, will be discussed in the following chapters.

CHAPTER 5
GROWING OLD: THE LIFE CYCLE

It takes a long time to gain the wisdom, knowledge and judgment that marks a grown-up. I became grown-up two years ago. I am proud to be old. (A 81 year old woman)

Stupid old-child, get out of my way! You know nothing and can do nothing, so why don't you go home where you belong? (A young bus driver to a 65 year old woman)

Who is defined as an "old person" in Botswana today? That is the question this chapter deals with. The indigenous life-cycle will be presented to add depth to the definition. I will discuss the generalized concepts progressing from infancy and childhood, through the phases of adulthood, and the return to childhood and infancy, with the infancy at the end of life united to the infancy at birth through ancestors and God. Tradition established the life-cycle. Modernization has changed the timing of its phases. This, too, will be discussed, first in a general manner and ending with sketches of two individuals of similar age.

The Attributes of Old Age

Tswana individuals living in one urban and four rural areas were interviewed prior to the in-depth study to obtain broad based information on attitudes involving the aged. Two methodological pathways were followed. The first was based on ethnographic interviewing for in-depth interpretations of aging. Respondents were asked to describe an old person, speak about their behaviors, the advantages and disadvantages of being old, and talk about their present involvement with the aged. Identical initial question were asked of 57 individuals, 27 males and 30 females, approached randomly within preselected sectors of one urban and four rural communities. Respondents were encouraged to expound upon their answers. Particular attention was paid to verbalization of emergent thoughts reflecting interpretations and expected behaviors of the aged. This type of ethnographic interviewing is believed to give general data about the population at large, while using a small numbers of respondents (Bernard, 1987).

Secondly, participant observation in family and community settings allowed me to view interpersonal interactions involving the aged. This supported many of the previous answers and provided additional insight into the varied aspects of life as an old person. Participant observation continued during the main study, adding increased depth and understanding. A card sort aided in defining the timing and signaling of life cycle phases.

Old Age as Defined by the Urban Sample

The urban sample, which included 23 people ranging in age from 18 to 78, unanimously defined an old person as one who had many years and required assistance, mainly financial. Old age was stressed as a time of economic dependency. There was nothing good about advancing age as it was a time when a person was incapable of meaningful activity, making them children. No one perceived themselves as having an old person as a friend. Fellow workers and neighbors were considered colleagues, not old people. Old people belonged in the villages, not the city.

All urbanites stressed the cultural tenets of aged adults returning to the village of birth and of adult children providing for aged parents. This care was interpreted as a need to provide remittances. The stated social trend was that remittances were monthly, although most individuals sent money home once to four times a year. Household remittances were viewed as parental support, without mention of other household members. Only individuals employed in higher paying professional positions claimed to provide monthly funds. None planned to terminate urban employment to fulfill obligations of providing physical care. Interestingly, the majority of both sexes planned to return to their village of origin with advanced age and felt sure their own children would provide them with comprehensive, traditional care.

No mention was made of family or social roles of the aged. Their activities were viewed as play as the aged did not produce income. Old people were called usurpers and parasites, taking but giving nothing in return. Only with questioning did it emerge that parents provided child care to the adult child's children or maintained the village home to which the urbanite returned.

<u>Old Age as Defined by the Rural Sample</u>

The four villages, including Ramotswa, were scattered in each of the four directions from Gaborone. Distances varied from 20 to 100 kilometers. All were of Tswana culture but of different tribes. Uniformity existed among villagers in their responses. All participants had at least one aged relative, usually several, living in the village.

Rural respondents stressed filial piety as a right of aging. Stress was on physical assistance. The needed economic assistance blended with household needs. Remittances were for household use, not for elder support.

In defining the aged, individuals divided old people into two categories: the active-old, or elders, and children. The active-old were not colleagues, as in the city, but elders. Childcare for women and political participation for men were identified roles for elders. Active aged who cared for grandchildren or participated in *kgotla* activities were not abstractly regarded as children,, although many young and middle-aged adults were quick to identify reasons why they should be classified as such. Chronological age had no bearing on the life cycle phase, although everyone was well aware of who was chronologically the oldest with seniority rights in a given situation.

Assignment of childhood came to the foreground with the aged having any visual, ambulatory, or memory recall limitations. Although not necessarily dependent on others for care, physical limitations prevented full participation in social and family activities. Additionally, slowness in thinking or occasional forgetfulness would be seen by younger generations as a valid reason for not including an old person in activities. Such aged were perceived as children in both activity and thought. Individuals had difficulty identifying specific examples of childlike thoughts and behaviors, other than "just like a child, they like to be with people while accomplishing nothing." Childlike communication practices were observed, with adult children answering for and altering the answers of their aged parents.

At no point was a chronological age ever assigned as an identification marker, nor were grandparental status or outward signs of aging. The assignment of elderhood and childhood to the aged reflected their position within the cultural life-cycle. (See Figure 5.1.)

The Life Cycle

Infancy and Childhood: The Ngwana

A child is born. Twelve hours later the mother walks home from the village maternity unit, carrying the newborn tied in a blanket around her chest. The *ngwana* (child) is given a name, usually a Tswana one, as Christian names are no longer required for baptism. Following custom, the name reflects whatever is on the mind of the mother. It may be "Joy" or "Good Fortune," especially for the first born. With large families, names of "That's enough" or "Oh, my Goodness" are common. If the family is rich in livestock, the new son may become "Donkey Man." Less popular are the Western biblical names that were once necessary for church membership.

For the first three months of life the mother is freed of all responsibility except infant care. Undivided attention is given to the baby's needs, nursing the child

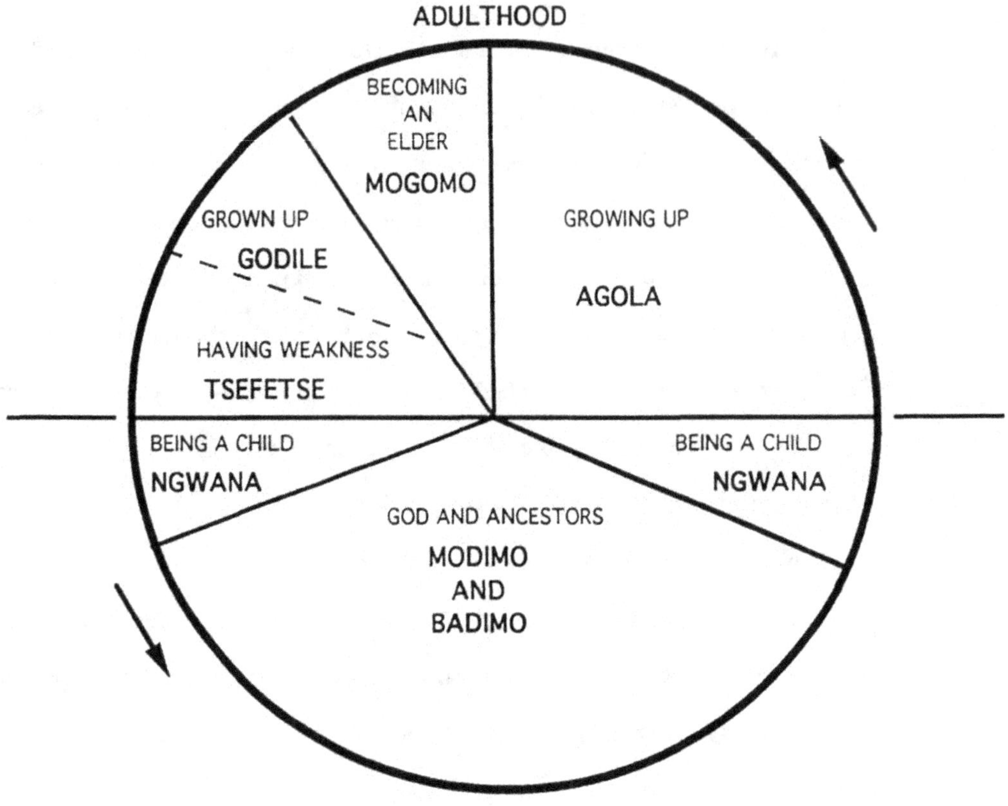

Figure 5.1: The Indigenous Life Cycle

frequently, singing and cooing as they stay together in one section of the house. Visitors are welcome to see this "Gift from God," who requires total care.

Childhood is entered as the infant learns to walk and obey simple commands. Overall, childhood is regarded as a carefree time for play. It is a rare child who owns a toy of any kind. A long piece of string is shared among girls for intricate games of jumping rope. A well worn soccer ball is the center of attention for older boys. When these two items are lacking and there is no chore to perform or school to attend, children sit or walk idly through the village. Imaginative play and play with adults have little place in this society.

The child is considered to be incapable of spontaneously recognizing important facets in life. Learning is through rote memorization, in school and at home. Facts are not to be altered and hence there is little encouragement to apply them to different situations. Reasoning is called cleverness, appreciated only if not disruptive. It, too, is not encouraged as a child is not expected to act independently of others. A child is to do as told. Activity is assigned through command. The form of command is "Go buy bread!" with never a please or thank you. Overt signs of appreciation for assistance are not part of the child's upbringing.

Outwardly there is an appearance of harshness and rigidity in childhood. Actually, it is a teaching that "all things come to those who wait." The minimal pressures for self-direction and decision-making reflect a learning of necessary concepts for gerontocracy (Rosenmayr, 1989). The goal is to instill respect for elders, with obedience seen as a necessity for household and social functioning. Beneath the directiveness is a security that, if true difficulty arises, there will be someone to help, be it a parent, older sibling or neighbor.

Ideally, the child must learn obedience to the law before becoming an adult. The child-beating used to control the behavior of earlier generations of children has been governmentally banned. It does not take a child long to learn that disobedience has no severe consequences. Refusal to carry water or fetch items may result in a few yells but little more. Respect and deference for elders can be thrown aside, for parents or guardians have little recourse. This is the one area where parents and grandparents severely resent governmental intervention. Fear of arrest for child abuse prevents even the mildest of spankings. Replacements for traditional discipline are absent. Concepts of withholding privileges or using a quiet corner are unknown. Shaming the child has little effect in today's world. The traditional action of having an older male talk to a disobedient child is used, but only if behavior becomes a severe problem.

Growing-Up: The Agola

Adulthood begins with a process of growing-up, or *agola*. *Agola* has always begun early in the village, and this continues in today's world. Traditionally, adulthood began with age-set initiation. Today, completion of primary school, usually at age 14, marks the end of childhood (Lee, 1985). The few who continue with secondary school are in a nebulous stage, neither children or adults. They are a child in that they are a student but an adult in their behavioral expectations of increasing responsibility.

Both teenage boys and girls are expected to wash and iron their own clothing, shop and cook meals and tend to all their personal needs. They are seen as capable of becoming economic contributors to the household, if opportunity arises. Therefore, it is not long before many young adults migrate to cities or mines seeking employment. Young women, many of whom continue to view themselves as producers, would rather produce money than traditional crops. Unemployed young men dream of work. Those without employment often perceive themselves as the managers or supervisors of the household members who are dependent on their physical presence, such as younger siblings, aging mothers or grandparents.

As in the rest of Botswana, there is a decreased emphasis on marriage as a marker of adulthood (Ingstad and Saugestad, 1987; Suggs, 1987). In itself, marriage today does not necessarily carry status among the young, being viewed more as an event than a commitment. Divorce, which has always been acceptable, is more so. The only change is that a judicial divorce proceeding has replaced the customary act of the woman placing her blankets and clothing on her back and leaving the house. As before, the children and all possessions become the man's property.

Producing children without marriage is the one way the woman can obtain parental rights and, as such, is becoming more common (Ingstad and Saugestad, 1987; Suggs, 1987). Parenthood does contain status, even though the single mother cannot claim the respected status of married women (Ingstad and Saugestad, 1987). I feel this is more true in regard to respect from elders, mainly the older adults and the aged who openly protest the increase in single parenthood. Among the young, single parenthood has become a statement of freedom from parental control, although the young mother usually continues to reside within the family compound. There is no prejudice against the child, as traditionally children would be produced by the single woman as insurance in old age. Unmarried fathers are expected, but not required, to contribute to child support. In reality, any child support becomes part of family support.

Young adulthood slowly blends into full adulthood, without ritual. Throughout these years, be it young or middle age, the process of growing continues. Unlike the physical maturity of childhood, growth in this stage involves learning, maturing, and gaining from interpersonal bonds with the extended family. Middle-aged women are freer to participate in community social affairs as her older children take over household chores. Lessening of household responsibility is not an increase in recreational time, as social interactions, be it informal visiting or community service, remain an integral segment of work. At all times, the person is learning about life. Respect and deference increases as one ages, and is greater for males than females.

<u>Elderhood: The Mogomo</u>

As the years pass, a person gains recognition of being a social elder, or *mogomo*. A social elder is one who is still growing-up but knows much. The woman becomes a *masadi mojolo* and the man a *manna mojolo*. These terms do not parallel our terms of grandparent for social aging. The elder may have had grandchildren but people may have grandchildren when relatively young, and the sixty year old woman may have a teenage child. Strands of grey hair or laugh lines around the eyes are frequently named as a marker of elderhood but a good number reach their seventies or more without dominant greying or massive wrinkles.

Some fifty year old women proudly consider themselves to be *masadi mojolo*. Men tend to wait until they are much older before labeling themselves as *manna mojolo*. Basically, the terms signify that the person has many years behind them, enough years to be proud of living a long time in a village where many continue to die young. Social elderhood has been reached, which reflects the presence of judgmental over physical strength. Many duties have been assigned to younger household members, but elders are not idle. Available time, not age, provides opportunity for leadership; the growth of wisdom contributes to the quality of leadership. The household is under their control. The village considers their decisions concerning social and service club matters.

Romanticism needs be avoided in describing this period of life, as all is not ideal. Conflicts exist between social fact and its application. Elders must actively strive to maintain their position, as deference is no longer automatic. The right to control through seniority in age must be proven with demonstrations of proper knowledge and contributions. No one claims special privileges or taboos associated with elderhood.

<u>Grown-Up: The Godile</u>

Older adults, although recognized as having reached their height of physical stamina, have not reached their height of personhood. In order to grow-up, or reach the apex of life, many years of family, social and economic related activity are required. One must be very old chronologically (*motsofe*) before becoming *godile*, or a grown-up. Only then has full adulthood been reached. Incorporated into the concept of *godile* are wisdom, proper judgment and knowledge of life. Says Monate of her 73 years:

> you cannot look at, or even talk with a person, and tell if they are *godile*.
> I had been *mogomo* for a long time but it wasn't until a few months ago
> I felt grown-up. I have learned a lot about life, how people behave,
> about what is really important to me, my family, and village. I may not
> be able to change things but I have gained the insights and knowledge I
> did not have when I became a *masadi mojolo*. This does not mean I
> have stopped learning or know the book learning of the young.

Being *godile* is a personal feeling, not assigned by others, or even discussed with others. Self-recognition of being *godile* seems to come after an informal conversation or a reflection of the day's events. It is not associated with any special event. Full adulthood is not recognized socially. No change in social rank or household status is expected. Instead it involves a sense of personal security that one has the ability to contribute wisely to provide direction for others and in decision-making. Independent decision-making continues to be shunned. The *godile*, like others, seek other family elders to provide input during important times. The *godile*, like any other aged who are the eldest in the family or who are alone, close their eyes and seek affirmation from ancestors.

Some aged feel they may not reach this stage, either in the near or far future. Dianna, a few years older than Monate, is still growing up. She says:

> I still don't understand this world I live in. Other people do not act like
> they should and I must always figure out what I should do. I think like
> an adult. Wisdom has not come to me yet although I know many things.

Dianna frowns as she says, "I am becoming *tsofetse, not godile*."

<u>The Physiologically Aged: The Tsofetse</u>

Tsofetse represents a loss of strength and energy. It refers to the physiological or functional state of aging and occurs independently of the mental state of growing-up. It does not preclude final maturity, as either may occur first or not at all if death comes first. There are no set conditions for being *tsofetse* as it comes slowly. The young assign *tsofetse* with a waning of energy and almost always assume mental ability is declining also. The old see the phase as the time when activities requiring excessive physical exertion become impossible. To them, mental decline is not associated with the physical change. They may occasionally forget but see themselves as competent thinkers.

In contrast to *godile*, which brings contentment, self-recognition of *tsofetse* carries trepidation. *Tsofetse*, with the overt loss of stamina and strength, is a transitional phase. It signifies the dimming of adulthood and the nearness of the final stage of the life cycle, that of being a child.

<u>Childhood and Infancy: The Ngwana</u>

Everyone, if they live long enough, will return to childhood and infancy (*ngwana*). Maria considers herself a child.

> I have many, many years so it is natural that I need help. There is no shame in needing others to help you. I am a child as my daughter must wash and dress me. Eventually, I will become like an infant, needing people to feed me. I see nothing wrong with being ngwana as it is a natural part of nature. When I will die my breath will go to the skies. That, too, is a natural part of life.

Ngwana is not regarded as abnormal or embarrassing to the old, as it reflects the natural progression with aging. The systematic process of becoming grown-up is reversed, with marked physical decline and some changes in mental acuity. Here one finds ambivalence. There is pride in years but acceptance of being *ngwana* is not the same as liking it. Like a child, assistance is needed for the basics of living. Although a few aged worry about future care with reversion to childlike physical needs, most fear

the ascription of being "a child in thought," as the two must come together. Concerns of
acceptance by others and meaningful incorporation into family and society abound.
Adulthood and *godile* are left behind. Like the child, the *ngwana* is regarded by others
as having minimal mental ability, incapable of having insights and of making decisions.
The *ngwana* is forced to relinquish any remaining status producing roles, such as
supervision of household function and input into community affairs, regardless of mental
abilities. As the child patiently waits for directions, these aged "are those that sit,"
incapable of achieving. Therefore, the aged want to delay assignment of *ngwana* as long
as possible.

What many gerontologists regard as normal mental changes with aging, such as
occasionally forgetting where something was put and difficulty in recall of a wanted
word, was widely admitted, even by some sixty year olds. Such individuals would not be
impaired in thought under the Western clinical interpretation of mental impairment. To
the Tswana, this is the first sign of impending childhood. Sometimes it is the only sign
used to ascribe childhood.

Interestingly, no cases of Alzheimer's Disease were found. Two females had
severe depression, medically documented as resulting from the tragic death of a family
member. Both of these women were oriented to person, time and place. Only one of
the aged had survived a stroke, inducing motor impairment only. Two others, both 101
year old females, required patience and time to obtain full interviews as they tired easily
and became distracted. They remembered my past visits and were eager to talk about
their life.

Younger Tswana say that the old-children talk with their heart. Emotion and
feeling overtake wisdom and knowledge, as they begin to unite with ancestors. This
nearness to ancestors can give them a power called *dikgabe*. *Dikgabe* can be used to
direct worries or illness to others. Naturally, *dikgabe* is feared by the young. Such

power is often thought of as stimulating respect, obedience, and care giving to the very old (Alverson, 1979; Ingstad et al., 1991). My observations strongly suggest this may be true with elderly care providers, although they most often mentioned love and caring as the basis of good care. Younger people, especially those under forty, use *dikgabe* as an excuse to separate themselves from the aged. They fear getting close to or speaking with an aged stranger. As long as they remain strangers, without names, no harm will come.

<u>Ancestors and God: Modimo and Badimo</u>

As the phases of adulthood connect the stages of childhood in the visible world, God (*Modimo*) and ancestors (*Badimo*) serve as the metaphysical connection in the invisible world. They are not separated from the living, as both directly affect life on earth. Therefore, life is not severed from those who came before, or those who will come after.

The traditional meaning of god, as supreme director of the universe with communication through prayer, parallels the introduced Christian god. The name is the same. *Modimo* is the one who controls the sun and rain, birth and death. Birth is a purposeful act and so is death. Witches must have God's blessing for disruptive or fatal acts. Physicians can treat disease but God controls the outcome. When individuals see "the cloud of death" hanging over them, no escape is possible.

Ancestors provide the necessary guidance for behavior and decision-making. Communication may be instigated by the person or ancestor. Messages arrive through dreams and intuitive cognition. Deceased parents and grandparents are especially active in directing earthly behavior and thought. In this manner, ancestors are vital in the continuity of the life-cycle. (See Figure 5.1).

<u>The Indigenous Life-Cycle and the Aged in Bechunaland</u>

Descriptions of colonized Botswana reflected a society based on a strong gerontocracy. Ruling by elders permeated social structure and functioning (Schapera, 1944, 1953, 1955; Campbell, 1971). The claim is that elders were active in the village, participating with others in the subsistence, political and social realm. Respect for one's own elders was a paramount aspect of the moral code within the village even when a breakdown in monetary support by migrant children occurred (Schapera, 1953).

Many traditional societies around the world distinguished between the active and decrepit old, expressing a differential attitude in assigned status and treatment according to the level of physical ability (Glascock and Feinman, 1981). I believe such a division occurred in the traditional Tswana culture. Schapera (1955) separated interpersonal relations with elders from interactions with the old in only one instance. In describing the old, Schapera quoted from the Tswana. "Because they are old and have no more strength they must be supported" (Schapera, 1955:179). I believe this quote reflects the giving of care in the *ngwana* stage of aging. The comments of the present aged strongly suggest this is the case, as respect to the decrepit old was shown mainly through provisioning of physical care. Utilization of knowledge and guidance from the decrepit was never mentioned. Happy Sound described her relationship with her aged mother very clearly.

> My mother became like a child as I had to bring her food, wash her clothes and bathe her body. That was our way of showing respect for the very old. She knew what was happening around her but had no power in the house. No, I didn't seek many of her thoughts for she was *ngwana*. Instead, I gave her the things she requested and she was well cared for, and she was happy.

Happy Sound was not the only one to emphasize the giving of physical care with an apparent withdrawal of inclusion in decision-making and decrease in authority. Discussions of aged parents in the past included comments like: "When my mother could no longer move about the house, she had no say in what happened." "My father had trouble walking so I told him what he could do and where he could not go. He was to do what I said."

I do not question the historical priority of physical assistance to the decrepit old, but I do question the decisiveness of written history regarding increased status with each year of life. Deference based on physical needs is quite different than deference arising from power. It appears that the traditional response when *ngwana* occurred was to delete the assigned position of stature. If it were possible to talk to the aged of 50 years ago, would they have the same complaints about loss of status that the aged experience today? We must keep in mind that information on the degree of life satisfaction for the aged of the past comes from the providers of care, not the receivers.

The comments of the aged indicate that loss of family stature was probably also accompanied with exclusion from family affairs. Being *ngwana*, with the needs of a child, is a contradiction to the previous life phase of being *godile*, where one has maturation in thought. Just as the child was and is excluded from family involvement during decision making, I believe it was traditionally so with the aged *ngwana*. The result was a form of social isolation, a situation known to exist in colonial Botswana with widowhood and unmarried motherhood (Comaroff, 1953:68). Thus, the aged with physical limitations were not in a position of control and status prior to modernization. In other words, there has always been a point when increasing years decreased one's ability to find access to the doors of status and power.

The Indigenous Life-Cycle in Contemporary Life

The traditional meanings and associated overtones of the process and phases of aging would be expected to undergo adjustment or direct change with the on-going transformation of Tswana society. Change can take the path of frequency and magnitude or a new direction in kind with a total change of attitude. The attitudes of the contemporary young, especially the urban, fail to portray an increase in wisdom and knowledge with aging. *Agola* quickly becomes *ngwana* as the young shorten the time frames, or overlook, *mogomo* and *godile*. The loss of stamina and vigor no longer accompanies to the later phases of adulthood. Instead, it marks an almost immediate entrance into childhood. (See Figure 5.2.)

Such feeling reflects more a change in magnitude than a new attitude. All phases of aging, from birth to death, remain events controlled by nature, which, in turn, dictates expected behaviors and social reactions (Alverson, 1978:138). Social change, compounded by absorption into a money market economy, has created a new habitat with increasing social pressures. This has reconstructed the meaning of the conduct during old age to emphasize monetary contributions with devaluation of roles providing abstract values for social functioning. The traditional strengths of knowledge and guidance are of little good in helping the family meet economic needs. Thus direct and indirect (i.e. child care) materialistic contributions become the basis for separating elders from old (Guillette, 1990).

As with the traditional interpretation of aging, when levels of "taking" become greater than the levels of "giving", one advances towards childhood and is treated accordingly. *Ngwana* occurs at a much earlier chronological age than before, as "giving" is increasingly interpreted on an economic level, with devaluation of the roles and contributions providing for community good. Traditional old age assets of land control for food production are also frequently devalued. Those who were perceived in the past as elders and grown-up with signs of decreasing strength and vigor are now considered children. Social fact, built on tradition, dictates that these aged children require attention and guidance for everyday activities. As with young children, they lack the capacities for judgement and decision making. Self-instigated behavior is not expected or desired. Their role is to follow and obey.

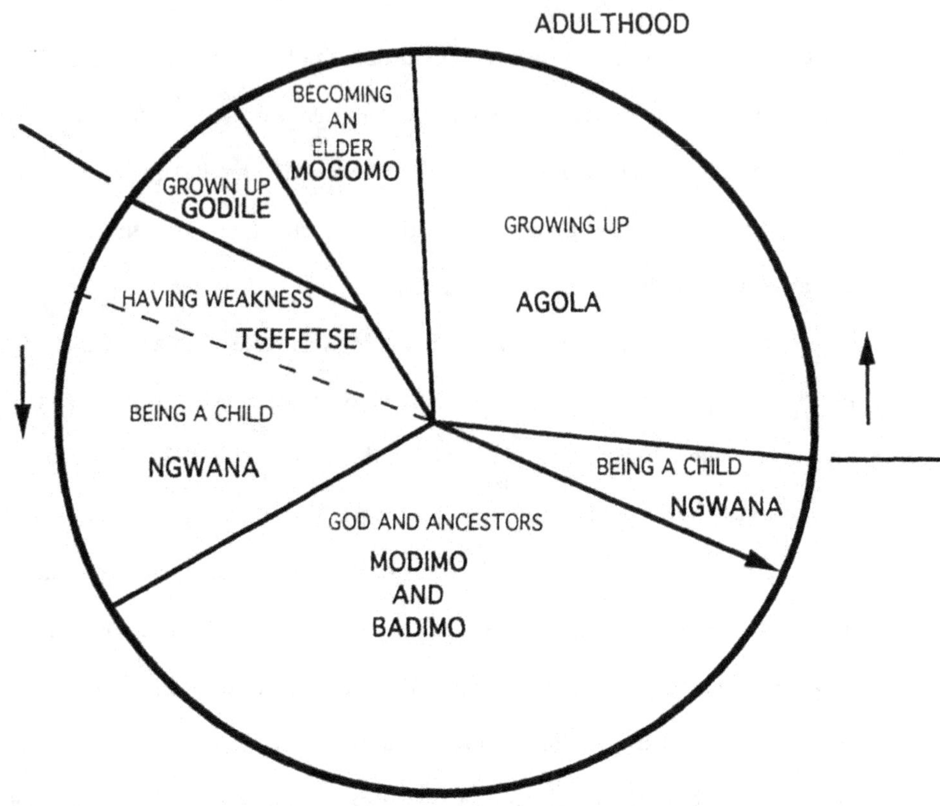

Figure 5.2: Applications of Indigenous Life Cycle to Contempory Life

The result is a psychic distress for the aged that arises from contradictions between self and society in the interpretation of phases of aging. It can only be assumed that some degree of psychic distress with aging always existed, as mentally alert aged were cut off from performing elder roles within the family. The change in definition of the aged is not so much in kind but in direction and degree. With changes in the definition of old, status loss occurs earlier. Disrespect for all aged is becoming verbal and open as more and more old people are considered old-children without status and control over others for obedience to the law.

The aged, who are past their prime in physical fitness, repeatedly expressed their anger in being treated like a child.

> Once the body begins to go, people treat you like the mind goes also. They say I walk with my head under my feet. I know many things and think clearly, but to them I am a child.

> When I was healthy, everyone would visit. Now they say I am a child and no one visits. I am *godile* but nobody cares. No one asks for advice as I am useless. Once the body goes, people act like the brains goes too. I am not a child!

<u>Corresponding Change in Deference and Respect</u>

Public expressions of old age status were of great concern to all aged, as all had at least one incident of disrespect. Most had multiple experiences of being teased and debased. This is one area where the ladder technique was most helpful in investigating social change.

The aged were asked to describe the reactions of others to the old, as the Setswana language has no special word for social status. The middle rung of the ladder was identified as representing the reactions of others to the aged in society prior to Independence. I asked, "How do the people of the village react to the aged today?" Only one of the 31 in-depth participants felt the aged had moved upward. His comment referred to self and family with overtones of disrespect on the social level.

> I know I am respected very much because my family gives me
> food and washes my clothes. I am still in charge of my house. But if I
> went to look for a job now, they would laugh and chase me away. Some
> old are cared for by the government and some by their children, but
> some are alone and no one cares.

Those three aged who retained the status of the aged on the third rung were all

indecisive.

> That could be up or down. Sometimes when I go out, people
> laugh and call me a worn-out woman. Other people say, 'Good morning,
> Mother'. Then some old people are always treated nice and others are
> not, even though I know of no difference between them.

The other 27 aged felt the social position had declined, with 19 of them placing the

aged on the lowest rung. The following quote are from three of these individual.

> Young people liked the old people before Independence.
> Civilization was not like it is today. When we were growing up, we knew
> we must assist the old as we were stronger than they. In exchange, they
> helped us with our problems and guided the household. Today there is
> no respect for the old. I am not treated as an elder. People don't see
> an elder as any different from a child. We are the same. Our children
> shout at us and do what they want, regardless of what we say.

> There is no respect, as in the old days. Now children do not help
> the old. Before children listened to their elders and asked them for
> advice. Today, when children go for a walk, all they think of are the
> paths that will take them to a bar or other young.

> Not many people have any manners that show the way old people
> should be treated. Some shout at parents, others laugh at what their
> parents say. The young don't like old people. They say the old must die,
> even our own children want us to die.

Schapera's (1955) definition of the old as helpless intones a decrease of social

status along with the loss of family power. For the young, the new definition of social

old age includes this decrease of social status. The change is that open disrespect has

replaced customary deference, which traditionally indicated respect.

Much of the psychic distress the aged feel today is actually status anguish. Status

anguish is created from an inconsistency between what the aged perceive as a deserved

elderhood ranking and the public's conveyance of childhood and dislike (Lauer, 1973).

This is particularly true with the active or young-aged who have minor limitations. They exist in two worlds. They see themselves as elders. Others see them as children.

These active aged experience unexpected disrespect as elders and a loss of status before they feel they deserve it. They know that social isolation with exclusion from family functioning is a very real possibility long before it is a necessity. Monate, who is still active in home and community affairs, says:

> when I leave the house I've heard people say, 'old woman, when will you die? You are too old.' I am still taking care of grandchildren and cooking at feasts. I am not useless. I still think well. To others I am the living dead, for they only want me to sit. Even my own children tell me to sit.

The situation creates concern in the aged, not just for themselves but for others who will become old in the years to come. Senatla, who has seen change over his 88 years says:

> the old are way down. In the old days the aged were still strong, in they had control of children. Today there is no control. In the old days the aged had duties of mixing mud for houses, roles in weddings, and they helped with the lands. Today there are no duties, no one seeks my help or advice, or asks about the past.
> Young children should learn about the traditional Tswana culture. I think the culture will be lost with development and the children will make their own culture. If a new culture arises, it will make for many problems When I was about seven years old, an old person's children cared for him, at least provided most things. We always said good morning. Today, the seven year old has little respect for the old and the bigger children do not help parents, but 70 years from now it will be worse.

Variables Underlying the Life-Cycle Phases

Thought regarding the life cycle must also include the concept of the continual process of achieving for family and social unity. Achieving, with goals of self-good leading to social-good, encompasses the Tswana maxim that self-sufficiency is actually for everyone's well-being (Alverson, 1978:137). This process is superimposed on the phases of aging, separate from the direct economic ramifications.

Decreasing strength reverses the process. The process of achieving begins to falter. Production and possession of trade items gradually decreases and possible errors in decision-making may occur. Yet being *tsofetse* does not negate the need to follow social rules, although one is increasingly separated from social function. Conceptually, self-sufficiency progressively decreases as childhood blends into infancy. Decision-making is not expected; knowledge of the rules may be questioned; and behaviors become those of play.

<u>Dependency</u>

Logically one may ask if the decrease of self-sufficiency is not the same as an increase in dependency. Unfortunately, the term "dependency" is used to describe a wide variety of types and degrees of factors, ranging from the individual to international level and reflecting anything from physical dependency and the dependency syndrome, through the economic dependency ratio, to dependency theory.

Hansen (1990) describes dependency as a complex concept with specific characteristics providing several layers of meaning. In its simplest form, dependency implies a relation between any two or more factors. The characteristic of directional flow between the two factors, which may be one or two way, is placed on the relationship to give it additional meaning. The characteristic of need comprises an new layer, with the term dependency reserved for those relationships in which one element or factor needs the other to maintain its state. The content of the relationship may be specified as well, such as substance or physical care. The final layer emerges from characteristics of value. A value judgment is placed on the relationship, its directional flow, need and content (Hansen, 1990).

Americans tend to view old age with a selective focus on the retention of physical characteristics. Such traits are highly valued and susceptible to decline with aging. In the majority of cases, the judgment value placed on the aged is negative when

At the same time, society incorporates the concept of adequacy in self-sufficiency. The physically well, middle income, full-time housewife who insists on others performing all household chores would not be considered self-sufficient, as performance of these duties is incorporated into the social definition of her responsibilities. For the Tswana, the receiving of service is a right of motherhood and has no bearing on self-sufficiency. Instead, socially defined adequacy is based on the ability to trade abstract organizational abilities for control over, not performance of, household activities.

The final step in self-sufficiency is social recognition that the previous steps are being followed in a laudable manner. The most laudable are performing behaviors that meet ideal social standards. Social acceptance is greater for the working father than the father who is supporting his family with unemployment benefits, although both follow the rules of conduct and legal laws. For the Tswana housewife, participation in cooking rituals, such as conversation with the cook, brings greater recognition than non-participation. Cooperation and sharing self with others are lauded social traits in demonstrating self-sufficiency.

When superimposing this concept of self-sufficiency on the Tswana life cycle, it is easy to see that in infancy and early childhood no self-sufficiency is expected in any form. Everything is given. As childhood progresses, elementary resources of energy and agility emerge. The child is told how to use these resources. Socialization for the rules with an awareness of social sanctions occurs. In early adulthood the acquisition of trade items and the opportunity to experiment with decision making occurs. The individual is expected to follow the rules and will feel increased positive feedback in social responses, with errors expected as a part of learning. Progression through the life cycle brings increasing self-sufficiency, with an interactive blend of self, family and social welfare. This process continues through elderhood. Total application, with understanding, comes with becoming completely grown-up.

care and support must be given. The old must be kept as independent of others as possible. If correction cannot be made, more dire consequences arising from increasingly dependent relationships are anticipated. These traits of dependency, so disliked in American society, are valued in Tswana relationships. The serving of food and performance of chores for the aged serve to stabilize the family and society. Tswana behaviors reflect interdependency, which is part of any dependent relationship (Munnichs, 1976).

A multitude of dependent relationships are found with the Tswana. These relationships, with both one and two directional flows, are between individuals, families and society, and exist at all age levels. Such relationships are necessary to provide for needs. Traditionally, the dependency in a required, unidirectional relationship is not judged negatively. Neediness is not associated with negativeness.. Providing food and economic support to the aged is socially valued. It appears that outside philosophical thought, with emphasis on the individual rather than group, has contributed to the emerging negativism with old age. Materialism, with monetization and modernization resulting from incorporation into a cash market economy, directs the judgments. Young people talk about the economic and physiological dependency of the Tswana aged without equal judgment placed on other younger members of society. Numerous households contain young adults that are supported along with the aged by remittances yet it is only the aged who are identified as dependent. Working age adults are excused from negative connotations. A similar example is found with Tswana families with disabled young adults, with negative judgment placed on the handicapped (Ingstad, 1989).

<u>Application to the Village Aged</u>

Self-directed, independent action, corresponding to the Western definition of autonomous individual action, is becoming more pronounced in village life. At the same

Self-Sufficiency

Self-sufficiency, in its most elementary meaning, is the ability to supply the basic necessities for life by oneself. The hermit, providing entirely for and by oneself, is self-sufficient by definition. Usually people are not hermits, nor do they want to be. In a social situation self-sufficiency includes the production and possession of items for trade in social exchange. According to social exchange theory, trade items are not limited to money, but include concrete in-kind goods, abstract resources of knowledge and guidance and physical contributions. Trade, therefore, becomes an important step in social self-sufficiency.

Successful trade requires a second step: that of applying reasoning and decision making skills. Trade items without the ability to instigate trade and to trade for one's own advantage negates the first step. The most extreme example of failure at this stage would be a mentally retarded person with money in a trust. The funds are adequate but the skill to use them for self-good are lacking. Competency for beneficial trade is learned within the setting where it is to be used.

Skills for rewarding trade are applied to promote self-sufficiency according to the individual's perception of self-good. The miser may live in chronic malnutrition, not because of lack of trade items, but because he wants to guarantee trade items if and when hard times ensue. The spender, on the other hand, may envision self-good in the present as more valuable than attempts to predict the questionable future.

What about the thief who steals enough to have it all and can trade freely? Is this self-sufficiency? No, as the social rules for self-sufficiency are being broken. Legitimacy in the possession and use of trade items is the third step. Different societies may have varying rules of ethics. For instance, the overcharging for jewelry during trade can be a legitimate action in the Trobriand Islands but would bring social sanctions against the American small town individual.

time, such action is not included in the Tswana concept of self-sufficiency. Perhaps this is best illustrated by the daily lives of two individuals, both similar in age and living alone. They resemble one another in that both consider themselves *tsofetse* but not *godile*, and both still strive to achieve.

<u>Diane</u>. At 78, Diane is slender, small in body, and with bright, expressive eyes. During her late thirties she recognized that she would never marry and migrated to South Africa for employment. Preparing for the future, she had two daughters. Both died from infections during their teenage years.

A few years ago, Diane returned home. She was becoming old. Most of her family are deceased, although a brother and several cousins live at varying distances from her house. She seldom interacts with them as "they have a different way of looking at life." The distant aunt next door is her closest confident. Most of her decisions are made without consultation although she sometimes relies on ancestors to provide some direction.

Diane generates her own income by making and selling *mokhankgalasi*, an alcoholic beverage. Each evening she employs a neighboring teenager, rather than asking her niece, to fetch the ingredients. Before retiring for the night she mixes yeast, brown sugar and water in several 20 gallon plastic pails. With the rising sun, Diane paces her chores of laundry, cooking, and errands according to the degree of fermentation that occurred during the night. When the beer is ready, she opens her front door. It is a rare day when the *shebeen* (bar) is closed. Even with funerals, she plans attendance according to brewing and selling times.

<u>Sego</u>. Not far away lives Sego, whom you have already met. His peppercorn hair and youthful actions belie his age of 78. He also worked in South Africa. He too lives alone as he lost his wife and two daughters during a 1950 measles epidemic. His

educated adult son lives in Gaborone. Monthly remittances, Sego's only source of income, provide for necessary purchases.

Sego is actively seeking a new wife. He firmly believes a man should be served and tended to by a young woman with him as the master. Until he finds the right lady, Sego relies on the woman who is living in one of the three rooms of his house. She cooks and cleans in exchange for room and board. When she is gone, Sego goes without food or seeks it elsewhere even though he can cook.

Sego talks about wanting the security and assistance with decision-making that come with marriage. To compensate he frequently visits various relatives in the village, seeking guidance, help, and companionship. During a wedding, he feels bound to proper visitation and required supervision for the slaughter of cattle. For the ten days of mourning prior to a burial, he pays proper respect to the dead by arriving at sun-up and staying until nightfall. During this time he often denies his own needs or desires.

By our standards, Diane is the success story representing Western independence and autonomy. To her neighbors, Diane is worthless and useless. She is seen as producing solely for her own good and not the good of others. Family has little meaning to her, as she neither takes nor receives. Above all, she allows nobody to serve her. Some say her behavior indicates she is a witch. Others fear she will use *dikgabe* against them. They purposely avoid her. Even her aunt tries to avoid her.

Diane publicly fails to demonstrate the ideology behind the process of achieving, even though she is economically productive and knowledgeable in decision-making. Her income is used only for her welfare, which is not seen as adequacy or regarded as laudable. Her social interactions are viewed as self-centered, contributing little to the good of society. Although she relies on others to purchase beer, there is no harmony between herself and the social environment. Such actions do not represent self-sufficiency within the rules of society or cultural confines. Society's judgment of her interpersonal behavior involved with income production, and the use of this income, is negative. Social opinion, combination with her age, make her as a child, best to be avoided if possible.

Sego portrays the Western picture of dependency and lack of autonomy. His urban son also pictures him in this manner. In the village, Sego is one of the few highly esteemed, social elders. What we perceive as dependency is actually the dynamic social process of interdependency. His actions involve others, both in the extended family and in the community. Not only does he gain from relying on others but others gain through serving him. His remittances, similar in size to Diane's income, are used to help others. Decisions are seen as wise, as they are made with group input. The high priority given to his extended family with much give and take is admired, and no one counts the lopsided directional flow of assistance with his physical decline. Although similar in age to Diane, he is not feared.

Autonomy extends beyond doing what one desires to include the maintenance of the integrity of the self (Munnichs, 1976:5). Society defines the values that should be maintained. Sego is autonomous, expressing integrity in his actions that are in harmony with the social context. Because of this, Sego is seen as the true elder, set apart from most others. He admits that he has to purposely continue in the process of achieving in order to maintain his status.

The term *motsofe*, once a positive term for old age, has now become pejorative and carries degrading connotations. The aged see living to sixty and beyond as a reason to be proud. Yet, in the public eye all aged are nearing or in childhood. It is better to avoid them than risk danger, now that no one knows who can be a witch. The resulting withdrawal of opportunities to engage in meaningful behaviors, and the avoidance of the aged from fear, places the aged in a precarious position. The active and the physically limited yet mentally alert, although viewed as useless, see themselves with much to give as good people. This contrast of perceptions influences the availability of doors, with age-children having the most difficulty in gaining access, regardless of its condition. If access to doors can be made, the door of acceptance, which has always had cracks and damaged areas, is now more severely broken. The following chapter will examine the assets the aged control to use in their attempts to deal with the existing doors.

CHAPTER 6
THE INDIVIDUAL: ASSETS FOR USE IN DAILY LIFE

> I am not quite as strong as before, but I am well and can do everything
> for myself. I still have my house, land, and one cow, which I am saving
> for my funeral. She is pregnant so soon I will have two. The important
> thing is that when people ask, I can give. (72 year old man)

> I am at home. Most of my relatives are dead. Before I had some money
> and possessions. Now I have very little. My cattle were lost with the
> droughts. I try to make more money but there is no way. I only sweep
> and one is not proud of that. I see myself as a worn-out woman. (88
> year old woman)

Most people do not go through life thinking, "I have this to get me that."

People do not generally think of the interrelationships between assets or the effect of

social change on what was once valuable or invaluable. This chapter begins with two

scenarios, briefly explaining the relationship of the past to the present while presenting

some of the assets the Tswana hold dear for making the good life. I then group assets

into the three resource dimensions and present data on the assets the aged do have and

do not have. I have tried to take an emic view, interpreting the meaning of health,

family and valued possessions as the Tswana do. At the same time, I try to relate

propriety to the setting in which they live.

Priscilla and Alfred

Priscilla and Alfred have been husband and wife almost 50 years. They were

children in "the days before civilization" and were married while the village was

"becoming civilized." Following the times, he worked in the mines while she raised the

family of five children. Occasionally, when there was no need to be at the lands,

Priscilla would work as a maid. Together, they had earned enough to build and furnish

a modest three-room house.

A high-school educated grown daughter shares the house with them. She is waiting to hear about a job, either at the hospital or the mill at Ramotswa Station, where she put in applications over a year ago. Meanwhile she does occasional laundry in the village. She feels she must continue to live at home because of her parents. If the situation were different, she would like to move to a city and work in a big store. The remittances sent by her brothers, although not large or regular, are sufficient for household support.

It is usually difficult to find Priscilla at home. She leaves early in the morning, sometimes at sunrise, when her energy levels are the highest. Where she goes depends on what is happening in the village. There is usually the necessity of cooking for a feast or visiting home-bound aunts or an ill nephew. Most of her activities center around her large network of kin.

> I don't have friends and never did. I know lots of people and am friendly, but I refuse to get close to them. That way I can avoid danger and trouble. My family is large enough to cause its own trouble. The one thing I miss most in my family is grandchildren. They live so far away I see them only when they come at Christmas. I do not have the health to spend 24 hours on a bus trip to visit them.

> The worst thing I ever did in my life was to let my daughter take over the household. Now I have no say in anything that happens. My daughter controls the money, the house and when we eat. Alfred is lucky, as a man he is Master. As my master, I have honor in serving him. I am proud I can still wash his clothes and make his bed.

Alfred laughs and breaks into the conversation. "I cannot afford to have my wife die before I do. She takes care of me as I am helpless in the house as I do not know how to do household things." Alfred rises and puts half a loaf of stale bread and a few kernels of maize into a burlap bag. With a proud grin, he slings the bag over his shoulder. He says:

> I have three hectares of land, past Ramotswa Station, down where the rivers meet. The walk is short, less than two hours. Priscilla is too old to

plow so I use the land for my one goat, two dogs, three cats and four chickens.

With a wink in his eye, he continues. "Oh, Priscilla, I will be late coming home today. I have some *thebe* {coins} so will stop and buy some *bajalwa* {traditional beer} on the way home." After Alfred's departure, Priscilla frowns and says:

> ever since we lost our two cows in the recent drought, he likes to drink. He buy beer with money from the chicken eggs. It's awful, the way he drinks. It's not every day but sometimes he takes too much. When he comes home, he accuses me of romancing with other men. Now, who would want a worn out woman like me? I hate beer so much I will not brew *bajalwa*, and that is the only way I can see to make my own money. People laugh because Alfred uses the field as a cattle post but it gives him something to do. In some ways I am lucky too. I have lots of clothes from the better days when Master was working, a daughter at home, and a nice house.

Maria and Martha

This morning I find Maria sitting on the ledge of her *rondaval*. Martha, her 63 year old daughter, has given her a basin of water for bathing. There is no embarrassment or covering of the body. Maria starts the feminine chit-chat.

> Do you remember I when told you about the varoom of cars? Those were the good days, when I could walk and do things. Now my bones ache. I only sit, too old to even pull grass from the yard. Even Martha is becoming old. Our husbands died long ago. There was no one to plow, so my land went idle. I think someone else uses it now.

Martha supplements the statement with:

> her nephew's family plows her land and we get some grain from them. Food is a heavy problem. It is so expensive, especially milk. Without milk, we do not have breakfast as black tea makes Mma sick. Sometimes my brothers and sisters send us money and sometimes I help a neighbor make *bajalwa*, but money is always dear.

Martha walks into the house to get her mother's clothes from a cardboard box. Each have a bed and several blankets. Otherwise the house is bare except for a box of laundry soap. While she is gone Maria tells me about the comfort and care she gets from her daughter.

> I can talk to her, especially about my sadness from everyone dying. Three people died in my family last year, including the one

daughter that Martha could talk to. Now there is no one to understand her. Even though people visit me, I get lonesome thinking about all that has happened. I worry about illness most of all. Martha has high blood {hypertension}. The hospital gives her pills but only the *Ngaka ya Setswana* {traditional healer} can remove the curse that caused it. His fees are too great so we must suffer.

Martha returns with clothes and a goat skin. Instead of dressing, Maria reaches out and grabs my breast. As she holds onto me, Martha says, "My mother is telling you that you are now seen part of our family. That is the way Tswana mothers show pride and affection for daughters." After helping her mother dress, Martha assists her to the front of the yard to sit on the skin. Martha shakes her head.

> Today is like every day. This afternoon I must visit other people and attend a meeting at the community center. When I leave there is no one to be with Mother. She is my responsibility so no one ever offers to help. Sometimes I must be gone a long time. She gets very upset when I leave but I must go out. I give Mother good care when I am home so I don't know why she gets angry when I leave. When I go I must walk. That takes time as I am also old. The bus would make it faster but who has money to spend that way? We don't. You can ask only for so much when you cannot give. Our younger kin who live in those two fancy houses across the foot path have money and lots of furniture. They tore down their deceased parents' house and built new ones.

Each of the above individuals stress the importance of physical well being, family and economics in their thoughts about themselves. Usually, the import of one area is linked to the import of another. As time affects the strength of one area, side effects are felt in another. For Martha, economics prevents the desired treatment of hypertension. Priscilla feels contact with her grandchildren is limited by her health. The proposed personal/physiological, social/familial, and fiduciary resources apply to the Tswana aged. The following section takes a close look at the specifics, which the aged see as important in seeking the good life.

The Gerontic Fund

Resources are generally considered to be assets that people use to gain access to desired goods and ends. The term "resources" has varying implications depending on the stance of the theorist. Gerontologists and anthropologists can vary in what is to be considered an asset. Care-providers and social agencies that are not under the direct control of the aged are frequently included. In recognition of the fact that resources can have such varied meanings and implications, ranging from owned assets through indirectly controlled assets to services received, I prefer to delineate self-controlled assets as a distinct unit known as the "Gerontic fund." It is the gerontic fund that the aged use for social exchange.

I believe that throughout life there is an accumulation and saving for the future, such as preserving health, having children, and assembling material goods. Some may have amassed more gerontic-wealth than others when entry into old age occurs. This accumulation of wealth, which includes the bodily self, is the gerontic fund. It is viewed as instrumental in promoting personal well-being. It serves as a source of material and non-material supply, to be used as an expedient in achieving particular ends in given circumstances. The arena for use is old age. Reflections on fund size and use can only occur at an abstract point in time, as becoming old is not automatic with turning sixty or sixty-five. The size of the gerontic fund does not remain constant.

The gerontic fund is distinct from personal wealth, as it is composed of bundles of assets based on the personal/physiological, social/familial and fiduciary dimensions, as proposed by Gubrium (1973). I believe specific contents are determined by the contextual setting, including the cultural, social, political and economic environment. Items that may be important in one setting may have little importance another. A simple example in the personal/physiological dimension is the ability to use electrical appliances. This has no meaning in communities lacking electric power.

Actual contents at a given time rest with the integrative action of past life style and life history, present context of life and the natural process of aging. Strength of the fund is based on the number of assets within content areas, and the way contents unite and interact with each other. Education, type of employment, and other life experiences contribute to the fund's strength but are not unique components in themselves. I assume necessary components are the same for both males and females, although they may use them differently during exchange.

Each aged person controls a gerontic fund, although contents vary in amount. Self and others place a value on the fund, and the congruency of values influences its purchasing strength. In turn, the contents and their value to others influence the degree of flexibility in deciding on a course of action for negotiation and expediency for spending, saving and budgeting. A large gerontic fund, in itself, can generate power without expenditure (Dowd, 1980:53). This power is not available to the fund-poor (Sen, 1981). Thus, a person is influenced by the knowledge of their own gerontic fund and how it may be used for interactional activity to promote well-being. The actual number of exchanges is not as important as the amount of resources required to meet physical and psychological needs. The division into dimensions and the necessity of asset ownership during the exchange process allow me to later specify where assistance for the aged is needed. It also allows for investigation into the impact of time on reaching goals through fund use.

The gerontic fund can be altered in shape and size with time. It may be invested to further increase its strength, such as improving a room in an owned house for rental income. It may be spent, as with the transfer of land control in exchange for part of the crops. At other times, parts of the fund are simply lost within the environment, as with movement of a family member or livestock death during drought. The ease of rejuvenation and the potential of permanent loss are considered in its use. As the Tswana so aptly commented in terms of physical actions, family communications, and monetary spending, "I can do it but I cannot afford to do it."

Meaningful research regarding the application of the gerontic fund concept requires that its contents, and the determination of its strength, be specific to the group under study. For the Tswana, the importance of functional interdependency within the extended family and the value placed on social interchange during the continuous process of achieving are vital parts of the three dimensions. The instrument developed to measure the fund, Appendix A, is situationally dependent for the group under study. It would need to be modified for use with other groups, especially non-African.

In this research, the tested gerontic fund consisted of twenty separate items, reflecting each of the three dimensions. Each item had a score range of 1 to 5, with 1 indicating no held assets and 5 representing a high degree of possession. Each dimension was scored by adding item scores within it. As unequal numbers of items were in each dimension, the individual's sub-totals were then weighted by dividing this score by the number of answered questions. This provided a statistical equality between dimensions. The individual's total fund was determined by dividing the total of the three weighted sub-totals by three, resulting in a number between 1 to 15, with 15 being the maximum possible score.

The rest of this chapter explains what specifics constitute a gerontic fund for the aged Tswana. Incorporated with this are comments that exemplify the relationship between desired fund and lifestyle. The findings reflect the size of the fund at the time of research. No attempt is made to document changes in the individual's assets with time. Variations in holdings, within and between age groups, are common. No person is the same, with some controlling large total funds and others striving to maintain the smallest, while others are rich in one dimension and poor in another. Diversity is so great that I avoid presenting a picture of the average old person. Instead, I emphasize the meaning of asset ownership, the modes, and the extremes.

<u>The Personal/Physiological Dimension</u>

The Personal-physiological dimension of the gerontic fund involves the levels of physical and mental health that merge to allow for the use of self as a participating agent in exchange. This area is evaluated in terms of abilities to perform daily activities that, in turn, allow for culturally laudable interaction with family and society. This differs from the usual Western style of functional assessment of the elderly, which focuses on impairments in relation to approaches for needed services to retain autonomy and independence. It is the "doing" and "being" in the achieving process, rather than remaining "autonomous" and "independent," that give social value to this category. No attempt is made to evaluate personality.

General descriptions of the aged frequently include the demarcation of the degree of decrepitude. These terms, such as active, frail and decrepit, can become pseudo labels, borrowed from the outside world, unless one knows how the terms are negotiated within the culture where they are applied. In Ramotswa, the person receiving large amounts of service from others is not necessarily frail. They are normal-old, receiving service as a right. Decrepitude, which implies a need for 24-hour supervision and care, is seen by the Tswana as a term applied only to the dying. Aged requiring major assistance are not seen as needing the continuous presence of others. Not infrequently, they are the only household member at home as these people are always seen as being capable of doing at least one or more things, such as feeding themselves or conversing.

The large discrepancy between my outsider's interpretation and the cultural interpretation resulted in much error when I attempted to place individuals in these categories. When the terms "frail" and "decrepit" are used in these writings, it is to indicate limited and extremely limited physical status, according to Western standards. For the research, I used the three generalized categories only to select representative individuals with varying health levels for in-depth studies .

A more valid measurement of the personal/physiological assets was grounded in terms of abilities to perform activities of daily living (ADLs). Activity is evaluated according to what a person can do for oneself. It involves self-maintenance activities performed daily, with the help of supportive devices if necessary, in order to retain personhood and to function as an individual. They range from basic activities of feeding oneself to those involving complex reasoning and movement.

I could not transfer the standard evaluation instruments for ADLs to the African setting. The usual supportive aids for activity maintenance were absent in the village, meaning a person could not be judged in relation to the use of crutches, dentures or such. In fact, many of the every day appliances, like a kitchen stove, were absent in homes, altering the meaning of meal preparation and house work. In addition, many of the factors that constitute assets with Euramerican did not always apply to the Botswana setting: the ability to make phone calls, the ability to drive. Therefore, the approach used in determining physical abilities, while seemingly simplistic, incorporates Tswana contextual and social reality. Other alterations in the standard ADL format may appear to place unreal expectation on the aged, such as the ability to walk long distances or to carry heavy weights. But these are the facts of life, necessary to perform in order to function as a member of the family and society.

While most researchers divide ADLs into low levels of functioning (toileting, grooming) and high levels of functioning (shopping, driving a car), I have separated them into self-care and social functioning abilities, unifying the process of achieving with cultural norms of interdependence. The self-care ADLs that involve bodily needs are similar in content and style to those used in developed nations. They measure the ability of the individual to physically care for oneself. Social ADLs are those functional abilities that allow the individual to be a contributing person within the home and community. While all these abilities do have significance regarding body integrity, the Tswana culture interprets the socializing abilities as having a unique functioning value.

With both socializing and self-care activities, the mitigating circumstances affecting the importance of abilities and their performance are highlighted.

<u>Self-care abilities</u>

Self-care, or tending to one's own bodily needs, was very important to the individual. This category contained items on cooking/eating, toileting, bathing/grooming and obtaining water, as outlined in Table 6.1. The majority (76%) could prepare and/or serve foods (doing so only if absolutely necessary) and 99% fed themselves. The 23% who could not serve themselves cooked foods relied completely on other. No special purees were needed for the numerous aged with dental problems and tooth loss, as the main food stay was a porridge of boiled sorghum, sometimes with well-boiled vegetables and occasional boiled meat.

The high roughage diet probably contributes to a lack of complaints about bowel elimination. Toileting is no special problem, other than some have occasional leakage of urine with laughing or coughing (31%). Only four, all with motor-neurological damage from accident or disease, are incontinent. The need to "rush" (31%) is controlled, in part, by external factors.

Latrines, while not part of the actual gerontic fund, are present in over 90% of the compounds. They are not always the preferred place for urination. It was perfectly acceptable for anyone to use the nearest bush, squat at the edge of the compound, or along a road. Latrines just make sitting easier. Hence, they are preferred for defecation. Construction of the building involves much hard work, so people tend to be careful with use. It does not matter if the building has a door or not; it is having a roof that is important! The lack of a latrine, either because it is full or never built, usually brings forth expressions of anger. Anger tends to be directed either at a son for not digging one, or at a neighbor for refusal to open their's to others. This latter reason caused an "Outhouse War" to erupt in one neighborhood.

TABLE 6.1 PERSONAL/PHYSIOLOGICAL SELF-CARE ASSETS. (N=105)

A. <u>Eating</u>: (Mean = 3.71)
    1. Must have food served and be fed. (1.0%)
    2. Must be served all foods, feeds self. (22%)
    3. Must have meals prepared, serves and feeds self. (24%)
    4. If food is present, can prepare and eat own meal. (11%
    5. Can shop, prepare and serve food. (42%)

B. <u>Toileting</u>: (Mean = 4.2)
    1. Complete incontinence. (5%)
    2. Urinary incontinence, bowel control. (4%)
    3. Must be taken to the toilet or given pan. (5%)
    4. Uses toilet by self, occasional loss of urine. (32%)
    5. Uses toilet by self, no soiling. (55%)

C. <u>Grooming</u>: (Mean = 4.0)
    1. Totally bathed and dressed by others. (7%)
    2. Can wash part of self: main bath and dressed by others. (2%)
    3. Gives self main bath, requires dressing by others. (16%)
    4. Keeps clean with minimal assistance with clothing. (30%)
    5. Keeps self clean and dresses by self. (45%)

D. <u>Carrying water</u>: (Mean = 2.9)
    1. Unable to transport cup of water any distance. (21%)
    2. Carries water by the cupful. (18%)
    3. Carries water by the liter. (21%)
    4. Carries water in 2 liter bucket. (29%)
    5. Carries water in large bucket. (11%)

Mary, a *shebeen* keeper, had worked hard to have a new latrine installed for herself and customers. She felt future replacement would be impossible, in recognition of her increasing limitations. The next-door family of six, including a physically-limited old woman, assumed the latrine was available for their use. One day, when the old woman walked towards the building, Mary chased her away with screams. The following day, Mary went after the old woman with fire in her eyes, additional anger in her voice, and a stick in her hands. The frail woman complained aloud bitterly, "There is no place for an old woman to go." Some area residents sided with Mary, saying she had a right to refuse outhouse use to the general public, as it would soon become full. The other group thought it should be open to all, for most people bought beer there at one time or another. Arguments came to a quick end a week later, when all concluded Mary was under the spell of witchcraft. The wild yelling and stick swinging, neither of which was acceptable behavior according to the law, was an expression of the witch's spell over which Mary had no control. Fear overrode desire, and the neighboring family went back to using an open area.

One might think that with the elementary living conditions, and ambient dust and dirt, that cleanliness and grooming have little import. The opposite is true, with strict adherence to high standards of personal cleanliness. This means clean clothes and bodies. Poverty may limit the wardrobe, but that doe not mean dirty clothes are acceptable. A general all-purpose granulated soap is used for everything. A limited soap supply causes as much worry as limited food.

Grooming is always done early in the morning, including the brushing of teeth (no toothpaste as a rule) and the combing of hair after a complete bath in a basin of water. As age progresses, it becomes increasingly necessary for others to bring the water, but very few (7%) rely on others to completely wash and dress them. No signs of skin breakdown or decubiti (bed-sores) are present in the bed-bound.

Several years ago, when the village increased the number of public water taps, households had the option of having a metered tap placed in the compound at subsidized installation prices. About 60% of the compounds have a private water tap, which is comparable to national census findings for the village (Republic of Botswana, 1982). Another 10% have a public tap within a 100 meters. The remaining households have to transport water from distances up to 500 meters. By custom, children should carry water for home use, but situations arise in which the aged must carry water for their own use. This ability implies a high level of overall strength and coordination. For this reason the ability to carry water is included in the self-care category.

Almost everyone, old and young alike, has to stop and rest when carrying a full, standard large bucket. The sixty to eighty year olds, as a rule, can carry a full 20 liter pail with minimal difficulty. Women usually carry this pail on their head. Men use their hands. As age-related weakness progresses, new ways of obtaining water are adopted. Either it is carried in smaller containers by hand or used directly at the tap. Some of the more affluent have rigged up running water to their house by connecting a hose to their private tap. Only those unable to walk the required distances were unable to any get water on their own. One should remember that the ability to carry water, as with cooking, does not always mean availability. Provisioning of food and water are considered rightful entitlements of age. Over one half refuse to perform these activities, even if it means doing without.

<u>Social functioning abilities</u>

Generally, the aged feel no embarrassment, or increased sense of dependency on others, if any self-care abilities are compromised. The loss of abilities necessary to promote interaction with others are those most strongly lamented. Such abilities as walking, hearing, seeing, doing (performing valued social acts), and knowing (remembering and contextual orientation) were identified by the aged as composing the core of personhood. These abilities were of significance, as the aged considered a degree of competency as a marker separating elders from old-children. Therefore, I separated them from the other physical abilities. (See Table 6.2.)

Walking is the major form of transport, a necessity in order to be included in household and community events and the reciprocity of household visitation. Individuals take great pride in the distances they can walk, with the mode for the 60 to 79 year olds as being able to walk over ten kilometers. They do so to get to the lands or to visit others living outside the village. Others in this age group limit their walking to the village, traveling up to five kilometers, often two or more times a day. If no severe physical problems (arthritis, limiting heart disease) are present, such activity continues until the nineties. The very old may stop along the way to rest or carry a "third leg" (stick) to help prevent falling on the rocky roads. The most common reason for being limited to the immediate yard and surrounding houses (19%) is severe degenerative arthritis. Those who cannot walk at all (6%) are pulled about the compound yard on an old blanket or burlap bag. Occasionally they are rolled into a wheelbarrow for transport to neighbors.

Hearing is a requisite for inclusion in the grapevine system of information sharing. The Euramerican forms of gaining knowledge (television, books, newspapers) have no place in the taxonomy of information exchange in the village. A normal or slightly raised voice could be understood by 66% of all aged. Excellent hearing, always found with the 60 to 69 year olds, remains the mode until age 90. Most of these very old can understand a raised voice without misinterpretation of words. Only three of the 105 interviews are limited because of deafness in the respondent. (I cannot help but wonder if deafness will increase, with the popularity of very loud radio music and the entrance of modern noise-making machinery into the village.)

TABLE 6.2 PERSONAL/PHYSIOLOGICAL SOCIAL FUNCTIONING ASSETS
(N=105).

A. <u>Participation in Family Activities</u>: (Mean = 3.1)
    1. Does not attend or participate in any area. (19%)
    2. Attends events, no participation in events. (18%)
    3. Able to do minor tasks. (17%)
    4. Able to participate except for heavy tasks. (25%)
    5. Full participation. (31%)

B. <u>Walking</u>: (Mean = 3.8)
    1. Can not move about or crawls. (3%)
    2. Moves about house/yard with assistance. (19%)
    3. Can walk in immediate neighborhood, less than one kilometer. (17%)
    4. Can walk around village, up to 4 Kilometers. (25%)
    5. Can walk any distance. (31%)

C. <u>Hearing</u>: (Mean = 4.2)
    1. Totally deaf or unable to recognize sounds. (3%)
    2. Understands loud voice with difficulty. (11%)
    3. Can understand raised voice without difficultly. (6%)
    4. Understands normal voice with occasional misinterpretation. (14%)
    5. Hears well. (66%)

D. <u>Vision</u>: (Mean = 3.7)
    1. Totally blind. (9%)
    2. Minimal vision, can see color or shapes. (8%)
    3. Can see to perform basic functions. (19%)
    4. Can either recognize people walking past the house or fine handwork. (26%)
    5. Can do both, recognize people and do fine hand work (38%)

E. <u>Memory</u>: (Mean = 4.2)
    1. Does not remember any of the past or present. (1%)
    2. Very forgetful, remembers only main events of past. (1%)
    3. Tends to forget present events, aware of past. (10%)
    4. Minimal difficulty with recent recall. (forgets where put item) (59%)
    5. Remembers well, knows where things are placed. (29%)

Visual problems are much more common than hearing problems. Cataracts and past eye damage from flying dust are common. Corrective lenses, which have to be purchased, are owned by very few, and are not updated with changes in vision. Like hearing, the measurement of acuity does not involve technological tests. Judgements are based on what the person said, which closely corresponded to observations on my part during the interview.

The Tswana view vision as "the core of being," the major ingredient of personhood. "The eyes make you a person. Without vision, one is no longer a real person. He does not see others and others do not see him." Poor vision, although increasing in intensity with age, was not thought of as a direct function of being old. It was interpreted as due to other long-standing, uncontrollable circumstances. Women attributed failing eyesight to a weakening in the uterus following menopause. "The womb controls all the ligaments of the body. My womb is now old and dropping out of place. It has pulled the ligaments to my eyes down so that I can no longer see." Men mentioned past abdominal surgery or injury to limbs as the cause. Blindness, in turn, created a "failure within the being" that tended to abort the process of achieving and hence, personhood.

Most 60 to 69 year olds (90%) have minimal problems; are able to recognize passing people on the lanes and identify pictures held at normal range. Seventy to 79 year olds retain the mode of good vision, although increasing numbers have difficulty in the sense that vision "wasn't as good as before." They may need help threading a needle, or have difficulty in seeing at night. Half of the 80 year olds do not visually recognize me before entering the yard, and also have severe problems with sewing or other fine handwork. All of those 90 and above are either totally blind or can identify only the basic shape of large objects. The visually impaired person has learned, in many ways, to compensate for the loss. The near blind still cook and take care of their own personal needs. The totally blind assist with household chores and walk about town with grandchildren leading the way, in order to participate in acts of "doing."

Acts of "doing" represent the expression of "being." This includes both doing with, and for, others. With doing, there is no concept of personal time or private space. One does not seek to escape or be alone, but wants to be with others. Thus, the ability to participate in group task activity, be it at home or in the village, is very important. Funerals, weddings, and neighborhood parties are not to attend as a guest to be entertained but as an opportunity to participate in the processes of life's achieving. The ideal is completeness in "doing." If the ideal cannot be met, one tries to do as much as one can. The old man sitting quietly at a funeral wake is "doing," maybe not as much physically as the those slaughtering the cow, but his presence has importance, and his giving of himself is recognized.

This aspect of doing is measured by the degree of physical contribution a person can make at a group event. Most of those aged under 80 years are able to participate fully except for maybe the very heavy tasks such as lifting large iron cooking pots or moving slaughtered animals. These tasks are shared with young adults. The average 80 to 89 year old is more limited in that they contribute through washing dishes, stirring small pots, greeting visitors. The oldest are restricted to "doing" at home, striving to help with chores or maybe just talking during discussions.

Participation in conversation requires remembering and contextual orientation. The Tswana encompass this ability under the term of "knowing." Everyone, with the exception of the two with diagnosed mental illness, is aware of time, space and people. This includes knowing where absent household members are, and their general activity at the present time, such as school, shopping, or work. Absence of recall difficulty is reported by 30%, usually among the younger aged. The major complaint regarding memory for all aged (59%) parallels the western bane of aging: that of occasional forgetfulness. This complaint occurred most frequently in the 70 and above age-groups. The stories are familiar: that of not being able to find a certain paper, or having money or keys in a pocket and not realizing it. This complaint increases in frequency with older ages, but not necessarily in severity. Individuals readily recall my previous visits and topics of conversation. The average score for those age 90 and above is like that of others; occasional forgetting. All have accurate memories of the past. Their historical stories are quickly confirmed by family members or other aged.

Overall, the personal/psychological bundle of the gerontic fund shows many assets for many people. For all ages, the weighted total mean was 3.8 out of a possible 5 (SD = .9, N = 105). This indicates that the typical aged could care for themselves and participate in life with minimal assistance. Like aged everywhere, increased years do take a toll, with a decrease in strength and stamina. There is a significant relationship between personal/psychological assets and age (p = .0001). The 60-69 year old age have weighted average scores of 4.2. This weighted average progressively drops in each ten-year age group, although it was not until reaching 90 years or above do scores routinely fall below 2.

Men fare no better or worse than women. For both sexes, if an individual's scores are weak in the self-care category, they tend to be weak in social functioning, reflecting a generalized decrease in body integrity. As a rule, it appears that major, all-encompassing, physical limitations do not appear until quite late in life. There are exceptions to this rule, especially for those with chronic debilitating illness.

The Social/Familial Dimension

Social/familial assets are based on the availability of interactional partners and social structures that can be used for the creation of service and support. As explained earlier, with meaningful others placed within the fund, they can be regarded as assets to be utilized to obtain goals through use, investment or trade. The value of this approach lies in the scientific ability to analyze the good life in terms of the social/familial assets, which provides a basis for success or failure in regards to fund applications. The death of a person is regarded as a lost, non-renewable asset while the birth of a grandchildren or great grandchildren rejuvenates this aspect of the fund. Trade can occur with individuals either within or outside of the fund.

This dimension includes assets of adult children, related children of school age and below, extended family, continuing contact with social agencies and access to motorized transportation. (See Table 6.3.) These assets, in reality, provide access to the good life, as care and satisfaction cannot be considered an automatic outcome. Measurement of size and density of one's social networks do not quantify the types of interactions. Outcomes of support and satisfaction depends on the adequacy of the relationships via the exchange process (Ryan and Austin, 1989). This is a slightly different approach than the more common concept of regarding children as automatic providers. For some readers, this may mean a switch of emphasis in the meaning of children. Also, the fund does not consider family functioning, in itself. Family functioning is considered later, as a setting for gerontic fund use.

The social/familial components vary in two distinct ways from developed nations. First, I have placed a stronger emphasis on grandchildren as an asset. Grandchildren from migratory workers are sent to live with aging parents to learn village customs and provide care in the later years. Numerous aged report having the child since infancy, with the expectation of having a care provider as the child matured. It is common to find infants and/or preschoolers under the direct care of grandparents. Very young grandchildren are viewed as an asset, in that "they create a reason to live." School age grandchildren are desired as they signify the availability of provisioning for old age

TABLE 6.3 SOCIAL/FAMILIAL ASSETS. (N = 105)

A. <u>Adult Children</u>: (Mean = 4)
    1. No adult children or has lost contact. (10%)
    2. See adult children less than once a week. (7%)
    3. Adult children in village, sees once a week or more. (11%)
    4. Contact with adult children in household only. (19%)
    5. Contact with adult children in and out of household. (53%)

B. <u>Grandchildren</u>: (Mean = 4.1)
    1. No grandchildren. (10%)
    2. Has grandchildren but not in household. (3%)
    3. Only children 14 years or below living in household. (11%)
    4. Has only children above the age of 14 in household. (13%)
    5. Has children of all ages in household. (62%)

C. <u>Other Meaningful Family</u>: (Mean = 3.7)
    1. No other meaningful family. (15%)
    2. Sees less than once a week. (6%)
    3. In village, Sees once a week or more. (9%)
    4. In household only. (34%)
    5. In household and village. (36%)

D. <u>Intimate Relationship</u>: (Mean = 2.9)
    1. No intimate relationship. (38%)
    2. Has intimate relationship, sees person less than once a month. (7%)
    3. Has intimate relationship, sees at least monthly. (10%)
    4. Has intimate relationship, sees at least weekly. (14%)
    5. Has intimate relationship, sees daily. (30%)

E. <u>Known to Service Agencies</u>: (Mean = 3.5)
    1. Has no contact with service agencies. (7%)
    2. Is known to one service agency. (18%)
    3. Is known to two service agencies. (22%)
    4. Is known to three service agencies. (26%)
    5. Is known to four or more service agencies. (28%)

F. <u>Access to Transportation</u>: (Mean = 2.7)
    1. No access to transportation. (18%)
    2. Access to bus only. (13%)
    3. Access to bus and car outside of village. (55%)
    4. Access to bus and car in village. (5%)
    5. Access to bus and car in household. (9%)

rights, such as having a child sleep with them. Thus, grandchildren's ages, as well as proximity, were important fund considerations

Secondly, there is no inclusion of friends as assets. This is in direct contrast to American theoretical stance where friendships are directly related to life satisfaction in old age (McKee; 1982; Lieberman and Tobin, 1983; Mumford, 1987; Keith, Fry and Ikles, 1990). True friendships with non-kin are rarely found among these Tswana. The aged considered acquaintances as having no place in the fund, as threat of harm overrode value. Therefore, in this section, those rare friends who act as fictive kin, are classified as an extended family member.

The Tswana consider adult children as one of, if not the most significant, resource in this dimension. The strength of this resource was not based so much on the numbers of adult children but on their geographic proximity and continuing contact. Offspring, in themselves, had no or minimal value if contact had been lost or distance and lack of communication prevented interaction. Thus, scores were determined by where adult children lived, with those having children residing locally receiving higher scores. Children residing in the home carried more weight than children in the village as availability was assumed more constant. Those with children in the home and village scored the highest.

All age groups reflect great variation in the number and location of living adult children. The majority (53%) have one or more adult children living in the home and the village, with another 19% having one or more adult child living in the compound without other children in the village. A sizeable number have scored low in this item with no children (10%) or no children within a day's commuting distance (7%).

While 72% of aged have adult children in the household, 87% have young grandchildren living with them. Preschoolers and school age children are present in 62% of the households. Another 22% have grandchildren of only one age group living with them. If grandchildren live locally they tend to reside with the grandparent, as only 3% were with local but nonresidential grandchildren. The 10% without any local grandchildren lamented this fact.

Adult grandchildren, with adult roles and duties, were considered as extended family members. Spouses, siblings, parent's siblings, even more aged parents, and fictive kin, were also classified as extended family members. They must have had positive significance in some manner to be considered an asset. The relative who was not desired to be seen, thus never seen, was excluded. Some extended family members were older than the individual under study, some were younger, but all the significant others mentioned by the aged were adults. Again, proximity was the key in determining the strength of the asset.

The aged are evenly divided between having significant others in the household and not in the village (34%), and significant others in household plus weekly contact with relatives in the village (36%). Very few (8%) have weekly contact with family members residing in the village with no extended family living in the household. About one/fifth of the aged (21%) are without any extended family or have less than weekly contact with them. It may be that such kin were never significant. More likely, in consideration of the social context, having kin in the household increases the likelihood of maintaining meaningful relations with other family members.

Age, gender, kin relationship, and social norms enter into the degree of permitted emotional intimacy. Intimate relationships are also affected by geographical distance. As Maria says, "Even though I have two daughters and many grandchildren, there is no one I can talk to, I mean tell my feelings and share the good and the bad." This type of intimate relationship is included as an asset.

Slightly under a third (30%) have an emotionally intimate relationship with someone seen daily. The person is usually an adult child of the same sex living in a household established by the old person. Another 32% have a person they could talk with, but did not see this person daily. Usually this relationship is with extended kin in the village seen weekly or monthly, or with a migrated child seen monthly or yearly. Over a third (38%) have no intimate relationship. Spouses, if present, are seldom mentioned in the context of intimate relationships. Married couples converse and rely on each other but it may be that the "master-wife" interpersonal relationship prevents intimacy.

The majority of men have a daily relationship that included intimacy. In contrast, the mode for women is to have no intimate relationship. The discrepancy between the genders is most probably related to the social laws. The elder man is generally master of any compound. Only this authoritative figure is normatively permitted to express dissatisfaction. This creates an ideal milieu for males establishing what they interpret as intimacy. In contrast, women are expected to be submissive and are confined in expressing thoughts. Female authority is limited to being the "woman of the house." This is not necessarily the eldest female, for with the assumption of chores, contemporary young women have also assumed control. Without authority, older women become more restricted in permissible expressions of desires and reactions. This thought is supported by the finding that both female intimacy and household control decrease as age groups progress upward.

At one time I thought tribal affiliation and length of village residency should be part of the fund. This was eliminated as all but two were Tswana, and all but four were either born in the village or married into the community as a young adult. These people were similar to other aged in assets and outcomes. This factor should not be belittled in cases where it may have significant importance, such as research with urban African aged where the custom of aging in place of birth is disrupted.

Everyone knows of the various social service agencies in the village. These include District Headquarters, the Tribal Administration, the hospital, churches, library, and community center where other non-governmental agencies such as burial societies and the Red Cross may meet. Few individuals (7%) do not have on-going contact with any agency, usually because of a disbelief in western medicine and religion and resistance to political intervention in daily lives. All other individuals attend, or are attended by, one or more agencies, on a regular basis. Contact with four or more agencies is the mode (28%). The only service provider never used by the aged is the library, although many could read and a good selection of adult books were offered. The library is avoided as it is regarded as "a place for children."

The final asset in this dimension is access to motorized transport when desired. None of the aged own a car. The presence of a car in the household is rare (8.5%) and even fewer have access to a car located outside of the household (5%). When a care is owned by a family member, it is regarded as a possession to be shared. It can be had "just for the asking"' if the need was justified. Justified needs usually mean transport for medical care, but not visiting, local shopping or church attendance. Therefore, the bus was important to all.

Bus service provides limited degrees of in-village transport, but includes the major points of congregation and the main shopping area. More important to the aged, bus service decreases walking distance to the lands and provides access to major towns where migratory children, and other kin, reside. Over half (68%) have access to the bus, which means being able to get to the bus stop, able to get on, and having the money for the fare. Bus rides are limited to "only when I have the money" for 13%, with another 18% feeling access is always limited. Besides fares and inability to get to the bus stop, individuals are limited by the initial high step on buses, some being 24 inches above the ground. Only a few women are willing to grin and bear it as a rider grabs her arms to pull her on board while another pushes her upward from behind.

The social/familial dimension, in terms of the discussed modes and a group weighed average of 3.6, makes one think that most people had moderate items in all areas. In reality, most individuals show great variation within the dimension. Some have many grandchildren, but no other family. Other have assorted family members, but are without intimacy or social assets. Some are strong in the social segment, but live alone. As a dimension, the social/familial segment creates the most variation in final total scores.

The social/familial dimension averages the same between sexes for similar age groups. The lower intimacy scores for women did not pull this dimensional strength downward. Although sex differences in other items were not significant, they were high enough to produce sexual equality in within the dimension.

Age is weakly correlated with overall scores in this dimension ($p = .014$). The average un-weighted social/familial score is 21, out of a maximum of 30. It is fairly consistent within most age groupings, except for the over-ninety group. With the very old, scores are in the low-teens with highs hovering around 20. Not only do the very old have the least variation as a group, but also the strongest tendency to have the same score for all items. They have outlived kin and significant others. This is the group, who in response to questions on social assets will most often say, "Yes, I can ride the bus, or go to church, but who would I visit and what would I do once there?" It must be keep in mind that other younger aged are in the same situation, Approximately one/fifth of the people in the each of the younger age groups have scores equal to those of the very old.

The Fiduciary Dimension

The fiduciary dimension is usually thought of as actual monetary resources of retirement income, investments and accrued property values. In Botswana, retirement pay is a new and limited event; investments tend to be in livestock instead of the bank or bonds; and land is not owned by the individual but the tribe. The household, in contrast to the individual, is considered as the source of economic production. Actual income is shared, with multiple family members participating in agriculture for food and cash crops, and husbandry. Blended together, these economic inputs reflect what is commonly known as household economic status.

The enigma of determining income and socioeconomic status is well known to researchers in developing countries (Sahlins, 1972; Brathwaite, 1986:13-14; Hill, 1986:39, 46-47). Botswana governmental officials are also well aware of the difficulty, giving various reasons: hesitancy of the population to reveal personal worth, personal inability to calculate value of land and livestock, the lack of accounting with remittances, in-kind and other income and a smattering of mistruths. Rather than try to do the impossible and arrive at a questionable set of figures on income, the fiduciary dimension was broadened to reflect personal direct control over money, land and livestock and the possession of material goods and housing. This is shown in Table 6.4.

Actual monetary assets for the aged were measured in terms of source and regularity. Most households functioned in some degree of poverty, so that regularity of income to pay expenses was important. Only with regularity was there the small possibility that the non-employed aged would be given spending money for personal use. The aged had direct control over sharing and/or keeping any self-earned income and pensions. Therefore, personal income provided additional strength for the fund. In all cases, little attention was paid to the amount as reports were unreliable. Other than

TABLE 6.4 FIDUCIARY ASSETS. (N = 105 unless noted)

A. <u>Income</u>: (Mean = 3.0)
    1. Registered destitute. (11%)
    2. Irregular remittances from others. (36%)
    3. Irregular employment and irregular remittances. (8%)
    4. Regular remittances. (32%)
    5. Regular source of income (employment, rental, pension). (12%)

B. <u>Investments in Cattle, Other Livestock, Animals</u>: (N = 73; Mean = 1.7)
    1. Owns no animals. (63%)
    2. Owns chickens or a dog. (21%)
    3. Owns less than 5 goats or pigs or has one cow. (4%)
    4. Owns 5+ goats or pigs or 2-4 cattle. (10%)
    5. Owns 5 or more cattle. (2%)

C. <u>House Ownership</u>: (Mean = 3.5)
    1. Temporary housing provided by government. (0%)
    2. Housing owned by other family member. (31%)
    3. Owns house, major repair needed. (16%)
    4. Owns house, minor repairs needed. (26%)
    5. Own house, in good repair. (27%)

D. <u>Furniture and Household Goods</u>: (Mean = 3.1)
    1. No bed, 2 or fewer blankets. (19%)
    2. No bed, 3 or more blankets. (24%)
    3. Owns bed, no furniture. (12%)
    4. Owns bed and either wardrobe or chest of drawers. (13%)
    5. Owns bed, wardrobe/chest and one other piece of furniture. (31%)

F. <u>Agricultural Land</u>: (Mean = 2.5, N = 103)
    1. No control over land. (42%)
    2. Controls one or less hectors. (7%)
    3. Controls two to three hectors. (26%)
    4. Controls four to five hectors. (9%)
    5. Controls six or more hectors. (17%)

recognition of the registered destitute, no attempt was made to pinpoint the individual's level on an economic scale.

The statistical mean indicates that the average aged relied on regular remittances. This was not reflected in reality, as only 7.6% of the aged fell into this group. Roughly one-third (36%) shared in the use of irregular household remittances and another third (32%) had irregular money making schemes providing money for personal use. Even with the provision of remittances and/or irregular money making schemes, adequate income was felt to be absent in many of these families The other age were equally divided between the two extreme ends of the scale, having a regular source of personal income, either through employment, rentals or pensions (12%) and that of being a registered destitute (11%).

A registered destitute is one who lives at 10% below the poverty line and is without family who have means to provide support. Numerous aged live below the poverty line, but because the family has the ability to provide support no government funds are given. It makes no difference in regulations if families do or do not provide money (Knudsen, 1988).

What proved to be a accurate way of assessing control over money was not discovered until quite late in the research. I mention it only because the method may be valuable in future work. The simple question of "When was the last time you had money in your pocket?" was so different from the usual questions asked in typical surveys that it took people completely by surprize. They responded without thought. Some would look in their pocket and show me what they had. Others would laugh and say, "I found 10 *thebe* (5 cents) on the path last month," or hang their heads and reply, "I haven't had my own money for over a year." This question was followed with probing into how money was obtained and how it was used. While the numbers asked these questions was relatively small, the 19 respondents did support my growing assumption that, unless the aged generated their own money, none was available for personal spending.

Men, compared to women, were more apt to be employed and to have pensions. The four men who held pensions, (20% of males) had experienced forced retirement from civil service or formal industry. With the females, which made up 80% of the study groups, only two held pensions. This discrepancy is due to, in part, the numbers of women in agriculture and past intermittent employment. In contrast, women were more apt to have irregular money-making schemes. The most common were making and selling *bajalwa*, and cutting purchased smoking tobacco into snuff.

All traditional beer in the village is made and sold by women, usually older women. As a group, women provide a constant supply, yet individuals use this money-making scheme only once or twice a year. The making of *bajalwa* is labor intensive and requires large amounts of either purchased or surplus grain. Frequently the initial cost, labor and profits are shared between two or more women.

Snuff is popular with women of all ages, with men preferring smoking. The making of snuff is an individual endeavor, usually performed by much older women on a routine basis, processing one or two bags of tobacco a week for a small profit. Smoking tobacco is purchased, then chopped at home with a large knife. Buyers bring their own containers.

The Tswana have long regarded investment in cattle as the major form of savings for old age (Schapera, 1954; Arntzen, 1984a). It must be noted that my initial questioning on livestock was not satisfactory, as the aged included cattle owned by extended family members in their answers. Hence, cattle ownership was omitted as a gerontic fund component for the first 32 respondents. The revised question of "Some people never owned cattle. Did you ever own cattle?" was more satisfactory. The implied assumption was that none were held at the present time, and there was no need "to prove oneself." The question was followed with ethnographic style questioning for more valid responses on livestock history, including reason for gain and loss, and present status.

The reason given for cattle loss was always, "My cattle all died." It was only with specific questioning that the causes of death were explained. Drought was the underlying reason for most cattle death. Some cattle did die directly from lack of water and food or from cattle disease accompanying drought. Others were slaughtered and used as food during droughts. Other cattle were slaughtered to use for burial and wedding feasts, unrelated to drought. Very few were sold for slaughter.

These 73 histories on cattle ownership highlight important points about gerontic fund accumulation and loss. Cattle, which can be owned by both sexes, were never owned by 22% of the aged. These were those who were not privileged to inherit family cattle because of position in birth, those who received no cattle inheritance because family cattle died in the 1930 drought, and women who were denied inheritance of her husband's cattle because of the absence of bride-wealth payment. Those who were adults during the 1930 drought lost their cattle at that time and were never able to replace them. For those who are now 60 to 89 years, 53% had no cattle by the end of the 1960 drought and never replaced them. Loss at this time was due mainly to dry environmental conditions. An additional 38% lost their herds before the end of the 1980 drought. For these aged, cattle loss was not necessarily due to lack of pasturage or water. Some were slaughtered for food and some were sold for needed money. Final loss occurred as cattle were used for funeral feasts and for payment to traditional doctors for needed cures. Limited finances was always the main reason for not replacing cattle.

Age correlated directly with cattle loss, with the oldest experiencing herd lost in the first drought, the 70-80 year olds having permanent loss during 1960 and the present

Table 6.5: History of Cattle Ownership and Herd Loss with Drought by Age Group.

Age Group	N	# Never own cattle	# Owned Cattle	Year 1930	Herd 1960	Lost 1980	Retain Cattle
60-69	31	8 (26%)	23 (74%)	0	7 (30%)	12 (52%)	4 (17%)
70-79	19	4 (21%)	15 (79%)	0	10 (67%)	4 (27%)	1 (6%)
80-89	19	3 (16%)	16 (84%)	0	13 (81%)	3 (19%)	0
90+	4	1 (25%)	3 (75%)	3 (100%)	0	0	0
Totals	73	16 (22%)	57 (78%)	3 (0.5%)	30 (53%)	19 (33%)	5 (9%)

60 year olds having permanent loss in the 1980s. (See Table 6.5.) Four of the five individuals presently owning any cattle are males, age 60-69. The fifth owner is a female in her low seventies, possessing one cow that she is saving for her funeral.

Using these 73 aged to determine total livestock assets, I found 2% own five or more cattle, with another 10% owning either 1-4 head of cattle (3 individuals) or a herd of 15 or more goats (9 individuals). By far, the majority (63%) controlled no animals. These people expounded on the fact of "not even having a dog or a chicken." The 21% with a dog or a chicken or two considered these animals as an arena for in-kind trade, with offspring or eggs exchanged for sugar or beer.

The majority of households in Botswana rely on agriculture to either provide, or supplement, food sources (Kerven, 1982:240-242). In Ramotswa, agriculture remains a family affair with the aged, children, and some adults participating. As land becomes more scarce due to population growth, it is quickly reassigned if not used, as land is insufficient for requests. Thus, the number of hectares an old person controls determines asset strength. Control over land extends beyond possession. It implies decision-making for production, and a say in the use of crops. Actual participation in agriculture is not necessary for control.

The average land size, slightly over 2 hectares, is not reflective of actuality. Forty-two percent of the aged do not control any land. Causes are varied and include such reasons as never acquiring land because of long term migration, loss of lands from disuse, and transfer of lands to extended family members. The remaining individuals (58%) have control over land, usually three to four hectares, with some having larger or smaller fields.

Those without land are mainly women, while the mode for men is to control three or more hectares. In one sense, this is surprising as land is shared by married couples, with land going to the widow or widower. Also, women are regarded as the mainstay in agriculture (Kerven, 1982:240). They are the ones who plant, weed and harvest. Men are responsible for the initial plowing, and then are traditionally free of agricultural duties until exerting control over harvest use (Schapera, 1953:27).

The gender imbalance found with land ownership, with nearly 50% of women owning no lands compared to 10% of men, is not startling when looked at in terms of family authority. The eldest man in the compound can demand or command agricultural service from the young, keeping the land under his control. The woman, with the transfer of household responsibility, loses power over daughters and sons. Fields are lost from non-use or transferred to family outside of the immediate household.

Social change also enters into the picture. With the more recent growth of the international money market, subsistence agriculture does not produce adequate and rewarding returns (Kerven, 1982:242). The young prefer wage labor and have little desire to either work or inherit the family lands. Older women, even if lands are plowed, must have enough other fund resources to command continuing agricultural assistance. As the physiological and family gerontic fund assets decrease with age, so does the loss of fields. All those after age 90 claim the loss of land from disuse.

Residential land, like agricultural land, in owned by the populace, not the individual. The aged of today built their original homes on land assigned by the chief. Giving up the home means that the land and hence the building reverts back to the government for reassignment. For this reason, even if the majority of time is spent at a daughter's or another relative's house, the aged, if at all possible, will return home for one night a week to retain ownership of this monetary and emotional investment.

Home ownership, regardless of the type of construction, comprised one item in the fiduciary dimension. Maximum credit was given if the aged person owned a home in a good state of repair, as a well maintained home created less worry and fewer problems with environmental elements than a house needing repairs. A home, even in poor condition, had more intrinsic value to the aged than a building provided by others. During my initial visit to Ramotswa I met a couple of aged individuals living in temporary housing. One was renting and the other had loan of a large tent. No such living arrangements are present with the sampled aged.

By far, the majority (69%) live in the home they constructed earlier in life. Of these 72 aged, 28 live in well maintained buildings and 27 live in homes needing minor repair such as broken windows or a cracked side-pole replaced. The remaining 17 live in homes where major repair is needed. Major repair includes the need to correct support beams eaten away by ants, walls eroded to a degree of potential danger, or large holes in roofing limiting protection from cold and rain.

A roof that does not leak in heavy rain is a rarity. Sound roofing is found only on more expensive buildings, which also were the only ones with ceilings. It does not matter if the roofs is thatch or corrugated iron, they leak. Almost everyone complains. During rain belongings are sometimes left in place with the knowledge that they will soon dry. At other times, it is necessary to move belongings. I, too, had to learn to sleep in contorted positions to avoid damp places in bed and to avoid using certain sections of the floor for storage during the rainy season. Based on the pervasiveness of this housing trait, leaking is not included in the state of repair.

The intrinsic importance of home ownership is shown by the reluctance of the aged to move when their housing conditions become substandard. They remain in potentially dangerous houses, although better living conditions are available. Other family members are usually living in the compound and many had offered to provide housing. With the 30% who are living in another person's house, the move tended to be to an adult child's house within the same compound. Those who move in with a more distant relative, almost always another aged individual or grandchild, usually have to change compounds.

The number of aged who had left the village to live with migrated children is unknown. Several aged are resisting pressures to do so, as they prefer living under poorer and more limited conditions to the breaking of the custom of living one's last days in the village of birth. The concept is similar to the Western desire of aging-in-place, resisting forced moves to new areas with the loss of friends and home.

I heard accounts of dissatisfaction from two who had made such moves and had returned. Others told stories of unhappiness, never satisfaction, regarding friends who experienced moves to the city. A similar dissatisfaction existed with those who had moved to a new location within the village. A change of residence was not truly accepted by the aged until external causes, such as "my house finally fell down," provided a valid reason. Even then, moving was something to be endured. For these Tswana, aging-in-place means living in one's own original home on family land. Second best is living in the village of birth.

Possessions owned by the aged have great meaning to them, making ownership of furniture an important asset. The meaning of possessions is compounded by the laws regulating the use of other's belongings. According to law, a table in the home can be used by all, as it was a necessity for meal preparation and open storage. Chairs are under the control of the owner, to be used with approval. Chairs are not an extremely high priority item as sitting on the ground, or a small wooden and leather stool, is often preferred for meals, conversations, and watching the sights go by. Other basic furniture is to be used by the owner only. This included beds, dressers, chest of drawers, and wardrobes. Personal goods. including blankets, radios, silverware, flashlights and other material objects, commonly shared in Western homes, are also restricted to owner-use. The regulations influence the purchase of items by individuals. The old state that during

their working years, they always purchased a bed first, then a storage unit. The third
and fourth purchases were a table and straight back chair.

A leading council official, in perusal of my research instrument, proudly stated
that no one in this village was without a bed. The days of sleeping on a goat skin or
mat had gone by. That may have been true for the younger generations, but not for the
aged. Their possessions reflect the stated preference in purchases, with many unable to
obtain the basics. Nearly half (43%) sleep on the floor with a blanket, mat or goat skin
under them. If only one piece of furniture is owned, it is always a bed. With two
pieces of furniture, the second is a storage unit, usually a chest of drawers or a
wardrobe. Only when a person owned these items do they add the table. Fewer than a
third (31%) own anything above this. This does not mean assorted furniture is not is
the house. It is, for most aged do not live alone.

I was quite interested in discovering the degree that laws regulating use of
furniture were practiced. The reply to, "Do you use your daughter's bed during the
day?" was always denial. I tried other methods to give permission to say "yes," and
always heard "no." Several aged, without their own house and with no or few
possessions, house-sat for migrated adult children. These homes were furnished with
beds owned by children who returned once a year. I expressed my feeling that if I knew
the owner would not be back for 11 or 12 months, I would probably use the bed if I
was sick. In all cases I was severely reprimanded for the thought. The same was true
with the use of another's blankets, even if it was cold. One man, living in his son's
house containing easy chairs, always sat on his sleeping skin. No permission had been
given for the use of the furniture.

In each of the fiduciary categories, individuals are fairly equally distributed within
item options. Patterns of distribution between options varies with age groups, with the
older groups consistently reporting fewer fiduciary assets than the youngest one (p =

.006). The average weighted score for the 60 to 69 age groups was 3, the exact middle of the possible score, with 11% scoring in the above average range. The 70 to 79 year olds and the 80 to 89 year olds are just about as apt to score in the mid-range as to score slightly below. It was the over 90 age groups that had minimal economic resources, with a mean of 2. The minimums of the oldest people were equal to the minimums found in other age groups, although their maximums were consistently and considerably lower. The oldest also had the least variation among scores. No one was totally without any contents in the fiduciary dimension, always having at least a little in two or more of the categories, usually support from irregular remittances, limited material goods or owned housing. (As one person commented, those who were without fiduciary resources were probably dead.)

Overall, the fiduciary dimension is the weakest of the three dimensions. The average weighted score for the fiduciary dimension is 2.78, or roughly a full numeral below the other fund dimensions (3.75 and 3.57). Is the evaluation of economic resources too harsh for the ambient setting of poverty? This I doubt. The casual observer can notice that the fancy housing in town is usually owned by younger adults. My initial errors in interviewing for livestock ownership pointed out that cattle is owned by sons, not aged parents. In my survey of household belongings, I find younger members own more than aged parents ($p = .0001$). Almost never did I enter a home that did not contain at least one bed, several chairs and a wardrobe controlled by adult children. They are also the ones to have radios, watches, paraffin stoves, and easy chairs. They also have direct control over household remittances. Such facts strongly support the assumption that an additional "poverty-of-old-age" exists within the generalized setting of poverty.

The poverty-of-old-age, although always present, became more pronounced as age groups progress. I believe history plays a role, in that the majority of aged never had the present day opportunities to establish strong fiduciary assets. Prior impetus for personal economic growth reflected desires for traditional social and familial sharing for equality, rather than individual personal materialism. Historic loss also enters the picture, with devaluation of assets and the actual loss of assets with calamity and drought. My questions on cattle ownership supported the claim that cattle ownership transfers to the younger elite who can afford purchase and grain during drought (Arntzen, 1984a; Arntzen, 1984b).

What value do the young place on the economic assets of the aged? With better wages and opportunities, they regard a broad assortment of furniture, which the aged do not have, as a necessity. Radios and watches are a mandatory luxury, with gas stoves and refrigerators quickly moving into this mandatory category. The changing emphasis on material goods has altered the meaning of the traditional inheritance. Younger members of society claim it is desired, but for more modern reasons. Farm land is supplemental to employment, the household compound is a place to erect a better house, and livestock is to add to a larger herd, if not sold with the money placed in the bank.

A Comprehensive Look at the Gerontic fund

All aged have at least a few assets in their gerontic fund. The majority have a fair amount, and a few are rich. All aged who are in good health retain strength at a level that appears to surpass Americans. All Tswana aged have some sort of access to family members. Although monetary deficits are numerous, all have a place to live. The average score, based on weighted averages for each dimensions, was 10.1 in a total of 15. (SD=1.9). Although this average is relatively high, the variations between individual funds are expansive. Of great importance is that extreme variation existed within all age groups, except for those over 90 who are always poor in assets.

One person scored 14.7 out of 15. Slightly over one/half (55%) controlled funds within one point of the average. However, more fell below one point of average (28%) than above (17%). Average scores were fairly equally distributed between 13.5 and 6. Three people scored between 5.9 and 4.5 These people were the very old. This is a very weak fund, as 3 was the lowest possible score. Maybe the small number of people over ninety (6) was not an adequate representation. Maybe it is. The person's comment that a lack of monetary resources brings death may extend to all dimensions. With an exceedingly small gerontic fund in combination with advanced age, does life become impossible? Does this relate to the lack of extremely old men in the village? It can only be assumed that they too would lack non-renewable assets. Do they require more numerous assets than women in order to survive?

It was the rare person who was relatively strong in one dimension and weak in the remaining two. More apt to occur was a relatively weak fiduciary dimension in comparison to moderate strength in the other two areas. Never did a person have an exceeding rich area(s) contrasting with an exceeding poor area(s). An interplay between dimensions appears to exist, with a lack of assets in one area affecting assets in the others.

Statistically, the average old person in the village is of satisfactory health, having more strength and stamina than what one would expect in an American setting. The average Tswana aged are active, involved with the family and village. They are economically poor, but often generalized poverty extends throughout the household. It would be a great disservice to the aged to paint this picture of the average old person onto all aged. Averages, as shown, distort the true picture. The amount of variation within age groups and between groups is too great. The young-old, as a group, have more personal/physiological and fiduciary resources than the very old. The young-old also have a tendency to have more social/familial assets. One must be cautious in assuming the young-old individual has sufficient assets, as this class of aged showed the most variation. Some relatively young aged have very few assets to provide access to the good life. At the same time, increasing age brings the warning that assets are diminishing ($p = .0001$). The correlation can be explained by losses in the personal/physiological dimension, smaller loss in the fiduciary dimension, and even smaller loss in the social/familial dimension.

The decreased assets with age is, in part, a reflection of the rapid growth of the money market economy, the corresponding social change, and the environmental disasters that have occurred in recent years. In addition, the decreased funds found with the oldest of the old stimulates that thought that once loss occurs it cannot be the rejuvenated, at least to the previous level. This is very true with cattle ownership. Neither can loss be compensated for with increase in other areas. This assumption should hold true with both the young and old aged, although it cannot be proved with facts as no such measurement over time occurred. It is strongly supported with comments from all aged reflecting that they have less now than before. The comments of 73 year old Monate are representative: "Since getting old, I do not have the strength of before, the family or the things that money can buy." Toto, at 94, says, "Now that I am very old, I have nothing left; no family other than a grandchild, no money, and no {physical} body that I can be proud of."

Daily life-style does have a strong influence on what is regarded as valuable within the gerontic fund. Life style, as a composition of individuals, families, social structures, functions, and values, also influences how the fund can be used. The following section takes a broad look at the family setting in which the aged function and the condition of the doors for social exchange.

I don't have friends as you can only trust family. My children are my main companions. If the children fight I settle it as I am Master of the house, although sometimes they advise me on what to do. (68 year old man)

A big difference from the past is that today we don't like each other, especially within the family. Today's children control themselves and do not hear their parents. (82 year old woman)

It is generally believed, that within the African family setting, traditional familial piety and oral tradition continue to protect the aged (Biesele and Howell, 1981; LeVine, 1965; Turnbull, 1983; LeVine and LeVine, 1985; Rosenberg, 1990). Old age care is provided as a result of culturally sanctioned behaviors that favor the debt of nurturing and care resulting from parenthood (Dowd, 1984). In southern Africa, payment of this debt is interlocked with a concept of mutual aid, with the aged teaching tradition to the young child and providing direction to the young adult through approval (Biesele and Howell, 1981; Colson and Scudder, 1981, Shostak, 1983).

When social conditions become unstable and the rate of culture change increases, exceptions to the guaranteed care may present themselves (Simmons, 1960). Age-related power, knowledge, and the demand of payment for old debts no longer provides the necessary resources for old age care. The emergence of a money market economy, and its multiple ramifications, is the most frequently mentioned avenue of change affecting the aged in southern Africa (Schapera, 1955; Hampson, 1982; Hampson, 1985; Hay et al., 1985; Tlou, 1986). The family experiences conflicts between means and desire, which promotes transference of attention from the old to the

170

wage earners. (Goldstein, Schuler and Ross, 1983). In turn, wage migration further thwarts intergenerational ties and removes family as an asset (Shostak, 1983; Khasiani, 1987).

Many African rural aged do reside in extended families, which have been diluted in size through wage migration and thinned in ability to provide adequate care due to economic pressures (Hampson, 1982; Hampson, 1985; Hay et al., 1985; Tlou, 1986; Khasiani, 1987). These new social conditions exist side by side with continuation of familial piety, thereby providing protection to the majority of rural aged (Biesele and Howell, 1981; Colson and Scudder, 1981, Shostak, 1983, Khasiani, 1987). As mentioned earlier, this pollyannish conceptual framework, and the assumption of normative care provisioning tempered with economic constraints, can have faults. No situation automatically fulfills the needs and expectations of the aged.

Thought needs to be given to the communications and behaviors found within the locally existing families of the aged, far beyond the economic angle. A hard look at the Tswana family will be taken in this chapter. The intent is to present the family of today as a setting for social exchange, asking if the multiple changes in family life have sealed or broken traditional doors and/or created new ones and thus effecting the value and use of the gerontic fund. I do not stress the gerontic fund as such. My intent is for the reader to realize how the individual and family interact. This includes the generalized give and take, the various options for exchange, and the way decisions are made within the family.

I have chosen two typical households and their extended kin to illustrate common threads reflecting culture and historical developments. Factors behind kin interactions, and behaviors, are discussed afterwards in terms of cultural continuity and social change. The chosen events and behaviors described are not unique, although I do admit that all families are inherently different. I have added the dimension of time by presenting the families as seen in different months over the research periods.

The Kerengs

September 15, 1988

Mogolokwane insists I use the English translation of her name, Happy Sound, as she likes to "feel modern." During her 69 years she has learned assorted English phrases. Idioms are liberally scattered throughout her speech. "Okey-dokey" is her favorite.

Widowhood came soon after the birth of her last child, Leru, 37 years ago. Agriculture was her way of life for many years. After widowhood, she relied on a large goat herd for income. Today, the goats are gone and Happy Sound no longer goes to the land. Time has not diminished her love of plants and beauty. She enjoys tending flowers scattered about her grass free courtyard. A simple, two room, "four-cornered" house and a traditional *rondaval* sit in the courtyard. Her daughter, Leru, lives in the more modern home. Leru, now divorced, is a typist for the local carpentry brigade. Her three children, ages 14, 10 and 11 months, sleep with Happy Sound in the *rondaval*. A married daughter lives in Gaborone. Another son lives in northern Botswana.

Happy Sound is washing the breakfast teacups in a pail of water. My arrival signals a good opportunity to divert herself from physical work and proceed with the equally important work of teaching me about old people. Taking the baby with us, we sit in her house to escape the cold, spring winds. The small fire adds little heat. Obviously chilly, Happy Sound pulls her own thin blanket around her, ignoring the warmer blankets on her 14 year old grandson's bed.

"Okey-dokey, let's begin. Young people do not ask about being old, so you must learn and tell others." She is explaining her many kin when the 14 year old grandson enters. Without a word, he sits with us on the mats. Happy Sound continues uninterrupted. "Sometimes I visit the chief's family. His father and my father were cousins. One day, not long ago, the chief brought me a bag of sugar, to show we were kin." The child speaks. "I did not know that you were related to the chief." Without waiting for a reply, he ambles outside to join his sister. The youngest child continues to sit quietly. Happy Sound, talking continuously, adds more wood to the fire to begin the noon meal.

> Leru will make the tea when she returns for lunch. She has no time to cook. She is a good daughter, cooking and cleaning when she is home. All my children are good, either working or raising families. One girl died as a child, and a son died last year. I knew he was going to die as he went to the bars and drank all the time. One morning the school children found him on the footpath. They thought he was asleep, but he wasn't. The beer killed him. Leru goes to the bar only on Saturday night. My other children drink also, but that is true of most young people. The bad thing is that needed money goes to alcohol.

With the meal now cooking, Happy Sound sits back down. She indicates she is satisfied by clapping her hand and saying another "Okey-dokey."

> I had to give up plowing when the children were small. It was because of Leru. She would not be born so the doctor opened me up to take her. That day the doctor told me not to lift heavy things and that plowing would kill me. Since then, we have had to buy all our food. My children have always helped with the money.

<u>October 7, 1990</u>

Happy Sound's household has increased by one, since last year. Her son, Kaizer, returned home from the mines at Orapa three weeks ago. Diagnosed with tuberculosis, he is to avoid intense activity. This morning he is tinkering with an old automobile part on the far side of the yard.

I have the ladder that I use for interviewing in my hand. I ask Happy Sound how she sees herself compared to the person who can walk everywhere and do all they desire. She ranks herself in the middle. "I am not the highest as it is the youth who are strong, I'm not the lowest as I walk and do chores in the house." The next words come more slowly.

>My children think I am the on the lowest rung, unable to think and do as they do. At the same time, they think I am on the highest, never running out of energy. I am expected to raise the children, tend to the yard and walk to the shops. The old no longer have the traditional roles in the household, like being in charge and having others do the work. I must make my own role according to what the {adult} children say.

Pointing to the middle rung I say, "This is how old people were treated by others in 1966, the time of Independence. Where are the old people now?" The reply is an empathic:

>on the bottom! I cannot talk for other families, only mine. Children do not have manners in the way they treat the old. There is no respect. They give money for food but they are the ones to buy a new dress. I only get the clothes no one else wants. My children go where they want without asking permission. Sometimes they even forget I am a person. It is my grandchildren who love me. Even they can get out of hand sometimes. Who will take care of me when I am old?

The appearance of Leru and the school age children means lunch time is near.

Leru takes me aside as I leave.

>It is very good Mother has someone to talk with. If she is not talking she is working. I tell her not to work so hard but she does so anyway. She worked very hard as a young lady to give her children a good life. Now it is time that I return that work, as mother is a child. Kaizer could do more to help, but he is sick. That does not stop him from spending the day with his buddies and fixing their old car. Now I must hurry back to Mother. Good-bye.

<u>November 14, 1989</u>

The winds have been exceptionally strong the last few days. I can see from a distance that Happy Sound has thatch missing from her roof. As I near the gate, angry voices of a man and woman explode in the air. Leru's voice bellows: "Mother's roof has been damaged for two days, as the son you should fix it!" Kaizer screams back. "I am sick, I will fix it when I get well." Leru is quick to on the retort, saying:

you always claim illness yet you run about town and get drunk every day. You are lazy. As a son, you are useless to Mother. How can I return to work this afternoon with such a shameful man in my mother's compound?

My entrance does nothing to temper the argument. The older children ignore the scene and continue their usual pastime of idly drawing lines in the dirt. Happy Sound, pretending deafness to the yells, indicates I should sit on the goat skin with her, under the tree . As we talk, Leru leaves for the brigade. Kaizer walks over to the children, bending to tie their shoes. He addresses his mother with the request to visit friends. She hesitates, and then grants permission, although pain in her voice is clear. Turning towards me she speaks softly.

> When I talk to you of disrespect, this is what I mean. The children scream at each other. The son does not fix the roof. The daughter acts like she directs the household. The children of today do not follow the law. Will they follow the laws of giving food, and bathing me, when I am an old parent?

Happy Sound smiles as she lifts the baby (now a toddler) in her arms. She looks at him while she says:

> I teach my grandchildren the laws, maybe they will keep them. I am lucky as I have only three. With any less, they could not do all that is expected. Any more would be too many. It is much work for the grandmother to raise more than three grandchildren.

December 18, 1989

Happy Sound and I meet on the footpath near her house. A blanket holds the toddler on her back. Her hands hold a dozen quart sized beverage containers that she has picked up along the road. Once used for *Shake-Shake*, a commercial traditional brew, she claims she has another use for them. She will show me, if I go home with her.

Lined up along the side of the *rondaval* are numerous containers, each with a plastic fork shading a small plant from the bright sun.

> When I last visited my daughter in Gaborone, her master had many of these forks that people had discarded near the take-away food

> store. When I told him I wanted to make money by selling plants but
> that the seedlings needed shade, he gave them to me. Last month,
> before Leru cooked vegetables, I saved the seeds. Look, the tomatoes,
> green peppers, and squash have come up. I am waiting for the orange
> seeds to sprout but it has been a long time. This month I sell the plants
> for 25 *thebe*. Next month, when the rains are good, they will cost 50
> *thebe* {25 U. S. cents}.

Kaizer joins us, with a pail of water in tow. Gently, he and Happy Sound water the plants. I promise to return.

<u>February 3, 1990</u>

I hurry to visit Happy Sound before the older children arrive from school. She sits under the tree, nibbling peanuts from a small cellophane bag. The toddler is sleeping on a goat skin near by. Kaizer walks towards his mother to make his wishes known: "Make me tea, now!" Happy Sounds moves to the fire, eating the nuts as she talks. "Kaizer is master of the house. If Leru were here, she would be making the tea. The women of this house have pride in serving the master." When she sees the children coming, the bag of nuts is quickly hidden in her apron.

> I bought these nuts with 25 *thebe* from the plants. I gave some
> nuts to Kaizer, because he is master. He does not get extra food at night
> by being the master. Sometimes I give some of my dinner to the
> children if they are very hungry. If they see my nuts, they will want some.
> These nuts are mine and I do not have to give them to others.

The children are in the house. Happy Sounds drops the now empty peanut bag on the ground to pour the tea. Kaizer takes the cup without a word of thanks and enters the square house. As I leave, Happy Sound says "I have a good family. I help them and they help me."

<u>March 6, 1991</u>

Today is Happy Sound's turn to build her preferred house on the felt board. The entire family comes over to watch. Past experiences has taught me that vast family discussions occur on what is the ideal, with the authority of the aged overruling the children's preferences during this pretend situation. It is also a time when the parent can openly compliment or criticize their offspring, for such comments should always made indirectly. I welcome the observers, knowing honest thought will prevail. Each young child holds one of the cardboard people. Kaizer holds the cattle and goats, while Leru watches her mother put the objects on the board.

Happy Sound begins her dream while placing her preferred building on the felt. "I want the round house. Thatch is quiet when it rains. It also makes me think of my husband. Wouldn't it be wonderful to have a bed inside the house." She puts the bed aside, choosing the bag of grain instead. "More important than a bed is food. I want lots of food, not like the limited grain we have now." Realizing she can put many things in the house, she adds the bed and a table. Leru suggests a chair so she would not have to ask to use one. Everyone nods in agreement. Mother should have her own chair.

Now it is time to design the yard. The garden is firmly placed by the side of the house. "I want a garden, where I can grow food, and feel that I am at the land. The garden will need water so I must put in the faucet near the vegetables, as no one carries water for me." The young children giggle, knowing they have ignored past requests. Kaizer suggests adding goats. "No, I've had goats and they are much work. They give us milk, but Leru is working and cannot care for them. Once you leave, there is no one to fix the fences." Looking towards me, she says "You know, the roof on my real house is still broken!" Leru looks at Kaizer, then turns her attention back to her mother, saying "What about the cooking fire?" Happy Sound shakes her head. "It would be much better if we saved money a bought a paraffin stove for inside the house. I will put a stove inside. Now I will add trees to the yard. Lots and lots of fruit trees, not like the ones we have now." The adult children shake their heads. "Fruit trees, who grows fruit trees?"

Everyone laughs as I ask if she wants to put herself in the picture. Gazing intently at the models representing women of different ages, Happy Sounds moves her hand from one to another. "People think I am this very old woman. I would like to be regarded as this young one. I will take this lady as she is like me, not too old and not too young. The young one will be my daughter." These pictures are placed by the house. "My daughter is cooking, while I am supervising." She select more people. "Here is my son, plowing the garden, and my grandson is hauling water. The granddaughter is sweeping and the baby is happy." Suddenly she shakes her head and jerks away all the people.

> Who will take care of me when I am old. Family just causes problems, too many problems. They argue and disobey the laws. They spend money on themselves and forget the old lady. My children do not obey. Having many people in the house causes fights. I want to be happy in my house. By law I must have someone to take care of me, but I would really be happiest if there was no one.

Slowly, she looks at the scattered facsimiles of people. She replaces the granddaughter and herself. "I will only have my granddaughter. She will obey the law." The other pieces are picked up. "I also want Leru and Kaizer in the picture. I will scream back when they yell at me, but that is breaking the law. I do not want to break the law."

The caustic finish to the picture creates nervous laughter. Leru and Kaizer assist Happy Sounds off the ground while commenting on her honest picture. The smiles become more genuine as we gather up the assorted game pieces. Happy Sound speaks. "Thank you; Thank you; Thank you. That was hard to do, but it was good."

On the way to the gate, both Kaizer and Leru comment on how little they understood their mother before I began my visits. After extending an invitation to return, Kaizer reflects on the afternoon. "I liked the picture. I never knew mother thought that way." I leave with thoughts on how my data collection may influence future family interactions. Do researchers have the right to interfere and make private thoughts open knowledge? I rationalize. Maybe, this time, the interview would strengthen an already loving family.

The Mosokos

<u>October 9, 1989</u>

Jacob and Betty Mosoko, both in their late seventies, live in a large traditional house nested in a small yard of flowering bushes and hedge. A traditional *rondaval*, which was once the main house, sits in the corner of the yard. Jacob's jubilance over retiring last month radiates about him as the two of us sit together on the ledge surrounding the building. I ask how life is different with his return to his village after so many years. He figures money would be no great problem as he is receiving a small pension, but life would be different.

> In South Africa I had electricity and running water in my house.
> There were many things that could be bought in the shops. None of that
> is here. Betty had a difficult time adjusting to the lack of goods and
> amenities when she returned years ago to raise the children. For me, the
> village has more important things to offer. My family and kin are here,
> but most of all is acceptance. Here I am a Tswana, not a Kaffir to be
> pushed around. I am a person, free to participate in the *kgotla* and
> direct my life as I see fit.

Jacob hears Betty preparing afternoon tea and asks me to join him. The couple's robustness and the sight of biscuits on the tray indicate money for food is not a problem. Betty serves Jacob first and then me. As she starts to retreat into the house, I suggest she join us, to which she replies, "Usually I have tea with Master, but with important conversations between Master and others I do other things." Betty quietly explains if I return when her master is gone she would be happy to talk with me.

Jacob explains to his wife that I am in Ramotswa to learn about the life of old people. This does not keep her from leaving. As she closes the door behind her, Jacob says:

the brain is different between men and women. The man's brain is made so he can be Master. He can see the important things in life and tell others about them. The woman's brain is more soft, so she can tend to children. She knows all about the house and how to make people happy. This doesn't mean she is less intelligent. It takes both kinds of brains to have a good family as decisions in the house are often made together with both man and wife saying what they see should happen.

An eight year old boy enters the gate in the hedge and quickly disappears behind the house. Jacob beams.

That is my grandson, Tobie. He is coming home from school. I wish my granddaughter was as trustworthy as Tobie. I have two sons, both live in South Africa. One was divorced three years ago, It is the law that with divorce all children go to the father. That is how we got Tobie, as my son could not take care of him. The other son was divorced a long time ago. At that time he gave us his infant daughter, Dorothy, to raise while he worked. She is out having fun in the village with her friends while her baby is inside. She thinks Betty should care for the baby. Betty is old and is raising Tobie so she says Dorothy should tend to him. Dorothy is too young for me to give her responsibility for the household. Like other young adults, she gets caught up with people in the village and thinks of herself. She is worthless as a grandchild. She does not try to achieve, but lets nature take it's way.

Tobie comes out with a bucket to get the evening supply of water. Betty pokes her head out the door, a four month girl in her arms. She asks if Dorothy is home yet. Jacob shakes his head. As we disperse, he calls out "Come back next year."

<u>December 19 1990</u>

I have seen Jacob and Betty many times since returning to Ramotswa. Both have lost weight and have become disillusioned with old age. The pension was only for twelve months, not life. Tobie's father has not sent money for three months. Dorothy's father has not been heard from for seven months. He still does not know Dorothy had a second baby in November. Both of her babies receive free supplemental formula, as Dorothy is unemployed and the children are underweight. Tobie gets a meal at school.

Jacob and Betty applied for destitute funding last month but were turned down. Betty said the social worker entered the house and saw the table and dresser. "This made her think we are rich, without thinking that we could have had money before but live with hunger now. We still have one cow that we want to save for the funeral of whoever dies first." The social worker claimed the sons were capable of supporting the household and that the cow represented available money. She expressed anger that so many aged must live in poverty because adult children think the government should support the aged and so do not send remittances.

This afternoon Jacob is at his cousin's house . Betty and Dorothy are cleaning for the Christmas season. The family expects tradition to hold true, with the two sons arriving sometime during the holidays. Lack of communication does not dampen their hopes. Betty makes it known that she wants the latrine spotless and leaves to clean it herself. Tobie, who has been sitting by himself, is told to carry the supplies. He continues to sit as Betty gathers rags and leaves the house.

Dorothy and I are alone in the house, except for the two babies sleeping on the floor and Tobie, now in another room. I use this opportunity to discuss her role as the care provider.

> My grandmother is never happy with the way I clean. She is always moving a chair slightly or rearranging the ashtray on the table. Both my grandparents are children, unable to think or do anything for themselves. They want me to do all the work but they will not turn the house over to me.

Using the ladder, she rates their physical abilities as in the middle. She explains that they can take care of their bodies but have many years. In ranking them in comparison to the person who has money and objects to give away versus the person with nothing, she places them near the top, yet she knows about the discontinued pension and infrequent, irregular remittances. "They are the ones with the money. They just won't give me any, except to buy food at the shops."

I put the ladder away. I ask how she would envision her life if she did not have to take care of her grandparents. She pauses before saying:

I have never thought about that. It is my duty to be here. I have no thoughts on the future. My future is caring for them. They direct my life. Old people can give good advice at times, but usually they are too old fashioned. I need my grandparents. They are the one I talked to about birth control and breast feeding the babies, but I did not breast feed for long. It was easier to get free formula at the hospital. That way we all get a little bit of milk.

Dorothy glances out the window and sees a pair of her discarded panties that Betty has put in the roadway to be burned with other trash later in the day. Visible shaken, she runs outside to retrieve them. Coming back inside, Dorothy loudly cries:

she shouldn't have done that! You never know when a witch may come along and take them. Then all sorts of bad curses could be placed on me. I have enough forces that work against me. Today began with the sun not waking me up in time to wash before the children needed breakfast. Then, as I was going to the shop for my grandmother, a rock got in my way and I hurt my foot.
My grandparents do nothing right. They are children. They yell at me, and I yell at them. All the time, we scream. They are so useless. Why can't grandparents behave like they should, sit and take life as it comes? Then my life would not be so hard.

<u>January 8, 1990</u>

Betty is taking me to her land today. She grins while saying "The free tractor plowing and seeds from the drought recovery program makes for much work now." She sends Tobie to school. He is wearing new shoes much to large. Dorothy is tending to the babies. Jacob says he has had enough visiting recently, and will help Betty and I weed.

The sun in hot after the 45 minute early morning bus ride and an equally long walk. The couple wonders if they will accomplish much, as there is no shade or drinking water. Betty hangs her head in shame as she says:

I used to be so proud of my lands as they were always clean of grass. Look at them now. Dorothy says it is old fashioned to grow our food, but we need food. We wouldn't be able to plant at all if it wasn't for the free plowing and seed from the drought relief program. I don't know how we will live next year when the program stops, for then we will be more hungry than we are now. The program was to stop this year. Because this is election year, the government was afraid we would change our votes if they stopped it.

As we work, the topic of conversation turns towards Christmas. Jacob begins by saying:

> we waited and waited for our sons. Tobie's father came two days
> afterward. It was good to see him but he had to leave that afternoon to
> visit people in Gaborone. He gave us a tin of powdered milk. You
> know, we've been using the babies' formula in tea so we don't get sick.
> This morning we had a fine breakfast of tea with lots of milk and a little
> sugar. Tobie was given the new shoes. While our son was here, he also
> gave us enough money for school expenses. The other son never came
> and never sent word.

Jacob leaves to search among the two hectares of maize for randomly planted squash plants. Betty speaks quietly.

> My heart is very sick today. The sickness is caused by Dorothy.
> She makes my heart very heavy. She does not try to achieve. She yells
> at us all the time. We have talked and talked to her, even Jacob's uncle
> spoke to her. Yesterday she slammed the door in my face. We took her
> to the headman, who is my distant uncle. He told her to obey her elders.
> She has behaved better since. It brings me great shame to say such
> things about my family. Jacob and I tried to be good parents. What
> made things go wrong?

<u>March 6, 1990</u>

Earlier this week Jacob was ill and told me he did not want to go to the hospital. The combination of problems with children and the sale of his last cow for food last month has made life "very heavy." His money is gone and the crops are doing poorly. Today, his softened voice indicates the depth of his depression.

> I feel the cloud of death. When I die I want my family with me.
> No one will talk to me about getting well. That is not our way. I will
> tell them what they should do in the future. To me, it is important that I
> die on my back and someone holds my legs straight. It is very bad to be
> curled up when death occurs. I do not know why. The church says there
> is a heaven with happiness, and a hell if you are evil. If people cannot
> be evil, as only evil forces, not evil men, exist, how can there be a hell?
> The same is so about the Christian heaven. It is better to be an ancestor
> (*Badimo*) with the breath in the skies.

Betty nods in agreement and presents her view. "A person decides for himself if he is sick and if medicine is wanted. I could force him to go to the hospital but I must respect his wishes." Dorothy's disagreement with Betty carries no weight, for she too agrees that God (*Modimo*), not doctors, controls life and death.

<u>March 11, 1990</u>

When Tobie entered our compound at sunrise I knew the message he was carrying. The funeral would be in 10 days, a week from Saturday, giving time for the dissemination of news and arrival of migratory family. Ignoring the increasing daytime temperatures, I dress properly for visitation, in long sleeves, a shawl, and a kerchief on my head.

The family had begun mourning rites by the time I arrived. Betty was secluded in the *rondaval*, face down on a bed of goat skins. Male cousins had arrived before day light for the customary guarding of the gate. A distant aunt was heating water for tea as she carefully stepped around the quiet children. She introduced herself as Auntie, saying that Jacob and Betty think of me as family and that makes me part of her family also. Other women kin were beginning the brewing of *bajalwa*. Dorothy, reached in the hut to remove Betty's breakfast dishes, being careful not to touch her, or her belongings, as direct contact with the widow is forbidden. I stand in the shade of the hedge. Other than Auntie, no one has spoken.

Later Dorothy tells her mother of visitors who have arrived. Betty requests to see her niece and her daughter (myself). We receive Betty's solemn, vacant gaze as we enter the *rondaval*. The door closes behind us, and a black cloth on the window isolates us from the living world. Betty is laying on a goat skin. She will not sit or stand until the time of burial.

Mourning is a time when all thoughts are centered on the dead. Only female kin may sit with the widow. Along with touch, speech is prohibited except for barely audible, needed commands and exchanges. Several hours of silence pass until Betty nods, indicating we may leave and other kin may enter. Before many days pass, her growth of pain will become so great that visitors will no longer be welcome. Dorothy will only place food and water in the room.

The silence continues in the courtyard. Women sit near the house. Men congregate by the gate, where a small table has been placed. Mourners drop a *Pula* or two into a tin can; money to help defray funeral expenses. Dorothy hands me a cup of tea as we slip into the back yard with the cooks, all kin. Only there can we converse.

Dorothy has assumed household decision-making, and feels a mix of security and uncertainty in directing events. Observations of previous family funerals has taught her must be done. Dorothy tells me all that she has arranged, as she feeds the smallest baby and keeps an eye on the other toddling among the women. Tobie is in school. Phone calls to the sons, and other kin, are being made. Some will come immediately; others later in the week. The village mortuary will soon bring a tent to provide shade. The burial society, of which Betty is a member, will be buying $500 Pula worth of grain, sugar, and powdered milk, as visitors must be fed to maintain strength. Dorothy stresses Auntie's wisdom and guidance as a big factor in getting through today's activities. No mention is made of Jacob or of her own grief.

<u>March 20, 1991</u>

It is the day before the funeral. The congregation of people at the Mosoko house has grown each day. The grain and most of the bajalwa have been consumed. Tobie's father arrived earlier in the week to assume responsibility for the purchase and delivery of a cow. It was slaughtered near the hedge earlier this afternoon, and is being butchered for tomorrow's funeral feast. A select piece of meat hangs in a tree for Jacob's brother. Various males plan how they will cook the meat. Young babies are tied to their mother's back. Otherwise, there are no children. Dorothy is in the hut with her grandmother. Adult women hurry with other food stuffs, while aged women, including Auntie, wash dishes. It is almost time for the "parade of the corpse."

A pick-up truck arrives from the mortuary. Women, sitting in the side yard, watch as men in the center yard step aside to made an aisle leading towards the hut. Jacob's closest male kin carry the coffin to a place of rest at the partially open hut door. Over a hundred people join in the service of prayers and songs lead by the Christian preacher. Afterwards, the pallbearers slowly move the coffin into the *rondaval*. Silently, the group departs.

Betty and Jacob will spend his last night on earth together, alone, and without interruption. The cooks will return to the open fires, some staying all night. The brother will sit by the gate long into the evening. All will be quiet with thoughts only on Jacob.

<u>March 21, 1991</u>

The final preparations for the burial feast begin with sunrise. The meat, already cooked, deboned, and shredded by male relatives, is made into *Tshwaiwa*. *Mosokwane* (sorghum boiled to a solid state) and *samp* (corn mixed with butter beans) are prepared by younger women. Older women are making tea and bread. The mourners arrive, slowly at first, and then in larger numbers. Many stand outside the hedge, as no room is left in the yard. Tobie and the babies are not seen.

The coffin is carried from the hut to the center of the courtyard. Ministers conduct a second, more elaborate service. Bouquets of vividly colored plastic flowers, wrapped in thick cellophane, are handed, one by one, from a table to the brother. He reads the attached cards. "Your passing leaves much pain in our hearts, your nephew." "You remain dear to us, rest in peace, your cousins." Men slowly circle the coffin for a final, somber glance through a small window over Jacob's face. Women slowly enter the circle as men return to their places. No sounds are made as Betty is escorted from the house by Auntie and Dorothy. As her weak body slumps down, she leans heavily on the two women. They follow the pallbearers towards the truck, which will lead the funeral procession to the church.

Only a few, more distant, kin remain behind. Out of respect for the dead, the yard has not been swept since mourning began. These women must now sweep the remnants of death from the houses and yard. The *rondaval* and ledges are to be smeared with fresh mud mixed with a heavy amount of dung, signifying continuing life.

The funeral procession increases in size during the four kilometer walk to the Lutheran Church. Four or five hundred people squeeze into the building and stand outside the doors. Today, as on most other Saturdays, it is a shared funeral service. This time the prayers, praises and songs are for Jacob, a six year old boy, and 37 year old man. As the Lord's Prayer draws the service to a close, the immediate families accompany the coffins to the trucks. Other follow on foot in the final procession towards the cemetery. Some of the very old and ill leave, as the cemetery is another twenty minute walk over rocks and stubble and the noon-time heat has become a burden.

The group hangs back until Jacob's coffin is centered over the grave and Betty is seated. A few more words are said by the minister as coffin is lowed into the ground. Jacob's brother begins shoveling. As he tires, other male relatives and then other mourners, take over. Two men, of more distant relationship, gather stones to mark the outline of the grave site. A large basin of water for the washing of the hands is shared by the men.   A few women and older men have stepped from the group to sit on the ground. Most stand during this hour of burial. When all is finished a wooden cross, bound with black plastic and colored ribbons, is placed over the head of the grave. There is no public weeping or tears, either in the preceding ten days or now.

Muted conversations begin as the congregation departs. The crowd forms into smaller groups., Some people begin the walk to their homes. Others join those from other funerals. More than one hundred kin begin the walk back to the Mosoko

compound for the funeral feast. Only the fast of foot manage to find a ride from the

few trucks parked nearby. The young women squeeze together to let Auntie sit on the

tailgate, but there is no room for her aged brother.

Dorothy and Auntie are silently sharing Betty's pain within the hut when I arrive

back at the compound. Guests are seeking a place to sit on the still damp mud. The

scent permeates the air. I sit with the women. The funeral feast begins. Holding out

my palm, a niece uses her fingers to give me a scoop of *sump*. Gradually all the food is

distributed, and is consumed slowly with the fingers. An occasional guest will shake her

head in refusal of a serving from one woman, an enemy, but willingly take the same

food from another. No liquids are served.

Some of the guests begin to leave. I move towards the cooking area. Dorothy's

father joins me, in order to introduce himself.

> There was much work to do in Johannesburg. I couldn't leave
> until the middle of last night. I was afraid I would not get here in time
> for the burial, but I made it to the church. Mother saw me come in and
> I know she was happy. I knew Father was sick and would die soon. I
> wanted to visit him but had no time. Instead I gave $600 Pula {300 U.
> S. Dollars} to the burial fund. That is much money and I can be proud
> of what I did.

> I have not yet talked with mother. Mother will stay in the hut
> today. She was fed the feast foods by Dorothy. It will probably be a
> couple of days before she can sit up and go outside, as she is very weak.
> Another difficult time for her will be in winter. She will have to take
> Father's belongings out of the house, and share them with the family.
> Long ago father designated certain things to certain people, the rest will
> be given away by Mother. The taking out of clothes is a very sad time,
> as Betty will picture Jacob and cry bitterly. She will wear a black dress
> for a year and should stay in the compound a month or more. I have
> black necklaces for the children to wear to demonstrate their mourning.
> I will wear a black cloth on my right arm to show I lost a father. The
> next year will be a very dangerous time for us and those who talk to us.
> We must warn others with the wearing of black.

The heat of the afternoon sun is beginning to wane as I walk toward home with Sego. Sego shared the same grandfather with Jacob, with friendship building on the kinship. I let him lead the conversation.

> Jacob was master of a fine family. He and Betty upheld the law in raising their children. I will miss Jacob, although I can see why he wanted to pass. As Tswana, we do not share our troubles with people but I could tell he was worried about hunger and poverty for his wife and the children. He never talked about his sons. I knew them as children, and they were good. One gave a daughter to help in his old age. The other son gave Tobie. Grandchildren should provide much joy, but outsiders never know what happens within the house.

> I don't understand funerals anymore. We used to serve the grains and beef from our lands, and bury the dead in the yard. Now there are fancy coffins, and much food to buy. Children want to show everyone how much they can spend, and how rich they are. Sometimes they rent a fancy hearse for their dead mother who lived life with nothing. Why don't children give to their parents while they are alive and need assistance, instead of waiting until they are dead? Then, with funerals, they spend too much. There are many hungry people in this village. For the next few months, there will more even more hunger. Yes, funerals make for too much unnecessary hunger, There are too many funerals causing too much hunger.

The Family as a Setting for Exchange

The Tswana family has been described as a happy family, with laughter and gaiety (Schapera, 1967). Walk into any household and you, too, will be met with warmth and an eagerness to assist. It is not until the cover of inhibition is replaced with security that the outsider has an opportunity to see the additional panhuman emotions of anger, sadness and worry; the same emotions shown in the previous portrayals.

Some of the situations stimulate empathy. Anyone who has lived with an aged parent can attest to frustrations with the rearranging of knickknacks, and the complaints that chores are not done correctly. Other portrayed situations may have created negative emotion, such as a desire to rebel against the induced pain with funeral practices, or the hoarding of time and belongings that lead to the suffering of others.

Particularly in this chapter, the observer's emotional reactions must be put aside. Tswana behaviors cannot be judged by Western tenets and called abusive.

Abuse, as perceived by the Tswana aged, is the purposeful withdrawal of available food, care or shelter. An unserved meal or being left alone occurs in all households and includes all family members, not just the aged. Usually it is attributed to the impact of outside, uncontrolable forces, as with Kaizer failing to repair the roof. In these cases, any accompanying suffering is not perceived as wrongful abuse. It is acceptable deprivation due to nature getting the upper hand over good people.

This does not mean people do not react when others behave inappropriately. Dorothy felt put-upon with the pantie incident. Betty felt put-upon being expected to tend to Dorothy's infant. Even so, these two individuals lacked what we consider feelings of abuse. There was no assigned guilt involved with the doing of wrong. Again, wrong is nature controlling life.

Ideally, suffering from wrong should not occur. When suffering does occur, it is because of the way things are. Laws are broken, such as with yelling. That is wrong, but the results of broken laws do not appear to be considered directed abuse. It is more like uncontrolable suffering for which no one is really at fault. Dorothy believed Betty had no control over the discarding of the panties, and Betty saw external events controlling Dorothy's time away from home. A person can try to achieve, but nature determines the outcome (Alverson, 1978). Dorothy was taken to the headman to create shame and stimulate her to work harder at the process of achieving, not as punishment for psychological or physical abuse.

My own concept of abuse had to be temporarily set aside as I learned more and more about the Tswana value system. When using the Tswana concepts, I found no culturally-defined abuse. Some of the observed deprivation resulted because the aged refused to alter laws to their own benefit. The refusal to use another's belongings created personal hardship in a household where it could have been avoided. This self-denial was part of the Tswana way of finding integrity in achieving.

No one denied the aged food if they asked during meals, or housing when a roof was blown off during high winds. I found that sick or extremely hungry aged, including those who lived alone, always found temporary assistance from kin if help was actively sought. (Hesitancy to ask was from the awareness that others also live in need, and/or from fear of sorcery.) It is not polite to constantly ask a particular individual or household for food or money everyday. A few of the very poor say they "have run out of kin to ask." Somehow, even these individuals find meager, but sometimes life-supporting, food.

No sexual misconduct towards the aged is ever mentioned. I also doubt if it was present, as laws regarding sexual activity were strictly adhered to. Rape is virtually unknown. No aged acknowledge ever being beaten as an old person. One woman's complaint is that the new law regarding beatings prevented her from knowing where she stood with her husband!

The Tswana perceive situationally-induced deprivation as quite different from abuse. Suffering induced by such lacks as material goods is seen as the usual way of life. Being cold from the lack of a blanket, or uncomfortable from a lack of a bed or chair, is interpreted by the aged as a result of self-poverty. Adult children and other family are not faulted. Additional goods may be desired, but they are not expected as gifts. Only the most affluent households may give the aged a blanket or cloth slippers at Christmas. Usual gifts are a bar of soap or a box of tea. Migratory children, returning for a visit, usually bring a a few food stuffs to be shared by the household. Such items are seen as gifts, not as part of the obligation of support.

Many people, of all ages, suffer from some sort of economic shortages. Whatever extra money is available is spent by the producer of that money, usually for his or her own good. This is a practice, not a law. Many present day aged report accumulating items for themselves over the purchasing gifts for others. They never mention that a law regarding income was gradually changed, even with questioning.

It seems that the present day working adults are building their own future gerontic fund, just as their parents did. Their first purchase continues to be a bed, then the dresser and other furniture. The middle-aged sense that if multiple blankets, radios, watches and such are not obtained soon such items will never be obtained. Like their parents, they do not expect such gifts in old age.

As each family varies in the degree of deprivation, so each family also varies in the breath and depth in the maintenance of traditional laws and the incorporation of Western thought. My purpose is to provide understanding of how these social practices unite in family structure and functioning, and thereby influence the value and use of the gerontic fund. Family structure is defined as at the communication network with a family, with family function encompassing family supportive behaviors (Ryan and Austin, 1989). Each are separate and important variables in the process of social exchange as family structure and function are the doors connecting household members with each other.

<u>Family Structure</u>

The two main influential variables in structure, repeatedly referred to by the aged, are discourse and decision-making patterns. These two factors transcend family size and ages of the members. Patterns are similar between household, always reflecting aspects of cultural continuity, and some degree of conflict between tradition and change.

<u>Family discourse</u>

The teaching of every young child begins with the cultural rules of discourse. The majority of such regulations promote power in the elder: only the elder may show displeasure in the face and tone of voice, one does not criticize an elder, one remains quiet when elders are talking, one obeys an elder without questioning. Additional rules govern all conversations: one must speak in a quiet voice, one must listen until the other is finished, one must not tease, or play word games in jest with another, one does not display pride or boast.

Each family then personalizes methods of communications between sexes, generations, and kin relationships. In some households, women do not partake in men's "important" discussions, as with Betty. In others, women listen and contribute. There is also much variation with expressions of pleasures and unhappiness within households. This may be done verbally, although custom prefers non-verbal expression with smiles, frowns, or a leaving of the area. Such traditional constraints do not constrict knowledge of another's feelings. As one young adult said, "We just know how others feel, if they like us or not, if they are happy or sad. We know by the way others act."

The aforementioned rule mandating the use of indirect criticism does not limit its use. Complaint discourse is an intrinsic part of conversation, although no names are mentioned. As such, complaint discourse allows for the rightful expression of power based on age. Its use is almost exclusively by eldest household member(s), who have a vivid flair in denouncing all and sundry behaviors in the household. It is not a time for argument or refutation. The comments of the elder are final, and no direct retort is permitted according to traditional discourse laws. This does not mean that response is lacking, as the listener participates non-verbally with head shaking and eye expressions. All know that the comments contain elements of truth and fiction, as with Happy Sound and her family. Complaint is her way of controlling behavior and stressing the need for adherence to traditional law.

Negative discourse helps seal family bonds, as each person can recognize themselves and others. The object in to promote introspection for all listeners, as statements are broad and harsh. Comments like, "No one gave me food today" are not to be taken literally. The daughter realizes she skimped in cooking time, the son thinks of the money spent on himself instead of the family. Happy Sound, with repeated questions on future care, was not implying a fear that she would be deserted. Instead, she was making known her family's failure to serve and defer, according to her perceptions of rights and proper treatment.

Disagreement, separate from complaint discourse, follows a different course. The aged report that, based on tradition, the eldest person rules with no back talk permitted. They say argument was unknown in the days before civilization. If this held true, why the reported use of uncles, headmen, and lastly the chief to settle household differences (Schapera, 1953:40)? No doubt dispute has always been present in household. The question is how verbal did such differences become.

Today's open expression of conflicts between siblings, and between children and parents, creates obvious anxiety in aged parents. To the aged, acts of yelling extend beyond the purposeful breaking of the laws. Open arguing between any family members presents a threat to their status as elders, as it overtly shows that the aged member lacks control over the situation. The yelling and fighting between adult children is the most mentioned cause of unhappiness by the old. Participation in yelling is perceived by the aged as an undesirable source of power, as it indicates a lack of self-control. In contrast, the discarding of other discourse regulations does not create as much anxiety. Many of the aged break the rules themselves, as they talk out of turn, interrupt, boast, accuse, and praise directly.

On the other hand, the younger generations accept open and loud disagreement as a valid method for anxiety reduction in times of conflict. To them, fighting between siblings is not associated with respect for elders, and argument with parents is viewed as a necessary a step for the now desired independence. It is not used as a purposeful status-reduction mechanism, directed towards the aged At times, it seems the aged can grow in status if they discard the old law and can prevail in the shouting match.

Unresolved disputes threatening household functioning do exist. Custom continues to be followed, with aged male kin outside of the immediate household, making a final decision. Both sides of the dispute are presented and weighed. Recommended action for involved parties is presented, and usually followed. The power conflict between Dorothy and her grandparents was partially resolved by asking the older uncle to decide on behavioral guidelines for achieving. When this failed to control shouting, the headman was used to bring about resolution of conflict.

What is said, and how it is said, remains private with those directly involved. Neighbors shy away from listening to loud conversations next door. All discourse regulations are applied more rigidly to extended kin outside the household, and even more rigidly to non-kin. Internal family disputes and difficulties are never shared. Sego, one of the more progressive aged, had never asked Jacob about his personal life, although they were kin with friendship. The refusal of Happy Sound to talk about the status and care of other old people was because she really did not know how any other families treat their aged. Gossip, if spread, is nameless, referring to a woman or a man. Only when shame is made public will others have a hint of what is occurring.

<u>Family Authority and Decision Making</u>

Emphasis on the subordinate role of women in African societies has promoted interpretations of male dominance and independence involving decision making. Historically, males controlled the family and compound, with feminine domestic duties performed under the guidance of the senior male (Schapera, 1953). This thought perseveres with the idea that the lack of male leadership is equally as threatening to household security as the lack of a bread-winner. (Ingstad and Saugestad, 1987). Only recently has consideration been given to the idea that African women can, and do, play an active and authoritative role in the management of comprehensive family well being (Coles, 1990).

The life histories of the aged indicate women's participation in decision-making is not a new phenomena. Jacob, like many others, found value in his wife's thoughts, with joint decisions made regarding cattle, agriculture, employment, style of housing, and education for children. Other aged report similar spouse interactions, with aged single men still seeking wives, reportedly to assist them in this arena. With some families, men nodded in agreement as women laughingly claim to be in total charge of the household, turning to men for the automatic stamp of approval.

The turning over of direct household responsibility to adult daughters and sons is a long standing tradition (Schapera, 1966). Adult children see this as including the broad scope of all decision-making. Priscilla's daughter, who was presented earlier, assumes all decision-making with the formal transference of domestic duties. Alfred, as a husband, continues having indirect power over his wife's daily activities, with neither having control household functions.

This transference of duties may also evolve without formal assignment. Leru, as the household bread winner, assumes the financial decision making, and extends her power to include other areas. Only in retrospect do the aged recognize the impacts of transference of duties. They do not foresee the congruent label of child and the accompanying family behaviors. It is the very rare household where the aged have supreme authority. This occurs only when a aged male is fully employed and no formal transfer of duties has occurred between mother and daughter.

The Mosokos are actively delaying the traditional "turning-over" act, saying that Dorothy is not yet of age for this to occur. Even so, the aged couple does not have complete authority. Some of Dorothy's antagonistic behavior reflects the decision-making power struggle commonly found between parent and young adults. Young women, in particular, are striving to make decisions independently of family input, with motherhood and employment providing excuses to do so (Suggs, 1986). Younger village men, especially if unemployed, seldom sought household authority. The deciding factor for the placement of control is often that of economics, independent of gender. This modern "household elder", who is not necessarily the oldest person in the family, controls and benefits from discourse regulations.

The one accepted exception to family control by the household elder is with sickness. Any adult is allowed to decide the course of action for personal illness. Others may be told, and advice sought, but the decision for action remains that of the individual. In contrast, the ill child is under the control of the head of the household. Jacob, although becoming old, was still the head of the household, and therefore, was not forced to seek medical treatment. This does not always hold true when the aged are perceived as children and lack aspects of elder authority.

<u>Family Function</u>

Descriptions of Batswana family function, in terms of supportive behaviors, usually center on the positive, or the help-giving and health-promoting behaviors of the members. (See Biesele and Howell, 1981, Suggs, 1986, Ingstad, 1989, Rosenberg, 1990.) Equal attention should be placed on the stress-producing and problematic behaviors, as these also influence the setting for social exchange. In the following I identify behavioral trends providing both support and stress from the emic point of view. The outcomes of these behaviors reflect the *goodness of fit*, or the degree of balance between the needs and demands of the aged person and the family environment (Ryan and Austin, 1989). A goodness of fit can be considered as working doors between the aged and the family. The degree of balance in relationships also influences the balance in exchange relationships. Not all households behaviors are mentioned, only those repeated mentioned by the aged. Much of the data comes from the felt board exercise, which expounded on praises and complaints mentioned in formal interviews, and the ladder questions that emphasized the degree of balance between the young and old.

Commonalties in all households exist, regardless of the degrees of balance. Meals are portioned equally between household members. Personal gifts and purchases of snack food belong to the individual, to share only if desired. Food entering the family either from child feeding programs or destitute funding for a specific individual is also shared equally. Some aged give part of their food to very young children in cases of food scarcity. This latter fact was noticed to occur mainly when all household members were underweight, contrasting with the overweight of affluent families.

Many families, regardless of household income, face problems related to high alcohol intake with one or more members. Unemployment also affects the majority of families. Blame for any problem is almost always placed on outside forces, either someone else or metaphysical powers. Dorothy reflects the commonness of fatalism that predominates in reacting to the present and thinking of the future. Few people actively plan for the future, considering it as unpredictable.

Family members consistently rank the aged higher in physical abilities than the ranking the aged gave themselves. Using the ladder, with the top rung as the person who can do everything and the bottom rung as the person who can do nothing, care providers usually place even the most physically limited parent on the middle rung or higher. Comments are, "She can feed herself if I leave food." and, "He can walk to the toilet." Most aged place themselves on the bottom two rungs. The reasons for these generational contrasts in perceptions are not clear, although two distinct probabilities exist. Either the adult children are giving praise to parents or are justifying their absence from the aged for long periods of time.

These commonalties among separate families has little influence in what Ryan and Austin (1989) call the "goodness-of-fit" between the aged and the young. It is the intermingling of strong points and faults of individualized family functions that determine the balance between generations. This balance will now be discussed, using the two family scenarios as a foundation.

<u>The aged in balance with the family</u>

The Kereng family exemplifies the assorted everyday problems found in Tswana households. For Happy Sound, the rewards generated by daily behaviors of the adult children and grandchildren outweighs stress-inducing events. She, like other aged in this group, see her family as one that works together. Their felt-board pictures contain actual family members performing chores necessary for household functioning. Usually children are playing, adult daughters are cooking or cleaning, and adult sons are going to work or tending cattle. The degree of personal decrepitude and household poverty are not major points, as they see themselves as having a place within the unified family. These aged feel accepted and needed. They have reasons for happiness. Some are happy to be loved or wanted, with others happy to see the sunshine, one day at a time.

The felt-board pictures parallel their own on-going lives. These aged men and women perceive themselves with definite roles and can be observed performing them. For some, the roles are more modern than traditional, with the performance of tasks previously assigned to the young. Others are helping children "grow-up", teaching the laws, and providing guidance. The personal process of continual achieving is contained in all these roles. Interdependency between household members is always highlighted.

The aged, of families in balance, describe family members as adding meaning to their lives. When asked to explain what the young children in their picture actually do, comments include aspects of emotional and physical caring. "He rubs my body when I am tired." "She listens to my stories of the days of becoming civilized." "The young children make me laugh." Adult family members are similarly seen. "My son cannot afford to give me things but he sits with me." "Her behaviors show me she really loves me." Such overt signs of acceptance, and being regarded as valued household member, contribute to old age security.

By custom, expressions of love are non-verbal and build on the sensual components. Evening visits to these families will often find them laying on mats, massaging each others backs and extremities. Conversation is not a necessity to the feelings of belonging and family unity. Kaizer's frequent quietness is not a sign of rejection but a symbol of respect for his mother. His seemingly disregard for others is a provisioning for personal space in an environment without much access for physical privacy. The grandson's entrances into the hut during the interview is a protective, rather than curiosity move, done so to ensure all is proceeding satisfactorily and safely. These are the families where private interviews were often an impossibility.

Stressful situations, which occur within the family in-balance, tend to be handled by the aged with coping mechanisms that reduce anxiety. Happy Sound interprets the delinquency of her son in repairing the roof as being due to illness, not the self-centeredness claimed by his sister. When sibling arguments occur these aged are just as apt to yell back or leave the compound as to separate themselves with pretend deafness, as Happy Sound did. Some sort of retribution is usually made afterwards by the guilty parties, such as Kaizer asking the traditional permission to leave. Other stressful situations, such as the use of alcohol, are not easily resolved, but the overall goodness-of-fit provides some protection.

The family-in-balance has accepted the fact that Westernization has altered the deliverance of traditional services for the aged. Wage migration has diluted the presence of family members. Time commitments for those remaining have increased. These changes did not negate traditional care-delivery. The family accepted its aged member, interweaving everyone's individualized, and diversified, version of the roles of modernization and tradition. Conflict was resolved near the fulcrum between the extremes, with family behaviors predominantly centered on buffering hardships in life.

In Happy Sound's eyes, she still has someone to cook for her. She also has grandchildren to serve her, although they are in school much of the day. Happy Sound is like other aged living in families-in-balance. She is able to place traditionalism in a modern setting and accept or resolve the conflicts between being an elder in her own eyes and frequently that of being a child in the eyes of her family.

<u>The aged out of balance with the family</u>

Being out-of-balance, or in a state of disparity is said to exist when the aged and the immediate family have multiple unresolved problems based on differences in the approach to life. Like all aged, the aged of these families stress cultural continuity, and the obedience of traditional laws in family life. The adult children are no more progressive in thought, or behaviors, than others. Young children accept, and refuse, commands with equal frequency as those in families in balance. Unlike the family in balance, these families are unable to mitigate differences.

Disparity was found in all types of households, large and small, rich and poor, educated and uneducated. It is at a minimum when the central care provider is also aged, as with Martha caring for Maria, and greatest when adult granddaughters are the central care providers. The numbers of households with disruptive disparity appears to be large. Of the 33 aged who created their ideal house, 70% either completely refuse to put in any family members (8), put in only one young grandchild (6) or one specific adult (8). In all these cases, multiple kin are residing in the household. The reasons given reflect the extent of tension arising from conflict. "If you have more than one, it just causes too many problems." "I want only my oldest daughter here, not the others. The others should get their own house, as they disobey the laws." "One person would take care of me, the others cause suffering and pay no attention to the laws." Visitors are sometimes added to the pictures. "I want to live by myself. I will have visitors and they will treat me nicely. They will stay a while, go home, and not ignore me or cause problems."

Nearly two/thirds of all aged feel disenfranchised from family affairs. "They refuse to listen to me as master, and pay no attention to what should be done." "They disregard me, and want to do things the new way." "Family makes a lot of noise and much work, then do nothing for me." When these individuals are asked how they see themselves as actual family members, the replies consistently reflect a lack of family cohesion. "I am just there." "I am nobody." "I am not free and cannot give of myself to them." "They treat me like the living dead." Other family members would reply to these comments. "Mother, you are too old fashioned." "You are out of date." "Mother, you are stupid and do not learn the new way."

Labor migration, the absence of family members, proliferation of small nuclear families, and the new diversified roles of women have influence on the balance the aged find with their families. All these factors influence the selection and availability of family care providers. Contrasting with balanced families, where someone willing assumes the position of care provider as an act of on-going tradition, families in disparity see old age care as a mandate stemming from outmoded tradition. Selection of the care provider tends to be haphazard and does not follow traditional norms. Some of these care providers are like Dorothy, given to the aged by migratory adult children for the express purpose of care. This releases the adult children, who no longer live locally, of direct care responsibility. Dorothy has no choice but to live with her grandparents and often resents having to serve them. Others are placed in the care provider position by default: "My brothers and sisters moved away and I was the only one left." "I am the only child who never married, so it is my job to take care of mother." In these cases, the aged who had centered nurturing on a specific child in anticipation of care provisioning, frequently have their plans disrupted. Resentment exists on both sides.

While an emotional bond exists between the family out-of-balance and its aged, it lacks closeness and understanding. The aged seek roles reflecting seniority and family interdependence. Younger family members perform roles and duties with the western epistemological stance of self-autonomy. Dorothy, like others her age, dislike the forced subserviency and dependency stemming from the seniority principle. Decision-making, aimed at self-good, directs many of her behaviors. At the same time, she admits a need for having grandparents, and had concern for their well-being.

Resentment does not prevent using the aged as advisors. Grandchildren, like Dorothy, often turn to the aged for assistance in personal matters, especially those involving interpersonal relationships. While the research does not specifically concentrate on grandchild/grandparent bonding, a special relationship involving closeness on the part of the grandchild is frequently seen. A grandchild will name the grandparent as a confidant, but the reverse never occurs. With families out-of-balance, the aged feel very insecure with the generation gap, questioning if grandparental advice was heeded. The degree that such insecurity is colored by generalized family tensions is unknown, but is surely present.

The aged of these families tend to internalize family conflict. This is the group who asked rhetorical questions such as "Did I go wrong?" and "What happened so they do not keep the laws?" Yelling is seen as a personal deficit, rather than disagreement between two individuals. Passive behavior is the norm when arguments occur, with the aged concentrating on the loud words without interfering or separating themselves from the scene. These aged also overlook their own requests when orders are ignored by others. There is usually no second, more authoritative ordering.

Independence and autonomy, desired in the Western-style family, conflict with shared functioning (Streib, 1972). Shared functioning, with family coordination of efforts in the process of achieving, is a hallmark of the traditional culture. The present day aged applaud individual achievements involving free thinking and self-motivation for social success. The difficulty lies in the fact that self-direction and decision-making have altered the interdependency occurring with shared function within the household. With families out-of-balance, the aged definitely become far removed from their traditional roles as gate-keepers for family decisions and actions. Family unification, if present, is controlled and directed by the young. It is not that the young have lost knowledge of the culture. Instead the pull of progress has distorted the push of traditionalism, creating a schism between generations.

<u>Extended Family Outside of the Household</u>

Extended family outside of the household must be considered in family functioning, as this is the unit that provides additional doors beyond the confines of the home. Discourse regulations may limit the depth of intimate knowledge about others, but not the scope of caring. The use of an extended family member as a confident is rare, the turning to kin for assistance is not. The reasons for requiring assistance need not be explained to them.

The interdependency found with shared functioning involving extended family is particularly notable with aged living alone. Sego, like other aged men, women, and aged couples without immediate household members, maintains self-sufficiency by utilizing extended family interactions. Some of these aged live alone by choice, others do not.

Many of these aged are satisfied with using extended family as informal care providers. They have an advantage in that the selection of kin to be included in their family function is based on congeniality and trust. All but two of the 13 without nuclear family feel secure in terms of continuity of care if it is needed. This is, in part, because kin regarded as enemies or non-caring are ignored, eliminating the common problems of fighting and household exclusion. In the selection process, little attention is paid to differences in economic, educational or social status. Observations of extended family exchanges, which include frequent visits and the giving of aid, support this claim.

The roles of extended family become paramount in times of interpersonal family crisis. Problems are openly shared among selected kin, with the opposing parties giving the two sides of the story. One or more elder male relatives make up the judicial body and pass judgement. The story told is kept secret. The judgement is usually made public to direct shame towards the guilty, leading to enforcement of retribution or correction of offending behavior.

This was the process used by Betty and Jacob to create shame for Dorothy. Betty felt shame only with the admittance of her own lack of control in achieving as a grandparent. She had no public shame. As the family viewed me as adaptive kin, I was in a position utilize public shame to place pressure on Dorothy to yield to seniority. Although I never spoke to Dorothy about the matter, it was obvious from the way she hung her head during the following visit that she knew the event had been shared.

If the internal family crisis increases in severity, additional extended family become involved. Marital disagreement threatening divorce is presented to a large board of elder male kin. This is more formal, with all extended family knowing in advance of the upcoming trial, but never the specifics. Afterwards, the extended family joins together to assist in promoting peace and insuring the decision. Only after this process fails are official courts visited.

The crisis of death immediately brings social support from male and female kin, both close and far in genealogy and geographical distance. Although supportive care during mourning does not parallel American or European behavior, support is there and it is meaningful. Sitting replaces touch. Silence replaces crying. It is a time to respect the need to be alone instead of showing respect with group involvement.

Betty, during her confinement, had continuous comfort in knowing that kin were performing the necessary rituals of greeting and guarding. Alone in thought she was never alone in the compound. To the Tswana, this was the way it should be. Confidence in Dorothy handling the situation was never questioned, as there was always someone, like Auntie, to encourage and indirectly direct.

The division of labor continues to depend on age, sex, and kin relationship during crisis with death. It is never questioned that work may not be done, or responsibilities never carried out. When a person becomes tired, another will take over without being asked. All kin contribute monetarily, physically and emotionally. This is the time when tradition reigns, with respect given to eldership and the principles of seniority followed. With the passing of the crisis, life returns as before.

The parting remarks of Sego after the funeral seem all so true. Why do fine, old-people fail to find emotional peace and security? Why is it that adult children give, or do not give, to their parents when they are in need? Adult children do have more money and material possessions than the aged, in spite of the generalized poverty. Many times they also fail to recognize the needs of the aged extend beyond that physical hunger.

Many aged function in a setting of hunger for continuity in customs, meaningful roles, and family interdependency. Some aged seek to fulfil their hunger in a liberalized territory where known rules for social exchange no longer apply. Others seek the traditional approach. The irony is that the segments of continuing traditional law also prevents fulfillment. Many adult children do care but are prevented from succeeding in helping through custom and poverty.

I have presented the family setting in which the aged must function. Changes in discourse patterns, the loci of power and authority, and the degree of shared functioning are but a few of the ways in which the doors to the good life have been broken. It is difficult, if not impossible, for the aged to find access through them. Some of the aged rely on tradition to keeps the doors working. Other aged modify their approach to the doors.

These same damaged doors of family structure and function also effect the behaviors of other generations. Some people believe modernization and Western thought should build new doors, with the government assuming some of the traditional family responsibility for care and economic support. Others want the aged to provide repair by adapting more Western and liberalized ways. For many, the conflicts between generations prohibit any repair of the damage.

This chapter has dealt with the doors that effect family interactions, There are other doors that the family, and its aged, must use. These doors operate within the setting of the community, the topic of the next chapter.

CHAPTER 8
THE AGED AND THE COMMUNITY

> The old people say we should listen. I say if there are opposing views
> between two people, each should go their own way as each individual is
> unique and has the right to be oneself. Let tomorrow take care of itself,
> as it will not be like today or yesterday. (21 year old woman)

> I have worked hard all my life to make Botswana a better country. I
> worked for the government and on the lands. Now they treat me like a
> dog. You know what a dog is? He gets kicked around. Lucky is the old
> person who can be somebody, for no one sees or hears us. (74 year old
> man)

During the past hundred years, modernization and Westernization have touched

every nook and cranny of the village. This was no accident, as new thoughts and

methods were disseminated from the top down by foreign and state governments and

carried upward by the young. (Harrison, 87:54). The aged of the present were and are

carriers of change, just as are the present day young.

Individuals are the mediators of social change, as they serve as the bearers of

culture (Social Science Research Council, 1953). Response to new ideas is set within

their perceptional framework as outlined by the past (Oliver-Smith, 1986:16).

Innovation incorporates tradition, with cultural continuity and social change existing side

by side. The old is negotiated with the new, which alters the approach to life. The

question is how much alien cultural material can a society incorporate and still preserve

basic values and patterns. As additional unfamiliarity is introduced, the new of the past

becomes the traditional and another renegotiation occurs.

One can assume that there will always be contradictions, opposition of interests,

plus real or latent conflict with negotiation (Social Science Research Council, 1953:977).

There can be no completely satisfying life to all members of a society. The aged, operating within a framework that carries the past back further, have historically larger perceptions of the paths that negotiation should follow.

The aged were socialized to society when sources of status were linked with ceremony, ritual and traditional power. They introduced new status sources of the Western kind. Employment has led to a conspicuous consumption of a modern life style and acquisition of necessary funds to abrogate the bonds of traditional obligations (Harrison, 1987:467). These easily adapted material forms of culture are gradually influencing the less amenable symbolic forms.

The village young are caught in the bind of social pluralism as they mediate new knowledge with old folklore, interpret modern beliefs in light of traditional world view, and negotiate new ethics with customary laws. Seeking satisfaction, the adoption of the latest Western fashions and lifestyles gives the young a symbol of supposed superiority, which they hold with contempt over the aged (Harrison, 1987:54). The resulting fission and fusion breaches the bond between being Tswana and being Batswana (the people of Botswana). The paradox is that, in seeking the new national Batswana ethnicity, reliance for direction stems from traditional Tswana tribal thought. The irony is, that with the forces of time, today's young are very apt to be without the what they hold dear at the present.

In much of rural Africa, the primary social struggles are not just between traditional and new ideas, but against a social injustice found with abject poverty and gross inequality in opportunity (Harrison, 1987:467). Social justice in not a simple matter of the distribution of goods. It includes the provisioning of social, political, economic and cultural rights. This chapter deals with the hope and despair arising from the struggles of change when interfaced with the continuation of assorted traditionally familiar meanings and symbols, and the disappearance and new acquisition of others.

The purpose is to enumerate aspects of a functioning, evolving society as it is at this point in time, and as it effects the aged. The contradictions and ambivalence found with change will also be addressed as I inquire into the interpretation of social justice and the criteria for status and success.

Social Interactions

In this section I present primary material by cases, or situationally united descriptions of events, conversations and behaviors experienced in the village. As with the family, I stress the setting as an arena for social exchange. My observations reflect my role at the moment of the event, which could be an anthropologist, a "granddaughter", or a member of the community. A discussion follows each scenario. When my feelings are put in, it is mainly to emphasize the differences between my society and the Tswana, and between myself and others. They are not to be interpreted as a judgmental statement.

Case 1: The Hungry People are the Lazy People

I am taking a short cut through the church yard, when one of the European missionaries asks how my work is proceeding. In response to my reply that I see much hunger, she boils out:

> there should be no hunger in the village this year! The rains were good, and the drought is over. It is the fault of the people if they do not have food! They are too lazy to plant. They want us to give them everything. Last year the destitute in the village of Otsi were given blankets. Do you know what they did? They sold them to get money for bottled beer.

I had heard similar comments before, from assorted expatriates and government workers and knew refutation would fall on deaf ears. I continue walking. A few minutes later I stop at an food shop that is off my usual beaten path. I asked the owner, a middle aged village woman, if she has any tomatoes or fruit. "There is only cabbage, that is all I've had all week. I always have cabbage as it is cheap. Many people are hungry in this village as food is so expensive." Wrapping up four loaves of bread, she says "Take this. I've heard of your work. Give this to the old people who are the hungriest."

Who are the hungriest? Choices for decision-making overwhelm me. My friend, Sego, is sometimes hungry, but he manages to get a meal at least once a day. Others are more in need. The loaf I give Elizabeth and her nephew is shared with the preschool great-grandchildren and quickly disappears without accompaniments or beverage. Then I see Anah, who shares her destitute rations meant for one with her six pre-school grandchildren. These young children spend most of the day sitting on the ground, obeying Anna only after persistent commands. They scramble quickly to the gate when they see Anah breaking the bread. Again, the food is gobbled immediately. I know I will find Sepia and his 88 year old mother at home. Both have told me they reason they sit so much is that they "cannot afford to walk." Their bread is gone in three minutes. My last stop is at the home of Dikeledi, an 82 year old man, who lives with his 91 year old sister. She is asleep and the way the bread is disappearing, I begin to doubt if she will get any. Dikeledi reassures me that he will save some for her. I do not worry about the four grandchildren at school, as they will get a large lunch of vegetables and *mealie meal* (ground corn). As I take my leave, Dikeledi holds me back.

> People in the city with their fine houses quickly forget how people in the village have to live. Even those with education, and thus work for the local government, do not want to think of what it is like to be a poor villager. Those who work see only themselves, and they have the money to be fat. They tell me I am a useless old man. On the streets, people call me 'old man' in a nasty way. They think that because I am destitute I should work. Ten years ago, the place where I worked said I was old and worn out. They made me quit work. They took away my independence and made me a child. Now they say 'work'. Who will give work to someone as old as I am. They don't see that I tend to my sister and the children and we have no land to plant.

Hunger is a multifaceted problem affecting young and old alike. A few families have abundance in food resources; many have none. Prior to the 1981 drought, national rates of malnutrition among preschool children hovered around 24%. With the drought, rates increased to 34% and would have climbed higher except that Botswana was one of the largest *per capita* recipients of food-aid in Africa (Morgan, 1988). The prevalence of severe malnutrition in children under age 5 now remains about 30%, and this may be a gross underestimate (Hay et al., 1988; Knudsen 1988:19). Similarly high rates are believed to present among older children and adults, as the basic cause creating hunger in young children also causes hunger in older individuals. That cause is chronic household poverty (Morgan, 1988).

Rural incomes are declining in real terms as inflation increases, with those in poverty affected the hardest (Hay et al., 1985:11). Survival becomes the objective. The need to survive forces participation in agriculture. Three/fourths of the studied households planted with the sole purpose of obtaining food. Those not in agriculture were mainly the aged destitute living alone or with very young children. They were also the ones who either lost or never owned land. Very few were like Priscilla and Isaac, who could gather enough resources for food without being rich.

If the rains were good last year, as claimed by the missionary, why are these farmers hungry? Substantial risks in planting accompany the risks in life. Expected production does not occur when scattered rains fail to fall on a segment of land, when plowing is late because of community demands on tractor time, or illness and absenteeism of sons and daughters prevents proper maintenance of the crop. An unexpected expense means the selling of intended food grain for a monetary obligation. For most, production is between two to five bags of grain. There is never enough to feed the average household of 6, much less support a family, for a year (Hay et al., 1985). Sorghum alone, even if sufficient, is not adequate nutrition. Meat, vegetables, and milk need to be purchased to round out the diet. Food security comes only with secure employment or regular remittances.

The assets of the affluent families for increased production are not available to the poor. The fact of being affluent, in itself, decreases potential for agricultural loss as the rich are better able to provide fertilizer, man-power, and mechanical power to increase yields. Risks are not as great as employed household members provide other resources. In addition, the grain they sell is surplus, available for sale at any time, while the poor are frequently forced to sell food-grain immediately when prices are lowest. (Hill, 1986:71-14).

The process of time adds another dimension to hunger. Sorghum, if not sufficient, is consumed before the next harvest. A hunger period occurs. For many in this village, this hunger period exasperates chronic under-nutrition. There is no time of plenty followed by want. In this particular year the hunger period came early for many, as massive village flooding in October destroyed stored grains harvested five or six months earlier.

Few, if any, of the village households suffered from hunger due to laziness, as ascribed by many an outsider. The aura of laziness appear because of hunger, as there is a relationship between activity and food intake. When intake of energy is below requirements, even slightly so, there is a downward adjustment of energy demand. Insufficient caloric intakes means working more slowly and taking more frequent rests. This same fatigue, either from mild undernutrition or slow starvation also leads to mental dullness and generalized apathy (Dirks, 1980). Sepia, and his mother, are correct when they say they cannot afford to walk; they do not have the calories to spare. Anah's grandchildren do not have the calories to play. It is difficult to say how many of the aged, who cannot walk in the village without frequent rests, suffer from malnutrition, and not the hospital diagnosed "old age." I found that one or two of the daily meals for many aged are tea with sugar. As much as we like to say sugar is empty calories, these calories are vital to life.

The problems of under-nutrition, in part, are recognized in the village but little is being done (Tlou, 1986; Knudsen, 1988) The school lunch program is invaluable for possible intervention of mental retardation in children, but only 75 of the approximate 3,400 village preschoolers receive supplemental foods (Central Statistical Office, 1982; Knudsen, 1988). The present destitute program, for those individuals living more than 10% below the Poverty Datum Line, provides for a 1,750 calories-per-day diet (Ministry of Local Government and Lands, 1980). In reality, part of that funding is spent for matches, soap and other household essentials. By policy decree, individuals, not households, are destitute. What food rations that are given to a given destitute individual are frequently shared with the household. In Anah's case, her absentee adult children are deemed adequate wage earners to provide for their children's nutritional needs. It does not matter that the children's parents do not do so. Only Anah receives rations.

If you were Anah, knowing that last years rags would provide warmth again this year, would you sell a new blanket? Would you buy refined beer? Money is spent on food. The few cents found on the village path buys candy, for energy as much as enjoyment. The poor rely on gifts of *bajalwa*, never overlooking the fact that the main ingredient of sorghum provides high nutritional value. People living so close to the edge cannot afford to overlook situations that promote life. They cannot follow the activity norms of the more affluent, or luxuriate in lazy idleness. To do so would mean not to survive.

It has been said that many of the rich, including the government workers and the outsiders, tend to live in their own cocoons and perceive others accordingly (Chambers, 1983:13, 104-106). It seemed to me that within the village the bodily reactions to hunger and the related problematical social impacts of school absenteeism, decreased production and the investment in the need to survive filtered through the perceptions of the more affluent. The government infrastructure and local expressions of that structure made the extent of poverty difficult to conceptualize. The village tarmac road passes

impressive service-providing buildings. Most households, by outward appearance, cannot

be distinguished as rich or poor. Equally hidden are the number of economic dependent

residents, and the vulnerability and powerlessness of the individual family. The same

infrastructure makes poverty difficult to conquer. Poverty is a social phenomena, caused

by social conditions (Sahlins, 1972:37).

Dikeledi's observations of rich are astute. Obesity is a sign of wealth. The

village rich, who are mainly government employees and well-educated entrepreneurs,

admit to being obese because of too much to eat. Many of them say the thin-poor

avoid hunger with the system of giving, as other have surplus to share. Like the notion

that no one sleeps on the ground, no one is chronically hungry. I must agree with

Dikeledi, and others, who say that the affluent have forgotten what it is like to be poor.

Accompanying this forgetfulness is a blindness to the conditions in which the majority

survive. A lone shop village keeper was able to admit all was not well.

<u>Case 2: The Loss of Social Treasures</u>

Several weeks have passed since I gave bread to Sepia and his mother. Sepia

had been supporting his mother with remittances from employment in South Africa. A

few months ago he became ill. This was soon after his sixty-first birthday. His worry

increased when South African physicians were unable to cure him. He then feared the

cause of his illness was sorcery, and that he would never see his mother and village

again. Silently, he returned home in the dark of night, breaking his work contract. This

was his personal secret, for such behavior is illegal. The nature of his return also meant

breaking ties with his own children and grandchildren in South Africa. This man, so full

of warmth and kindness, still felt the loss of fatherhood.

The illness forced him to visit the village hospital, where a permanent catheter to

allow for urine flow was inserted. Sepia says he lost his "manhood" that day. Repeated illness, diagnosed under the heading of "old age," has limited his involvement with the village and his extended family.

Sepia's main worry is not himself but his 88 year old mother, who is also chronically ill. Each live in their own hut, bare of furniture except for a bed and multiple blankets in Sepia's house. These belong to an absent brother. His sister visits every couple of days to do the laundry but his mother must do the cooking and general cleaning. The failure of his mother's other migratory children and grandchildren to adequately support the household means money is very scarce. Again, because some of the adult children are employed, Sepia and his mother do not quality for destitute funding.

Today, Sepia is laying on his bed of an old burlap bag. A ragged blanket is his cover, with his trousers serving as a pillow. His weakness and pain are obvious as he tries to pull himself into a sitting position. Unthinkingly, I grab a blanket from the brother's bed to prop him against the wall. In mortification, Sepia screams out "No! Stop!" His bulging eyes and expressions of fear emphasize the severity of my error. His mother runs into the hut. After seeing her son free of danger, her only statement is "Give me food. We had no tea this morning. Tea is our food."

Tears roll out of Sepia's eyes as he asks me for 15 thebe (8 U. S cents) for bus fare to the hospital. I search my pockets in vain, feeling even worse as I do not have my usual hard candy. Sepia speaks words of comfort, for himself and for me. "Someone will come later. I will ask them for coins. If they have, they will give." I ask whom he expects. "I don't know. Someone will come."

An evening walk leads me back to Sepia's compound. He and his mother are sitting by a dying fire. They are sharing a small bowl of porridge, brought by the sister who assists with laundry. A look of discomfort crosses Sepia's face. Immediately his sister's hands reach out to massage the sick man's back. She explains that her visit must

be short, as she must return home to her own family. She then continues:

> it is awful the way my brother and mother live. There is much suffering.
> I love my mother and brother even through they are useless. I bring
> food when I have extra. I help with chores when I have I time. I try to
> help, but there are limits on what I can do. I must serve my own master
> {husband} first.

The food is gone. The mother complains how the neighbors do not give her sugar or

tea, although she is sure they have some. Sepia nods and says:

> today, when children go for a walk, all the think of are paths leading to
> fun with other young people. No one came today. I went out, but no
> one had coins that they would give me. I walked to the hospital. The
> walk took a long time and when I got there I had to wait. They told me
> I was just worn-out, that nothing was wrong, and there was no need to
> come back. I was given some pills for pain so I do feel better.

There are many diverse elements that contribute to the feeling of personhood.

Sepia lost two of these elements: fatherhood and manhood. He describes the loss in

terms of events. I agree that it was a time when his particular needs did not blend with

the offerings of the employment and medical systems, but more is involved than the

breaking of a labor contract and the insertion of a catheter.

Sepia does not feel guilt or the pain of sin with his return to the village, as

neither exist in his culture. He does feel shame in that he did was unable to continue in

the process of achieving for the good of his South African and Tswana kin. He also

feels agony for the way his mother must now live. She must give him care, whereas he

should be providing for her. This inner turmoil clashes with his concept of personhood

and produces a loss that cannot be overcome in conditions of poverty and poor health.

Sepia is experiencing other unexpected loss. He can accept the poverty. He

knows he is useless, and that is acceptable for others to call him so. It is the loss of

acceptance that bothers him. No Tswana cared that he was in extreme need and

deserving of food and bus fare. He had been treated as a undeserving and unwanted entity all day. He had found the doors for community acceptance broken and sealed. Only his sister and mother care.

The one way Sepia can maintain any degree of integrity as an old person is to uphold any traditional laws he can. The use of the make-shift bed, over the use of his brother's, was his way of demonstrating the importance of law in finding personhood. My touching the brother's blanket was a threat to his ability to maintain the law in his own home. Like many aged, he finds pride in maintaining laws, even though it means deprivation of needed comfort.

Many relatives care about their aged, and love them as parent. However, ability to give care and express love must be considered in relation to social rules and the boundaries of the present. Body massage and breast holding (between kin-related females) or genital grabbing (between kin-related males) is often a more meaningful expression of love than direct service. Direct service is influenced by law. Since Sepia and his mother live in one household and the sister lives in another, the sister must put her master and children first. The aged and others say this is proper, a part of the traditional law. The assistance given to Sepia by his sister fulfills his and village expectations, although they may appear inadequate or lacking by Western standards.

Sepia's story raises various thoughts considering the outsiders when they work in one cultural system and come from another. I will use the social process of sharing as an illustrative example. The custom of asking and giving for equality applies to kin and non-kin. Anyone who has signs of wealth is frequently asked for money, as they have more than the asker. Many an outsider visualizes the system as begging, or as being dependent on aid, or as another sign of laziness. Outsiders frequently fail to see the inherent social value of the system.

Chambers (1983:13-16) states that outsiders praise their own belief system and degrade the equally valid local system. I found the same was frequently true with the highly Westernized native urbanite, many of whom complained about village "begging." This bias is easily recognized by the village members, and interferes with trust.

My inclusion in society meant I was expected to give. Requests were so numerous that I quickly recognized my work would be compromised, both financially and in anthropological value, if all requests were fulfilled. My own Western ethics would not allow me to lie to escape giving. I seldom carried money to protect my work and to insulate myself. A pleasant "I do not have money today" was sufficient and acceptable. Instead I carried small candies to fulfill requests. *Life Savers* lived up to their name!

The Tswana accept "honest" lying for negating social expectations. It was perfectly acceptable to claim "I do not have money" and show an empty pocket and have money elsewhere on the body. I found the contrasts between the moral behaviors were burdensome. In addition, I experienced secondary conflicts when not having hidden money for when I felt giving was justified. In this case, a few coins would have benefitted Sepia. (Yet, if I had interfered by giving coins, data would have been lost). This sketch demonstrates that biases an outsider carries cannot always be resolved.

Sepia acted according to traditional social behavior with his shaming of me for touching the blanket, asking for money, and expecting someone to visit. His personal security was undermined as he experienced clashes between the old and new behaviors, between himself and the outsider, and between old age and youth. Villagers expect agreement on proper social behaviors, only to collide and tumble when tradition collapses. Sepia managed to hold onto some of his social treasures, while losing others, as he groped along the unwritten path from South Africa to our evening together. He was able to still use a multitude of broken doors. Jacob, unable to manipulate the broken doors in the darkness of insecurity, fell and died. Maladies take many forms.

<u>Case 3: My Arm is Sick</u>

The usual habit in the village is to drop all unwanted items directly on the ground. The Chief asks that the villagers spend a day each year to clean and improve the village environment. This morning the Village Development Committee has been discussing the annual "Clean-Up Day." Three Thursdays from now, everyone in the village is to pick up the papers, glass, tin cans and plastic that litter the village streets and compounds. After the planning meeting, a group of us walk together on our way home.

Monate begins coughing. I ask her if she has been sick. She replies, "My cough became very bad and would not stop, I went to the hospital. The nurse gave me pills to make the pain in my arm go away." Feeling confused, I inquire for more information.

> A cough is normal. I always cough. Last week I was coughing so much that it made pain in my arms and shoulders. That is not normal. So I told the nurse my arms were sick. My arms are well now, and I am not sick.

The other women tell her that it was good that she went to the hospital. Happy Sound tells her why, using her Caesarean delivery for Leru as the source of her knowledge.

> It is important to follow their directions, especially with an operation. An operation is the beginning of the end of your life. With each operation a piece of life is removed and you will never be as strong as before. I had an operation a long time ago and still wash that place every day with boiled water, just like they said. They told me not to do heavy work or it would kill me. I haven't been to the lands since then.

The others agree that what Happy Sound says is true. Martha tells Happy Sound to be careful in the rain, and not get her feet wet.

> You know if you stand in water it will travel up your legs to the operation and cause much sickness. Maybe the sickness will be in your eyes or kidneys, or it may be high blood (hypertension) like I have. Operations are good as they keep you alive, but you must be very careful forever after.

The hospital is a good place, but many times the nurses do not understand what it is like to be old. They say we are wearing out and our disease is in our head. Sometimes the disrespect for the old worries me more than any disease, even high blood.

Sometimes I fear the hospital. I have heard they sometimes do things that make people die. Other times, they do things that keep people alive when they should die. I've heard that when people reach the end of high blood disease {have a stroke}, the doctors can put in a tube for food, even though the person cannot move or talk. It is good they are alive but there is no way for them to live in village. All they can do is stay on a blanket and be pulled along the ground. That is not the life that God wants us to have. The meaning of life has been lost.

Sego has been listening to the women talk. He tells us of one old man who has

sick legs.

Other people and witches started the sickness. He went to the hospital where they treated him like they wanted him to die. He went to the traditional doctor. That doctor said he could dig a hole and put medicine in it that would cure him. My friend didn't have a cow for payment so he is still sick and will be sick until the curse is removed or he dies. No one will let him live happily.

The body is the same as grass. Grass grows, goes down and dies. It becomes green again after a rain. A person goes down without the water of life but can become green again after death. He is a new person in the skies. It is strange. When we want to live as elders, people act like they want us to go down and die. When we are children and want to die, they make us live.

For family survival, good health and food come first. Disease is approached with just concern, as health provides the necessary buffer of labor against food insecurity (Chambers, 1983:142, 144). In Botswana's favor, emphasis on preventive medicine promotes health, even in the poorest and most uneducated village households. Clinic and hospital walls have multiple posters promoting cleanliness, sound infant care, good nutrition, and safe sex. The general public strives to comply with preventive measures, as far as economics, tradition and social behaviors allow them. The ambient social and political settings promote compliance with maternal-child health programs. This is not so true with other programs. Knowledge of prevention is present. The means is absent, either in finances, material goods or beliefs.

I will just briefly mention that AIDS is present in the village, although only one reported death has occurred in the last five years. Sex is like hunger, an openly accepted human need to be fulfilled. Multiple partners are the norm, even after marriage. Condoms are available without charge but are seldom requested. Knowledge and abstract fear of AIDS is present; self-motivation for prevention is not. In this way, the village is similar to the developed world.

The disease of greatest concern to health care providers is hypertension. According to the hospital physicians, the government spends more money on hypertensive drugs than on any other medication, excluding aspirin. Blood pressures are routine for all patients, with the goal of identifying and treating all hypertensives. The thirty-day supply of medication is replenished at monthly follow-up visits, along with on-going recommendations to control weight and salt intake.

The aged consider hypertension as a major medical threat, as it almost always affects either themselves or someone in their immediate family. Of the 105 study participants, 102 want me to take their blood pressure. This procedure is not new to them, and many told me in advance if they had "high blood" or not. The diagnosis of hypertension, as with other diagnoses and "old-age syndrome," is readily confirmed with the individual's out-patient chart. Each person keeps their own medical records at home.

The incidence of hypertension, as defined as a blood pressure greater than 140 systolic over 90 diastolic, is 41%. The majority of the hypertensive readings are 160/100 or greater. Five more have borderline readings where either the systolic or diastolic pressure is greater than normal, and are not included in the above figure. A high incidence of hypertension is also present in younger household members. In a control group of 26 of the aged's adult children, with an average age of 38.7 years, 27% are hypertensive.

A similar high incidence of hypertension exists in African Americans, with black males having a higher incidence than females (Anderson, 1988:191). This is not true with the Tswana aged, where only one aged man is hypertensive, in the study group involving 23 males . With the control group of off-spring, males have rates similar to the females. As with African Americans, rates are highest in the 60-69 year old group. Frequency decreases in the higher age group. Two things may be happening. One is the very old, by placement in history, avoided the dietary changes, stress conditions, and affluency leading to obesity. More likely, only those without the disease live to be very old. The sex differences raises the questions about sex differences in ages for coronary death. and about the availability or acceptance of treatment for males.

The effectiveness of the hospital hypertensive identification program is reflected in the fact that 30 of the 42 hypertensives had received treatment. Thirteen remain under treatment. The remaining seventeen give various reasons for noncompliance. The most common reason is "I ran out of pills and I could not afford to go for more." The meaning of "afford" is time, energy, and emotional expenditures. The walking distance, time in waiting, and inability to sit until one's turn are all mentioned. Equally frequent is mention of emotional expenditures, which run deep. Says one, "I went to the hospital and was told 'Old woman, it is time for you to die, you are too old'. I am old but I have the right to say when I want to die." Others complain of rudeness and a lack of understanding of what it means to old. They feel shuffled between nurses, seldom seeing a physician. Concern for specific, age-related needs is mentioned as lacking. "They say 'Don't eat salt.' How can I not eat salt. They won't tell me the ways it is possible not to eat it." "I am old and do not understand all the things they say. They say 'lose weight', 'do not get angry'. Tell me, how is a person to do those things?" "When I ask questions, they laugh and say I am too old to understand." Free health care and medications do not seem to guarantee availability, as other indirect expenses are too great.

Earlier, I supplemented Sepia's expense of illness through my desire to provide comfort with a blanket. The error was acting on my cultural care-giving heritage instead of his, bringing shame to me and disquietude to Sepia. Similar unconscious errors were observed repeatedly with foreign health-care providers. The difference was that the cultural conflicts were not recognized. Instead of embarrassment, there was anger for impropriety of behavior or refusal to comply. This anger was directed towards the Tswana patient, adding timidity to already present trepidation.

Health-care workers also make errors in failing to seek depth for the interpretation of symptoms. Monate knows that total wellness does not include coughing, but similar to undernutrition, malady is an accepted inclusion in everyday life. Coughs and weight loss are to be expected with aging in poverty. In Monate's case, arm pain is the major concern, as it prevents work. In terms of health, the cough is the most important symptom to treat. In the same vein, individuals seek help for dizziness and headaches during the hunger season, without mentioning weight loss or a lack of food. Yes, the proper treatment of the presented symptom is pain-relief. Although correct in theory, such treatment falls short from alleviating the true cause of symptoms and complaints. I openly admit providing comprehensive medical care based on what patients' say, and how they perceive health, can be a challenge. The reality of the effects of poverty, and belief in sorcery, should be incorporated in meeting this challenge.

In opposition to misinterpretation, the Tswana frequently fall overboard in interpreting the path for health. This too must be considered in the total scheme of health care. The drive for health unnecessarily limits the ability to meet the labor demands of poverty. The life-long washing of surgical scars, and life-long avoidance of lifting and plowing after surgery, are but two of the many instances of carrying out instructions to an extreme. Hence, surgery is ambiguously accepted as a cure, as consideration is given to the long term consequences.

The hospital staff provides outstanding care in the majority of situations. Gerontology, with an emphasis on providing specialized care based on the separation of the normal changes with aging from the abnormal, is a relatively new field of medicine. Misguided care is not limited to this African village. As a nurse, I have heard aged patients in America, and other countries, state the identical complaints of the Tswana aged. Many of the problems stem from the lack of information and knowledge of gerontology superimposed on misinformation. I found a strong desire from the staff to acquire sound knowledge and methods for application. Methods included the establishment of workable out-reach programs.

The problem of adjusting the mechanics of care delivery to the given cultural socioeconomic framework cannot be easily overcome. The present models for the delivery of medical care come from developed nations, and include life-support systems. Often times, contrasts exist between the guidelines of Western medicine and patient needs reflecting the local sociocultural environment. The ethical dilemmas of life-support already exist for both the health profession and the recipient of care. Feeding tubes for stroke victims are already in use in Ramotswa. The village is rapidly hearing of more advanced techniques to maintain life. Some individuals are for their use; others are against it. Arguments revolve around the meaning of life and death, debating if medical interference removes God's control over life. This growing debate contains philosophical similarities found in developed areas.

The local European physicians also ponder if their outside Western models, promoting quantity rather than quality of life, are always the ideal model in the village context. Part of the medical debate is financial: Is the limited money best spent on supporting one life, or providing improved quality of life for many people? In addition, the locally underdeveloped service-providing infrastructure, and the scope of poverty, prohibit access to the necessary equipment and support programs for rehabilitation and later survival. What may be prolonging life in a developed nation may be prolonging the act of dying in other situations. What is viewed by one group as a new door is seen by the other as a broken door.

<u>Case 4: Elisa's Teeth</u>

I realize that I had overlooked an interview question on dental health for one of the female participants, Elisa. Socially, Elisa is a child, as she has limited mobility and is recognized for her nearness to ancestors. Passing by her house with Agony, a 26 year old friend, we see the old woman in her yard with her late-middle-aged cousin. We are invited in.

I see no harm in asking Elisa if she has problems with her teeth, as personal health is openly discussed in detail with others. Elisa responded openly. "My teeth? I have no problem as I am on my third set. When I was working in South Africa, the dentist pulled all my teeth and I now have new ones."

She smiles proudly as dentures are virtually unknown in the village. Evidently, this modern advancement has never been covered in my friend's education. Agony turns to me, saying, "No one can grow three sets of teeth. It is impossible for a person to grow a third set of teeth, isn't it?" The old woman prevents me from explaining, as she continues talking. "These teeth are very special. It is possible only in South Africa." Seeing bewilderment and increasing fear on Agony's face, I try to interrupt. Elisa insists on explaining the features of dentures. "What is special about these teeth is that I can take them out." As the dentures gradually protrude from her mouth, Agony lets out a tremendous scream and continues yelling as she runs out the door. She stops only when she reaches the footpath. Elisa and her cousin, at first startled, now laugh heartily. The cousin is asking more about the teeth, as I quickly close the interview.

Back on the footpath, the trembling Agony refuses to let me explain. I see how the removable teeth has only reinforced her earlier learning about life.

> That woman is a witch. I always thought she was a witch. I can tell because of the look in her eyes. This just proves it! There is no way a person can take their teeth out! Old people are the worst kind of witches. They go to church to get God's blessing to perform their crafts. That old woman is a witch, and I'll never go back there again.

Agony never allows me to speak of the miracles of modern dentistry.

New technology, from fancy earth movers to computers, is constantly introduced to the village. The aged have extensive experience with new and unusual objects entering their lives. Dentures are but one more thing the modern world has created for better living. Much to the new originates from outside the area. Elisa, in a position to obtain dentures, did so. Surprize to her cousin arises from the uniqueness of this marvelous new appliance, as it is just one more item she had not seen before.

For those young adults, who have never migrated, familiarity with technology stems from what has been in the village for some time (telephones, radios and electricity), or from the things their age-group has brought in (portable tape recorders and digital watches). The aged are believed to be "out-of-date" and "out-of-touch" with material goods symbolizing progressiveness. In Agony's eyes, this belief is particularly applicable to Elisa, an old, child-like woman. Explanation could only come from what Agony already knew: the metaphysical world. The metaphysical provided a sound and valid basis for understanding the event. How else could a woman remove teeth, other than being supernatural herself? The explanation is that much more plausible as Agony knows that aging and the powers of witch-craft coincide. Tradition, which the young try so hard to evict from their life style, continues to intersect their daily lives. Old beliefs infiltrate thought, sway reasoning, and tinge the direction of behavior.

The aforementioned assumed superiority of the young adult, as the forerunners of up-to-date material goods, leaves no room for acceptable innovation by aged. Proposed innovation by the aged does not have to be technological, as in Elisa's case to be met with rejection and devaluation of the person. Alfred, the old man who kept small animals at his farm land, instead of the more distant cattle post, is a source of laughter among the young. Happy Sound is scorned for wanting to grow fruit trees in her yard. To the aged, these are ideas to promote adulthood achieving. To the young, these are ideas that demonstrate childhood play. The exceptions are with successful innovations for money-making schemes, as with Happy Sound and her plants. Money is always valued.

Case 5: Grandmother Has Fallen

A young gentleman approaches to tell me that our grandmother fell yesterday, and that she wants to see me. It takes me a second to connect his face with an area on the other side of the village. I realize that Grandmother is Khuma, a strong-willed, and determined, 84 year old woman who lives by herself. Her only immediate family, a sister, lives in a new home by the village edge, although a distant aunt lives three houses away.

Khuma is not the young man's grandmother, but without relatives himself, he likes to talk to the old woman. The relationship is strained as he asks for money and drinks excessively, much to the disgust of Khuma. Through her directed and determined use of complaint discourse, it is no neighborhood secret that she feels the young man takes from her and never gives. Khuma does not consider him to be a grandson!

A detour on the way home for lunch brings me to Khuma's hut. The neighborhood seems excessively quiet as we sit alone on the front stoop. Khuma, in no apparent acute distress, tells me her story in an increasingly loud voice.

Yesterday, I asked a passing teenager to help me with some
lifting. He did not want to assist and we began arguing. He pushed me,
and I fell down and hurt my hip. I still cannot walk well, and I am in
need of water. I haven't had water since last night, How is an old
woman to wash and drink if no one brings her water?

I know the statement is one of complaint discourse directed towards the

neighbors rather than a direct request for my assistance. I offer to help her. Taking

her bucket, I make the 300 meter trip to the public tap. Khuma hobbles along beside

me, all the time repeating loudly "An English woman shouldn't have to do this for an

old lady. It is shameful that an old woman can get no water." Doors close quietly as

we pass homes.

As I leave Khuma's compound, the cousin calls me into her house. She closes

the door behind us. Handing me a bowl of porridge, she says "This is for helping the

old lady. You helped her when no one else would." Stunned, I ask why no one would

assist.

Khuma has friends and she has enemies. Some of her enemies
are my friends, and some of my enemies are her friends. I know who my
friends and enemies are, but not all of hers. If I got water the friends
and enemies would get all mixed up, and then they would say bad things
about me or about her. It makes too many problems to do such things
when everyone will see and talk. It is a shame she has no children, other
than you.

That evening another neighbor, a young mother, brings me a carrot from her

garden. She explains the gift.

This is to say thank you for getting water for Khuma. It is bad to
be alone and get sick. Everyone has enemies. They know who they are
but may not know why they are an enemy. If I got water for Khuma I
could get unknown enemies who are more dangerous than the ones I can
identify. A friend can help a friend but enemies make doing so very, very
dangerous.

Everyone has friends and enemies. The term friends, in this case, refers to

people that another has interactions with on a social level. They may be neighbors,

extended family, club members, or government officials. While considered friends, the usual elements of friendship such as closeness, trust and openness are missing. Friends are a "safe" person, as opposed to a enemy. The enemy is someone who is feared, as they spread rumor, tell falsehoods and know of witches for assigning doom or death. Overt malevolence towards enemies is to be avoided at all costs, as the enemy's retaliation will make the threatened a reality through witchcraft. Although no one had ever personally sought out a witch, everyone was sure their enemies had personal contact with one.

References about enemies tended to be unidirectional, with a person having multiple enemies but seldom seeing oneself as an enemy to many. As with other interpersonal conflicts, the naming of enemies to others is not usually done. Sometimes the enemies of a individual can be identified, such as with the refusal of food as demonstrated during the funeral feast. Other times, the sudden departure of a one person with the arrival of another signals a lack of friendship.

Allegiance with friends and alienation between enemies is not a constant. Change from friend to enemy can occur with a simple behavior, harshly spoken word, or siding with friend's enemies. More time is required for the reverse conversion, as a feeling of safety has to be generated. The regaining of trust seldom occurs as enemies tend to be avoided. As extended family usually provides sufficient trustees, there is no great motivation for improving the status quo of village acquaintances.

The system of friends and enemies places limits on the extent of possible interactions, and also upon the type of interactions, as the unknowns of who is friend or foe to whom is not constant or bilateral. In particular, the system complicates the law of asking and giving. In practice, this law applies only to friends and strangers, as strangers are friends until proven otherwise. One does not ask a enemy. Also, if one asks too often, or for too much, one is apt to become the enemy.

The directional flow of friendship does not have to be shared. Khuma is threatened by the young man's asking for money for drink, and therefor, considers him as an enemy. On the other hand, the young man sees Khuma as a grandmother. He avoids many of the risks in the friend/enemy issue by labeling her as adaptive kin. He could always deny any rumor or falsehood Khuma spread to others because socially Khuma is regarded as an aged-child, incapable of accurate memories.

Within this type of setting, individuals are constantly aware of their movement in the village. It prevents some from seeking medical assistance at the clinics, as the crowded waiting rooms may contain an enemy. Others avoid certain shopping areas or sections of town. Certain events seem to be immune from enemy harm, among them the *kgotla* meetings, church services, and funerals. Weddings, parties, governmental meetings and simple visitation require vigilance, and one must constantly evaluate the consequences of action. The result is that a generalized lack of assistance given to those who are not immediate family. (Surprisingly, there is no fear than enemy will enter an open house unless he, or she, is an actual thief.)

The neighbors did care about Khuma but could not afford to upset the delicate balance in relationships. It is not that they did not recognize a need or lacked emotional involvement. The neighborhood doors shut to prevent the giving of shame for not assisting. The fact of caring is shown through the gratitude for my assistance as a "granddaughter." However, as the "anthropologist", I remain exempt from the area's internal processes of friends and enemies. Why did not help come from the man who considered Khuma his grandmother? "Carrying water is the work of women and children."

<u>Case 6: The Wedding Dances</u>

Delight spread over the neighborhood this month. Ida and Mmanato Ramasu's youngest daughter is getting married. It is expected to be a grand wedding, reflecting the family economic status. Mmanato is arranging for one of his finest cows to be brought into the village. His brothers, John (Monate's husband) and Masimi, are assisting in planning. Monate, Priscilla and Happy Sound are helping Ida prepare *bajalwa*. Betty, still in mourning, sends Dorothy in her place. The older women are happy to have her strength and thoughts. It is a scene of mixed ages, with the broad, extended family working as one.

The ceremony begins early Thursday morning. Ida's daughter is taken in tow by a group of women dressed in finery. Led along the village paths, the bride-to-be sings out "I need a man to cook and clean for. I need a master to take of me." The women search every nook and cranny of the village, looking for the proper man. The lack of success does not deter them from parading the bride to her mother's house. Here she receives the laws of marriage, in the form of secrets passed on by the elder married-women kinfolk.

The fifty or so women from the morning's parade gather for the bride's party. Mmanato, John and Masimi sit, as guards, outside the compound gate. Only adult women may enter. Children play in open public area, going to the men if assistance is needed.

Calabashes (gourds) of *bajalwa* are passed among and between women. Soon the mortar, used for pounding grain, is turned upside down. It is carved from a tree trunk, with a flat top tapering to rounded-oval base. Monate giggles and squirms her hips as she points to the mortar, now a 2 foot tall phallic symbol. The turning of the mortar signals it is time to begin the dancing.

The women raise their voice in song. A line is formed, each woman holding the chest of the person in front. Intricate foot work keeps time with the song's beat. Winding between the mortar and compound buildings, the women dance until they tire. New songs are started. Each woman takes a turn dancing around the upturned mortar.

With each dancer, the movement gets closer, and closer, to the carved wood. The dancers now hike their skirts as they shake their hips and lift their legs over the piece of wood. As the number of songs increase, the number of voices declines. After an hour, only the older women are singing, with the other clapping in rhythm. This does not decrease the intensity and spirit of the dancing.

Priscilla decides that it is her turn to dance. As the older women sing, she begins by imitating the steps of previous dancers. Suddenly, she bends down to lick the mortar, and with quick gyrations turns and sits on it. Monate, Ida and Happy Sound laugh, and cheer her on. The young women boo. "What a crazy old woman." "Old women don't know how to do such things." A young woman gently guides Priscilla from the dance floor and takes over herself. Her dance, more graphic than ever, brings howls of laughter, and loud approval, from her young friends.

Almost imperceptibly, the older women move to a corner of the yard. Only one person is singing. Gradually the song fades out, and the dancing ends. I ask the young woman next to me to explain what is occurring. "The old women think they know everything. It makes me mad. It is true that only the old women know the words to the many songs. The young people have not bothered to learn them, so if the old do not sing, we do not dance." I ask what will happen in the future. She replies:

> somebody will always know the songs, but hopefully the old customs that
> have no use will be gone. The old say 'Don't sew when you are pregnant
> or the pain will be worse', 'Don't listen to music or the child will cry'.
> Kgotla meetings and funerals are the play for old people. They should
> not be telling us what to do. They tell us nonsense, for what do the old
> people know about life.

When I leave I spend a minute with the old men by the gate. They have been talking of marriage. John is anxious to tell me their conclusions.

> The ideal is to have ten years between husband and wife. Others
> appreciate a man more if he married an young woman. It allows a man
> to have a woman to care for him in old age. A husband should be
> master of the house. He shows this by bringing home food and money.

The wife has learned the laws of marriage, such as where to go and when to be home, so the man just has to remind her. She already knows what will happen at night. In the old days, people didn't know about sex. The man had to teach the wife, and if she refused, he threatened to tell her parents. Sex is always important, with any age.

One of the younger women is leaving in my direction along the path. She, too,

talks of marriage.

> Few girls today want to get married. Most husbands don't want their wife to work, but farm the lands. The food you get from the lands is not worth the effort. I refuse to plow. It is a silly custom, to do as parents did. I want to get married after I do all I want to do: make money and build me a house.
> Also, women won't spend many years with a husband as one or the other leaves. You can leave at any time. Sometimes the woman leaves if she makes the most money. Sometimes divorce is over love, like having a boyfriend or girl friend. It is okay to have boyfriend when you are married, but it can be a menace. As long as the man doesn't see him, he doesn't care. Keep lovemaking a secret and all is well.

A parallel event for the groom occurs the following day. He, too, must call out

his need for a woman to hunt and provide food for, a woman to take care of his house.

The male laws of marriage are passed onto him on top of the rock pile, not to far from

the *kgotla*. His party is for men only. Meanwhile, the bride's family slaughters and

butchers the cow. John and Masimi, as close relatives, hang their choice parts in a tree,

to be taken home after dusk and the completion of chores. The women pause in the

cooking when Monate arrives in a huff.

> As I was walking, some teenagers called me a worn-out, old woman. They laughed at my clothes and said I wasn't fit to be seen. They told me go home where I belonged. I wanted to turn around and go home, but I held my head high and kept walking. Why shouldn't I come to help? It makes me mad that those same teenagers will come tomorrow and eat the food I cook.

Saturday brings the wedding. The groom, dressed in a new suit with the labels

still attached, and the bride, regal in a white wedding gown and carrying a bouquet of

white plastic flowers, climb from the open pick-up truck at the church. The bride's

maids assist. Dressed in sheer yellow dresses, their tiger stripped bras and bright red panties show through. The groom's brother is the best man. He is finely dressed in a brown jacket with a black bow-tie. The short Christian ceremony ends with the couple scrambling back into the truck, to be driven to the compound of her childhood. Before guests are able to arrive, the newlyweds, with the wedding party, are cloistered in the hut.

The feast begins, with the honored guests served rice, potatoes and meat on china plates. Others use their fingers to eat from paper plates. The fanciest foods of iced fruit and select beef are served to the cloistered wedding party. The radio, connected to a car battery, plays the latest hit songs at ear shattering level.

Conversation is impossible as the dancing begins. Women move near the fence, and sway in soft, rhythmic motion. They dance alone, lost in thought. Men begin to dance alone, or opposite others, near the huts. Their movement is more forceful and angular, as they act out their thoughts. The men's dance is as much an expression of feelings as an actual pantomime.

A few young people snicker as an aged man, dressed in torn and patched overalls and high rubber boots, begins his dance. Soon he becomes the center of attention. With movements symbolizing a walk down the street, he catches the eye of a pretty girl in the crowd. His face, shimmering like instant coffee crystals, portrays instant love and happiness as he dances in front of her.. He moves on, revolving around the circle of onlookers. He comes upon a group of younger men. First, his face shows bewilderment, indicating he does not know what to expect from them. The boys begin to point and laugh. The dancer's features change to fear and repugnance. He quickly moves away with sadness in his eyes. He retreats into himself, ignoring others as the dance transforms itself into a reflection on his life. He starts off as a care-free child, grows into a masculine adult and ends up as an feeble old man, always expressing life's hurdles and enjoyments though movement and countenance. His eyes become aglow with satisfaction during the time of adulthood. A downcast smile shows his sadness with aging.

A young man begins to dance, and interacts with the old man. A battle between the generations emerges. Each show their respective strength in the pretend fight. The young man emphasizes muscles and agility. He grabs a stick to show his strength. The old man stresses the knowledge and wisdom of age, He places a finger next to his eye, thinking of ways to outwit the lad and his stick. The two men circle each other, lunge out, and retreat. The young man still has energy, but the old man is tired. The twenty minute interplay ends with the old man slumping to the ground and showing defeat by the younger generation. Hanging his head, he leaves the dance area.

Rumor spreads quickly through the crowd. We will see the married couple, as it is time for the bridal parade. The best man, and bride's maids, lead the bride and groom from the hut. As they solemnly wind through the crowd, the parents join the march, then others. Only the guests smile. The procession exits through the compound gate and proceeds around the village as uninvited onlookers clap. Returning home, the couple again enters the hut, to stay until the party's end in the wee hours of the morning.

Sunday is another party, this time at the groom's compound. As before, there is feasting, dancing and the drinking of *bajalwa*. The second slaughtering and serving of beef signifies complete approval of the union and legitimacy for offspring. It is dark when I give my departure blessing to the couple's parents. I notice two young gentlemen arguing with Alfred and Priscilla. The older couple feel I should be properly escorted home. The young men finally consent. I arrive home, tired but happy.

The amalgamation of the customary and modern in the wedding represents but one segment of the integration between the old and new in daily life The eyes of the beholder determines where the emphasis is placed. Aged individuals perceive the wedding as traditional, with the slaughtering of cattle, passing down of laws, and public announcement through feast and parade. The young see a modern wedding, with western wedding finery, plates for food and loud radio music..

I wonder if it was so different years ago when the Christian church service and cloth clothing became part of the rite. Did some of the present aged exalt in "civilized" weddings to the dismay of their parents? Now, like the Christian prayers with funerals, change has become tradition, serving to provide unity and state legitimacy to birth, marriage, and death. Change and tradition has interwoven to provide meaning.

It is much easier for youth to accept the process of modernization over the performance of much of the traditional. Tradition is perceived as a separate entity, used by the aged as a method for reasoning and the basis for behavior. Many customary practices that provide depth and meaning to village life are discounted, such as *kgotla* meetings and the role of the aged in preparing feasts. The fragility of valued practices, such as the dependency of the pre-wedding dancing on the songs, is unrecognized. The mortality of the aged is not associated with the possible mortality of the presently valued process for a meaningful, modernized life. Only the "useless" traditional behaviors will die.

The social situation creates conflict concerning which aspects of tradition have value, especially in regards to the gerontocratic principles. Schisms between the old and young abound. This was graphically and publicly proclaimed with the dance between the young and old men. The symbolic battle drained the energy of both men, as do the skirmishes in real life. The aged tire first during the interplay of daily interactions. Independent action for self well-being replaces family and village welfare, as the strength of youth frequently wins over wisdom of age. *Kgotla* meetings and funerals, sitting and death, are for the old. Sexuality and vitality is assumed to be only for the young. The aged are accepted as having limited learning from past experience, but should be far removed from the impervious knowledge and activity reserved for youth. Priscilla's dance confronted this assumption. The dance that followed discredited the thought that the aged resemble youth in desire and thought. "It is impossible for the aged be alive and modern."

The aged, exposed to public ridicule and degradation, sometimes bow out of potential skirmishes. The older women, when faced with doubt of their sexuality during the bride's party, withdrew from the festivity. Other aged pretend to ignore the actions of youth, although the hurt is deep with teasing and ridicule. Very few would purposely face confrontation, as with Elisa in her attempt to show superiority with the dentures. Most aged retreat.

The young do not retreat. Changes in discipline influence public, as well as family, behaviors. The lack of deference and respect no longer results in public shame among cohorts. The power of group-support promotes ridicule, and attempts to disengage the aged from public activities. The aged use their age-group for solace, instead of power. Priscilla and her cohorts effectively ended the derision caused by the dance, but failed to win. The few opportunities to use group action are compromised with threats of losing the whatever support they have from middle-aged adults, who lean both right and left.

The majority of schisms are representative of those found in our society. Martin Martel (1968:55-56) claims that with American social advancement, a shift in the sociological prime of life from middle age to young adulthood has occurred. Life after age 40 becomes "anti-climatic", with the aged shut out from meaningful interactions (Martel, 1968:56). The devaluation of the aged is well known in developed society. We might ponder if it is the social advancement that creates change in people, or does socio-biological human behavior become more pronounced with social advancement.

I have shown previously where the aged, as maturing children in a traditional society, exhibited actions demonstrating their belief in the priority of youth: the sneaking into the woods to be with a boyfriend against parental wishes, the rush for employment and material goods, the desire to attend academic over initiation schools. The young, with quickness of foot and mind, have always taken advantage of the new, probably with as much parental discomfiture as is found today. The change is that opportunities for new explorations, and permissible ways of self-expression, have increased. More and more aspects of daily life are affected as youth follow their instinct to broaden their horizons. The growth of the aged population and rapid social change stemming from technological advances enhance long-standing situations of conflict.

The schisms are heightened in intensity as the age groups compartmentalize themselves. The youth do not visualize themselves as anything but young. The aged see only the present, forgetting their own participation in rebellion and change. Middle age adults take bits and pieces from both sides, as with Happy Sound's daughter and son. Often, life for the middle generations exists on a fulcrum, where they are unable to balance the two sides.

Why are the present day aged ridiculed? It could very well be that some of the lack of social respect for aged revolves around traditional reactions to those who fail in the continual process of achieving through work. Any adult who plays is teased and taunted, as I once experienced. One day I stopped along the path to join in a youngsters' game. Very soon I was experiencing social shame with my failure to be achieving and learning about old age. What surprised me the most was that it was the aged who instigated the taunting by pointing out my play to others, and making rude comments. Are the young only doing what they have been taught when they tease the aged who slowly walk the footpaths? In the eyes of children and younger adults, the aged are engaged in social play. Early assignment of social old age, denied access to the expression of youth's values, and degradation of activity amplify the issue.

Social Concurrence and Collision

Each of the scenarios contain two important elements of contemporary life: those of Tswana tradition and those introduced with modernization. All ages base their behavior on what they consider is the correct interplay between these elements. Social emphasis is placed on change, with a failure to place this change in the many arenas of on-going cultural continuity. The ethnic components of achieving, and a world view centered on the metaphysical, continues. Dorothy's blaming outside forces as a cause for her failures, and Agony explaining the unknown through witchcraft -- both serve as examples. These two women also strive to achieve, oftentimes thinking of others as well as themselves.

The process of achieving has changed in that the goal is located in the economic realm. Young children have economic potential. In turn, as they mature, they are gradually given opportunities that carry social value and reward. The aged, many of whom are unable to generate tangible income, are scorned. The poverty of old age underlines their uselessness. The indirect economic value provided by the aged through their social contributions is belittled, and represents play. Only children play. Traditionally children, either old or young, are not part of village functioning.

The social assignment of all aged as children has broken the door for continuing achieving in old age. The aged feel that they are denied social access to the rewards that should come from their activity. Many times the door is unmanageable. Sepia could not make it work with tradition. Elisa could not make it work with modern technology.

The traditional fear of sorcery, with divisions between friends and enemies arising from unknown, has probably always been a broken door for help and mental support with all age groups. The need to survive with increased in rural poverty has given to reason to magnify the fear of unknowns, separating individuals even further. Kin must rely on kin, but even then fear may prevent help. The possible outcomes for the cousin, if she obtained the water Khuma needed, were too great to risk.

The unification of tradition and social change has damaged additional doors. Many relatives care but are constrained in the opportunity to provide care because of the interweaving of continuity and change. Sepia lives with his mother on the extended family compound. The sister lives in a nuclear family. Her master must be served foremost. This limits time she can be away from her own home. Health is compromised when new and traditional beliefs are superimposed on each other. Access to the essence of being is further damaged as outsiders, in their drive to improve village life, fail to accept the meaning and value of Tswana custom. Poverty increases the damage to all doors.

The next chapter will investigate which aged can manipulate the broken doors that they face. We have seen that Sego manages them fairly well. Others do not.

CHAPTER 9
FINDING THE GOOD LIFE

> I am happy as my family takes care of me and respects me. What will
> happen in the future no one knows. (A 78 year old man)

> I live with my family but they act like they don't care. Who will take
> care of me when I am old? (A 68 year old woman)

Not a day passed when I did not hear the serious asking of the rhetorical question of "Who will take care of me?" The presence of family and riches appeared to have little to do with the asking. The question seemed to be more of a pondering of the continuation of the present degree of life satisfaction. No one wanted to lose whatever segments of the good life were presently available.

The components of the good life involve freedom from physiological want, security in care and service, and being regarded as a worthwhile person. This chapter delves into who finds the good life, and the role of the gerontic fund in success or failure. First, time must be devoted to clarifying basic principles of the good life in Botswana.

Core Elements in The Good Life

Historically, the basics of the good life during old age in Africa have been thought of as respect and power. An additional right of elderhood was physical care when needed (Schapera, 1955). The good life appeared guaranteed. The social and family emphasis on seniority with aging gave the aged undivided power and authority (LeVine, 1965; Alverson, 1978; Meillassoux, 1981; Turnbull, 1983).

242

<u>Sources of Respect and Power: Yesterday and Today</u>

Control over the good life in traditional societies comes from the power of knowledge (Cowgill, 1979). For the Tswana, the majority of knowledge has been, and is, open knowledge. It is given freely for the asking. There have always been secret sacred laws taught with rites of passage. These rites occur relatively early in adult life, placing the aged in a position to enforce the laws. The Tswana aged have never been regarded an distinct hoarders of traditional knowledge. Instead they were perceived as having knowledge resulting from varied and unique life experiences. This gave them value to others but not necessarily power.

Power and status came to the Tswana aged through social organization emphasizing eldership (Schapera, 1955; 1966). Laws could be used advantageously. Those who were eldest in age were the recipient of the required giving of services and social deferments. Some laws that support elderhood still hold true to varying degrees. Daughters continue to cook and grandchildren run errands. Other laws continue to hold true, but the power associated with them has changed hands. Adult children, who control family economic assets, control the use of material goods. Other laws are being ignored as social change promotes self-serving action and wealth in the younger generations. The aged recognize these social changes, yet seek the promises of status and power found in the earlier gerontocratic society to which they were socialized.

African gerontological researchers tends to use the gerontocratic promises of respect, power, and provisioning as a pivotal point for describing the trials and tribulations experienced with aging today. (Osman, 1983; Rosenberg, 1990, Ingstad et al., 1991). My conversations with the aged indicate that power and status are a reflection of the finding the good life, but not the core. The core for successful aging is based on a validation of life through continuous processes involving self, family, and society.

<u>Similarities and Contrasts with Developed Nations</u>

The aged of developed countries validate their life by examining the past. Earlier successes are evaluated and used to provide meaning to the future (Rutter, 1991). Reminiscence and life narratives are freely shared in this effort for self-understanding and social acceptance. In contrast, the Tswana validate life according to the present. "What I do" is more important than "what I did." Memories of past encounters and achievements are important to define "who I was." "Who I am" is defined by today's achievements, unconnected to the past or future.

The future is considered by the Tswana aged, but not in terms of the usual long-range goals of increasing, or maintaining, goods and services found in the West. The future is seen in terms of being, with wants fulfilled. Wants are conceived as what is good, proper and moral in behavior.(Alverson, 1979:121). Thus, much of the good life is based on the process of striving and achieving, in contrast to actual material needs and purchases. Failure in the process of having wants fulfilled does not mean failure as an old person. Instead, the aged find disappointment from circumstances that interfere with the process.

The study of aging in developed areas places great emphasis on independence, self-sufficiency and autonomy (Clark, 1972; Gubrium, 1973; Lawton, 1983). The customary Tswana emphasis on reliance on others, family corporacy and joint decision-making appears to negate self-sustained individuality. This is not so. One cannot assume Western-based terms have the same intonations across cultures. Independence, self-sufficiency and autonomy are important to the aged Tswana. As explained earlier, differences in expression exist. Although I do not use these concepts directly in analysis, they play a very important part in finding the good life.

The contractual nature of Tswana human relationships cannot be denied. Dependency is a respected expectation within the traditional Tswana social system

(Chapter 5). It is found with the requirements for service, the input from others for decision-making, and actions leading to the betterment of others during achieving. Within such contractual social relationships are opportunities for valued independence. The Tswana see independence as the power to negotiate favorable contracts that result in reliance on others. In this sense, independence for the Tswana contrasts markedly with the typically applied Western markers of physical, mental, and economic freedoms.

In modern nations, self-sufficiency usually encompasses residential and economic competence. The aged of developed countries strive to remain in their own households and adopt their standard of living to fit pensions and social security plans (Clark, 1972). Governmental policy reflects this definition of self-sufficiency, in that great attempts are made to keep the aged economically and residentially independent of adult children (Achenbaum, 1987). Living with an adult child is regarded as a step backwards.

Living with an adult child is a step forward for the Tswana aged. Self-sufficiency is a process of participating in behavior that has value. Behavior that involves family carries the most value. Behavior that involves coping with adversity is also valued (Alverson, 1978:118-119). Adversity is expected with achieving. Achieving, in turn, is valued in relation to one's present abilities. Caring for one's body is the most fundamental aspect of "doing" for achieving for self-sufficiency, as the body is the source all other valuable behavior. Maria, the old woman who could only wash herself, saw this action as reflecting self-sufficiency. Like achieving, there is no beginning or end of goals to be met. One always tries to be self-sufficient, although arenas for self-sufficiency decrease with physical limitations.

In the Western world, autonomy is usually thought of as liberty to make decisions for oneself, free from outside control. The extreme opposite is considered to be coercion by others. For example, nursing home residents lose autonomy when they must follow an imposed, but unnecessary life style (Shield, 1990). The meaning of autonomy is frequently blended with independence, implying freedom in action is unlimited by the choice of others. Autonomy is doing what one wants when one wants to do it (Munnichs, 1976).

Autonomous decision-making can occur within contractual relationships (Rosenberg, 1990:35-36). The Tswana act within a framework of interpersonal relationships bound by law and kinship. The ideal is stay within these bounds of cultural restraints. The aged do not view their cultural norms as limiting autonomy. The aged see themselves as "thinkers." As decision-makers, they must utilize the direction given from specialists, advisors and authoritative figures. In addition, decisions should definitely include the needs and feelings of others. Thus, expression of autonomy is incongruent in some ways western expectations, but it is present.

<u>Finding the Specifics of the Good Life</u>

Each society has its own standards for conduct towards the age, and by the aged. What is honorable, respectful, and proper in one society may not be so in another. It is necessary to view the good life for the Tswana through the eyes of the Tswana. This means accepting the Tswana tenet that all people are continuously in need of subsistence, security, and recognition, with such needs being met through the continual process of doing and achieving (Alverson, 1978). Thus process becomes the central core element of the good life. This process undergoes change with aging, as eldership rights enter into the laudable methods of execution. Being served meals is as important as the presence of food. Receiving from others becomes more important than the giving of oneself. Material wealth is not a major factor in itself.

The measurement of the good life is designed to reflect three major areas valued by the aged Tswana. These areas are freedom from physiological need deprivation, security in being served as an elder, and recognition as an achieving individual (personhood). The measurement of physiological need-fulfillment and security is based on four items for both of the two categories, using a five point scale per item. Achieving is the most subjective of three variables. It is also the area where the confines of language and world view affected data gathering the most. The respondents were unable to provide gradations for responses. This resulted in specific yes or no answers for the five tested items. The total score is the addition of scores for the three areas, as differences in question design among the areas prevents valid weighting of categories. Hence, the total scores de-emphasizes the area of personhood. I limit the use of total scores for this reason.

<u>Freedom From Physiological Deprivation</u>

Freedom from physiological deprivation encompasses the areas of having adequate food, the necessary expendable goods for life maintenance, health provisioning and protection from the environment. (See Table 9.1.) The aged stress, that while there can be disappointment in not having advantages of living a middle or upper class life style, it is not a necessary dimension of success. There is no self-judgement or shame involved with the lack of material goods. The aged define the minimum as having enough freedom from deprivation to allow adequate calories for health and activity in a safe, health-promoting environment.

Freedom from hunger is mentioned repeatedly as a prime component in the good life. The frequency of meat and vegetables intake is not included as a variable, as the group considers obtaining any food the prime objective. Three solid meals a day are obtained by one/fourth of the aged. Another fourth are at the other polar end, having one meal a day, either with or without milk. Milk is the foodstuff singled out by the aged as it balances the grain and prevents nausea with tea. Approximately half eat twice a day, with slightly over half of this group having milk. Eating between meals is very rare, as snack foods are not generally affordable.

TABLE 9.1: AREAS OF PHYSIOLOGICAL DEPRIVATION IN THE GOOD LIFE
(N=105, unless noted)

1. Frequency of Meals (Mean = 3.47)
    1. One (1) meal a day without milk. (13.33%)
    2. One (1) meal a day with milk. (10.48%)
    3. Two (2) meals a day without milk. (17.14%)
    4. Two (2) meals a day, with milk. (34.29%)
    5. Three meals a day, with or without milk. (24.76%)

2. Basic Expendable Goods in Household (Mean = 3.62)
    1. Without food for next meal, matches and soap. (12.38%)
    2. Without food, has either matches or soap. (10.48%)
    3. Has food, without matches or soap. (19.05%)
    4. Has food, either matches or soap. (19.05%)
    5. Has food, matches and soap. (39.05%)

3. Health Provisioning (Mean = 4.19) (N-104)
    1. Health care of any sort is not available. (0.95%)
    2. Difficult to get formal/informal health care when well. (6.67%)
    3. Home assistance with medications, denied formal care when ill. (20%)
    4. Can get general health care as desired but not specialized care (cost of
        traditional treatment, eye glasses). (17.14%)
    5. Can get all preferred services. (55.24%)

4. Protection from the Environment (Mean = 3.56)
    1. Is without all of the following: a coat, shoes without holes, dry room during
        rain and freedom from worry about community safety. (3.81%)
    2. Has 1 of the above. (14.29%)
    3. Has 2 of the above. (22.86%)
    4. Has 3 of the above. (40%)
    5. Has all of the above. (19.04%)

Totals for Physiological Deprivation. Range = 5-20; Mean = 14.93; Mode = 16.

Extreme hunger, lasting for a full day, occurs during the week for 67% of the group, with 52% contributing hunger occurring due to lack of purchasing power, one or more times. The very poor regarded tea with sugar as their main food. More affluent households served meat and vegetables twice a week, with the poorer households feeling lucky if these foodstuffs were obtained at all during the past month.

Food, matches and soap are considered as the most important basic expendable goods for functioning. Almost one/fourth of the aged have no food in the house and half of these have no food, soap or matches. Another 20% have only food in the home. Slightly over a third (39%) have all three items.

Health care is judged by access to formal services and assistance with medications or treatments within the home. The majority (55%) felt that they obtained all preferred provisioning. I believe this statistic is influenced by free clinic and hospital care, along with an acceptance of western medicine resulting from the historical presence of a village hospital. Those 17% who see traditional care as the cure for disease (although symptoms can be treated with western medicine), and those presently requiring specialized services such as eye glasses, felt comprehensive care was beyond their reach. The remaining 27% feel they are denied aspects of care due to the lack of required transport, or other assistance. Less than one percent feel they cannot get health care of any sort. (Satisfaction with health care, as presented in the previous chapter, is not included.)

Freedom from deprivation includes protection from the environment. Four conditions have to be met to feel adequately protected. These include ownership of a coat or sweater, shoes without holes, having a dry area when it rains and a lack of worry about generalized safety. All four conditions are met by one/fifth of the group. The mode (40%) is having 3 of the 4 conditions met. Usually the missing item is a coat or cold weather clothing. Others lack shoes and coats (22%) or shoes, coats and either a dry area or feelings of safety (14%). Only 4% lack all protective factors. Safety is the least mention item on the scale, but is a major concern for those who feel being bumped or pushed when walking on the footpaths is a real possibility.

Freedom from deprivation is the segment of the good life where the aged scored the highest, with a mean of 14.93 out of 20. Scores of 17 to 20 are held by 40%. Only 4% score below half of the possible points. Sex and age have no bearing on the these scores. The total gerontic fund correlates strongly with the amount of deprivation (p = 0.0001). Both the physiological and fiduciary dimensions had a strong influence (p = 0.0004 and p = 0.0001, respectively). Those who controlled money-producing assets in conjunction with good health were the least deprived. These findings suggest that as health and finances decrease the amount of deprivation increases. This is not what one would expect in a society where care of the frail and economically dependent aged is mandated by social mores. In fact, the social/familial assets correlated very little with freedom from deprivation (p = 0.046). Two reasons come to mind: the situation of generalized family poverty creates deprivation for everyone, and secondly, that with the wage-earner making decisions for expenditures and labor and the coinciding displacement of the aged from family affairs, physiological need fulfillment is overlooked. I believe both play a role.

<u>Security in Being Served as an Old Person</u>

Security in being treated as an old person concerns the receiving the customary rights of service as an elder, regardless of the degree of decrepitude. This category involves both subjective and objective data. The reader should keep in mind that the concept of security is under evaluation, not the actual care delivery. Responses do not always reflect what may be reality. Several aged are sure a migratory daughter would quit a job to return home, and provide continuous care, if chronic illness occurred. My earlier research suggests employed adult daughters do not intend to return to the village to provide care for ill aged-parents (Guillette, 1990).

Assistance and care-giving begins long before the onset of frailty, as duties are delegated to growing children by parents. Later responsibility for household management is given to adult children who remain in the village, thus establishing the concept of family acceptance of old age disability and provisioning for more complete care when the need arises. The care the aged seek includes emotional support, as well as physical care and normative assistive behaviors. (See Table 9.2.)

Within this category falls the earlier mentioned norms of being served cooked food, having water supplied for personal use, assistance with chores, and home maintenance. Reliance on others is something to be proud of, contrasting with the negative judgment of dependency found in America. Many aged can, and do, perform these acts. Performance is viewed as part of achieving, but only when such services are available. For example, Happy Sound did not see herself denied service although she did the majority of household cooking and other chores. Many were not constantly "served", but approached the problem like Happy Sound and her chores. Outside forces, such as employment and school created the need for performing the act oneself.

Other aged are sometimes forced into doing these activities, which should be provided as a right of aging, when assistance is not available or refused. This lack of assistance and/or refusal brings shame. Idleness, play, or self-centered activity on the part of service provider are not justifiable grounds for omission of service. This is frequently the case with the Mosoko family, where Betty resents having to clean and cook.

One/third of the group consider themselves as receiving all the rights to service on a regular basis. Availability and access to prepared food is the most common missing service for all those not receiving full rights. Of the 67 aged who were hungry in the

TABLE 9.2: AREAS OF SECURITY AS AN EDLER IN THE GOOD LIFE.
(N=105)

1. <u>Security in Having Normative Services Provided</u> (Mean = 3.33)
    1. Does not have food served regularly, water provided, home repairs done and assistance with chores. (19.05%)
    2. Without 3 of the above. (14,29%)
    3. Without 2 of the above. (14.29%)
    4. Without 1 of the above. (19.05%)
    5. None of the above. (33.33%)

2. <u>Being Included as a Family Member</u> (Mean = 2.65)
    1. Feels 'lonely' more than once a day. (41.35%)
    2. Feels 'lonely' once a day. (10.58%)
    3. Feels 'lonely' four to six time a week. (11.54%)
    4. Feels 'lonely' one to three times a week. (14.42%)
    5. Feels 'lonely' less than once a week. (22.12%)

3. <u>Availability of Care if Ill</u> (Mean = 3.58)
    1. Would be alone, no responsible care provider. (14.29%)
    2. Someone would be present during night only. (9.53%)
    3. Someone would be present during night and part of day. (21.91%)
    4. Someone would usually be present, with short absences. (12.38%)
    5. Someone would provide constant care until well. (41.91%)

4. <u>Support During Death</u> (Mean = 3.41)
    1. Fears process of dying will not be recognized by others and their death will not be known. (10.58%)
    2. Fears family will not respect wishes about locale for dying and will die without family present. (20.19%)
    3. Fears family will not respect wishes for locale but will be with them most of the time. (22.12%)
    4. Feels that family will respect wishes and be with them most of the time. (11.54%)
    5. Anticipates full time care and support in desired locale. (35.58%)

<u>Totals for Security In Being Served</u>. Range = 5-20; Mean = 2.81; Mode = 16.

past week, 45 of them (67%) say is because there was at least one instance of no available prepared food. This condition is up and beyond the inability to purchase food.

Fairly equal numbers mention lack of service in the remaining areas of providing water, assistance with chores, and housing repair. Within this service sub-grouping, the aged consider the lack of providing water to be the most serious. Water, like food, directly affects health. Twenty-one percent see themselves denied water.

Objectively, one can say that half of the aged are secure in that all, or all but one, of the rights of elderhood will be provided. The seriousness of lacks raises serious thoughts for the other half. Who are the aged who receive none (19%) or only one (14%) of the most basic of services? The lack of water and food is mentioned by all individuals who were house-bound, or confined to the yard. My observations during visits confirm such lacks, and many of these individuals are without access to food and water for the entire day.

Another factor reflecting security as a family member involves being incorporated into family activity. This is the feeling of being loved and that someone cares, independent of social value and contribution. Linguistically, the lack of such acceptance can best be stated as "feeling lonely even when someone is around." Being *ikhutsafalo* (lonely with someone present) is the one category where the lowest possible score is the mode (41%). This means almost half feel lonely two or more times a day. These individuals state they are ignored by others, have others disappear without reason, and that others "create sad thoughts" unnecessarily. Another 11% are lonely at least once a day. No one is ever forever free of loneliness, although 22% say the feeling occurs less than once a week.

The evaluation of security in care when ill differs from service, in that answers cannot be based on recent recall of the past week. Consideration of the past and future is involved. The ideal of having someone available 24 hours a day is anticipated by

41%. Another 34% feel someone would be with them either the majority or limited part of the day and during the night. This leaves one/fourth of aged who believe they would be by themselves during the day, with 14% of these people saying they would also be alone during the night.

A complementary component to care during illness is support during death. Death does not have to occur in the home, or even the village, but should occur with kin. Reassurances that one will recover are not given. Visitations are to say good-by with the dying person giving final commands for the living. Someone must be present at the time of death to assure correct positioning. Death should occur in a flat position on one's back, even it means being constrained during pain. No reasons for this mandatory action can be given, other than this is the way it should be.

The aged of today recognize that family may not be available to provide care during death. They desire to die where they perceive the availability of people to be the greatest. Some believe this would be in the hospital. Others believe home-care would be most adequate. Only a third (36%) anticipate full time care in the desired locale. (This number approximates the number who anticipate care when ill.) An almost equal number (31%) fear they would die without the family present. Some believe that the family will not care if they are near death and the process would occur without recognition (11%), while others believe that family will show their uncaring attitude by insisting on an undesired locale and not remaining with them (20%).

In review, the component of being served as an elder includes being involved with kin who provide the social rights of eldership. These rights include assistance during wellness, and care during illness and death. Social change has done little in the modification of the definition of these rights as the young discuss "what should be done" in the same framework as the aged. Agreement on rights does not guarantee provisioning of rights. This is the area with the lowest scores. Possible scores range from 5 to 20. While very few (4%) received the minimum, 16% scored no more than four points above the minimum, with half the group (49%) receiving less than half the possible points.

A correlation existed between those who received high scores for security and a large gerontic fund ($p = .009$). The social/familial component had the strongest correlation ($p = .0007$). In contrast to freedom from deprivation, there is no direct economic cost to providing security. I suspect correlation is due to the availability of assorted family members who care and love the aged parent(s) and are willing to make social exchange a payment of past debts. Fiduciary assets are weakly correlated with security ($p = .04$), hinting that material wealth and potential inheritance and do not strongly contribute to the gaining of service and psychological support. This is understandable as many of the fiduciary assets of the aged are generally not valued by their adult children, and frequently adult children have more fiduciary holdings than the aged parent. Only a hint of possible relationship between physiological assets ($p = .08$) and security existed. This suggests service and support do not alter excessively with changes in health and vigor. There was one exception in the correlations between physiological assets and security, if individual categories are considered. Security in the single area of delivery of rightfully due services actually decreases with increased physical limitations ($p = .002$), supporting my observations that when social exchanges involving labor cannot be made, care is not always provided. This involves the aged who are the most in need of physical-supportive services.

<u>Recognition of Personhood</u>

One main reason for the Tswana wanting to grow old is to continue the "great works" of creating a unified family whose activities perpetuate household and kin welfare (Alverson, 1978). This last category of the good life is called personhood. Personhood consists of the web of social relationships that provide self-identity. It reflects the opportunities and rewards associated with the Tswana process of achieving as a person and as an elder. Both participation in relationships and the reactions of others have importance. (See Table 9.3.) The finding of personhood through achieving is conceptually possible until death. Therefore, I adapt items to include the physically frail.

The first item involves participation in meaningful events, which promotes feeling of social contribution. Activity, such as having a small church service in the home or the sharing of food brought home from a from a feast, opens this category to all aged. Some of the more frail aged see meaningful participation as just being told, and then talking about, events that they could not attend. The group is fairly equally divided between having opportunity to participate for social contribution (57%) and not having the opportunity (43%). The aged who express denied access to opportunities for social contribution gave reasons for exclusion. This includes not being told in advance of events, lack of transportation, and feelings of prejudice against the aged if they did attend. This was true of both active and frail aged. Home-bound aged stated they were not told of marriages and deaths, or not given an opportunity to express their thoughts.

A large majority perceive themselves as actuating self-identity through kinship gatherings. This involves being visited and/or visiting others.. This area has import beyond having company, as visiting elders is mandated by the law. Thus, the act of visiting reflects respect and concern for the elder. This allotment of respect is not commanded by power or prestige, but symbolizes the value of one individual to another. Seventy-one percent are involved with visitations. The rest are not visited and/or have no one to visit with.

Direct questions on the perceptions of status in the community and at home were a dismal failure. There is no specific Setswana word for social status. People know their place in the scheme of village life with everyone being recognized for their

TABLE 9.3 RECOGNITION OF PERSONHOOD. (N=105 unless noted)

1. <u>Participates in Events Meaningful to the Individual</u>. (Mean = 1.57)
    1. No participation in meaningful events. (42.86%)
    2. Does participate in meaningful events. (57.14%)

2. <u>Participates in Visiting</u> (Mean = 1.7; N = 104))
    1. Does not visit or have visitors. (28.85%)
    2. Visits and/or has visitors. (71.15%)

3. <u>Provides Direction in Family Life</u> (Mean = 1.39)
    1. Family members "do not listen". (60.95%)
    2. Family "listens". (39.05%)

4. <u>Perception of Functional Utility</u> (Mean = 1.43)
    1. Is useless. (57.14%)
    2. Is useful. (42.86%)

5. <u>Perception of Life</u> (Mean = 1.43)
    1. Is unhappy. (56.19%)
    2. Is happy. (43.81%)

<u>Totals for Recognitions of Personhood</u>. Range = 5-10; Mean = 7.5; Mode = 8.

role in the schema. The more recent association of economic success with social status is recognized by all. In addition, the aged emphasize a social ranking based on the traditional social structure of society. Structurally, age in itself determines the level of status in any verbal exchange. All aged should be given respect, independently of other markers of status. For data in this area, I took advantage of the two concepts involved with respect that are presented repeatedly by the aged: that of being "listened to" and of being "useful."

The concept of being "listened to" involves obedience to orders, respect for advice, and subservience to the elder's authority. Almost two/thirds (61%) of the aged said they were not "listened to" by the family, indicating that the aged not only lack family respect but also power. Being useful implies making valid contributions for the welfare of other. Sweeping to avoid idleness is not being useful. Sweeping in respect of the needs of others is useful. Over half (57%) see themselves as having no use-value to the family or society.

The final question concerns the view one has towards one's present life. Life is described as either happy or unhappy. Happiness comes from on-going recognition, by self and others, of achieving and of doing. Happiness is present, or it is not. Various events could make one cheerful or depressed at the time, but the aged insist on separating specific events from the overall feeling. No gradations of contentedness with life is possible. Two-fifths (44%) of the aged were happy, with the remaining 56% seeing life as unhappy.

Sex and age have no bearing on perceptions of personhood. Full personhood is obtained by 14%, with another 16% finding personhood in all but one area. Nearly one/third of the total group find no aspect of personhood (13%), or only one segment (18%). All aspects of the gerontic fund are related to the finding of personhood, although the social/familial dimension had the lowest correlation (p = .03). Again, this is not what one would expect in a setting where family is regarded as insurance for a good life in old age. Does one have to buy happiness, respect, and power with physiological and fiduciary holdings?

<u>The Gerontic Fund and the Good Life</u>

All aspects of the gerontic fund, including the total score, strongly correlated with the total good life score. All aged with strong holdings found a relatively good life. The fiduciary dimension had the strongest correlation ($p = .0002$). Everyone who scored in the top quarter of the good life also are in the top quarter for fiduciary assets, although a few with minimal economic assets are able find various aspects representing success. Strong correlations are also found with the social/familial dimension ($p = .0006$) and physiological/personal dimension ($p = .0008$). The majority, with social/familial holdings in the upper quarter, tend to find the good life, while very few of those in the lower quarter can succeed. The same is true with physiological/personal assets. Age and sex played no role, yet the very old are the group that stand out as fund-poor age group. One must keep in mind that there are numerous individuals in the other age groups that are equally fund-poor.

The diversity in life satisfaction cannot be accounted for in terms of gerontic funds alone. There is no statistical threshold in held assets below which a good life becomes impossible. A few of fund-poor, of any age, are able to exceed the midpoint in life satisfaction, although they never reach top scores. In additions, some aged, with a relatively strong fund, are unable to find any aspect of the good life.. Some old people with extremely limited assets may be skilled in the substitution of one asset for another to obtain what is desired. The fund-rich may not know how to use their fund effectively. I think other factors are also involved.

It seems likely that a larger household size, with its increased availability of benevolence, would increase life satisfaction. This concept is false ($p. = .42$). Neither does it make a difference if the household contains a second aged individual, as I thought that another aged household member would assist in keeping doors open. Living in a three generation household slightly increases the probability of obtaining the good life (p. = .048), but the number of household members make no difference.. The failure to find a good life probably exists because the setting prohibits affordable trade, either with the present gerontic fund or the calling upon past debt for rewarding benevolence among family members.

I believe much of the diversity in finding life-satisfaction occurs from success or failure in the exchange process is related to access to doors, their condition and sway. Household relationships have become competitive in nature (Comaroff, 1953). Social relationships involve mistrust and antagonism. Success in finding the good life depends, in part, on the success of finding willing and cooperative significant others in a supportive setting. Only then can the aged find esteem through service, security and personhood. as status cannot be self-generated.

Why do fund-rich score high in obtaining service and security and low in personhood. One must keep in mind the division between esteem or respect (holding in high estimation with the ingratiating regard for another's wishes) and deference (the yielding or submitting to superior authority with the possible absence of esteem). Power breeds deference but not necessarily respect. The fund-rich obtain service and security from the power that the fund generates for trade. This does not automatically provide respect for the individual as a person. Many of the doors for respect have been broken, and can only be transcended with the retention of middle-age values of activity and economic contribution.

CHAPTER 10
AGED CHILDREN: SOCIAL ELDERS

Do other old people live like I do? I see myself as alive, with much to
give, but others see me as the living dead. It's like the door to the future
has been broken and no one lets me through. (69 year old woman)

What we need is a place where we can be somebody, our own place
where we can find respect and be thought of as valuable people. The old
are no longer a part of the family or village. (78 year old man)

Anthropology is more than a study of social change as the anthropologist is a

facet in that change. My work had increased the villagers awareness of the difficulties

associated with aging and among the aged and their families. Frequent meetings with

local, and central government, officials had increased formal recognition that the poverty

of old age extended beyond income, to include deficits in social acceptance and

meaningful roles. This chapter presents the outcomes of this change in knowledge.

The Demand For Innovation

The interviews were almost complete as I began my last month in Ramotswa. I

had purposely planned sufficient time for the sharing of my preliminary findings and

thoughts with various agencies in Ramotswa and Gaborone. The hospital had set aside

two afternoons for staff in-service education regarding my findings and

recommendations. The Social Service Department and Health Department at the

District Council also requested classes for their staff. Additional requests for classes and

public seminars had been made by the University of Botswana in Gaborone.

My first priority was to have the public presentation of the preliminary results of

the study, fulfilling my earlier promise to the tribal administration and the people. Chief

Mokgosi and I set the date. There would be two meetings, one at the *kgotla*, the second at the Community Center. The second meeting, held nearer the other end of the village, would allow attendance by those who could not walk across town to the tribal headquarters. Over two hundred village aged and their families attended my community reports of the strengths and needs of the aged.

The general findings of the study were echoed by the crowd, with specific requests for better nutrition, treatment of hypertension and "something to make people like us." The absence of models did not prevent their recognition of a need for a place for old people to share common concerns and receive understanding of age-specific needs. Throughout the presentation came pleas. "We Need Help Now!" "We don't have years for the government to plan programs, for by then we will be dead." The present determination and past cooperation of the aged suggested the time was ripe for immediate intervention.

<u>Existing Plans and Projects in Botswana</u>

Botswana has recognized there is a need to institute programs for the aged. This is a welcome reversal attitude since the beginning of this research in 1988. The desire is to create policy that avoids the pitfalls of past public support programs. These programs, usually associated with drought, promote the public concept that it is the duty of the government to provide economic support during adversity. Anger arises as not all individuals receive equally (Guillette, 1990).

The most wide-sweeping program under consideration is tax relief for households containing aged. This will certainly relieve some of the economic burden of support. I cannot help but wonder if the aged member will directly benefit, other than receiving their share of purchased food. My research has shown the use of money is controlled by the person receiving it. Expenditures for family nutrition, and for the education of young children, take priority. Remaining funds, in this case in the hands of the tax-payer, are usually used for personal goods, either from necessity or the desire for luxury.

I have grave doubts if such income would be used for clothing, bedding, or traditional health care for the aged. Yet, this potential program indicates that the government is concerned and is trying to divide equally.

The second program under consideration is the construction of a government supported 36 bed nursing home, to be located in the heavily populated area of Francistown, some distance to the northeast. This will provide care to a select few aged, to be determined by policy relating to economic and health needs. Several drawbacks can be identified. A few aged receive benefits from a large expense. Their families receive "something for nothing", an ambient attitude created by past drought relief program that the government desires to combat. Equally important, this transfer of Western ways and means of gerontological care acts in direct opposition to the desired promotion of traditional family-centered care in the home. In addition, it is concrete evidence to the population that separation of the aged from the rest of society is acceptable.

The provision of care services is a growing problem in Botswana, and does need to be addressed. A novel approach to the problem was taken by the Maru-A-Pula High School in Gaborone, as part of their community outreach program. A housing unit of one elongated building with several bedrooms and a kitchen area was constructed in the middle of a small rural village. Aged who were living alone were invited to become residents. A mix of talents and abilities in residents resulted, ranging between the very active and alert to the care-dependent bedridden. Chores were divided according to strengths and desires. Care was given to the bedridden. Pride in achieving replaced fears of being alone when in need or at death. The house location allowed two-way interaction with the village, with visits from kin and to kin. The school was proud of their innovation. The community was proud the aged who lived there.

<u>Self-Help Programs</u>

Self-help programs for the aged in developing countries are drawing interest, but are few in number. Planned intervention usually originates with an outsider's selective perception of need and solution (Chambers, 1983:28-31). Most programs provide for direct employment by the teaching of new skills or establishing gardening and husbandry programs for the active aged (Tout, 1989). These programs, while successful in promoting income-producing opportunities, usually involve a small number of individuals, exclude the more frail aged, and require extensive external financial and manpower support. Many fail as models, as internal cultural strengths and inhibitors are not considered (Chambers, 1983).

Self-help programs in Botswana have been associated with drought relief. The government provided either income or food in exchange for labor for village development. The aged have been active participants in these programs, with the goal of stimulating social prestige and acceptance rather than becoming economically independent (Hay et al., 1985; Guillette, 1990). The last government sponsored self-help program ceased in 1989.

<u>Defining Gerontological Problems</u>

In gerontology, the manner in which problematical social trends are defined becomes the basis for social policy development. In Western nations, concentration is focused on how modern life styles have affected the previous support systems for the aged. Gender, age, family, and income have become the important issues (Stanford and Yee, 1991). African gerontology is no exception. The claim is that migratory employment and breakdown of the extended family places the aged-poor, the childless aged, the widow, and the very old in jeopardy (Osman, 1983; Tout, 1989; Ingstad et al., 1991; Thomas, 1992). They should become the targets for policy intervention (Khasiani, 1987; Thomas, 1992).

The identification of trends stressing the specific vulnerability of particular aged tends to override that which is generic and common to all older persons of the society. The present cultural and psycho-social forces that could be used to promote well-being are forgotten in both the definition of risk and the approach for planning.

The real issue with the aged is not one of age, living conditions, or traditional ethnicity, but one of poverty for social exchange. The deprivation trap (Chambers, 1983:111-114), which pulls a household below any possible threshold for economic recovery, also exists on the individual level. Aged individuals are the poorest of the poor: economically, socially and physically. The same interlocking clusters of poverty, physical weakness, social isolation, powerless and vulnerability exist for the aged. Many of these clusters are age-related, and intensify as age advances. Each economic, social and physical loss become a ratchet, increasing old age isolation, powerlessness and vulnerability. This always makes any doors to the good life harder and harder to manage.

The aged of Ramotswa, many of whom lack numerous gerontic fund assets, can no longer manage the present-day doors to care, support or acceptance with the process of exchange. The impacts of the resulting gerontic-deprivation trap are shown repeatedly in family function and communications, and in social interactions. The loss of rights is a by-product of the resulting social detachment (Chambers, 1983:115). Vulnerability is great, for the end-product is the loss of personhood. No one hears them. Their activities are play. It does not matter if the aged maintain the customs of ceremony or sweep from idleness. They are the valueless, and frequently the unwanted, burdens of society. This is why all the aged cry out for help.

<u>Aged Children Become Wise Elders</u>

I feel this call for help should not go unanswered. I am aware that standard gerontological approaches of service delivery, or monetary assistance, will not promote escape from the gerontic deprivation trap, which I see as the commonalty unifying the aged. Culture and need has to be united to establish any potential program direction and content.

Identified individual, family and social needs fall into two distinct, but complimentary categories: that of promoting physical and mental well being, and that of providing sources for status generation. Traditional culture dictates that the aged needed recognized roles of contribution involving power, authority, and status. Contemporary society suggests action that addresses the needs associated with family poverty. I also have to consider the basic facts of life in a Botswana village. The physical environment means any program has to be adjusted to the absence of modern conveniences and facilities. Motorized transport is also limited.

I still have many unanswered questions as a result of my research. I have only touched the surface of adjustment after retirement from employment in more developed areas, and how this relates to finding the good life. The combination of positive and negative interactions and the flow of affinities between grandparents and grandchildren still creates confusion. How does one approach the needed education for resolving generational conflicts in a culture without quilt or belief in self-control over life? I wonder if the teaching of newer Western discipline techniques and the addition of positive feed-back would restore any control and authority by adults and elders. Lastly, can one expect the Tswana, who have reached old aged, to have an extended life expectancy as found with African-Americans?

I do know that many of the problems the aged were experiencing were not unique to the village (Osman, 1983; Hay et al., 1985; Ingstad, 1987; Guillette, 1990). Everywhere, there is a strong need to enhance the filial obligations associated with the extended family (Hampson, 1982, 1985; United Nations, 1985; Tout, 1989; Guillette, 1990; Ingstad et al., 1991). In addition, the economic problems of this village's households containing aged are similar to those of other villages in Botswana and in Africa (Hampson, 1982; Guillette, 1990; Ingstad, 1991; Thomas, 1992). Material deprivation and the loss of social position with aging seem to be a problem in all developing nations, and even within developed nations (Gubrium, 1973; Cowgill, 1979; Goldstein et al., 1983; Braithwaite, 1986; Khasiani, 1987). With this knowledge, I take it upon myself to help the village aged, and hopefully develop a program that would have catholic implications and applications.

The Wise Elders

With less than a month remaining, a group of aged and myself set the groundwork for the requested "special place for old people." The "place" becomes the concept of a service organization. The organization is unique in Botswana, as it would be for old people only, who would give of themselves in return for the help they wanted. A name is chosen to signify membership: *The Wise Elders*.

My role is to be a provider of ideas and of resources that could be tapped. This places leadership immediately in the hands of the local aged. Plans for the development of a constitution and board of officers are made, in order to officially register the group as an approved, formal, non-profit organization with the Botswana government. Three types of services are discussed: a centralized monthly clinic for hypertension and health education, a small loans program for equipment or tools to establish income, and a food program.

The need for better family nutrition quickly become the main priority. The plan is for the organization to purchase the basic grains, tea, sugar, and dry milk wholesale in the capital, and resell them locally at wholesale prices for family use. Only members, anyone 60 years or above, are qualified to buy. (The home-bound aged could join the organization and have a designated family member make purchases.) Chief Mokgosi agrees to using the kgotla grounds for the twice monthly sales. Unsold food is to be stored in the silos, maintained but empty since 1935. The South-East District Council agrees to transport the food on the weekly service truck coming from Gaborone. The Embassy of the United Sates of America in Gaborone agrees to provide limited funding for the initial purchases, through its special Development Assistance Program .

A short but relevant note must go here. Several externally-based volunteer and grass-root organizations denied the group financial assistance. Careful presentations of needs and environmental constraints could not overcome ethnocentrism. "Senior Citizen Programs should provide dances and parties so the aged can have fun." "Why aren't you including field trips in the program?" "You should establish Meals-on-Wheels to provide cooked meals."

Excitement about the program increases during my remaining days. The aged see *The Wise Elders* as a means of achieving in their own right. Through the purchasing program, the aged can obtain food economically, for the benefit of the family. Families begin to think of the aged as providers of food, not an empty mouth to feed. The public makes comments that imply that the aged are becoming recognized as capable of valid and valuable contributions. More remarkable to the public is the fact that the activity of the aged is being supported by the government and tribal administration.

Repairs To The Broken Doors

The Wise Elders purposely overlooks the often stressed Western goal of creating independence and stimulating self-reliance through self-help programs. The desired outcome is built on the traditional goals of family interdependency and cohesiveness. Program design emphasizes that the aged are not the recipients of service or care, but are the providers for enhanced family nutrition and contributors for economic savings.

The Wise Elders program is unique in that it placed the aged in a position to begin repairs on the broken doors and thus increase their access to sources of social acceptance. The one problem of adequate food supply is but a small area of needed change in the overall setting. A small magnitude of change has a great significance when social worth is involved (Rossie and Freeman, 1982:68). The aged, by leading themselves within the organization, become traditional elders concerned about family and social good. Acts of achieving become recognized. The necessary group interactions pave the way for turning enemies into friends. The aged can see others were living in need, without speaking of their own problems and breaking discourse regulations. All these actions reflect traditional roles and social structure. Such actions have a better chance of inducing further change than if the program was planned abstractly, and administered by others (Foster, 1973:164-165).

The formation of the Wise Elders also serves to undo some of the door-sealing effects of gerontic deprivation ratchet. Traditionally, the Tswana elder acts as the door-keeper for family and social welfare. Change has opened new exits for the family, with a closing of the traditional doors. This program recreates value in using the traditional door, as families need the aged for access to the desired food. Although other doors to successful aging continues to have severe cracks, and has hinges with a tendency to work the wrong way, a small opening exists. These minor repairs promote positive social interactions and a reason for being. With this comes a twinkling of hope that the aged children of Ramotswa will once again be seen as the wise elders.

CHAPTER 11
CONCLUSIONS

You should write a book and tell the world how we, the old people, must
live. Tell them the people are good but the way people act is not good.
(78 year old woman)

No one allows us to live happily. It is the society that makes us live like
this. I think my family cares but it is hard for them to show it. (67 year
old man)

I began my field experience with thoughts on the contextual value of gerontic

fund. As I became more and more involved with the villagers, I realized the fund could

not be isolated from the setting and the way of life. Everyone, the young and old alike,

attempted to manipulate identical doors; defined by the early culture, constructed on

tradition and modified with time. Gradually, doors became kaleidoscopic, with each

person placing their own values, knowledge, and goals on door selection and use.

The aged also built their approach to life on their culture, experienced life

according to tradition, and modified life with time. Only when the past, and its doors to

the aging process, is comprehensively interwoven with the present can one evaluate the

diversity of the aged. The past also provides commonalties among all aged. The shared

commonalties are as important as the diversity (Standford and Yee, 1991). This chapter

concentrates on the commonalties and diversity among the individual Tswana aged,

between the aged and the village population, and between the Tswana aged and other

aged in the world.

<u>The Aged</u>

Diversity among the aged of Ramotswa exists, with widows, childless aged, the physically limited, the extremely poor, and the extremely aged contrasting with the economic elite, aged in large families, and the physically strong. All placed emphasis on tradition, but varied in the extent that Westernization played a role in their behaviors and thoughts. Individuals varied markedly in the strength of their fund, but no particular group stood out as having the best or worst in the good life, either by age, sex, family finances, or marital status.

The Tswana aged seek a good life by using the self-controlled assets in their gerontic fund. Physical, family, social, and economic resources are used in a process of social exchange to obtain service, care, security and recognition. The amount of held assets varies between individuals, with some aged entering old age with very weak dowries. Natural disasters, changing patterns of employment, limited education, and increased demands on income prevented them from entering old age with the needed riches. Others report having lost their assets since becoming old. Repeated drought, death and migration of family members, new expenses, and illness tend to erode the funds as age progresses.

Strong personal-physiological, social-familial and fiduciary resources creates power and command, allowing the aged individual to overcome the sociocultural and geo-physical barriers to a good life. Few aged had such strong gerontic funds. The poverty of old age, within a setting of generalize poverty, places many of the aged in a gerontic deprivation trap. Any small loss in health, family, or savings increases their vulnerability and decreases their already waning power over others. They must be able to use the remaining resources effectively in order to find any life satisfaction. Demand and compliance may provide the needed basics for life, but not satisfaction. Therefore, I see that the real issue with the aged is not one of age, or traditional ethnicity abutting modernization, but one of poverty for successful social exchange.

No point exists in which the gerontic fund automatically provides a good life, or negates its occurrence. Much depends on the openness and the swing of the doors that are used during exchange. Some fund-poor, regardless of age, found various segments of life satisfaction. All fund-rich reported some area of life dissatisfaction. Personhood, or being regarded as a valuable, wanted, and respected individual was the most difficult to obtain, regardless of resources. Most can obtain life-supportive care from others, although some fund-poor cannot acquire basic needed services. Benevolence for satisfying care cannot always be relied upon, although the traditions of giving goods and assistance continues.

The diversity between individuals in finding the good life of care, security and personhood, in conjunction with the diverse amounts of fund assets, strongly suggests that the assorted process of social exchange is as important as the items to be traded. These processes involve the aged, the family, and the community placed together in a cultural context.

The Context for Exchange

Ramotswa, as a village, is full of diversity. A mix of old cultural patterns abuts, clashes and collides with Western technology, architecture, approaches to health care, and religions. People of all ages are puzzled as to why traditional beliefs, and behaviors, do not blend smoothly with new thought and actions. Conflict creeps into everyday life as new pressures overtake obedience to the traditional laws. The young are guilty and so are the aged. Diane ignores others in her quest for the good life. Betty screams at Dorothy. Alfred never gives Priscilla a portion of the egg money. Naturally, the following of the basic laws, for and during social exchange, is more important to the aged, for this is what they learned as proper and necessary for continuing life.

<u>Cultural Duality</u>

Rebellion against traditional behavioral expressions of the law is not new. It was noted when the present aged were children and young adults (Schapera, 1944:265-266). Behaviors arising from laws continued to be questioned, mitigated, and sometimes discarded, throughout their lives. Each person attempted to pick out what was materially and morally superior when faced with two distinct modes of life. This has resulted in a cultural dualism (Harrison, 1987:46).

Cultural dualism is found among the aged. There is also a generational duality as each succeeding generation digs deeper into the traditional culture. The rebellion of the youth of today is towards what is left of the original culture, that which was originally deemed worth saving (Harrison, 1987:48). The self-identity of each generation is now threatened with the conflicts involving moral obligations between self and others. Contradictions also exist between self and society.

This is the area where the aged feel the pain of social change the most. Many of the doors to self-identity, and personal integrating with aging, have been severely broken and are difficult to manage. What few material resources that some aged use to promote self-identity, such as agricultural land and housing, are not valued by the job-centered young. Acts of achieving for the betterment of others do not always generate social recognition. Traditional definitions of a social elder have all but disappeared. For some aged, the doors to self-identity are too damaged to move. Jacob is not alone in feeling ancestorhood is the last open door.

<u>Cultural Continuity</u>

The flow towards cultural duality is counteracted with a strong current of cultural continuity. The diversity in obeying the laws and applications of the seniority principles does not prevent a generalized consistency in shared beliefs. Traditional philosophical values continue to influence the direction behavior should take. Mankind is good.

Goodness is expressed through achieving in the present. The products of achieving are to be shared. The major change is that sharing by the youth is directed more by choice than moral obligation, with the direction of flow going to other young. They believe the goodness of mankind is exemplified by being strong, active and achieving for self-advancement. The aged believe the goodness of mankind is shown through the wisdom of experience and achieving for the good of others. This goodness grows in strength as one grows in eldership.

The aged have always been divided into groups: social elders, those becoming limited, and the decrepit aged or children who will soon become ancestors influencing life from the sky. Before modernization became as pronounced, the aged child was one nearing death and had a relative short time of managing the broken doors, which I believe have always accompanied extreme old age. These traditionally broken doors prevented access to continuing achievement, as the decrepit aged were expected to think and react like a child. Limitations on acts of "doing" limited opportunity for meaningful exchange and increased their disengagement from the family and society. These same divisions of the aging process continued, except that the timing has been advanced. The periods of elderhood and "becoming old" have been shortened with a premature assignment of childhood.

Traditional metaphysical beliefs continue to influence behavior. Achieving is thwarted by outside forces, separate from the individual. Potential unwanted ill forces arising from enemies and sorcery continue to be avoided by limiting interpersonal relationships. Kin are regarded as safer than non-kin, but even then, not all can be trusted. Social change compounds the situation. The result is that people are leery of social exchange with those not considered safe. The imperfections in the doors leading assistance have enlarged, preventing entrance or exit for both the aged and others.

Many traditional patterns of behavior continue. Work and active leisure are not divided. Visiting, talking and feasting are integral parts of life's work. Life centers on the present. Therefore, one is, or is not, involved with "doing." "Doing" and "not-doing" separates achieving from idleness. It also separates the elder from the child, for without providing the desired contributions, one is useless and has no value. The past is ignored, making it difficult to rely on past accomplishments and experiences for establishing usefulness or value. The future is uncertain and unpredictable, directed by forces over which one has limited, if any, control. There is no quilt with failure, as it is forces beyond one's control that prevents achievement. One tries to achieve with continuous doing, but one must also accept the situation.

Those individuals perceived as not productively "doing" are shamed and teased for acting like a child. The aged have difficulty finding ways to be productive in a recognized manner, as economic and technological change has damaged the door of usefulness. The aged are degraded, as contributing has more social value than being the recipient of contribution and the rights of elderhood.

Continuity in many of the traditional laws also limits opportunities for the aged to be of value in the household. The daughter is to cook, the son is to make repairs, the child is to carry water. If an individual fails in duties, the duties are not assumed by others. Often, everyone does without, unless the individual responsible arranges otherwise. Many times the resulting deprivation comes from choice and not the physical inability to act. Participation by the aged in household activities, because of the absence and failure of others, is just as apt to bring shame instead of reward.

I disagree with Ingstad et al. (1991), who places the fault of lack of service on the adult children of the aged. The claim is that because they are "the losers of modern society," they are unwilling to serve the aged. I admit that many assume the position of care-provider for the aged by default, but nearly all are ambitious to achieve and "want to be somebody." The problem lies in the new social structure, which has altered the overt expression of some the traditional behavioral laws, not the elimination of them. Western education has given the younger generation knowledge and desire for achieving through skilled employment rather than participation in agriculture. This employment is limited in the cities and more so in rural areas. Adult children now want to live in nuclear families. This means that the daughters must serve husbands first, rather than fathers and mothers. The desire is present, the opportunity to fulfill valid desires is not.

The generalized poverty and unemployment means life satisfaction if difficult for anyone to find. Younger rural men turn to alcohol, and rural women turn to single motherhood in order to find satisfaction. These younger adults do not perceive their behavior as deleterious to their future, as they maintain the belief in traditional gerontocratic principles. "All good things come to those who wait." "One has children to care for them in old age."

This may never be. What used to be advantages of the gerontocratic principles are becoming liabilities. Some behavioral laws that remain intact work against the aged. Valued items, once controlled by the aged, are now controlled by adult children and grandchildren, and are not available for use. Social and economic demands on care-providers have taken on new dimensions. They are not always available to provide food, water and companionship. Service and security are frequently lost because conflicts between time and duty cannot always be resolved by the care-provider. The care-provider is also limited in performance, as other laws regulate the manner in which service can be provided. The care-providers often prefer to maintain household unity by upholding these laws rather than disregarding them and alleviating discomfort. All these things place additional cracks in the doors to old age rights and security. Both the care-provider and the aged care-receiver are trying to balance the precarious positions of these doors.

It is true that many generational gaps exist, with the young feeling independent of the aged. At the same time, the many of the principles of eldership are recognized by many of the adult children and grandchildren. Much of the present conflict stems from differences between the aged and others as to when the label of childhood is assigned. Childhood negates parts of the seniority principles. There is no need to give respect or ascribed status to old-children, for they have never been so rewarded. Old-children continue to be separated from household function and community recognition. The cracks in this original main door for establishing and maintaining elderhood have been enlarged to an extent where even the younger, active-old are prevented access to the good life.

This original broken door, now almost sealed, exists in tandem with the other doors to achieving and social participation, which have also been damaged. The interweaving of cultural continuity and social change present a situation where it is difficult to use one's assets for effective social exchange for either group or individual good. Care and service is available in the village. Assets alone do not provide access to freedom from physiological deprivation, security in be served, and recognition as a person. The individual aged must know how to use the broken doors, and find people willing to help in keeping them open. Some aged can manage the various doors, using choice, knowledge and assets. Very few can manage all.

<u>A Comparison to World Aging</u>

The Environmental Exchange Theory used in this research was developed in America and designed for primarily for use in developed nations (Gubrium, 1973). The three dimensions of resources necessary for successful aging have been shown to be equally important in Botswana as in America, although some of the actual components do not apply cross-culturally. The correlation between assets and the good life does not account for as much variability as the theory suggests. I believe that the process of finding partners willing to make trade that provides benefit to the aged is equally necessary in both settings, as the geo-physical and sociocultural barriers to the good life are rigid and unforgiving.

The assumption that assets decrease with age appears to be true. The aging process can account for physical losses. Other loss occurs without reference to age, such as that caused by calamity over which the aged have no control. Ambient economic inflation and changing values also takes a toll.

The universal goals of a guarantee for food, shelter and care while remaining respected, useful and healthy participants in society until the advantages of death outweigh the hardships of life (Chapter 1) hold true for the Tswana. The behaviors involved with goal attainment differs from the West. Fulfillment of basic biological needs should come from others, as a right of age. Respect comes from being served and usefulness comes from serving others. Independence, self-sufficiency and autonomy are important, but lay below the surface in the process of achieving. Contribution, as determined by Tswana culture, involves group good rather than promoting oneself. Death is an honorable step forward in the circle of life, not to be feared or unnecessarily delayed.

There are many indications that the Tswana aged resemble aged of the developing world, in that, as a group, they are the poorest of the poor; economically, socially and emotionally (Goldstein et al., 1983; Diouf, 1985; Brathwaite, 1986; Tout, 1989). Others tend to regard them as a human liability to the family and society (Hendricks, 1982). Isolation, powerlessness, and a loss of rights is a by-product of the resulting social detachment (Chambers, 1983:115). For the aged-poor, wherever they may be, the psycho-social implications of poverty are greater than the monetary implications, as the end-product is the loss of personhood (Streib, 1990). No one hears them. They are the valueless, and frequently the unwanted burdens of society.

Incongruity between generations, which involve definitions of a successful person, results in old age status-anguish (Dressler, 1988). This is one of the major problems faced by the aged of the world. (Hendricks, 1982; Tout, 1989). Both the assets, and the process of asset use, are ineffective in overcoming this anguish, as aged are prevented in becoming involved and participating citizen in society (Nusberg, 1988).

The Tswana aged claim they are "the living dead" and are without a place in life. These same complaints are heard in many places around the world, including the developed countries (Hendricks, 1982; Goldstein et al., 1983; Barker, 1990; Sokolovsky, 1990:289-291). These feelings are no different from those of the displaced disaster victim or war refugee (Oliver-Smith, 1986; Hansen, 1990). The main difference is that the aged are involuntarily displaced-in-place. The involuntary-displaced are disenfranchise from the host society, and find frustration with achieving and remaining valuable members of society (Oliver-Smith and Hansen, 1982:2-6). The difference with the displaced aged is that their "host" is their own community. The problems are the same.

Refugee workers attempt to reintegrate the displaced into the host society. Group commonalties form the foundation for reintegration. Group commonalties should also form the foundation for reintegrating the aged in their host society (Stanford and Yee, 1991). Segments of broken doors can be mended when tradition and change are used together for repairs, as with the Wise Elders. Traditional roles, past experience and accumulated knowledge can make a continuing contribution in developing society, with proper planning. (Nusberg, 1988). The strengths of the aged can assist with meeting the needs of society, while the resources of the society assist the aged.

All societies change. The society entered at birth is never the same as the society one exits with death, although the differences may not be perceived as great. Old age goals do not change. Change occurs with the doors needed to meet goals. It is the group characteristics that allow one to see which doors have become broken and prevent the aged from using their resources effectively. Broken doors are not unique to the Tswana. The types of doors may vary and the damage may be caused by other elements. No matter what resources the aged hold, unless doors that displace the aged from society are repaired, many individuals will be unable to open them.

APPENDIX A
GERONTOLOGICAL ASSESSMENT FORM

Code___________Census Tract________
Name______________________________Sex_____Year of Birth________Age_______
Village of birth___________________________Tribe________________
How long have you been living in this house?________ Village_______
Did you go in school?__________What grade?___________________
Religion_________________________________How long (date)?____________
Marital Status: (Circle) Married (Times_____) Married but spouse away
          Widowed Divorced

1. How do you see yourself? (elder, grown-up, with weakness, child)?

2. What sort of things make you feel that way?

3. Who lives in this household with you?
    number   ages/sex
    ____No one
    ____Husband/ Wife
    ____Children
    ____Grandchildren
    ____Parent(s)
    ____Brother/sister
    ____Other Kin
    ____Other

4. How many children did you have and where do they live?

5. How often do you see them?
    5. Daily contact with adult children in and out of household.
    4. Daily contact with adult children in household only.
    3. Adult children in village, sees once a week or more.
    2. Sees adult children less than once a week.
    1. No adult children or has lost contact.

6. Do you have grandchildren?_____ What are their ages/sex_________

7. Where do the grandchildren live?
    5. Has grandchildren of all ages in household.
    4. Has only grandchildren above the age of 14 in household.
    3. Has only grandchildren 14 years or below living in household.
    2. Has grandchildren in village but not in household.
    1. No grandchildren in village.

8. Do you have any other kin in the village who are important to you?____
Who?__________ How often do you see them?______
   5. In household and village.
   4. In household only.
   3. Live outside of household, sees once a week or more.
   2. Live outside of household, sees less than once a week.
   1. No other meaningful family.

9. Does the house in which you live in belong to you, or someone else?
   5. Self.
   4. Provided by adult child for use.
   3. Provided by other relative.
   2. Moved in with adult child, other relative or friend. (circle)
   1. Rental or owned by Chief/council. (circle)

10. Type of house resides in:
   5. Brick/wood with tile/tin roof.
   4. Brick/wood with thatch roof.
   3. Adobe with tin roof.
   2. Adobe with thatch roof.
   1. Other.________________

11. What repairs does your house need? (explain)____________________
   5. Well maintained, no repairs needed.
   4. Minor repairs needed, no inconvenience in type of repair needed.
   3. Minor repair, presents inconvenience. (broken window)
   2. Major repairs but house is structurally safe. (open holes in roof)
   1. Major repairs, may cause harm. (damaged support beams)

12. Type of toilet
   5. Indoor toilet.
   4. Enclosed pit with roof.
   3. Enclosed pit without roof.
   2. Open pit.
   1. Open area.

13. Where is the closest water tap?
   5. In yard.
   4. Within 100 meters.
   3. 100 meters, less than 300 meters.
   2. 300 meters, less than 500 meters.
   1. 500 meters or over. (Give distance)____________________

14. I will read a list of items found in houses. Tell me if it is in your house, and if you, or someone else owns it.

In the house   Owns item   Number owned   Condition

____Shoes
____Coat
____Blankets
____Bed with mattress
____Mat or Foam bed
____Watch or Clock
____Radio
____Chair
____Table
____Inside Stove
____Dresser/wardrobe
____Working Car
____Other

15. Do you have soap, matches, and food in the house now?
   5. Has soap, matches and food for next meal.
   4. Has food for next meal, either soap or matches.
   3. Has soap and matches, no food.
   2. Has either soap or matches, no food.
   1. No soap, no matches, no food.

Lets talk about your health now.
16. In general, how is your health?________________

17. Is there anything about your health that worries you?___ Explain_____

18. What sort of health care is available that you can use? What would your like to be available, including help from your family?
   5. Can get all preferred services.
   4. Can get general health care but not specialized care. (eye glasses, traditional
         medicine).
   3. Can get regular assistance/care at home, denied aspects of formal care if ill.
   2. Difficult to get formal and informal care.
   1. Health care of any sort is not available.

19. Which of the following people did you get help from in the last 6 months.. If you saw more than one person or clinic, tell me. Also tell me if there are any you would like to see but cannot.

                         Times seen   Reason
1. Western doctor____
2. Clinic nurse____
3. Family Health Educator _________________________________
4. Ngaka ya Setswana ______________________________________
5. Moprofiti____
8. Other____
9. None (Why)____

20. In what way does your health prevent you from doing the things you want to do?____

21.   Can you walk to the nearest health post/clinic?________How do you get there when ill?______________________________

22. Do you take, or should you take any medicine?____Why?______________ What difficulties do you have in getting or taking medicine. ____________

Now I would like to ask you about some of the activities that are part of our daily lives. I would like to know if you can do these activities without any help at all, or if you need some help to do them, or if you cannot do them at all.

23. What difficulties do you have with eating?
 5. Can shop, prepare, and serve food.
 4. If food is present, can prepare and eat own meal.
 3. Must have meal prepared, serves and feeds self.
 2. Must be served all foods, feeds self.
 1. Must be fed.

24. Do you have any problems with your teeth? (Explain)___________

25. Do you have difficulty getting to the toilet on time, or with your water leaking?
 5. Always uses toilet by self, no soiling.
 4. Uses toilet by self, occasional loss of urine with rushing, laughing, coughing. (circle)
 3. Must be taken to toilet/given pan, can control urine and bowels.
 2. Urinary incontinence, bowel control.
 1. Complete incontinence.

26. Do others help you with bathing and dressing?
 5. Keeps self clean, and dresses by self.
 4. Keeps self clean, minimal assistance with clothing. (buttons)
 3. Gives self main bath, requires dressing by others.
 2. Can wash part of self, main bath and dressing done by others.
 1 Totally bathed and dressed by others.

27. Do you ever feel you do not look nice?___Why?______________

28. Tell me how you can obtain water, if necessary.
 5. Carries in a 5 liter bucket. (Large sized)
 4. Carries in a 2 liter bucket. (Medium sized)
 3. Carries in a 1 liter bucket. (Small sized)
 2. Carries in a cup or uses at tap.
 1. Unable to get water.

29. Do you ever go without water?________Why?__________________

30. How far can you walk?
 5. Can walk any desired distance.
 4. Can walk around village, up to 4 Kilometers.
 3. Can walk in immediate neighborhood, less than 1 Kilometer.
 2. Moves about in house/yard with assistance.
 1. Crawls or cannot move about.

31. Are you able to attend and participate in feasts and parties?
    5. Full participation.
    4. Participates except for heavy tasks. (lift large pots, chop wood)
    3. Able to do minor tasks (sit to wash dishes, stir small pots).
    2. Attends events, no participation in events.
    1. Cannot attend.

32. How well can you see with/without glasses? (Circle)
    5. Sees well, recognize people and can do handwork.
    4. Can either recognize people outside the compound or do handwork. (Circle)
    3. Restricted vision, can see to perform basic functions only.
    2. Minimal vision, can see color or shapes.
    1. Totally blind.

33. How well can you hear?
    5. Hears well.
    4. Understands normal voice, occasional misinterpretation of words.
    3. Understands loud voice without difficulty.
    2. Understands loud voice with difficulty.
    1. Totally deaf or unable to recognize sounds.

34. How is you memory?
    5. Remembers well, knows where things are placed.
    4. Minimal difficulty with recent recall. (forgets where item was placed)
    3. Tends to forget present events, aware of past.
    2. Very forgetful, remembers only main events of past.
    1. Does not remember any of the past or present.

Now I have some questions about life in general.

35. How would you describe your life now (happy or unhappy). Why?__________

36. Do you think of yourself as useful or useless? Why?____________
    2. Is useful.
    1. Is not useful.

37. Do you visit others and do others visit you?
    2. Has visitors and/or visits others.
    1. Does not visit or have visitors.

38. Many old people are lonely. How often do you feel lonely?
    5. Very Seldom or never.
    4. Once or twice week.
    3. Three to six times a week.
    2. Once a day.
    1. More than once a day.

39. When you give advice or talk with others, how do they respond.
    2. Listen.
    1. Do not listen.

40. What sort of things do you worry about?____________________________

41. Do you have someone you can to talk with and who understands your problems?
Who?________________
  5. Has intimate relationship, sees daily.
  4. Has intimate relationship, sees person at least weekly.
  3. Has intimate relationship, sees person at least monthly.
  2. Has intimate relationship, sees person less than once a month.
  1. No intimate relationship.

42. Do you feel you are included in meaningful events, like family talks, or village
feasts? Explain_____________________________________
  2. Yes.
  1. No.

43. In the village, it is possible to become cold and wet because of the weather, or to
hurt your feet on the village paths. Do you have a coat, shoes without holes, and a dry
room during rain? Do you worry about your own safety, like someone hurting you or
taking something which is yours? (Circle those which are present)
  5. Has a coat, shoes without holes, dry room and freedom from worry about safety.
  4. Has 3 of the above.
  3. Has 2 of the above.
  2. Has 1 of the above.
  1. Has none of the above.

44. It is the custom for the family to provide water, serve food, help with chores and fix
the house for the aged parent. Which things are usually done for you? (Circle
performed activities)
  5. All (4) of the above.
  4. 3 of the above.
  3. 2 of the above.
  2. 1 of the above.
  1. None of the above.

45. Listen to this story. A lady, let's call her Mary, has become ill and the doctor is
sure she will get well with the new medicine. It will take several months in bed before
she will be able to work in the house again. Mary talked this over with her daughter
who works in the village. The daughter will be there at night and arranged for her
niece to help during the day.   Now, here's my question.

46. Suppose this happened to you. Who would be able to care for you? (Relationship
of Main person)________Others______

47. What times during the day and night do you think someone would be with
you?_______________________________
  5. Someone would provide constant care until well.
  4. Someone would be present majority of the time, except for short   absences.
  3. Someone would be present during the night and part of the day.
  2. Someone would be present during night only.
  1. Would be alone, without a care-provider.

The next few question are about what you eat. I want you to think carefully about the answers.

48. What did you eat for breakfast yesterday?____________ Today?________

49. For lunch yesterday?_______________ Today?_________________

50. For supper yesterday?________________ Today?____________

51. Did you have anything between breakfast and lunch, such as beer, tea, or food?_________________________________

52. What did you eat or drink between lunch and supper?___________

53. What did you eat or drink after supper or at bedtime?________

54. How many spoons of sugar do you put in your tea?____________

55. How often do you drink beer?_______; can of beer?___________

How often do you eat or drink something made with:
56. vegetables?____
57. eggs?____
58. meat____
59. milk?____

60. I will read a list of reasons people are hungry. Tell me if any are true for you.
   l. No one brings me food and I am unable to prepare it myself.
   2 I cannot afford to buy enough food for myself.
   3 I have no one to go to the store for me.
   4 It is better to let someone else eat the food. (Who)_____
   5 I/my family was unable to grow enough food this season.
   6. I do not have firewood or matches.
   7. I want to loose weight.
   8. I cannot chew or my mouth hurts.
   9. Everyone in this household is hungry.
   10. I cannot get water for cooking.
   11. Do you have another reason for staying hungry. (State)____
   12. I never go hungry.

61. Which is the <u>main</u> reason you stay hungry? (Mark one with an X)

62. How many times in the past week did you have a day without food?___Why__

63 I have a list of places people go. Tell me if you went there regularly before, if you go there now, and how often.

   Visited in past Visits in present Frequency

____	Community Center
____	Kgotla
____	School PTA
____	Church
____	District Council
____	Clubs/Organizations
____	Library
____	Bank
____	Other

64. How do you usually get to other places in the village?
  5. Walks by self.
  4. Walks with others.
  3. Dependent on ride, usually gets one if wanted.
  2. Dependent on ride, may be unable to get one.
  1. Never attends because of distance/transportation.

65. How do you get transportation when you need a ride?
  5. Access to bus and a car in the household.
  4. Access to bus and a car in the village.
  3. Access to bus and a car outside of the village.
  2. Access to bus only.
  1. No access to transportation.

66. What difficulties/problems would you have if you wanted to ride a bus?___________
  5. No problems, walks to bus stop and gets on by self.
  4. Rides bus, walks to bus stop, needs help getting on.
  3. Can walk to bus stop, cannot get on.
  2. Does not ride bus because of a lack of money.
  1. Can not get to bus stop or ride because of health limitations.

I have two more stories. Here is the first one. Mr. and Mr. Jonas were born in 1920. Their roof started to leak a couple of weeks age. Their son, John, said he would fix it. They were expecting John next Saturday. He sent word to tell them he did not know if he could make it because a friend at work was getting married.
67. What do you think John should do?

68. Using your knowledge of how people behave, what do think really happened?

Mrs. Ntseane is a very old woman who has been unable to walk outside for a long time. She lives with her daughter. One morning Mrs. Ntseane does not get out of bed and her daughter sees that she is very sick with a fever.
69. What do you think the daughter should do?

70. What do you think families actually do when an old person is dying?

71. What do you think would happen if you were dying?
   5. Anticipates full time care and support in desired locale.
   4. Feels that family will respect wishes and be present most of time.
   3. Feels family will not respect wishes for locale but will be with them.
   2. Feels family will not respect wishes about locale and will not be present.
   1. Feels dying will not be recognized by others, and death will not be known.
*I would like to ask you questions about your work and money. Please remember I am
not from the government and this information is between you and me.*

72. Type of work (include agriculture): Past_____________ Present________

73. In terms of work, what are you doing now?
   1. Employed full-time by someone else.
   2. Employed part time by someone else.
   3. Self employed full-time.
   4. Self employed part-time.
   5. Employed at a temporary job.
   6. Earns limited money with crafts, snuff, bajalwa.
   7. Not producing income.

74. Where does your money come from? (Check yes or no for each of the following,
and if yes, enter amount if known, frequency and regularity.
Yes/No Amount Frequency Regularity

____Present employment
____Spouse employment
____Money from family (Who) ___
____Agriculture
____Cattle
____Destitute Program
____Investments
____Retirement Pension
____Savings account

76. When was the last time you had money in your pocket?_______

77. How did you get the money._____________________

78. What did you do with the money._________________

79. Some people have never owned cattle. Did you ever own cattle?__

80. When was the last time you had cattle?_____________________________

81. What happened to your cattle? (ask reason for deaths)_____________

82. Which of the following do you, as a person, have today.
   Own   Number
   _______________________________________Cattle
   _______________________________________Goats
   _______________________________________Other livestock
   _______________________________________Chickens/fowl
   _______________________________________Dog
   _______________________________________Land/Fields

83. This area had a recent drought. How did the drought change the way you
live?____

84. In what way did the drought affect your family? (Losses, Migration, unemployment,
sale of cattle)__________________

85. Did you use any of the drought relief programs? (explain)_______

86. I have asked you many things. Is there anything else you would like to tell me, or
think I should know about old people?

87. Is there anything you want me to tell the government, or others, about old people
and what they need?

*Before I leave, I would like to take your blood pressure. I would also like to see your
clinic chart, if you have one.*

87. Blood pressure:

88. Weight is
   1. Appropriate for height.
   2. Heavy for height.
   3. Very heavy for height.
   4. Thin for height.
   5. Exceedingly thin for height.

89. Clinical record: Weight______Height_______Diagnoses________________

Date_____Time____Length of interview____Language(s) used_____Translator____

Comments.:

APPENDIX B
THE LADDER INSTRUMENT

<u>The Ladder</u>
The ladder is approximately twelve inches high. It is constructed with five rungs, each about three inches across.

<u>Opening Sentences</u>
This is a ladder. Each rung represents different people. (Point to rungs.) Sometimes I will hold it so that is sideways, instead of up and down. I will ask which rung represents you. (Move hand up and down rungs.) There is no correct or wrong answer. I want to know how you feel about yourself.

Questions: (Hold ladder on its side)
1. The rungs are five people. (Point to right-hand rung) This person is the stranger in the village, he/she is all alone as he/she knows no one. No one visits or talks with him/her. (Point to left-hand rung) This is the person who knows everyone. He/she has many friends and people come to visit and to ask for advice. Which rung is you? Why?

2. (Point to right-hand rung) This person has nothing that other's want; no food, no knowledge, no house and no physical body. (Point to left-hand rung) This person has everything and gives things, and assistance, to other people. Which rung are you? Why?

3. (Point to right-hand rung) This person can do nothing for him/herself. He/she cannot walk, must have food and water put in their mouth by someone else and someone must dress him/her. (Point to left-hand rung) This person can walk as far as they want without getting tired, can plow or earn money, and is never sick. Which rung are you? Why?

(Questions 4-7: Say "The questions change a little now". Hold ladder upright. Point to middle rung to begin each question.)

4. This is where you were at the time of Independence in 1966. Where are you now? Why?

5. This is where Botswana was at the time of Independence in 1966. Where is the country now? Why is this so?

6. This is where the people of Botswana were at the time of Independence in 1966. Where are the people now? Why did this happen?

7. This represents the position old people had in society and how important they were to others at the time of Independence in 1966. Where are the old people now? Why is this so? Which rung are you? Why do you put yourself there?

<u>Adaptations to Use With Care-Providers</u>

Questions 1-3: Use the same statements. Change questions to: Which one is most like the old person you take care of? Why?

REFERENCES CITED

Achenbaum, W. Andrew. 1987. Modernization Theory. In *Encyclopedia of Aging*. Springer Publication Co., New York, NY.

Alverson, Hoyt. 1978. *Mind in the Heart of Darkness. Value and Self Identity Among the Tswana of Southern Africa*. Yale University Press, New Haven, CT.

Amoss, Pamela T. 1981. Cultural Centrality and Prestige for the Elderly: The Coast Salish Case. In *Dimensions: Aging, Culture and Health*. Christine L. Fry, Ed. Bergin & Garvey Publishers, Inc., South Hadley, MA.

Anderson, Norman B. 1988. Aging and Hypertension among Blacks: A Multidimensional Perspective. In *The Black American Elderly: Research on Physical and Psychosocial Health*. Ed. James S. Jackson. Springer Publishing Company, New York, NY.

Arntzen, Jaap. 1984a. *Rural Agricultural Activities and Resource Utilization in Mmathubudkwane During a Period of Drought*. National Institute for Development Research and Documentation, Gaborone, Botswana.

Arntzen, Jaap. 1984b. *Changes in Rural Activities and Utilization of Natural Resources in the Period 1979-1983: The Case of Malolwane, Kgatleng District*. NIR Research Notes No. 14. University of Botswana, Gaborone, Botswana.

Atchley, Robert C. 1987a Disengagement. In *Encyclopedia of Aging*. Springer Publishing Co., New York, NY.

Atchley, Robert C. 1987b. Activity Theory. In *Encyclopedia of Aging*. Springer Publishing Co., New York, NY.

Barker, Judith C. 1990. Between Humans and Ghosts: The Decrepit Elderly in a Polynesian Society. In *The Cultural Context of Aging*. Jay Sokolovsky, Ed. Bergin & Garvey Publishers, New York, NY.

Bell, Bill D., Gail Kara and Constance Batterson. 1978. Service Utilization and Adjustment Pattern of Elderly Tornado Victims in an American Disaster. *Mass Emergencies* 3:71-81.

Bernard, H. Russell. 1988 *Research Methods in Cultural Anthropology*. Sage Publications, Beverly Hills, CA.

Bhila, Hoyidi. 1984. The Impact of the Second World War on the Development of Peasant Agriculture in Botswana, 1939-1956. *Botswana Notes and Records* 16:63-71.

Biesele, Megan and Nancy Howell. 1981. The Old People Give You Life: Aging Among !Hung Hunter Gatherers. In *Other Ways of Growing Old*. Pamela T. Amoss and Stevan Harrel, Eds. Stanford University Press, Stanford, CA.

Blau, Z.S. 1973. *Old Age in a Changing Society*. New Viewpoints, a division of Franklin Watts, Inc., New York, NY.

Botswana Central Statistics Office. 1974. *A Social and Economic Survey of Threee Peri-Urban Areas in Botswana*. Ministry of Finance and Development Planning, Government Printer, Gaborone.

Brathwaite, Farley. 1986. *The Elderly in Barbados*. Carib Research & Publications Inc., Bridgetown, Barbados.

Campbell, Alex C. 1971. The Rural Economy: A Sociological Perspective. *Botswana Notes and Records* 13:192-194.

Campbell, A. C. 1979. The 1960's Drought in Botswana. In *Proceedings of the Symposium on Drought in Botswana*. Madalon T. Hinchey, Ed. The Botswana Society, Gaborone, Botswana.

Campbell, A. C. 1982. Notes on the Prehistoric Background to 1980. In *Settlement in Botswana*. R. Renee Hitchcock and Mary R. Smith, Eds. Heinemann Educational Books, Ltd., Lesotho, South Africa..

Central Statistical Office, MFDP. 1982. *1981 Population and Housing Census, Summary Statistics on Small Areas (for settlements of 500 or more people)*. Government Printer, Gaborone, Botswana.

Chambers, Robert. 1983. *Rural Development: Putting the Last First*. Longman Scientific & Technical, Essex, England.

Chambers, R. and D. Feldman. 1973. *National Policy for Rural Development The Government's Decisions of the Report on Rural Development, Government Paper No. 2 of 1973*. Government Printer, Gaborone, Botswana.

Chatters, Linda M. 1988. Subjective Well-Being Among Older Black Adults: Past Trends and Current Perspectives. In *The Black American Elderly*. James S. Jackson, Ed. Springer Publishing Company, New York, NY.

Clark, Margaret. 1972. Cultural Values and Dependency in Later Life. In *Aging and Modernization*. Donald Cowgill and Lowell D. Thomas, Eds. Appleton-Century-Crofts, New York, NY.

Cohen, Ronald. 1986. Traditional Social Formations. In *Food in Sub-Saharan Africa*. Art Hansen and Della E. McMillan, Eds. Lynne Rienner Publishers, Inc., Boulder, CO.

Cohen, Carl I. and Jay Sokolovsky. 1989. *Old Men of the Bowery: Strategies for Survival Among the Homeless*. The Guilford Press, New York, NY.

Coles, Catherine. 1990. The Older Woman in Hausa Society: Power and Authority in Urban Nigeria. In *The Cultural Context of Aging*. Jay Sokolovsky, Ed. Bergin & Garvey Publishers, New York, NY.

Colson, Elizabeth and Thayer Scudder. 1981. Old Age in Gwembe District, Zambia. In *Other Ways of Growing Old, Anthropological Perspectives*. Pamela T. Amoss and Stevan Harrell, Eds. Stanford University Press, Stanford, CA.

Comaroff, John L. 1953 Tswana Transformation, 1953-1975. Supplementary Chapter. In *The Tswana*. I. Schapera, Ed. International African Institute, London, England.

Cook, Susan. 1987. *Social Exchange Theory*. Sage Publications, Beverly Hills, CA.

Cowgill, D. O. 1979. Aging and Modernization: A Revision of the Theory. In *Dimensions of Aging*. Jon Hendricks and C. D. Hendricks, Eds. Wentrhrop Publishers, Inc., Cambridge, MA.

Cowgill, Donald O. and Lowell D. Holmes. 1972. Summary and Conclusions: The Theory in Review. In *Aging and Modernization*. Donald O. Cowgill and Lowell D. Holmes, Eds. Appleton-Century-Crofts, New York, N. Y.

Cummings, E. and W. E. Henry. 1961. *Growing Old: The Process of Disengagement*. Basic Books, Inc., New York, NY.

Diouf, Abdou. 1985. *African Conference on Gerontology*. Government of Senegal and International Center of Social Gerontology, Dakar, Senegal.

Dirks, Robert. 1980. Social Responses During Severe Food Shortages and Famine. *Current Anthropology* 21:1:21-32.

Dowd, James J. 1975. Aging as Exchange: A Preface to Theory. *Journal of Gerontology* 30:5:584-594.

Dowd, James J. 1980. *Stratification Among the Aged*. Brooks/Cole Publishing Company, Monterey, CA.

Dowd, James J. 1984. Beneficence and the Aged. *Journal of Gerontology* 39:1:102-108.

Dressler, William W. 1985. Psychosomatic Symptoms, Stress, and Modernization: A Model. *Culture, Medicine and Psychiatry* 9:3:257-286.

Ellenberger, Vivien. 1937. History of the BA-GA-MALETE of Ramotswa (Bechuanaland Protectorate). *Transafrican Royal Society* 25:1-71.

Foster, George M. 1973. *Traditional Societies and Technological Change*, 2nd Edition. Harper and Row Publishers, New York, NY.

Fry, Christine. 1986. Emics and Age: Age Differentiation and Cognitive Strategies. In *New Methods for Old Age Research*. Christine Fry and Jennie Keith, Eds. Bergin and Garvey Publishers, Inc., South Hadley, MA.

Gage, Timothy B. 1991. Human Variation in the Age Patterns of Mortality. *Association for Anthropology and Genontology Newsletter* 12:3:6-7.

Glascock, Anthony P. and Susan L. Feinman. 1981. Social Asset or Social Burden: Treatment of the Aged in Non-Industrial Societies. In *Dimensions: Aging, Culture and Health*. Christine L. Fry, Ed. Bergin & Garvey Publishers, Inc., South Hadley, MA.

Goldstein, Melvyn, Sidney Schuler and James Ross. 1983. Social and Economic Forces Affecting Intergenerational Relations in Extended Families in a Third World Country: A Cautionary Tale from South Asia. *Journal of Gerontology* 38:6:716-724.

Goodenough, Ward Hunt. 1963. *Cooperation in Change*. Russell Sage Foundation, New York, NY.

Gubrium, Jaber F. 1973. *The Myth of the Golden Years, A Socio-Environmental Theory of Aging*. Charles C. Thomas Publisher, Springfield, IL.

Guillette, Elizabeth. 1990. Socio-Economic Change and Cultural Continuity in the Lives of the Older Tswana. *Journal of Cross-Cultural Gerontology* 5:191-204.

Guillette, Elizabeth A. 1991. *The Impact of Recurrent Disaster on the Aged of Botswana*. Paper presented at the 50th Annual Meeting of The Society for Applied Anthropology. March 18, 1991. Charleston, SC.

Hampson, Joe. 1982. *Old Age: A Study of Aging in Zimbabwe*. Mambo Press, Gwere, Zimbabwe.

Hampson, Joe 1985. Elderly People and Social Welfare in Zimbabwe. *Aging and Society* 5:39-67.

Hansen, Art. 1990. *Dependency and the Dependency Syndrome.* Paper presented at the American Association of Anthropology. November 28 - December 2, 1990. New Orleans, LA.

Harrison, Paul. 1987. Inside the Third World, The Anatomy of Poverty. Penguin Books, London, England.

Havighurst, Robert J., Bernice L. Neugarten and Sheldon S. Tobin. 1963. Disengagement and Patterns of Aging. In *Middle Age and Aging*. Bernice L. Neuggarten, Ed. The University of Chicago Press, Chicago, IL.

Hay, Roger W., Susan Burke and D. Y. Dako. 1985. *A Socio-Economic Assessment of Drought Relief in Botswana.* Botswana Co-operative Union, Gaborone, Botswana.

Hendricks, Jon. 1982. The Elderly in Society: Beyond Modernization. *Social Science History* 6:3:32-45.

Hendricks, Jon and C. Davis Hendricks. 1986. *Aging in Mass Society, Myths and Realities*. Little, Brown and Company, Boston, MA.

Hill, Polly. 1986. *Development Economics on Trial, The Anthropological Case for a Prosecution*. Cambridge University Press, Cambridge, England.

Hitchcock, Robert K. 1989. Settlement, Seasonality and Subsistence Stress Among the Tyua of Northern Botswana. In *Coping With Seasonal Constraints*. R. Huss-Ashmore with J. Curry and R. Hitchcock, Eds. MASCA Research Papers Vol. 15. The University Museum, University of Pennsylvania, PA.

Hoover, Sally L. and Jacob S. Siegel. 1986. International Demographic Trends and Perspectives on Aging. *Journal of Cross-Cultural Gerontology* 1:5-30.

Hudson, Derek J. 1977. Rural Incomes in Botswana. *Botswana Notes and Records* 9:101-108

Ingstad, Benedicte. 1989. *The Disabled and The Community in Botswana.* Paper presented at Organization for International Health Cooperation, Feb. 24-26. Bologna.

Ingstad, Benedicts, Frank Brunn, Edwin Sandberg and Sheila Tlou. 1991. *Care for the Elderly--Care by the Elderly: The Role of Elderly Women in Changing Tswana Society.* Paper presented at American Anthropological Association. Nov. 20-24, 1991. Chicago, IL.

Ingstad, Benedicte and Sidsel Saugestad. 1987. Unmarried Mothers in Changing Tswana Society:Implications for Household Form and Viability. *Forum for Utviklingsstudier* 4:3-25.

Jacobs, Jerry. 1974. *Fun City: An Ethnographic Study of a Retirement Community*. Holt, Rinehard and Winston, New York, NY.

Jacome, Eduardo Garcia. 1988. *Provida: A Non-Governmental Organization for the Care of the Elderly in Columbia and Latin America.* Paper presented at Aging, Demography and Well-Being in Latin America. February 23-25, 1988. University of Florida, Gainesville, FL.

Keith, Jennie, Christine L. Fry and Charlotte Ikels. 1990. Community as Context for Successful Aging. In *The Cultural Context of Aging.* Jay Sokolovsky, Ed. Bergin & Garvey Publishers, New York, NY.

Kerven, Carol. 1982. Rural-Urban Interdependence and Agricultural Development. In *Settlement In Botswana.* R. Renee Hitchcock and Mary R. Smith, Eds. Heinemann Educational Books, Lesotho, South Africa.

Khasiani, Shanyisa A. 1987. The Role of the Family in Meeting the Social and Economic Needs of the Aging Population in Kenya. *Genus* 43:1-2:103-117.

Knudsen, Thutego. 1988. *Family Welfare Educators in Botswana. Can They Be More Community Oriented?* Health Research Unit, Ministry of Health, Republic of Botswana, Gaborone, Botswana.

Lauer, Robert H. 1973. *Perspectives on Social Change*. Allyn and Bacon, Inc., Boston, MA.

Lawton, M. Powell. 1983. Environment and Other Determinants of Well-Being in Older People. *The Gerontologist* 23:4:349-357.

LeVine, Robert A. 1965. Intergenerational Tensions and Extended Family Structures in Africa. In *Social Structure and The Family: Generational Relations*. Ethel Shanas and Gordon F. Streib, Eds. Prentice-Hall, Inc., Englewood Cliffs, NJ.

LeVine, Sarah and Robert A. LeVine. 1985. Age, Gender and the Demographic Transition: The Life Course in Agarian Societies. In *Gender and the Life Course*. Alice Rossi, Ed. Aldine Publishing Co., New York, NY.

Lee, Richard B. 1985. Work, Sexuality and Aging Among !Kung Women. In *In Her Prime: A New View of Middle Aged Women*. Judith K. Brown and Virginia Kerns, Eds. Bergin and Garvey Publishers, Inc., South Hadley, MA.

Lieberman, Morton A. and Sheldon S. Tobin. 1983. *The Experience of Old Age; Stress, Coping and Survival*. Basic Books, Inc. Publishers, New York, NY.

Martel, Martin U. 1968. Age-Sex Roles in American Magazine Fiction (1890-1955). In *Middle Age and Aging, A Reader in Social Psychology*. Bernice L. Neugarten, Ed. The University of Chicago Press, Chicago, IL.

McKee, Patrick L. 1982. *Philosophical Foundations of Gerontology*. Human Sciences Press, Inc., New York, NY.

Meillassoux, Claude. 1981. *Maidens, Meal and Money Capitalism and the Domestic Community*. Cambridge University Press, Cambridge.

Morgan, Richard. 1986. From Drought Relief to Post-Disaster Recovery: The Case of Botswana. *Disasters* 10:30-34.

Mumford, Lewis. 1987. For Older People: -Not Segregation But Intergration. In *Housing and the Elderly*. Judith Ann Hancock, Ed. Center for Urban Policy Research, New Brunswick, NJ.

Munnichs, Joep. 1976. Dependency, Interdependency and Autonomy. In *Dependency or Interdependency In Old Age*. Joep Munnichs and Wim Van Den Heuvel, Eds. Intercontinental Graphics, Amsterdam, Netherlands.

Myerhoff, Barbara. 1978. *Number Our Days*. Simon and Schuster Publishing. New York, NY.

Ngcongco, L. 1982. Precolonial Migration in South-Eastern Botswana. In: *Settlement In Botswana*, R. Renee Hitchcock and Mary R. Smith, Eds. Heinemann Educational Books, Ltd., Lesotho, South Africa.

Nusberg, Charlotte. 1988. The Role of the Elderly in Development. *Ageing International* (December). Pp. 9-10.

Oden, Bertil. 1981. *The Macroeconomic Position of Botswana, Research Report No. 60.* Scandinavian Institute of African Studies. Uppsala Offsetcenter AB, Uppsala, Sweden.

Oliver-Smith, Anthony. 1982. Here There Is Life. In *Involuntary Migration and Resettlement, The Problems and Responses of Dislocated People.* Art Hansen and Anthony Oliver-Smith, Eds. Westview Press, Boulder, CO.

Oliver-Smith, Anthony. 1986. *The Martyred City, Death and Rebirth in the Andes.* University of New Mexico Press, Albuquerque, NM.

Oliver-Smith, Anthony and Art Hansen. 1982. Introduction Involuntary Migration and Resettlement: Causes and Contexts. In *Involuntary Migration and Resettlement, The Problems and Responses of Dislocated People.* Art Hansen and Anthony Oliver-Smith, Eds. Westview Press, Boulder, CO.

Osman, Sheilla. 1983. *An Investigation into the Conditions Under Which the Aged Live in Southern Botswana.* Unpublished B.S. Thesis, University of Botswana Gaborone, Botswana.

Parr, Arnold R. 1987. Disaster and Disabled Persons: An Examination of the Safety Needs of a Neglected Minority. *Disasters* 11:148-153.

Parson, Jack. 1984. *Botswana, Liberal Democracy and the Labor Reserve in Southern Africa.* Westview Press, Boulder, CO.

Parsons, T. 1964. Evolutionary Universals. *American Sociological Review* 29:339-357.

Peterson, Jane W. 1990. Age of Wisdom: Elderly Black Women in Family and Church. In *The Cultural Context of Aging, Worldwide Perspectives.* Jay Sokolovsky, Ed. Bergin & Garvey Publishers, New York, NY.

Picard, Louis A. 1987. *The Politics of Development in Botswana, A Model for Success?* Lynne Rienner Publishers, Boulder, CO.

Prah, K. K. 1979. Some Sociological Aspects of Drought. In *Proceedings of The Symposium on Drought in Botswana.* Madalon T. Hinchey, Ed. The Botswana Society, Gaborone Botswana.

Republic of Botswana. 1981. *Summary Statistics on Small Areas, 1981 Population and Housing Census.* Central Statistical Office. Government Printer, Gaborone, Botswana.

Rosenberg, Harriet G. 1990. Complaint Discourse, Aging and Caregiving among the !Kung San of Botswana. In *The Cultural Context of Aging, World Wide Perspectives.* Jay Sokolovsky, Ed. Bergin & Garvey Publishers, New York, NY.

Rosenmayr, Leopold. 1989. *Health of the Old as a Problem of Communal Intergration in Rural West African Society.* Paper presented at the XIV International Congress of Gerontology. June 19-23, 1989. Acapulco, Mexico.

Rossi, Peter H. and Freeman, Howard E. 1982. *Evaluation: A Systematic Approach,* 2nd Edition. Sage Publications, Beverly Hills, CA.

Rutter, Virginia. 1992. New Therapy Directions for Aging Families. *Aging Today* 13:2:5.

Ryan, Maura C. and Adriana G. Austin. 1989. Social Supports and Social Networks in the Aged. *Image: Journal of Nursing Scholarship* 21:176-179.

Sahlins, Marshall. 1972. *Stone Age Economics*. Aldine De Gruyter, New York, NY.

Schapera, I. 1944. *Married Life in an African Tribe*. Sheridan House, New York, NY.

Schapera, I. 1953. *The Tswana. Ethnographic Survey of Africa, South Africa, Part III.* International African Institute, London, England.

Schapera, I. 1955. *Handbook of Tswana Law and Custom,* 2nd ed. International African Institute Oxford University Press, London, England.

Schapera, I. 1967. Present Day Life in the Native Reserves. In *Western Civilization and the Natives of South Africa*. I. Schapera, Ed. Routledge and Kegan Paul, LTD, London, England.

Schapera, I. 1970. *Tribal Innovators, Tswana Chiefs and Social Change 1795-1940.* London School of Economics Monographs on Social Anthropology No. 43. The Athlone Press, University of London, London, England.

Sen, Amartya. 1981. *Poverty and Famines, An Essay on Entitlement and Deprivation.* Clarendon Press, Oxford, England.

Shield, Renee Rose. 1990. Liminality in an American Nursing Home: The Endless Transition. In *The Cultural Context of Aging.* Jay Sokolovsky, Ed. Bergin & Garvey, Publishers, New York, NY.

Shostak, Marjorie. 1983. *Nisa, The Life and Words of a !Kung Woman*. Vintage Books, New York, NY.

Silitshena, R. M. K. 1979. Chiefly Authority and the Organization of Space in Botswana: Towards an Exploration of Nucleated Settlement amoung the Tswana. *Botswana Notes and Records* 11:55-67.

Simic, Andrei. 1990. Aging, World View, and Intergenerational Relations in America and Yugoslavia. In *The Cultural Context of Aging*. Jay Sokolovsky, Ed. Bergin & Garvey, New York, NY.

Simmons, Leo W. 1945. *The Role of The Aged In Primitive Society*. Yale University Press, New Haven, CT.

Simmons, Leo. 1960. Aging in Preindustrial Societies. In *Handbook of Social Gerontology*. C. Tibbitts, Ed. University of Chicago Press, Chicago, IL.

Social Science Reasearch Council. 1953. Acculturation: An Exploratory Formulation. *American Anthropologist* 56:973-1001.

Sokolovsky, Jay. 1990. *The Cultural Context of Aging, Worldwide Perspectives.* Bergin & Garvey Publishers, New York, NY.

Spradley, James. 1979. *The Ethnographic Interview.* Holt, Rinehart and Winston, New York, NY.

Stack, Carol B. 1974. *All Our Kin, Strategies for Survival in a Black Community.* Harper Torchbooks, New York, NY.

Stanford, E. Percial and Donna L. Yee. 1991. Gerontology and the Relevance of Diversity. *Gernerations* 15:4:11-15.

Stanford, E. Percial and Fernando M. Torres-Gil. 1991. Diversity and Beyond: A Commentary. *Generations* 14:4:5-6.

Staugard, Frants. 1985. *Traditional Healers, Traditional Medicine in Botswana.* Ipelegeng Publishers, Gaborone, Botswana.

Streib, Gordon F. 1972. Older Families and Their Troubles: Familial and Social Responses. *The Family Coordinator* 21:1:5-19.

Streib, Gordon F. 1990. *There's No Place Like Home.* Paper presented at Aging Well in Place Lecture Series. October. 1990, Center for Gerontological Studies, University of Florida, Gainesville, FL.

Suggs, David N. 1987. Female Status and Role Transition in the Tswana Life Cycle. *Ethnology* 26:2:107-120.

Thema, B. C. 1972. The Changing Pattern of Tswana Social and Family Relations. *Botswana Notes and Records* 4:39-42.

Third World Guide. 1988/90. *Facts, Figures, Opinions.* Neal-Schuman Publishers, Inc., Montevideo, Uruguay.

Thomas, Samuel P. 1992. *Old Age in Meru, Kenya: Adaptive Reciprocity in a Changing Rural Community.* PhD. Dissertation. University of Florida, Gainesville, FL.

Tlou, Shelia Dinotshe. 1986. Gender Issues in the Health States of the Elderly in Botswana. In *Gender Dimensions of Development Research.* Ulla Kann, Godisang Mookodi and Jocelyn Snyder, Eds. National Institute for Development Research and Documentation, Gaborone, Botswana.

Tlou, Thomas and Alec Campbell. 1984. *History of Botswana.* Macmillan Botswana Publishing Company, Gaborone, Botswana.

Tout, Ken. 1989. *Ageing in Developing Countries.* Oxford University Press, Oxford, England.

Turnbull, Colin M. 1983. *The Human Cycle.* Simon and Schuster, New York, NY.

United Nations. 1985. *The World Aging Situation: Strategies and Policies.* Department of International and Social Affairs, United Nations, New York, NY.

Wolensky, Robert P. and Kenneth C. Wolensky. 1990. Local Government's Problem with Disaster Management: A Literature Review and Structural Analysis. *Policy Studies Review* 90:4:703-725.

BIOGRAPHICAL SKETCH

Elizabeth A. Guillette received her B.S. degree in nursing and her Masters degree in nursing education from the University of Rochester in Rochester, NY. Her career in nursing centered on preventive medicine and the maintenance of wellness. She taught both gerontology and family-centered nursing at various collegiate schools of nursing. She began graduate study in anthropology at the University of Florida in 1986 and simultaneously earned certification in gerontological studies.

I certify that I have read this study and that in my opinion it conforms to acceptable standards of scholarly presentation and is fully adequate, in scope and quality, as a dissertation for the degree of Doctor of Philosophy.

Art Hansen, Chair
Associate Professor of Anthropology

I certify that I have read this study and that in my opinion it conforms to acceptable standards of scholarly presentation and is fully adequate, in scope and quality, as a dissertation for the degree of Doctor of Philosophy.

Gordon F. Streib
Graduate Research Professor of Sociology
Emeritus

I certify that I have read this study and that in my opinion it conforms to acceptable standards of scholarly presentation and is fully adequate, in scope and quality, as a dissertation for the degree of Doctor of Philosophy.

Anthony R. Oliver-Smith
Professor of Anthropology

I certify that I have read this study and that in my opinion it conforms to acceptable standards of scholarly presentation and is fully adequate, in scope and quality, as a dissertation for the degree of Doctor of Philosophy.

Leslie Sue Lieberman
Associate Professor of Anthropology

I certify that I have read this study and that in my opinion it conforms to acceptable standards of scholarly presentation and is fully adequate, in scope and quality, as a dissertation for the degree of Doctor of Philosophy.

Allan F. Burns
Professor of Anthropology

This dissertation was submitted to the Graduate Faculty of the Department of Anthropology in the College of Liberal Arts and Sciences and to the Graduate School and was accepted as partial fulfillment of the requirement for the degree of Doctor of Philosophy.

August, 1992

Dean, Graduate School